New Perspectives on the History of Life Sciences and Agriculture

Archimedes

NEW STUDIES IN THE HISTORY AND PHILOSOPHY OF SCIENCE AND TECHNOLOGY

VOLUME 40

Archimedes has three fundamental goals; to further the integration of the histories of science and technology with one another: to investigate the technical, social and practical histories of specific developments in science and technology; and finally, where possible and desirable, to bring the histories of science and technology into closer contact with the philosophy of science. To these ends, each volume will have its own theme and title and will be planned by one or more members of the Advisory Board in consultation with the editor. Although the volumes have specific themes, the series itself will not be limited to one or even to a few particular areas. Its subjects include any of the sciences, ranging from biology through physics, all aspects of technology, broadly construed, as well as historically-engaged philosophy of science or technology. Taken as a whole, Archimedes will be of interest to historians, philosophers, and scientists, as well as to those in business and industry who seek to understand how science and industry have come to be so strongly linked..

More information about this series at
http://www.springer.com/series/5644

Denise Phillips • Sharon Kingsland
Editors

New Perspectives on the History of Life Sciences and Agriculture

Springer

Editors
Denise Phillips
Department of History
University of Tennessee
Knoxville
Tennessee
USA

Sharon Kingsland
History of Science and Technology
Johns Hopkins University
Baltimore
Maryland
USA

ISSN 1385-0180 ISSN 2215-0064 (electronic)
Archimedes
ISBN 978-3-319-12184-0 ISBN 978-3-319-12185-7 (eBook)
DOI 10.1007/978-3-319-12185-7

Library of Congress Control Number: 2014955545

Springer Cham Heidelberg New York Dordrecht London

Printed on acid-free paper

Springer International Publishing is a brand of Springer

Springer is part of Springer Science+Business Media (www.springer.com)

Contents

Chapter 1
Introduction

Denise Phillips and Sharon Kingsland

The essays in this volume explore problems in the history of science at the intersection of life sciences and agriculture, from the mid-eighteenth to the mid-twentieth century. We interpret agricultural practices in a broad sense, including the practices and disciplines devoted to land management, forestry, soil science, and the improvement and management of crops and livestock. Our purpose is to show that investigation of this border zone raises many interesting questions about how science develops. In particular, it challenges us to reexamine and take seriously the intimate connection between scientific development and the practical goals of managing and improving—perhaps even recreating—the living world to serve human ends. Without close attention to this zone it is not possible to understand the emergence of new disciplines and transformation of old disciplines, to evaluate the role and impact of such major figures of science as Humboldt and Mendel, or to appreciate how much of the history of modern biology has been driven by national ambitions and imperialist expansion in competition with rival nations. Focusing on agricultural practices also leads to new insights about how life sciences have interacted with economics and politics.

A few prescient historians of biology have recognized the importance of looking at the agricultural context to understand the emergence of new disciplines, such as genetics. Their pioneering contributions are amply recognized in the references to the essays in this volume. But on the whole, despite the promptings of these scholars since the 1980s, historians of biology have devoted most of their attention to academic biology and have continued to neglect the agricultural context. Historians

S. Kingsland (✉)
Department of History of Science and Technology,
Johns Hopkins University, 3400 N. Charles St., Baltimore, MD 21218, USA
e-mail: sharon@jhu.edu

D. Phillips
Department of History, University of Tennessee,
915 Volunteer Blvd., Knoxville, TN 37996, USA
e-mail: aphill13@utk.edu

D. Phillips, S. Kingsland (eds.), *New Perspectives on the History of Life Sciences and Agriculture,* Archimedes 40, DOI 10.1007/978-3-319-12185-7_1

of agriculture on the other hand concentrate on social and economic history, neglecting the science. We hope to engage both of these communities of historians and bring them into closer conversation with each other.

In so doing, we are building on the momentum of recent scholarly work in our field that suggests that scholars are keen to explore the complex intellectual, social, and economic problems that are raised at the intersection of life sciences and agriculture. In 2006, Jonathan Harwood, a contributor to this volume, edited a special issue on biology and agriculture for the *Journal of the History of Biology*. He noted that the long neglect of agricultural topics was changing, perhaps in light of the emergence of modern concerns about food safety, environmentalism, and the recent controversies over genetically modified organisms.[1] Contributors to that issue (Barbara Kimmelman, Christophe Bonneuil, Thomas Wieland, Karin Matchett, and Lloyd Ackert) not only expanded scholarship on history of life science and agriculture but also embraced an international perspective by examining topics in American, French, German, Mexican, and Russian history of science and agriculture. In this volume, we build on the momentum created by these and other scholars with an expanded international comparative approach that, in addition to the USA, France, Germany, the Netherlands, and Russia, also extends to Indonesia, China, and Japan. Once we adopt a longer-term perspective spanning three centuries, and take a broad multinational view, we notice many areas of connection and continuity, linking past to present and also linking the history of one nation to another. Our introduction will serve to indicate some of the synergistic connections between themes developed by contributors to this volume. There is, to be sure, much more to be done, especially in the second half of the twentieth century, but we hope these essays will prompt interest not just in the connections between life sciences and agriculture but also in cross-national comparisons.

Indeed, one significant theme that runs through numerous contributions to the volume is agriculture's historical importance in conceptions of national strength and wealth. Agriculture was a nexus of concern that brought together the resources of state, civil society, and science, all in the interest of transforming natural or cultivated landscapes. Several of the chapters on the eighteenth and nineteenth century show government officials, voluntary associations, and scientific experts working together (or sometimes working at cross-purposes), trying to bring natural knowledge to bear on problems of pressing agricultural concern. Joseph Horan examines Napoleon's attempts to make cotton a successful cash crop, an effort, Horan argues, that was part of the regime's broader political strategies for dominance in Europe. Like Napoleon's continental blockade, homegrown French and Italian cotton was supposed to strike a blow to the economic strength of national rival Great Britain. Anastasia Fedotova and Marina Loskutova's chapter on forestry in the Russian Empire looks at another ambitious expansionist dream, the attempt to increase rainfall on Russia's drought-prone southern steppes by building up the region's forest cover.

In addition to being an important arena for practical intervention, agriculture also played a significant role in political discourse about the nation. In the eighteenth

[1] Harwood (2006).

and nineteenth centuries, agriculture was a primary source of wealth and power for Europe and Asia's landed elites, and as a result even people with no experience of the hard physical labor of farming often took an avid interest in agricultural improvement. Examining the Japanese context, Jakobina Arch's chapter highlights another major issue of concern—the threat of famine. In Europe, too, food shortages could lead to widespread suffering and discontent. European historians have called the eighteenth century the golden age of food riots, and in the nineteenth century, political elites were well aware that public order was tied to the stability of the food supply.

Agricultural debates also provided a vehicle through which ideas from the life sciences could become integrated into broader political discussions, much as was the case in the better-studied instance of eugenics. Corinna Treitel's essay examines the place of farming reform in late-nineteenth and early-twentieth-century German political thought. In the mid-nineteenth century, political liberals advocated "natural" agriculture (which for them, interestingly, included the use of chemical fertilizers) because they believed that it would help produce not just healthy crops but strong, healthy Germans as well. By the early twentieth century, this rhetoric had been adopted and transformed by figures on the political right, who now advocated a "biological" approach to farming that was linked with new racial understandings of the nation.

Treitel's connection between German agricultural science and nationalist programs of racial renewal, pointing toward the racial and imperial designs of the Third Reich, is echoed in Sander Gliboff's study of the career of Erich Tschermak, the Austrian co-rediscoverer of Mendel's work. Tschermak successfully adapted to the Nazi regime while promoting his career and his research program. One consequence was his attempt to extend the logic of Mendelism to formulate a naïve and crude anti-Semitic eugenic argument. Treitel's narrative, perhaps more surprisingly, also links up with Mark Finlay's discussion of soil bacteriology and the promotion of legume inoculants in early-twentieth-century America. Finlay identifies concerns about food supply and the strength of the nation in Britain and the USA that are not unlike those voiced by Germans. American scientists in fact drew on German discoveries of the symbiotic work of bacterial microbes living on the nodules of legumes, which helped to fix atmospheric nitrogen. These historical narratives have their particular trajectories, reflecting each author's choice of focus and theme, but they are linked by common concerns about national expansion and maintaining the health of the population or race. We see these themes also in the chapters by Robert-Jan Wille on botany in the Dutch East Indies, and Kaori Iida on Japanese genetics.

An international comparative perspective automatically raises questions about how ideas and methods travel around the world. Victoria Lee discusses the development of pure culture techniques in microbiology in relation to the Japanese sake and soy fermentation industry. Her perspective enables her to balance an analysis of how foreign ideas entered Meiji era Japan, with the observation that traditional local brewing industries continued to have an important role in shaping scientific approaches and concepts. She argues that the concept of microbes as "living workers" having a complex physiology, a perspective characteristic of traditional brewing culture, influenced the development of microbiology in Japan. There was synergistic interaction between Western-trained Japanese scientists in technical colleges and universities and expert workers in the brewing industry.

This theme—the two-way conversations between scientific experts and practition ers—emerges in several essays in this volume. Farms and forests had multifaceted and complex relationships with formal scientific locales such as botanical gardens and experiment stations, and exchanges ran in both directions. For example, Horan shows that Parisian academicians and other French botanists knew the cotton plant reasonably well, and they could grow it with some success in botanical gardens. Turning the plant into an economically viable field crop was another story entirely, and despite a significant initial investment from the regime and a number of provincial agricultural academies, cotton did not conquer Italy or southern France.

Similarly, Loskutova and Fedotova's examination of Russian applied entomology traces the relationship between professionalizing experts and a broader lay community of observers. In this case, leading naturalists put significant effort into trying to train a competent network of entomological observers, but provincial officials and landowners also worked to solicit and create the kinds of expertise they needed to manage the growing threat posed by insect pests. Brendan Matz's chapter follows the development of scientific research into animal nutrition in the second half of the nineteenth century, and simultaneously shows the persistence of practical, localized know-how as an important voice in animal husbandry. In any particular instance of development in scientific agriculture, there are multiple interacting communities, and knowledge does not flow one way from experts to practitioners, but moves back and forth between these various communities. This point is also central to Jonathan Harwood's and Margaret Derry's analyses of plant and animal breeding practices in relation to Mendelism.

Engagement with agriculture also shaped the disciplinary and intellectual development of the life sciences in numerous ways. As Denise Phillips and Nils Güttler show, agricultural discussions played a key role in the emergence of plant geography. In a period that saw the introduction of many new crops to new places around the globe, learned naturalists and agricultural authors began to inquire more extensively into how differing physical conditions affected plant growth. As naturalists began to work out these relationships in greater detail, the practical literature on agricultural improvement provided them with visual traditions and empirical reference points for understanding the spatial dimensions of botanical diversity. Plant geography was not the only area in which the nineteenth-century life sciences overlapped with agricultural interests. Debates about evolution and heredity also touched on phenomena that were the objects of agricultural interests. Cristiana Oghina-Pavie's contribution to this volume analyzes how plant breeders and savants wrestled with the nuances of variation in rose bushes and pear trees, often using a shared vocabulary and drawing from the same empirical well. Breeders' findings and theories connected up in complicated ways with broader scientific questions about evolution, geographic variation, and patterns of inheritance.

The chapters by Robert-Jan Wille, Jeremy Vetter, and Adam Sowards all involve analysis of the link between agriculture and the transformation of existing disciplines or the emergence and maturation of new disciplines. Wille's study of the Dutch botanical garden at Buitenzorg (Bogor), Java, reveals that in response to the needs of large-scale colonial research enterprises like this, the traditional science of morphology was not abandoned (as is commonly argued) but was transformed and reinvented. The Dutch botanical garden operated much like a department of agriculture for the Indonesian colony, and scientists successfully tapped into the financial resources of the colonial plantation economy. The result was to reinvent morphology as a broader science that was more ecologically oriented than the traditional approach taken in the European university. Wille shows us how a modern ecological orientation can evolve from traditional disciplines that must adapt to the demands of different societies in different locations. The economic context of science is important in understanding how disciplines evolve.

In much the same vein, Vetter and Sowards develop the theme that the emergence of American ecology as a distinct discipline depended crucially on American agricultural interests, debates about land use, and the growth of new agricultural institutions. Vetter focuses on the US Department of Agriculture's Office of Dryland Agriculture, under the leadership of Ellery C. Chilcott in the early twentieth century. The problem that Chilcott faced was how to synthesize information pouring in from local field stations and produce knowledge that was applicable to the whole region. Chilcott deftly addressed this problem by insisting on standardized practices for measuring and collecting data, while avoiding overgeneralizing from those data. Those data showed that the Great Plains environment was variable and diverse, and this variability was important to recognize. Vetter identifies Chilcott's approach as truly "ecological" in both questions asked and the rigorous approach to getting answers. Even if scientists did not identify themselves as "ecologists," their work was driving the development of a distinct ecological research program. We should therefore acknowledge the profound impact of agricultural developments on the emergence of this perspective and this discipline.

Sowards carries Vetter's themes into the mid-twentieth century, by which time American ecology was acquiring guiding principles, theories, and a common core of research questions. Sowards focuses on Rexford Daubenmire's work on the Pacific northwest ecosystems of grasslands and forests. Because Daubenmire was concerned that science should provide management strategies that would reduce inefficient use of resources, it was important to him to assess the past state of a given landscape and to predict its future as population pressure increased. To this end, the concept of the ecological community, or later the ecosystem, served as an indispensable organizing principle. However, his assumptions about the reality of the community (and later the ecosystem) generated opposition with the Wisconsin-based school of thought that denied the objective reality of such ecological units. Sowards's analysis of this controversy reinforces the thesis that practical or managerial outcomes grounded in specific regions and land uses have profoundly shaped the intellectual directions of ecology. His chapter suggests a provocative question: Can the goals of "prediction and control" of nature, which are inherent to much of ecology as well as to other fields of science, be reconciled with the notion that scientists "seek truth" or objective understanding of what nature is really like?

How institutions shape the goals and character of scientific research is another running theme of this volume. Peter Lavelle's study of early experimental stations in China, especially in Beijing, assesses their importance for forestry, sericulture, and agriculture. Chinese scientists took Japan and the USA as their models, especially Japan at first, and used experiment stations to figure out which plants might grow best in certain environments, and how to improve soils with fertilizers. An interesting observation is that breeding was not part of their activity, and only later did breeding become important, coinciding with the growth of genetics in universities. Lavelle points out that Chinese researchers knew about and discussed Mendel's work even though genetics was not institutionalized as early in China as in other countries. One wonders whether, had the institutional agenda been more focused on breeding, Mendelism would have had greater impact in China, and would we expect genetics to have arisen in the context of these experimental stations? Did the institutional focus on matching plants to environment and on improving soil delay the development of genetics in China?

Harwood's, Derry's, and Gliboff's complementary analyses of plant and animal breeding and Mendelism suggest that that answer to this question is by no means obvious. Gliboff shows that for Erich Tschermak, Mendelism provided some guiding principles for his breeding experiments, although he did not adopt many of the ideas that we commonly identify as Mendelian. In addition, Mendelism provided a useful rhetorical tool, for it was an emblem of professional status to have been one of the few re-discoverers of Mendel's work. But rhetoric notwithstanding, Mendelism's capacity to transform breeding practices was far less certain.

Harwood subtly explores the question of whether there was a Mendelian revolution, using new scholarship that has shed light on breeding practices. Although he finds that Mendelism affected breeding in specific and limited ways, these fall short of revolutionizing breeding. Derry's study of animal breeding, focusing on the poultry industry, reinforces much of Harwood's argument. She too finds that the new Mendelian science did not persuade poultry breeders to change their practices at first, and that the scientists were often to blame because they did not understand breeding practices well. Her larger point is that it is necessary also to look closely at industrial practices and the culture of breeding. Changes in the structure of the industry—for instance, in the relationship between breeders and producer/growers—could provide a rationale for change in breeding strategies. When the poultry industry adopted the practice of marketing "hybrid chickens," following the model of hybrid corn, the stage was set for greater corporate investment that ushered in dramatic change in the chicken breeding industry.

Both Harwood's and Derry's chapters contribute to a new scholarly literature exploring the multiple contexts of the science of heredity, or what Staffan Müller-Wille and Hans-Jörg Rheinberger have called a cultural history of heredity.[2] One feels that a new synthesis or perhaps a complete reevaluation of the history of Mendelism and of genetics is in sight, one that will propel the history of biology in new directions. Not only will we look more closely at institutional contexts, at the

[2] Müller-Wille and Rheinberger (2012).

relationship of theories to practices, and at the structure of industry but we will also, as Harwood suggests, need to pay closer attention to the relationship of science and technology. In these and other chapters, contributors to this volume are questioning the pat distinctions between "pure" or "basic" and "applied science" and insisting that we need to evaluate in greater depth our conception of "applied science," with respect to both its historical and contemporary usage.

We would also benefit from a comparative study of Mendelism, breeding, and genetics. In this regard, East Asia provides interesting comparisons to British, European, and America-centered stories of the history of genetics. Both Lisa Onaga and Kaori Iida, in keeping with the arguments already developed in Arch's and Lee's chapters, show that in Japan there were very close links between Mendelism, genetics, and practical studies in sericulture and agriculture. All four papers dealing with Japan show that it was characteristic of Japanese biologists to take a broad approach to their research, one that could range across different disciplines. This breadth almost certainly reflected the way specific practical goals were driving scientific inquiry. As Onaga shows, Japanese interest in the "working silkworm" led scientists such as Toyama Kametarō into a variety of problems that included genetics, environmental effects, physiology and sex, and non-Mendelian inheritance. Iida's study of plant genetics, focused on Kihara Hitoshi, likewise emphasizes the lack of separation of genetics from other biological disciplines. Both Onaga and Iida mention the close relationship between Japanese biologists and German-American biologist Richard Goldschmidt, who advocated a distinctive physiological approach to genetics, one that appeared unorthodox in the American context but was compatible with Japanese approaches to biology. As Iida argues, genetics in Japan would be more accurately described as the effort to create a "science of breeding" comparable to what Nikolai Vavilov was doing in the Soviet Union, rather than a "discipline of genetics" that grew increasingly isolated from other disciplines (as occurred in the American university).

Helen Curry's chapter brings us back to the question of how interest in experimenting with the evolution of species drove many scientific and popular enthusiasms, such as the use of X-rays to speed up evolution by generating new mutations. The enthusiasm for "revolutionizing" agriculture reminds us of Finlay's themes concerning legume inoculations; Americans succumbed easily to exaggerated expectations when it came to the creation of new life forms. Again we see, as Harwood reminds us, that enthusiasm for new technology and the technological fix (today's biotechnology and genetic engineering) carries right through the entire twentieth century and up to the present. As Harwood notes, inflated claims for the importance of Mendelism lent support to the notion that scientific theory plays a decisive role in technological innovation in the early twentieth century—an assumption that is also seen in molecular biologists' claims about the revolutionary impact of biotechnology in the early twenty-first century.

The volume ends with a recent case study, indeed a story that is not yet over, concerning efforts to develop a model organism that is suited to agricultural problems. Christopher Lyons and Karen-Beth Scholthof argue that the inadequacy of the first model plant to have its genome sequenced, thale cress (*Arabidopsis thaliana*), led scientists to develop a different model organism, the grass *Brachypodium distachyon,* which had links to the plants critical for food security—rice, maize, and wheat. Their chapter discusses how a community of scientists came to support Brachypodium's role as a model organism, how scientists developed the tools expected of a model, and how President George W. Bush ensured that resources would be available for this work by making the development of biofuels a national priority. Here again we return to earlier themes about the links between science, agriculture, and a nation's security: in this case, how food and fuel security created resources that aided the promotion and development of new model organisms that would serve as better tools for the job.

We hope that these essays will stimulate further research at the intersection of life science and agriculture, especially in relation to these issues:

- The interaction of life sciences with economics and politics, including the ways in which national goals and the desire for economic growth drive scientific development
- The transformation and emergence of disciplines
- The material basis of disciplines
- The role and impact of such "founding fathers" of new disciplines as Humboldt, Mendel, and Tschermak
- The effect of the two-way interactions between communities of practitioners and scientists on scientific fields and discourses
- The meaning of "applied science" in both its historical and contemporary uses
- The significance of cross-national comparisons in drawing out common elements as well as crucial differences in the material basis and culture of science

References

Harwood, Jonathan. 2006. Introduction to the special issue on biology and agriculture. *Journal of the History of Biology* 39:237–39.

Müller-Wille, Staffan, and Hans-Jörg Rheinberger. 2012. *A cultural history of heredity*. Chicago: University of Chicago Press.

Denise Phillips is an associate professor in the Department of History at the University of Tennessee, USA. Her research deals with the cultural history of science in eighteenth- and nineteenth-century German-speaking Europe. She is the author of *Acolytes of Nature: Defining Natural Science in Germany*, 1770–1850 (2012).

Sharon Kingsland is a professor in the History of Science and Technology Department at the Johns Hopkins University, USA. Her research deals with the history of modern life sciences with focus on ecological sciences, including population ecology, ecosystem ecology, physiological ecology, and urban ecology. She is coeditor of the Johns Hopkins University Introductory Series in the History of Science.

Chapter 2
Plants and Places: Agricultural Knowledge and Plant Geography in Germany, 1750–1810

Denise Phillips

Introduction

Over the last 20 years, a rich body of literature has explored the ways in which natural history functioned as a science of resources in the eighteenth century. In this period, natural historical inquiry was intimately bound up with debates about national wealth and luxury, and also linked with the integration of novel or exotic products into European markets. Eighteenth-century natural history was a body of knowledge constituted within networks of global exchange.[1]

In the following chapter, I would like to continue to explore the connections between natural history and practical knowledge in the eighteenth century, but with a shift in emphasis. Most past work has focused on the quest for expensive or rare colonial plants and medicines; it has also devoted a great deal of attention to botanical gardens, the sites where such plants could be classified and grown.[2] I would like to consider plants that were grown in less rarified soil, in the fields, meadows, and forests of German-speaking Europe's countryside. In particular, I want to examine how elite interest in agricultural improvement fueled the development of a stronger geographical perspective among botanists in the decades around 1800.

In past work on the history of plant geography, scholars have described the decades between 1790 and 1820 as a key transition period. During these years, a handful of botanists began to analyze more systematically how patterns of plant distribution related to the varying physical conditions present in different places. European commercial and colonial expansion threw up many of the questions that early plant geographers sought to answer, but historians have also identified other intellectual traditions that purportedly infused geographical perspectives into botanical

[1] See, e.g., Koerner (1999), Schiebinger and Swan (2005), Spary (1996), Spary (2003), Müller-Wille (1999), Müller-Wille (2003).

[2] For example, Spary (2000), Schiebinger (2004).

D. Phillips (✉)
Department of History, University of Tennessee, 915 Volunteer Blvd., Knoxville TN 37996, USA
e-mail: aphill13@utk.edu

D. Phillips, S. Kingsland (eds.), *New Perspectives on the History of Life Sciences and Agriculture,* Archimedes 40, DOI 10.1007/978-3-319-12185-7_2

research. Many of the central figures in this new, more geographical strain within botany were German, men like Carl Willdenow and Alexander von Humboldt. In tracing the roots of "Humboldtian" plant geography, scholars have pointed to several features within the German intellectual landscape that encouraged geographical thinking: eighteenth-century state statistics, Wernerian geology, the Kantian geographical tradition, and, last but not least, German Romanticism, whose holistic and aestheticized approach to the study of nature has been seen as particularly important to the thought of Humboldt.[3]

In each of these cases, historians have identified various external examples of geographical thinking, and then argued that these other forms of geographical analysis prompted similar developments within botany. It is certainly correct to think of eighteenth-century botany as part of a wider complex of disciplines devoted to mapping, understanding, and managing territory, a grouping to which fields like Wernerian geology and cameralist state statistics also belonged. There is also little doubt that practices and concepts travelled freely among these different fields. But eighteenth-century scholars and landowners interested in plants had their own particular motivations for reflecting on geographical variation, and it is these motivations that the following chapter explores. In the final third of the eighteenth century, one can find discussions about the complex interrelationships between plants and places in a variety of authors who were working at the intersection of botany with *Oekonomie,* the eighteenth-century field that dealt (among other things) with the study of agriculture.

While most treatments of eighteenth-century economic science have focused on global trade and the promotion of national wealth, the practical management of local agrarian landscapes was an equally important focus of this literature. Historians of agriculture have often argued that a real relationship between science and farming began only in the mid-nineteenth century.[4] According to the knowledge categories that prevailed in the German Enlightenment, however, there had long been a science—*Oekonomie*—whose aim was the improvement of agricultural productivity. This field had well-developed connections to natural history and natural philosophy, connections that were real, not merely rhetorical.[5] Practical, regional agricultural concerns formed an important part of the history of fields like plant geography, which in their Humboldtian form owed a clear debt to the literature on agricultural improvement.

Before moving on to address these claims directly, let me start by describing the science that eighteenth and early nineteenth-century Germans called *Oekonomie.* This category of knowledge was a very old one, dating back to antiquity. In the classical tradition, *Oekonomie* was the body of knowledge that dealt with the running of

[3] Browne (1983, pp. 32–57), Nicolson (1990), Nicolson (1987), Steigerwald (2000), Cooper (2007).

[4] Haushofer (1963), Klemm (1992), Uekoetter (2006), Uekötter (2010). A few authors have described this transition as a more gradual process; see, e.g., Abel (1967).

[5] On connections between learned natural history and agricultural improvement, see Ambrosoli (1997), Koerner (1999).

a landed estate. Previous scholars have argued that this older tradition had died out by the middle of the eighteenth century, giving way to a new science of *Oekonomie* whose primary concern was the economy of the state rather than the household.[6] This transition, they have argued, was caused by the rise of cameralism, a new kind of state science introduced into German universities in the eighteenth century. Central European cameralists wanted to improve state administration and increase state revenues by better managing economic life, and they were particularly interested in getting the most out of a state's natural riches—its mines, forests, and agricultural land.[7]

In fact, the cameralist perspective added to, but did not replace, many older questions that had characterized *Oekonomie* in preceding centuries. While the expansion of cameralism certainly introduced new, more state-centered concerns into *Oekonomie,* the older, household-oriented version of the science also persisted into the nineteenth century. Mid-eighteenth-century authors like Christian Reichart and Johann Gottlieb von Eckhart still used *Oekonomie* in its older, narrower sense in the 1750s, as did writers like Friedrich von Rochow in the 1790s.[8] As late as the mid-1830s, Jena's influential *Allgemeine Literaturzeitung* reviewed books on *Oekonomie* under a joint heading with books on other aspects of household management (cookbooks, for example).[9] Throughout the eighteenth century, a primary concern of *Oekonomie* remained the running of individual estates.

Both of these orientations—towards the dynastic state on the one hand, and the rural estate on the other—meant that enlightened *Oekonomie* was very much a science that raised concrete questions about how plants related to specific places. What could be grown, where, and why? What explained the varied success of different crops in different locations, even within a given region? In this respect, eighteenth-century *Oekonomie* generated just the sorts of questions that nineteenth-century plant geography would try to answer.

Science and Agriculture in the German Enlightenment

In writing the history of German agricultural science, scholars have often organized their narratives around supposed watersheds when the study of agriculture became more closely linked with the practices of natural science. A number of eighteenth- and early nineteenth-century authors have been heralded as early advocates of this shift, most prominently the Göttingen professor Johann Beckmann and the Berlin

[6] Richarz (1991), Tribe (1988). Marion Gray has recognized the continued importance of the household-level of analysis in his work. Gray (2000).

[7] On the aspirations and failures of cameralism, see Wakefield (2009), on cameralist economic thought, Tribe (1988).

[8] Reichart (1753), Eckhart (1754), Rochow (1794).

[9] For example, this grouping was used throughout the 1835 volume of the *Jenaische Allgemeine Literaturzeitung.*

professor Albrecht Thaer. There has been a consensus, however, that a truly robust link between science and agriculture first appeared only in the middle of the nineteenth century with the advent of modern agricultural chemistry. For many scholars, this conjuncture also represented the moment when scientific *expertise* as such became a serious force within German agriculture.[10]

Eighteenth-century elite Germans, however, already thought that they had a body of authoritative knowledge, albeit not a perfect one, that could guide the practice of agriculture.[11] Indeed, in the 1750s and 1760s, it was common to find people saying that so much had already been written on *Oekonomie* that it might seem pointless to publish more. The authors of economic treatises saw themselves as contributing to a well-established field with an ancient and venerable pedigree. The eighteenth-century science of *Oekonomie* differed, of course, from later nineteenth-century concepts of "agricultural science." As I will discuss shortly, the former included a number of topics that were later siphoned off to other fields. *Oekonomie* was nonetheless a coherent and widely cultivated field, and one that was accorded significant cultural and epistemic authority.

Most eighteenth-century agricultural writers also thought that the field of *Oekonomie* had significant connections to natural history and natural philosophy. One can find a range of views on how exactly *Oekonomie* should rely on these other fields, but the claim that it ought to get *something* from them—that sentiment was widespread, and by no means the provenance of a few farsighted professors, as the previous secondary literature on German agricultural science would seem to suggest. It was, rather, part of the standard understandings of the field.[12]

Oekonomie, as mentioned above, had somewhat different boundaries than the nineteenth-century discipline known as agricultural science. It focused on core areas of agricultural production like the growing of field crops and animal husbandry, and also included things like fruit and vegetable gardening and sometimes even ornamental gardening (the latter had its own, separate literature, but got included in handbooks of *Oekonomie* as well). It frequently stretched to include rural crafts and manufacturing activities.[13] In the first half of the nineteenth century, in contrast, discussions of gardening would become more strongly disaggregated from discussions of agriculture proper, and a new science of *Technologie* would take all of manufacturing under its wing.

In addition to having a broader purview than its nineteenth-century analog, *Oekonomie* also had a broader audience. It dealt with topics like fertilizing practices or the growing of rye that were primarily of interest to people who owned or worked large plots of rural land, but it also encompassed discussions of fruit and vegetable gardening that were relevant to wealthy urban burghers. The authors of texts on *Oekonomie,* and, from what we can reconstruct, their audiences, drew from both of these groups. Affluent urban citizens who owned garden plots at their town's edges,

---

[10] See note 4, as well as Finlay (1991a), Finlay (1991b), Rossiter (1975), Borscheid (1976).

[11] Popplow (2010).

[12] See, e.g., Reichart (1758), Flurl (1799), Reuss (1777).

[13] See e.g., Reichart (1753), Eckhart (1754), Münchhausen (1773).

pastors who grew fruit trees in their church yards, or Prussian noblemen planting large amounts of grain for the international market—all of these could find something to interest them in *Oekonomie.*[14]

Eighteenth-century *Oekonomie,* in other words, was a body of knowledge that dealt with a variety of interventions into rural (and sometimes also urban) landscapes. Its practitioners did not just concern themselves with bounded spaces like fields. They wrote extensively about how wooded areas ought to be managed on an estate, and how fruit trees should be planted in orchards or along the sides of roads. They suggested garden designs and pondered the improvement of pastures and meadows.

These varied sites of plant cultivation provided the backdrop for *Oekonomie*'s intersections with the science of botany. To give one telling example, in the early years of the nineteenth century, the Prussian noblewoman Helene Charlotte von Friedland had the botanist Carl Willdenow produce a flora of her estate, and this flora included both the wild and the cultivated plants on her land: the fruits, vegetables, and flowers in her gardens, the crops in her fields, the grasses (both seeded and wild) that grew in her meadows and pastures, and the trees that grew in her woods. Willdenow's Flora recorded the wild and the sown, including carefully tended garden plants alongside weeds as part of one composite description.[15]

Historians of plants geography have often mentioned in passing that figures like Humboldt and Willdenow frequently discussed cultivated plants in their writings and showed a keen interest in agriculture. In what follows, I would like to look more closely at the connections, both intellectual and social, between eighteenth-century botany and *Oekonomie,* with an eye to illustrating how issues central to plant geography emerged at the interstices of these two fields. First, however, I would like to look at why questions about how plants fit with particular places came to seem so pressing to a wide audience of Central European elites.

Oekonomie and the Geography of Plants

By the 1790s, when Alexander von Humboldt or Carl Willdenow wrote their earliest reflections on plant geography, German-speaking Europe had a wider field of authors interested in broadly similar questions. Over the course of the second half of the eighteenth century, several intertwined developments had made the issue of how plants fit with particular places one that was on many people's minds. Historians of botany have already explored how global botanical exploration helped to spark these kinds of interests, but there was also an active intra-European trade in domesticated plants (some from the Old World, some from the New) that prompted

[14] The leading review journal for this field reviewed works in all of these areas, see Beckmann (1770–1806).

[15] Willdenow (1815). Willdenow's original preface from the 1803 first edition was reprinted in this later version.

similar kinds of discussions. On the one hand, this interest in new crops and garden plants was part of the development of a more market-oriented, commercial agriculture in the second half of the eighteenth century. Many German landlords, particularly in Brandenburg-Prussia, were pushing to get more revenue out of their estates, and experimenting with new crops and techniques to this end.[16] On the other hand, an interest in new fruits and vegetables was also part of the growth of consumer society in the German lands. By the later eighteenth century, for example, many villages around cities like Frankfurt am Main had switched to producing fruits and vegetables for sale at nearby urban markets, where a clientele eager for culinary novelty would offer a good price for their wares.[17] Many urban elites also owned leisure gardens by the end of the eighteenth century, and grew flowers, fruits, and vegetables on their own plots.[18]

The publishing networks of *Oekonomie* provided one important forum in which the benefits and perils of new crops or garden plants were vetted. Clover, for example, was one of the most widely discussed agricultural plants of the eighteenth century, and numerous pamphlets, books, and articles were composed singing its praises.[19] Clover's most famous German advocate, Johann Christian Schubart, received a title for his efforts on the plant's behalf. Joseph II raised him into the imperial nobility with the moniker Edler von Kleefeld [literally translated, "noble of the clover field"]. As we now know, clover is a legume that fixes nitrogen to the soil. Though this mechanism was not understood in the eighteenth century, clover was already widely celebrated for its ability to restore fertility to tired land. Its advocates argued that it allowed the farmer to skip the fallow year without exhausting the soil, providing useful fodder for animals in the meantime.[20] Other food and fodder crops—from turnips and potatoes to new kinds of grasses—were also exhaustively discussed in print.

Within this practical literature, the fit between soil, moisture level, region, and plant was a common topic of discussion. Johann Gottlieb Gleditsch warned his readers that fodder crops had to be appropriately matched to the climate [*Klima*] of a place to succeed.[21] Johann Friedrich Mayer advised landowners who wanted to reseed their meadows that grasses grown in dry, heavy soil would be much tastier and more nutritious (to cattle, that is) than those grown in lighter soils.[22] Johann Christian Schubart's career as a clover enthusiast began after he realized that the species of clover then being grown in his region was a poor fit for the area; only after finding a better-suited plant did his endeavors succeed. One always had to

[16] See, e.g., Wunder (1996), Hagen (2002). On the broader European context, see Ambrosoli (1997), pp. 337–398.

[17] Schuricht (2011).

[18] Dülmen (1999).

[19] To name just two examples, Schimper (1780), Schimper (1792).

[20] Rockstroh (1841).

[21] Gleditsch (1782).

[22] Mayer (1792, p. 281).

take the specificities of a place into account when deciding how a plant was likely to behave.[23]

This lesson was learned locally many times in the eighteenth century, as landowners and gardeners experimented with Ray grass, turnips, potatoes, lucerne, and many other domesticated plants. Indeed, enlightened learned societies often devoted considerable effort to testing whether or not novel plants would work in their particular region. The Economic Society of Bern, for example, repeated agricultural experiments undertaken elsewhere to see if they would work in the region around Bern.[24] Furthermore, people who tried out plants in a new place were seen as creating new knowledge worthy of commemoration. The Zedler lexicon, mid-eighteenth-century Germany's most important encyclopedia, argued in its article on experiment that the first person who proved that a crop would grow in a specific area ought to be credited with a new discovery.[25] Much in the same way that a botanist who found a new species deserved recognition, the discovery that something could be grown (or grown better) in a given area was also an important contribution to the storehouse of general knowledge. Christian Reichart, for example, was celebrated in Erfurt and elsewhere for perfecting the cultivation of watercress in his region; he was also known for being the first person in Central Europe to coax cauliflower into seed.[26]

Driven by this widespread interest in new crops and gardening plants, the German seed trade expanded in the second half of the eighteenth century, and the mechanics of buying and selling seeds also raised questions about the complex relationships between plants and places. Alongside a widespread interest in introducing novel food plants came an accompanying concern with knowing whether or not the seeds you were buying (sometimes from another region or nation) would actually flourish in your area.[27]

Over the course of the eighteenth century, German gardeners started growing a much wider range of fruits and vegetables than they had in preceding centuries. A number of edible plants that would have been rare novelties in the early eighteenth century had become widespread by the end of the century.[28] The most famous case, of course, was the potato, but there were many other examples as well. The previously mentioned cauliflower, for example, went from being an exotic novelty to a much more common food, one that deserved multiple recipes in cookbooks like Christian Heinrich Steinbeck's *Neues bürgerliches Kochbuch.*[29] From the careers of people like the seed merchant Christian Reichart, one can reconstruct the growing customer base of nobles and urban elites interested in trying out new plants. Reichart authored one of the most successful eighteenth-century handbooks on *Oekonomie,*

23 Rockstroh (1841, pp. 59–68).

24 Gerber-Visser (2010).

25 [1746] "Versuch-Kunst," *Grosses vollständiges Universal-Lexicon,* p. 2176.

26 Czekalla and Prass (2011).

27 For an example of seed merchants' advertisements, see [1794] "Kauf-und Handels-Sachen."

28 Schuricht (2011).

29 Steinbeck (1826).

and much of his six-volume work can be read as advice to his customers. He reported on the conditions under which certain kinds of seeds would flourish, and advised his readers on the circumstances under which a plant's failure to grow was the fault of the purchaser, not the seed merchant.[30] In other publications of the period, one can see a similar fusion between *Oekonomie* and the commercial trade in plants. The Flora of the Friedland estates, for example, placed a special mark beside plants (mostly newly introduced kinds of trees) that were available for sale.[31]

Europeans' exploitation and exploration of the New World is by now a familiar feature of the history of early modern science.[32] Running parallel to the dramatic and better-studied colonial exchanges of this period, however, was a lively intra-European exchange in food and fodder crops. Many of the plants of interest to eighteenth-century improvers were not from the far corners of the globe; they were imported from another part of Europe. Europeans saw a steady stream of new plants coming in from Asia, the Pacific, and the Americas, many of them rare and exotic foreigners that stayed predominantly in the nurturing confines of carefully tended botanical gardens. But there was also intense interest in more humble kinds of plants: new grasses to reseed meadows, or new vegetables to plant in the kitchen garden.

As these less glamorous fruit varietals, vegetables, legumes, and grains moved around Europe, they sparked discussions of why certain plants flourished in certain locations but not in others. These questions were of obvious practical importance. If a landowner or gardener knew what could successfully be grown in a given place, he could avoid wasting money experimenting with crops that were doomed to fail. By the 1780s and 1790s, a large practical literature dealt with the exigencies of farming, fruit tending, and gardening. Countless eighteenth-century pamphlets and handbooks wrestled with specific, practical questions about what kinds of plants would grow where.[33] More generalized reflections on this topic, however, appeared primarily in works with the explicit aim of forging a closer relationship between botany and *Oekonomie*.

Botany and *Oekonomie*

In 1784, Berlin's Society of Nature-Researching Friends announced a prize competition, soliciting an answer to the following question:

[30] Reichart's correspondence unfortunately does not survive, but he published letters from landowners in his handbook, and often presented his comments as being for the benefit of affluent garden owners. See Reichart (1758).

[31] Willdenow (1815).

[32] See, e.g., Schiebinger and Swan (2005), Drayton (2008), Delbourgo and Dew (2008).

[33] Friedrich Weber began compiling this literature into a bibliography early in the nineteenth century. Weber (1803–1842).

> What kind of economic knowledge about plants taken from botany [*Gewächskunde*] as a whole is actually the kind of knowledge that will put us in a position to accurately identify the natural state, fertility and flaws of land in forests, fields, meadows and so on when we are assessing the worth of parcels of land?

A physician from Montpellier, Pierre Joseph Amoreux, submitted the winning answer, and the society published his essay in German translation in its 1785 proceedings.[34]

The society's prize question addressed an issue of obvious practical importance to contemporary agriculture. What could botany do to help landowners figure out the best uses for their land? How could it guide their decisions about what new plants to try out in their fields, forests, gardens, and meadows? In his winning essay, Amoreaux discussed a number of ways in which botany was useful to *Oekonomie.* He boasted, for example, that the discovery of plant gender had allowed gardeners to figure out why certain trees had been infertile when a female plant was grown in isolation from the male of its species. The central argument of Amoreux's essay, however, dealt with plant geography, what Amoreux (in German translation) called the "*Erdkunde der Pflanzen.*"

The most important contribution botany could make to *Oekonomie,* he argued, was to provide knowledge about the original conditions in which a plant had grown, and then to help *Oekonomen* understand the complex conditions that prevailed in the place that a plant was to be introduced. Amoreux emphasized that someone dealing with this conundrum needed to take multiple factors into consideration. They needed to consider not just the comparative temperatures of the two places in question but also features like elevation and soil type. Indeed, one needed to be very particular in one's observations. Every plot of land was subject to a variety of physical influences that needed to be kept in view.[35] He advised that one quick way to get a sense of what could be grown on a given plot of land was to observe the wild plants that grew nearby, and then continued:

> This distinguishing feature is naturally a good one, but it needs to be based on observations that the country resident [*Landmann*] has occasion to make in every region and even in every area, every corner of the earth: in meadows, fields, in woods, and so on. For these different positions and situations all make up many different climates [*Klimate*].[36]

The *Klima* of a given piece of land determined which plants would grow there, and in this context *Klima* was a very specific term that referred to all the influences that might prevail in a precise location.[37]

The Society of Nature-Researching Friends' question, and Amoreux's answer, belonged to a wider European discussion about how natural history could aid the cause of agricultural improvement. In the German states, one can find a number of handbooks addressing this conjoined set of interests, works like Georg Suchow's

[34] Amoreux (1785). The text of the original prize question is reproduced in the table of contents of the 1785 *Schriften* of the society.

[35] Amoreux (1785, pp. 19–51).

[36] Amoreux (1785, p. 53, quoted on p. 54).

[37] For the broader history of this term, see the introduction to Fleming (2011) and Glacken (1967).

1777 *Ökonomische Botanik* or Heinrich Christoph Moser's 1796 *Deutschlands Oekonomische Flora*.[38] Eighteenth-century educated Germans usually spoke of botany and *Oekonomie* as distinct sciences with distinct practitioners, despite the fact that there were many people whose interests stretched across both fields. For example, Johann Beckmann, probably the most important academic figure in *Oekonomie* in the second half of the eighteenth century, had a correspondence network that included learned naturalists but also a large number of landowners.[39]

Beckmann, as a university professor of cameralism, had a strong institutional investment in asserting natural history and natural philosophy's relevance to practical economic life.[40] There were also other settings in which defending the utility of natural history took on particular strategic importance. Heinrich Christoph Moser, author of one of the textbooks mentioned above, was a professor at a forestry academy, while Georg Suchow was on the faculty of the cameralist academy in Kaiserslautern. In other words, both of these men worked at educational institutions where students received an education that joined together natural history, natural philosophy and the practical sciences.[41]

The practical sciences were typically seen as low-ranking subjects in learned contexts like the universities, but in German society as a whole, *Oekonomie* had considerable clout. Successful authors like Otto von Munchhausen and Friedrich von Rochow were noblemen with political connections and significant social and material resources, and they borrowed from their more learned contemporaries while still possessing great confidence in the powers of their own judgment.[42] For learned naturalists, attempting to explain plants' varying success in different locales provided a good way to generate interest in natural history among a socially and politically influential clientele.

Like Amoreux, other authors on economic botany thought that an understanding of geographic variation was one of the most important services that natural history could provide to agriculture. In 1791, Heinrich Christoph Moser published *Ueber Feld- und Gartenprodukte, mit Rücksicht auf das Klima in Deutschland* [On Field and Garden Products, with Attention to the Climate in Germany], and this book, as one might expect from the title, included an extended discussion of *Klima,* or climate. Moser started his discussion of climate by differentiating geographers' use of the term from its meaning in *Oekonomie*. "The geographer pays attention only to the length of the longest day, while the *Oekonom,*" he wrote, "pays attention to plants."[43] In *Oekonomie,* the defining features of a given climate were the plants

---

[38] Suchow (1777)and Moser (1796).

[39] Much of his correspondence was reprinted in the *Beyträge* he published through the 1770s. See, e.g., Beckmann (1779).

[40] On the importance of the natural sciences to cameralism's public image, see Wakefield (2009).

[41] Lowood (1991).

[42] On Rochow, see Tosch (2010). In the sixth volume of his *Hausvater,* Münchhausen claimed that his decades of experience running his estate had prepared him to create an entirely new natural philosophy; he presented himself as a new Aristotle. Münchhausen (1773).

[43] Moser (1791, p. 148).

that would self-reproduce in that area, and a variety of physical and human influences acted together to determine which plants that would be. The latitude and temperature of a place mattered, but so did the nearness of the sea, forests, or mountains, or the strength of winds, elevation, and amount of rainfall.[44]

In discussing climatic variation, Moser used several examples of the large-scale global variations familiar to historians of botany from past scholarship on this period. He also pointed out, however, that significant variations could be observed even within more limited geographic areas. Within Germany, he noted, there were large differences in how well plants thrived. A particular variety of cherry, for example, produced exemplary fruit when grown in the area around Erfurt. Planted elsewhere, this same variety produced fruit of only middling quality. Moser went on to discuss several unsuccessful attempts to introduce certain kinds of fruit trees to new areas, projects that had failed due to a lack of attention to the peculiarities of a particular German region.[45]

Both Amoreux and Moser claimed to be making novel contributions to botany and *Oekonomie,* but also presented those contributions as part of an ongoing discussion about plants and geographic variation. Both assumed that questions about geographic variation and the growth of plants were of wide practical interest, and both also cited similar authors writing at the borders of natural history and *Oekonomie* who had previously concerned themselves with this issue.[46]

Alexander von Humboldt and Carl Willdenow, generally recognized as the most important German writers on plant geography in this period, had multiple points of connections to the discussions described above. The young Alexander von Humboldt studied cameralism at Frankfurt an der Oder, and in his 1807 essay on plant geography he drew many examples from Friedrich Karl Ludwig Sickler's work on the history of domesticated fruit trees (the origins of domesticated plants was a topic of interest to both Amoreux and Moser, and also to other writers working in economic botany).[47] Willdenow was the leading botanist in the Society of Nature-Researching Friends in the 1780s, when the group announced the prize question with which this section started. He also published several works aimed at gardeners or estate owners. One of these was the Flora of the Friedland estates mentioned earlier; another was a 1796 work on the trees and vines that could be grown outdoors in the area around Berlin.[48]

In his textbook *Grundriss der Kräuterkunde,* Willdenow chose to justify one of his own favored topics of study—mosses—in terms that would have appealed to contemporary *Oekonomen.* It was important to understand the cryptogams, Willdenow argued, because they were the pioneers of the plant kingdom. They were the first to take over rocky ground where soil had been washed away, and through their patient work, they would eventually help rebuild the soil and make the land fertile

[44] Moser (1791, p. 150).

[45] Moser (1791, pp. 151–155).

[46] Amoreux (1785, pp. 39–40) and Moser (1791, p. 155, 157).

[47] Humboldt and Bonpland (1807).

[48] Willdenow (1815) and Willdenow (1796).

once again. Bringing new, previously marginal lands into cultivation was an important component of agricultural improvement in the late eighteenth century, and this image of mosses slowly changing rock to agricultural land would have been a potent plea for the humble crypotogams.[49]

Oekonomie and "Ecological" Perspectives

Willdenow's description of mosses' slow, transformative work is only one example of the ways in which economic botany raised questions about how natural and cultivated landscapes changed. Many of the authors discussed above assumed that the *Klima* of a given region was subject to significant human alteration. "However raw and unfriendly a climate might be," wrote Moser, "it can nonetheless be completely transformed through art and human industry."[50]

Given the prominent place accorded Humboldtian geography in the roots of modern ecological thought, it seems worth pausing to consider the ways in which *Oekonomie* framed the interaction between humans and the natural world. Günter Bayerl has argued that the enlightened practical sciences cultivated a "techno-economic gaze," conceptualizing nature as a warehouse to be exploited. According to Bayerl, this approach to the natural world helped lay the groundwork for modern ecological crises by introducing a new level of human rational rapaciousness into dealings with the environment.[51] It might be tempting to read Moser's claim about humans' ability to transform climate through this lens, as an example of heedless and exploitative confidence. When we set *Oekonomie* in its specific local setting, however, a somewhat different picture emerges. Indeed, *Oekonomie* incorporated many of the habits of thinking about human–nature relationships that have been labeled as novel and innovative in the work of thinkers like Humboldt. Laura Dassow Walls and others have described Humboldt as a prescient voice who merged the human and the natural to form an early version of an aesthetically infused ecological consciousness. Gregory Cushman has credited Humboldt with being the modern discoverer of human-induced climate change.[52] If one looks forward from the eighteenth century, rather than backward from the twenty-first, these claims to novelty seem overdrawn.

At least in a general form, each of the views described above were common in the eighteenth-century practical sciences. First of all, *Oekonomie* was a science that, by definition, fused a consideration of human activity with a discussion of natural processes. It joined together the study of nature and the study of human labor. Economic handbooks considered the total process of production, including how to

[49] Willdenow (1798, p. 438).

[50] Moser (1791, pp. 162–163). For similar convictions in a colonial context, see Golinski (2007), Grove (1996).

[51] Bayerl (1994) and Bayerl (2001).

[52] Walls (2009), Sachs (2003), Cushman (2011).

manage the labor of the people who planted, tended and harvested the crops that elite Germans wanted to see grown.[53] For example, in an article that looked at the influence of "locale" on agriculture experiments, the Prussian nobleman Friedrich von Rochow thought of a locale as defined not just by weather or the type of soil but also by the type of people who lived there. Their customs, laws, physical traits, and traditions influenced what was possible in a given place. This concept of place had philosophical roots in thinkers like Montesquieu and in even older, classical sources, but for figures like Rochow, such ideas also formed part of practical deliberations about what was possible on a given estate. In his own improvement efforts, he wrote, he had learned that things could not be thoughtlessly transported from region to region; both Nature and human actors placed limits on what was possible.[54]

Second, *Oekonomie* was a science that cared deeply about nature's productivity, but also celebrated its beauty. Rational calculation to improve productivity was certainly a part of this science, but so were many sentiments that fit into the prehistory of modern environmentalism. For eighteenth-century practitioners of *Oekonomie,* the desire to turn natural riches into the coin of the realm was tempered by other, equally important desires. Handbooks on *Oekonomie* emphasized the importance of preserving the long-term fertility of the land, an essential task for noble families whose status and security depended on the continued productivity of their estates. The landscapes of a given region were not warehouses of raw material to be plundered, but trusts to be tended for future seasons and coming generations. Christian Reichart, for example, ended his six-volume handbook with a poem, asking God to preserve his garden for his children and his children's children.[55] Many German landowners in this period also wanted their estates to be as pleasing to the eye as they were to the pocketbook. In the pages of agricultural advice manuals, the "ornamental estate" was the goal of agricultural reform—an estate where beauty and productivity were seamlessly merged. There were also numerous examples of landowners who tried to put these principles into practice. Noble estates, after all, were not just spaces for generating profit, but locations for the display of status-appropriate taste and knowledge.[56]

Conclusion

When looking at the early work of Humboldt and Willdenow on plant geography, it would be taking the point too far to argue that these works were written only as responses to local and regional agricultural concerns. Many of the questions these thinkers addressed came out of debates within natural history itself, as Janet

[53] On enlightened economic writings aimed at peasants, see Böning (1989) and Böning (2004).

[54] Rochow (1785).

[55] Reichart (1755).

[56] Düselder (2009).

Browne has shown.[57] Agricultural discussion, however, deserves significant credit for the growing importance of geographical questions within eighteenth-century German botany. Both Willdenow and Humboldt used many domesticated plants as examples in their essays, and this fact is not surprising.[58] These plants were the ones Europeans knew best. Thanks to the exchange networks of agricultural improvers and gardening enthusiasts, these were also the plants whose success or failure in different climes had been most carefully documented. It is also worth emphasizing that questions about plant geography emerged not just in the context of European colonial expansion, important though that was. Of comparable importance was a new, more intensive intra-European exchange of cultivated plants between regions and nations of the Old World.

Many of the circumstances described above were not unique to German-speaking Europe in this period. The fascination with agricultural improvement that gripped so many German noblemen was also to be found among French and British elites.[59] Authors in these different linguistic contexts also read each other's work. Indeed, one of the authors discussed above, Pierre Joseph Amoreux, was a Frenchman who submitted an essay to the Society of Nature-Researching Friends, a Berlin-based society whose membership stretched across Europe.[60] But given the importance accorded the Humboldtian strain of plant geography, it is worth emphasizing some particular features that shaped discussions of plant geography in German-speaking Europe. In the late eighteenth century, Germany's practical academies of mining, forestry and cameralism, as well as its university chairs of cameralism, provided a context in which bringing natural historical traditions together with practical questions took on particular programmatic importance. Here, we find a group of men, people like Moser and Suchow, who had a concrete interest in forging connections between natural history and practical economic concerns, in part to bolster the profile of their novel educational institutions.

If we want to understand the reasons for botany's increased regional focus in this period, agriculture debate is one essential place to look. Several other intellectual traditions have previously been suggested as important sources for this new geographically inflected form of botany; Kantian geography, Werner's geology, and state statistics have all been put forward as possible models for plant geography. In each of these cases, however, scholars describe a situation where a geographical mode of thinking that originated elsewhere later began to inform thinking about botany as well. There is another German intellectual tradition, however, in which questions of plant geography emerged directly, part and parcel of *Oekonomie's* practical debates about the growing of plants. All of these related fields—state statistics, geography, and geology—were part of broader, interlocking discussions of place, governance, and economic improvement in eighteenth-century Central Europe. It

[57] Browne (1983).

[58] Jackson (2009, p. 21, 28).

[59] Drayton (2008), Spary (1996), Spary (2000). On eighteenth-century economic societies more generally, see Stapelbroek and Marjanen (2012). See also McClellan (1985), Lowood (1991).

[60] Heesen (2001) and Böhme-Kassler (2005).

was in agricultural discussions, however, that questions about plants and places figured most prominently and explicitly. As Heinrich Moser suggested, thinking about climate from the perspective of *Oekonomie* meant thinking about plants.

References

1746. Versuch-Kunst. *Grosses vollständiges Universal-Lexicon aller Wissenschaften und Künste*, vol. 47. Leipzig: Johann Heinrich Zedler.

1794. Kauf- und Handels-Sachen. Bekanntmachung den Verkauf von Sämereyen betreffend. *Kaiserlich privilegirter Reichs-Anzeiger* 1:180.

Abel, Wilhelm. 1967. *Geschichte der deutschen Landwirtschaft: vom frühen Mittelalter bis zum 19. Jahrhundert*. Stuttgart: Eugen Ulmer.

Ambrosoli, Mauro. 1997. *The wild and the sown: Botany and agriculture in Western Europe, 1350–1850*. Cambridge: Cambridge University Press.

Amoreux, Pierre Joseph. 1785. Physikalische-botanische Abhandlung zur Beantwortung der Preissfrage der Gesellschaft naturforschender Freunde für das Jahr 1784. *Schriften der Gesellschaft naturforschender Freunde* 6:1–71.

Bayerl, Günter. 1994. Prolegomenon der ‘Grossen Industrie’: der technisch-ökonomische Blick auf die Natur im 18. Jahrhundert. In *Umwelt-Geschichte: Umweltverträgliches Wirtschaften in Historischer Perspektive,* ed. Wener Abelshauser, 29–56. Göttingen: Vandenhoeck & Ruprecht.

Bayerl, Günter. 2001. Die Natur als Warenhaus: der technisch-ökonomische Blick auf die Natur in der Frühen Neuzeit. In *Umwelt-Geschichte: Arbeitsfelder-Forschungsansätze-Perspectiven,* ed. Sylvia Hahn and Reinhold Reith, 33–52. Wien: Verlag für Geschichte und Politik.

Beckmann, Johann. 1779. *Beyträge zur Oekonomie, Technologie, Polizey und Cameralwissenschaft*. Göttingen: W. Vandenhoeck.

Beckmann, Johann. 1770–1806. *Physikalisch-ökonomische Bibliothek worinn von den neuesten Büchern, welche die Naturgeschichte, Naturlehre und die Land- und Stadtwirthschaft betreffen*. Göttingen: W. Vandenhoeck.

Böhme-Kassler, Katrin. 2005. *Gemeinschaftsunternehmen Naturforschung: Modifikation und Tradition in der Gesellschaft Naturforschender Freunde zu Berlin 1773-1906*. Stuttgart: Franz Steiner.

Böning, Holger. 1989. Der ‘Gemeine Mann’ als Adressat aufklärerischen Gedankengutes: ein Forschungsbericht zur Volksaufklärung. *Das 18. Jahrhundert* 12 (1): 52–82.

Böning, Holger. 2004. Popularaufklärung-Volksaufklärung. In *Macht des Wissens: Die Entstehung der Modernen Wissensgesellschaft,* ed. Richard van Dülmen and Sina Rauschenbach, 563–585. Cologne: Böhlau.

Borscheid, Peter. 1976. *Naturwissenschaft, Staat und Industrie in Baden, 1848-1914*. Stuttgart: Klett.

Browne, Janet. 1983. *The secular ark: Studies in the history of biogeography*. New Haven: Yale University Press.

Cooper, Alix. 2007. *Inventing the indigenous: Local knowledge and natural history in early modern Europe*. Cambridge: Cambridge University Press.

Cushman, Gregory T. 2011. Humboldtian science, Creole meteorology, and the discovery of human-caused climate change in South America. *Osiris* 26:19–44.

Czekalla, Eberhard, and Reiner Prass. 2011. Christian Reichart und der Erwerbsgartenbau im 18. Jahrhundert. In *Blumenstadt Erfurt: Waid-Gartenbau-iga/egapark*, ed. Martin Baumann and Steffen Raßloff, 49–74. Erfurt: Sutton.

Delbourgo, James, and Nicholas Dew, ed. 2008. *Science and empire in the Atlantic world*. New York: Routledge.

Drayton, Richard Harry. 2008. *Nature's government: Science, imperial Britain, and the 'improvement' of the world*. New Haven: Yale University Press.
Dülmen, Andrea van. 1999. *Das irdische Paradies: Bürgerliche Gartenkultur der Goethezeit*. Köln: Böhlau.
Düselder, Heike. 2009. Von den angenehmen ländlichen Beschäftigungen. Die Bedeutung der Gärten des Adels für Gartenkultur und Wirtschaft. *Zeitschrift für Agrargeschichte und Agrarsoziologie* 57 (1): 10–26.
Eckhart, Johann Gottlieb. 1754. *Experimental-Oekonomie*. Jena: Hartung.
Finlay, Mark. 1991a. The rehabilitation of an agricultural chemist: Justus von Liebig and the seventh edition. *Ambix* 38:155–166.
Finlay, Mark. 1991b. Science and practice in German agriculture: Justus von Liebig, Hermann von Liebig, and the agricultural experimental stations. In *Worldviews and scientific discipline formation,* ed. W. R. Woodward and R. S. Cohen, 309–320. Dordrecht: Kluwer.
Fleming, James, ed. 2011. *Klima*. Osiris 26.
Flurl, Mattias. 1799. *Rede von dem Einflusse der Wissenschaften insbesondere der Naturkunde auf die Kultur einer Nation*. Munich: Joseph Lindauer.
Gerber-Visser, Gerrendina. 2010. 'Die Beschaffenheit des Landes': topographische Beschreibungen der Oekonomischen Gesellschaft Bern. In *Landschaften Agrarisch-ökonomischen Wissens,* ed. Marcus Popplow, 261–275. Münster: Waxmann.
Glacken, Clarence J. 1967. *Traces on the Rhodian shore: Nature and culture in western thought from ancient times to the end of the eighteenth century*. Berkeley: University of California Press.
Gleditsch, Johann. 1782. Beytrag zur Geschichte der einheimischen Futterkräuter in der Mark Brandenburg. *Schriften der Gesellschaft Naturforschender Freunde* 3:42–83.
Golinski, Jan. 2007. American climate and the civilization of nature. In *Science and empire in the Atlantic world,* ed. James Delbourgo and Nicholas Dew, 153–174. New York: Routledge.
Gray, Marion W. 2000. *Productive men, reproductive women: The agrarian household and the emergence of separate spheres during the German enlightenment*. New York: Berghahn Books.
Grove, Richard H. 1996. *Green imperialism: Colonial expansion, tropical island Edens and the origins of environmentalism, 1600–1860*. Cambridge: Cambridge University Press.
Hagen, William W. 2002. *Ordinary Prussians: Brandenburg Junkers and villagers, 1500-1840*. Cambridge: Cambridge University Press.
Haushofer, Hans. 1963. *Die deutsche Landwirtschaft im technischen Zeitalter*. Stuttgart: Eugen Ulmer.
Heesen, Anke te. 2001. Vom Naturgeschichtlichen Investor zum Staatsdiener: Sammler und Sammlungen der Gesellschaft Naturforschender Freunde zu Berlin um 1800. In *Sammeln als Wissen: Das Sammeln und seine Wissenschaftsgeschichtliche Bedeutung,* ed. Anke te Heesen and E. C Spary, 62–84. Göttingen: Wallstein.
Humboldt, Alexander von, and Aimé Bonpland. 1807. *Ideen zu einer Geographie der Pflanzen*. Tübingen: F. G. Cotta.
Jackson, Stephen T. 2009. Introduction: Humboldt, ecology, and the cosmos. In *Essay on the geography of plants,* Trans. S. Romanowski, ed. A. von Humboldt and A. Bonpland, 1–46. Chicago: University of Chicago Press.
Klemm, Volker. 1992. *Agrarwissenschaften in Deutschland: Geschichte-Tradition*. St. Katharinen: Scripta-Mercaturae-Verlag.
Koerner, Lisbet. 1999. *Linnaeus: Nature and nation*. Cambridge: Harvard University Press.
Lowood, Henry. 1991. *Patriotism, profit, and the promotion of science in the German enlightenment: The economic and scientific societies, 1760-1815*. New York: Garland.
Mayer, Johann Friedrich. 1792. *Das Ganze der Landwirthschaft*. Nürnberg: Johann Eberhard Zeh.
McClellan, James E. 1985. *Science reorganized: Scientific societies in the eighteenth century*. New York: Columbia University Press.
Moser, Heinrich Christoph. 1791. *Ueber Feld- und Gartenprodukte, mit Rücksicht auf das Klima in Deutschland: für den Naturliebhaber, denkenden Oekonom und Gartenfreund*. Leipzig: S. L. Crusius.

Moser, Heinrich Christoph. 1796. *Deutschlands ökonomische Flora*. Leipzig: in der von Kleefeldischen Buchhandlung.
Müller-Wille, Staffan. 1999. *Botanik und Weltweiter Handel: Zur Begründung eines natürlichen Systems der Pflanzen durch Carl von Linné (1707–78)*. Berlin: VWB.
Müller-Wille, Staffan. 2003. Nature as a marketplace: The political economy of Linnaean botany. *History of Political Economy* 35:154–172.
Münchhausen, Otto von. 1773. *Der Hausvater*, vol. 6. Hannover: Im Verlag seel. Nic. Försters und Sohns Erben.
Nicolson, Malcolm. 1987. Alexander von Humboldt, Humboldtian science, and the origins of the study of vegetation. *History of Science* 25:167–194.
Nicolson, Malcolm.1990. Alexander von Humboldt and the geography of vegetation. In *Romanticism and the sciences,* ed. Andrew Cunningham and Nicholas Jardine, 169–185. Cambridge: Cambridge University Press.
Popplow, Marcus. 2010. Die ökonomische Aufklärung als Innovationskultur des 18. Jahrhunderts zur optimierten Nutzung natürlicher Ressourcen. In *Landschaften Agrarisch-ökonomischen Wissens: Strategien innovativer Ressourcennutzung in Zeitschriften und Sozietäten des 18. Jahrhunderts,* ed. M. Popplow, 3–48. Münster: Waxmann.
Reichart, Christian. 1753. *Land- und Garten-Schatz. Erster Theil.* Erfurt: Johann Heinrich Nonnen.
Reichart, Christian. 1755. *Land- und Garten-Schatz*, vol. 6. Erfurt: Johann Heinrich Nonnen.
Reichart, Christian. 1758. *Einleitung in den Garten- und Acker-Bau.* Erfurt: Johann Heinrich Nonnens sel. Erben.
Reuss, Christian Friedrich. 1777. Abhandlungen wie die Naturkunde der Grund zu einer wohleingerichteten Oekonomie, und wie gross der Einfluss derselben in dieser Wissenschaft ist. *Beschäftigungen der Berlinischen Gesellschaft naturforschender Freunde* 3:3–28.
Richarz, Irmintraut. 1991. *Oikos, Haus und Haushalt.* Göttingen: Vandenhoeck & Ruprecht.
Rochow, Friedrich Eberhard von. 1785. Ueber den Einfluss die Lokalen in die Anwendung allgemeiner ökonomischen Grundsätze. *Annalen der Märkischen Oekonomischen Gesellschaft zu Potsdam* 1 (3): 58–75.
Rochow, Friedrich Eberhard von. 1794. Ueber die Schwierigkeit, solche ökonomische Versuche zu machen, die, als richtige Erfahrungen, allgemeine Verhaltungsregeln geben. *Annalen der Märkischen Oekonomischen Gesellschaft zu Potsdam* 1 (2): 60–71.
Rockstroh, Christian. 1841. *Johann Christian Schubart, Edler von Kleefeld.* Dresden: Arnold.
Rossiter, Margaret W. 1975. *The emergence of agricultural science: Justus Liebig and the Americans, 1840-1880.* New Haven: Yale University Press.
Sachs, Aaron. 2003. The ultimate 'other': post-colonialism and Alexander von Humboldt's ecological relationship with nature. *History and Theory* 42 (4): 111–135.
Schiebinger, Londa. 2004. *Plants and empire: Colonial bioprospecting in the Atlantic world.* Cambridge: Harvard University Press.
Schiebinger, Londa L. and Claudia Swan, ed. 2005. *Colonial botany : Science, commerce, and politics in the early modern world.* Philadelphia: University of Pennsylvania Press.
Schimper, Johann Conrad. 1780. *Kurzgefasste jedoch vollständige Anweisung zum Kleebau aller Art*. Zweybrücken: Hallanzy.
Schimper, Johann Conrad. 1792. *Meine Erfahrungen über den Kleebau*. Ulm: Stettin.
Schuricht, Werner. 2011. *Zur Geschichte des Obstbaus in Mitteldeutschland vom Mittelalter bis zum Jahre 1945*. Erfurt: Deutsches Gartenbaumuseum.
Spary, E. C. 1996. Political, natural and bodily economies. In *Cultures of natural history,* ed. Nicholas Jardine, James A. Secord, and E. C. Spary, 178–196. Cambridge: Cambridge University Press.
Spary, E. C. 2000. *Utopia's garden: French natural history from old regime to revolution*. Chicago: University of Chicago Press.
Spary, E. C. 2003. 'Peaches which the patriarchs lacked': Natural history, natural resources, and the natural economy in France. *History of Political Economy* 35:14–41.

Stapelbroek, Koen, and Jani Marjanen, ed. 2012. *The rise of economic societies in the eighteenth century*. New York: Palgrave Macmillan.
Steigerwald, Joan. 2000. The cultural enframing of nature: Environmental histories in the German romantic period. *Environment and History* 6:451–496.
Steinbeck, Christian Heinrich. 1826. *Neues bürgerliches Kochbuch, oder gründliche Anweisung zur Kochkunst für alle Stände*. Nonneburg: im literarischen Comptoir.
Suchow, Georg Adolph. 1777. *Ökonomische Botanik*. Mannheim: C. F. Schwan.
Tosch, Frank. 2010. Der Aufklärertypus Friedrich Eberhard von Rochow. In *Landschaften Agrarisch-Ökonomischen Wissens,* ed. M. Popplow, 155-173. Münster: Waxmann.
Tribe, Keith. 1988. *Governing economy*. Cambridge: Cambridge University Press.
Uekoetter, Frank. 2006. Know your soil: Transitions in farmers' and scientists' knowledge in Germany. In *Soils and societies: Perspectives from environmental history,* ed. J. R. McNeill and Verena Winiwarter, 322–340. Isle of Harris: White Horse Press.
Uekötter, Frank. 2010. *Die Wahrheit ist auf dem Feld: Eine Wissensgeschichte der deutschen Landwirtschaft*. Göttingen: Vandenhoeck & Ruprecht.
Wakefield, Andre. 2009. *The disordered police state: German cameralism as science and practice*. Chicago: University of Chicago Press.
Walls, Laura Dassow. 2009. *The passage to cosmos: Alexander von Humboldt and the shaping of America*. Chicago: University of Chicago Press.
Weber, Friedrich Benedict. 1803–1842. *Handbuch der ökonomischen Literatur, oder, systematische Anleitung zur Kenntniss der deutschen ökonomischen Schriften*. Berlin: H. Frölich.
Willdenow, Karl Ludwig. 1796. *Berlinische Baumzucht: oder, Beschreibung der in den Gärten um Berlin im freien ausdauernden Baume und Straucher fur Gartenliebhaber und Freunde der Botanik*. Berlin: G. C. Nauk.
Willdenow, Karl Ludwig. 1798. *Grundriss der Kräuterkunde zu Vorlesungen entworfen,* 2nd ed. Berlin: Haude und Spener.
Willdenow, Karl Ludwig. 1815. *Verzeichniss der auf den Friedländschen Gütern cultivirten Gewächse. Nebst einem Beitrage zur Flora der Mittelmark*, 3rd ed.
Wunder, Heide. 1996. Agriculture and agarian society. In *Germany: A new social and economic history,* ed. Sheilagh Ogilvie, vol. 2, 64–99. London: Arnold.

Denise Phillips is an associate professor in the Department of History at the University of Tennessee, USA. Her research deals with the cultural history of science in eighteenth and nineteenth century German-speaking Europe. She is the author of *Acolytes of Nature: Defining Natural Science in Germany*, 1770–1850 (2012).

Chapter 3
Drawing the Line: Mapping Cultivated Plants and Seeing Nature in Nineteenth-Century Plant Geography

Nils Güttler

Introduction

In summarizing early efforts at botanical distribution mapping, cartographer Max Eckert had only a few positive things to say. His comments on this topic in his mid-1920s reference work *Die Kartenwissenschaft* (*Cartographical Science*) described a history full of failures and misunderstandings. By 1800, Alexander von Humboldt and other key figures of early plant geography had invented botanical distribution maps, but in Eckert's eyes, not much else had happened in the first decades of the discipline. Without a doubt, he admitted, Humboldt's famous cartographical cross sections of the South American continent (Fig. 3.1) had been "picturesque," but they basically matched the expectations of "naive minds" and could hardly "satisfy" the needs of "science."[1] The same was true for the botanical section in Heinrich Berghaus' *Physical Atlas,* which appeared in the mid-1840s and is today often regarded as a visual compendium of Humboldt's *Cosmos* and a landmark of early distribution mapping.[2] Here again, Eckert found the method of most maps either strange or completely unscientific. Vegetation maps of this period, he concluded with deep disappointment, turned out to be a "stillborn child" (*totgeborenes Kind*).[3]

But Eckert made a significant exception. For him, one of the maps in the *Physical Atlas* stood out: the world map on the distribution of cultivated plants (*Kulturpflanzen*) (Fig. 3.2). In this map, the reader could not only compare the distribution areas of particular crops to average temperatures (represented through lines, so-called

[1] Eckert (1921–1925, quotes vol. 2, p. 388).

[2] Berghaus (1845–1848), sect. V (botanical geography). This atlas was soon reissued in Great Britain, see Johnston (1848). For the history of these atlases: Camerini (1993), Espenhorst (2003–2008, vol. 1, pp. 365–390).

[3] Eckert (1921–1925, vol. 2, p. 389).

N. Güttler (✉)
Chair for Science Studies, ETH Zurich Gebäude RZ G 9.1,
Clausiusstrasse 59, 8092 Zurich, Switzerland
e-mail: nils.guettler@wiss.gess.ethz.ch

D. Phillips, S. Kingsland (eds.), *New Perspectives on the History of Life Sciences and Agriculture,* Archimedes 40, DOI 10.1007/978-3-319-12185-7_3

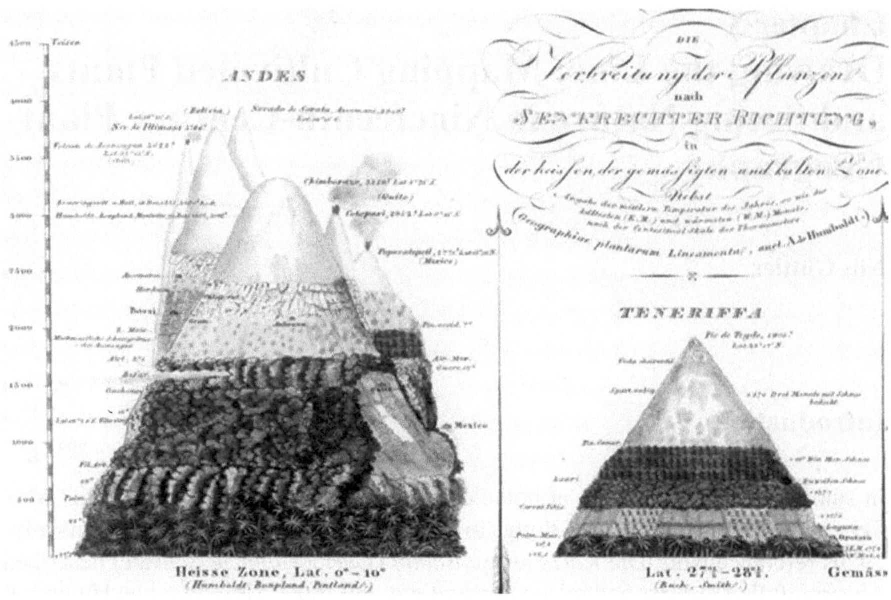

Fig. 3.1 Humboldtian cross sections. (Extract from Berghaus 1845–1848, "Umrisse der Pflanzengeographie," (*originally colored*), Sect. 5, No. 1)

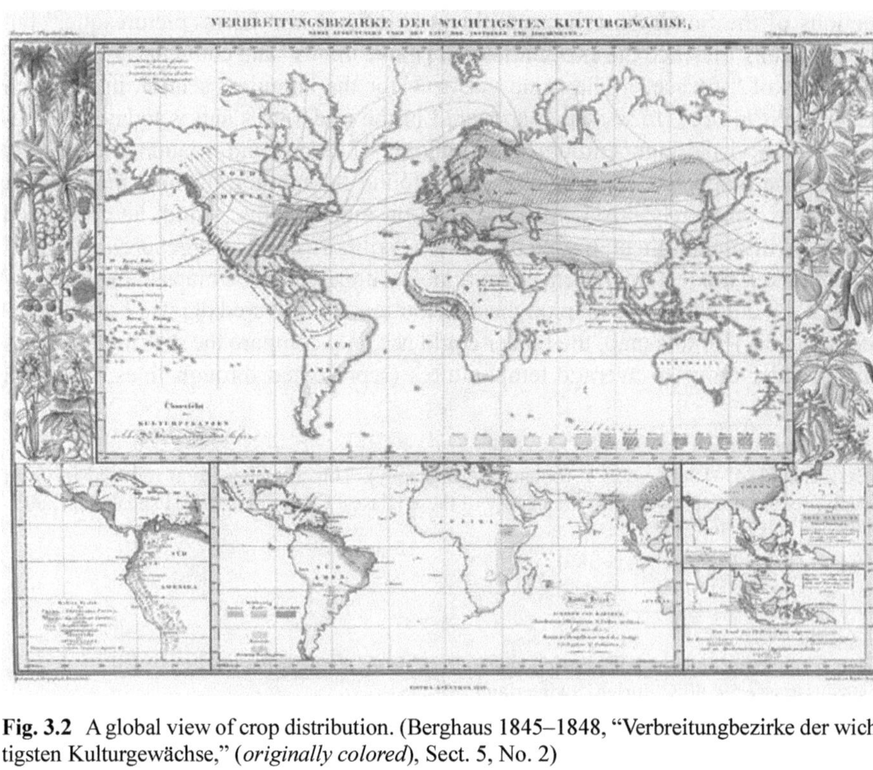

Fig. 3.2 A global view of crop distribution. (Berghaus 1845–1848, "Verbreitungbezirke der wichtigsten Kulturgewächse," (*originally colored*), Sect. 5, No. 2)

isotherms and isochimens); smaller maps also contained information about particular "tropical plants," showing the distribution of coffee, cocoa, or cinchona. Because it included a wide range of observational data and depicted a complex set of relationships, this map "pointed in the right direction, for which the time was not yet ripe."[4]

Despite his harshly polemical tone and explicitly teleological framework, Eckert made a notable observation. Even if one can easily challenge his ahistorical decision to judge the early nineteenth century by the scientific and iconographic standards of his own time, it is true that for a long time the cartographical depiction of the distribution of wild plants—in contrast to their cultivated companions—was an extremely difficult task.[5] In fact, not many maps on the distribution of wild plants were published until the mid-nineteenth century, and naturalists regularly complained about the limited applicability of maps to botany. In the 1830s, Victorian naturalist Richard B. Hinds was even skeptical that his colleagues would ever be able to use maps as a research tool. After having discussed all the botanical maps he knew, he remarked: "[I]t is not to be expected that any maps with which we may be furnished would convey much information. The few that exist are, therefore, very bare of facts, containing merely the names of some plants.... They seem framed more to meet and please the general reader, than for any benefit for the advancement of science."[6]

Whereas botanists like Hinds or his colleague Hewett C. Watson[7] were not sure what exactly to depict in maps of wild plants (landscapes? the distribution of singular species? statistical relationships?), maps of cultivated plants and particularly crops were apparently easier to design. Particularly when it came to publishing, maps on cultivated plants seemed to be more promising. They soon became widespread through atlases and journals; they were also presented to the public in other forums, for instance at the great exhibitions.[8] Thus, when naturalists thought cartographically about plant distribution in the first half of the nineteenth century, they primarily had crop maps in front of their eyes.

Starting from these observations, my chapter will discuss the practice of mapping crops from the late eighteenth to the mid-nineteenth century with the aim of exploring its impact on the conceptual framework of early plant geography. My goal is to show that in this period, when it came to mapping, plant geographers fundamentally relied on knowledge of applied botany, especially knowledge of agriculture. Crop

---

[4] "Damit wird ein kartographischer Weg gezeigt, den zu beschreiten die Zeit noch nicht reif war." Eckert (1921–1925, vol. 2, p. 389).

[5] A comprehensive history of early botanical distribution maps does not exist yet. For the broader context of "thematic" (distribution) mapping, see Robinson (1982, pp. 100–108). For the second half of the nineteenth century, Güttler (2011).

[6] Hinds (1835, p. 498).

[7] Watson (1836).

[8] A significant example for the popularity of "crop mapping" is the most widely distributed geographical journal in this time, so-called *Petermanns Geographische Mittheilungen*. Here, observations on the distributions of crops dominated observations on wild plants. See, for instance, the index volume of the journal's first decade: Anonymous (1865, pp. 43–44).

maps thus became an important reference point for maps of wild plants. I argue that the practical and conceptual importance of crops in early plant geography can be explained by two factors. First, mapping as a practice itself came out of a milieu of knowledge production that was mainly driven by applied and economic interest in plant distribution. Early distribution mapping was closely linked to discourses on the inventory of regions and states that took place all over Europe. Economists' and agriculturists' techniques of visual "data management" had a sustained epistemic impact on early plant geography. Many economists were keen to cartographically simulate a general view, a so-called *coup d'œil,* of their collected data. This cartographical practice of looking at plant distribution at a glance was soon adapted by the first generation of plant geographers.

Second, for early plant geographers, the distribution borders of crops gave an indication of how wild plants might also be distributed in space. As I will show, many botanists acted on the assumption that farmers would always push the environmental limits of particular crops as far as nature allowed. Thus, the borders of crops gave a hint to the invisible borders of nature. By cartographically drawing the line of crop distribution, botanists hoped to approximate the geographical dispositions of the plant kingdom in general.

With my focus on the entangled history of "pure" and "applied" distribution mapping, I follow a marked shift within recent scholarship on nineteenth-century botany.[9] For the last two decades, historians have convincingly uncovered how deeply nineteenth-century plant geographers were embedded in political, economic, and colonial infrastructures.[10] Yet, as soon as it comes to mapping, historians still tend to focus on the intellectual or epistemic functions of botanical maps, especially because "mapping" played a key role in the framework of "Humboldtian science."[11] This chapter has a more skeptical view on the actual role of Humboldt and "his" science, although it goes without saying that Humboldt had a sustained influence on plant geographers. But if one looks at the practice of mapping, it was clearly the emergence of "lay" audiences that allowed maps to become widely used. These audiences, consisting of amateur botanists with an applied interest in plant distribution as well as wider "botanophile" audiences on the private cartographical market, were the driving force that made maps a widely used research tool in botany.

The chapter is divided into four parts. The first part begins with the *Carte botanique de France,* designed by the Parisian botanist Augustin-Pyramus de Candolle in 1805; it explores the broader institutional context of early plant geography, particularly with respect to applied botany. In the second part, I link Candolle's map to

9 For the history of plant geography more generally, see Nelson (1978), (Browne 1983), Nicolson (1996).

10 The literature on colonial botany has grown enormously after the pioneering work of Brockway (1979). See especially: Browne (1992), Browne (1996), Drayton (2000), Schiebinger and Swan (2005), Endersby (2008). I will refer to more literature concerning the role of botanical gardens later in this chapter.

11 Dettelbach (1996) and Dettelbach (1999).

particular forms of travel writing and to discourses on inventories of states and their cartographical visualization, especially those developed by German cameralists. The next part will concentrate on the widespread assumption that studying crop distribution could reveal more general laws of nature, and the last part will describe the role of botanical distribution mapping on the cartographical market. Following crop maps throughout the nineteenth century shows that mapping was basically a pan-European practice until well into the second half of nineteenth century. Due to the particularly booming cartographical market in Central Europe, however, the German states became a stronghold for this visualization technique from the 1840s onwards.

Candolle's "Carte botanique de France" and Post-revolutionary Botany

In 1805, Jean-Baptiste Lamarck published the third edition of his *Flore Francaise.*[12] It enumerated all plants that had been found in the state's territory, and it could easily be read as a political statement: Society changes, nature does not. The first edition of the flora appeared one year before the storming of the Bastille, the second in the midst of the revolutionary struggles in 1795. When the third edition was printed, Napoleon was abroad fighting against the Habsburg Empire. Despite revolution, war, and economic struggles, all editions of *Flore Francaise* became bestsellers.[13] Encouraged by the work's public success, Lamarck never changed the flora's general structure, and only added new observations. But the third edition had quite an unusual feature. It was supplemented by a map of France displaying the general patterns of plant distribution on French soil: the *Carte botanique de France* (Fig. 3.3).

The map was designed by Augustin-Pyramus de Candolle, one of Lamarck's colleagues at the Museum d'Histoire Naturelle in Paris and coeditor of the third edition. It was one of the first plant geographical maps in history.[14] Its visual complexity was already remarkable. With the exception of France's national borders, Candolle had deleted all political and administrative boundaries. Contour lines (a recently developed method for displaying different heights above the sea level) allowed the reader to compare plant distribution with the state's topography. Drawing upon a large number of local observations, Candolle divided the French territory into five, entangled "botanical regions" which he signified with colors.

There was more botanical information in this map, although it was less visible at first glance. Three black lines crossed the "botanical regions" as parallels reaching from the southwest to the northeast (Fig. 3.3). These lines displayed the northern boundary of the distribution areas of several cultivated plants: olive, maize,

[12] Lamarck and Candolle (1805). For the history of this flora in the context of French botany, see Williams (2001, pp. 61–68).

[13] Stafleu and Cowan (1976–1988, vol. 1, p. 442 and vol. 2, pp. 731–732).

[14] Ebach and Goujet (2006).

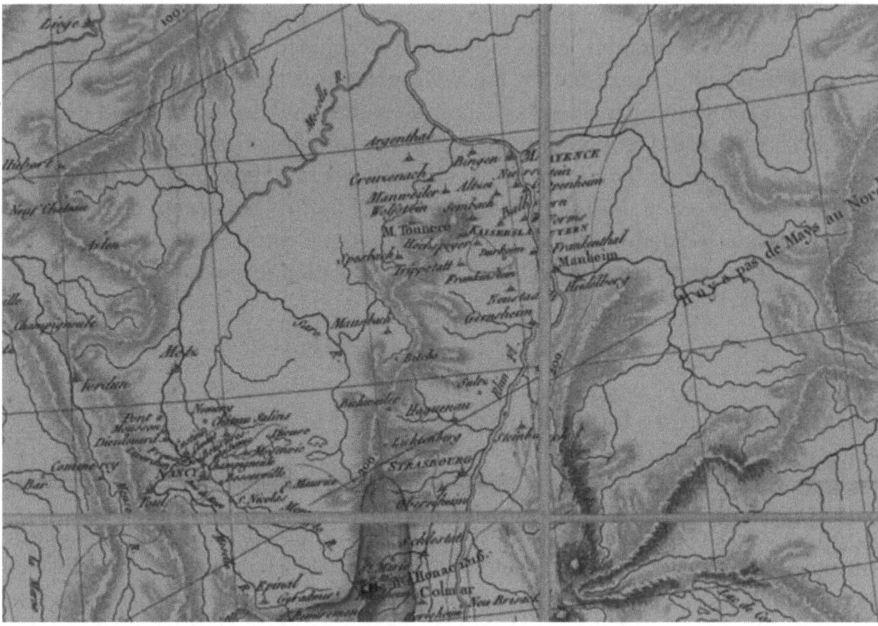

Fig. 3.3 Depicting post-revolutionary botany. (Extract from Lamarck and Candolle's "Carte botanique de France" (*originally colored*), 3rd ed., vol. 2, 1805)

and wine. In the text that accompanied his map, Candolle noted that he had drawn these lines based on the observations of Arthur Young, one of the most well-known European writers on agriculture: "This esteemed traveler, who has attentively studied cultivated plants, remarked that if lines are traced between the northern points where olive trees, grapes and maize are grown, three nearly parallel lines are obtained that all tend to meet to the north on the eastern side."[15] Indeed, Arthur Young's account of French agriculture—his *Travels, during the Years 1787, 1788, and 1789* (first published in 1792)—contained a map that was accompanied by a long descriptive passage.[16] Before Young travelled to France, he had studied the agriculture of English counties like Suffolk or Hertfordshire, and had cartographically depicted the geology and botany of these regions.[17] The similarity of Candolle's and Young's map is striking: Young's *New Map on the Climate and Navigation of France* showed exactly those lines of cultivated plants that Candolle copied later. Indeed, one could easily mistake Young's map for an uncolored version of Candolle's *Carte botanique de France.*

[15] Lamarck and Candolle (1805, vol. 2, p. VIII). Quoted from the translation of this text in Ebach and Goujet (2006, p. 767).

[16] Young (1792, pp. 293–301).

[17] See, for instance, Young (1794), containing a colored "Map of the soil of Suffolk."

"[T]here is something very remarkable in this [...]",[18] Young had said about the parallel boundary lines. He also described how astonished he was when he first observed their parallelism "at a glance" (*coup d'œil*[19]) on the paper. Once he had become aware of this phenomenon, he spent much of his time tracing it in the field. Regarding the northern boundary of the olive, for instance, he remembered: "In travelling [sic] south of Lyons, we see [sic] them first at Montelimart; and, in going from Beziers to the Pyrenees, I lost them in Carcassone [...]."[20] Young's interest in the distribution of crops and his willingness to spend so much energy on their observation was linked to his interest in the French climate; for him, the boundary lines indicated the meteorological and geological dispositions of the nation's territory. With the help of a map, Young was able to divide France into distinct zones, and he supposed that they were highly relevant for national economics.[21]

At the same time, the map allowed Young to speculate on the line's course beyond the borders of France. Concerning the cultivation of grape vines, for instance, he noted: "The line [...] which I have drawn as the boundary of vines in France, may be continued into Germany, and will probably be found to ascertain the vine-climate in that country, as well as in France."[22] In this regard, France was a testing ground that allowed Young to apply the technique of mapping to all kinds of agricultural phenomena. He also encouraged his readers to participate in the accumulation of cartographical knowledge on this topic: "A great many repeated observations must be made, and with more attention than is in the power of a traveller, before such a subject, apparently very curious, can be thoroughly ascertained."[23] The result of this collective research, Young believed, would be of true practical value. Even in countries like England, it could possibly "enable the farmer" to make plants like wine and maize "a common culture."[24]

Beyond the striking iconic similarity of Young's and Candolle's map, both enterprises were conceptually linked in multiple ways. Candolle's argument for using a map was exactly the same as that of Young. Both understood their map as a tool that allowed them to mobilize and coordinate their heterogeneous networks of observers and contributors. The only difference was that Candolle's audience, instead of consisting of farmers, was made up of amateur botanists (although there was most likely a significant intersection between these two groups).[25] With his *Carte botanique,* Candolle explicitly invited nonacademic naturalists, lay observers, to

[18] Young (1792, p. 293).

[19] Young (1792, p. 294).

[20] Young (1792, pp. 294–295).

[21] Young (1792, p. 294).

[22] Young (1792, p. 294). With respect to the lines, Young also discussed the possibilities of cultivating maize and wine in England, see (1792, p. 295).

[23] Young (1792, p. 295).

[24] Young (1792, p. 295).

[25] Williams (2001, p. 110) emphasizes how much Lamarck and Candolle relied on the local floras designed by lay observers all over France.

participate in the improvement of botanical knowledge.[26] By marking the names of cities and villages that had been covered by local floras, Candolle made visible gaps in the existing network. Thus, local botanists were able to explore botanically uncharted regions as well as to test whether or not the borders on Candolle's map matched with their experience.[27]

At the same time, Young's research focus—cultivated plants—fit perfectly within the framework of the institution where Candolle was employed: the botanical garden in Paris.[28] As Emma Spary has shown in detail, by the turn of the century, the Jardin des Plantes' particular way of practicing natural history attracted naturalists from all over Europe and the garden soon became an institutional role model for many botanists abroad, especially for early plant geographers.[29] From the 1760s onwards, the garden had been incorporated into the French "colonial machinery."[30] During the revolution, it was integrated into the Museum d'Histoire Naturelle and was linked with multiple scientific and administrative institutions that were intended to support a continuous reform of French society and economy through the reform of French agriculture. Because the garden soon became a hub for the acclimatization of plants in France, the consequences of this reform for the everyday work of Parisian botanists were far-reaching. Confronted with many more practical demands than elsewhere in Europe, naturalists at the Jardin des Plantes programmatically entered into scientific discussions with social groups that had originally been on botany's periphery: amateurs, gardeners, economists, physicians, landowners, pharmacists, and farmers.[31]

Opening up scientific discourse to lay observers was not new when Candolle published his map in 1805, although the revolution had amplified this prerevolutionary trend. An early example of a lay observer (and one that Candolle referred to in his writings) was the autodidactic naturalist and priest Abbé Jean-Louis Giraud-Soulavie. Already a decade before Arthur Young published his remarks on the distribution of cultivated plants in France, Giraud-Soulavie observed the distribution of exactly the same objects as Young did (except the maize) in the southern province of Vivarais in the French Alps.[32] His botanical research was embedded in a broader study of the geology and geography of this region, and he published his results in a two-volume *Histoire naturelle de la France méridionale*.[33] In one of the maps supplementing these volumes, Giraud-Soulavie charted this region in a manner that Humboldt would apply to the South American continent 20 years later. In a *Coupe*

[26] For the broader history of lay observation within the history of science, see the special issue in *Science in Context* edited by Vetter (2011).

[27] Lamarck and Candolle (1805, vol. 2: pp. V–VI).

[28] For the close links between Candolle's botanical work and the politics of the Napoleon administration, see the paper by Joseph Horan, chap. 5 in this volume.

[29] Spary (2000).

[30] McClellan and Regourd (2000).

[31] Spary (2000, esp. 23 and 125).

[32] For the broader context of Giraud-Soulavie's research, see Bourguet (2002).

[33] Giraud-Soulavie (1780–1784).

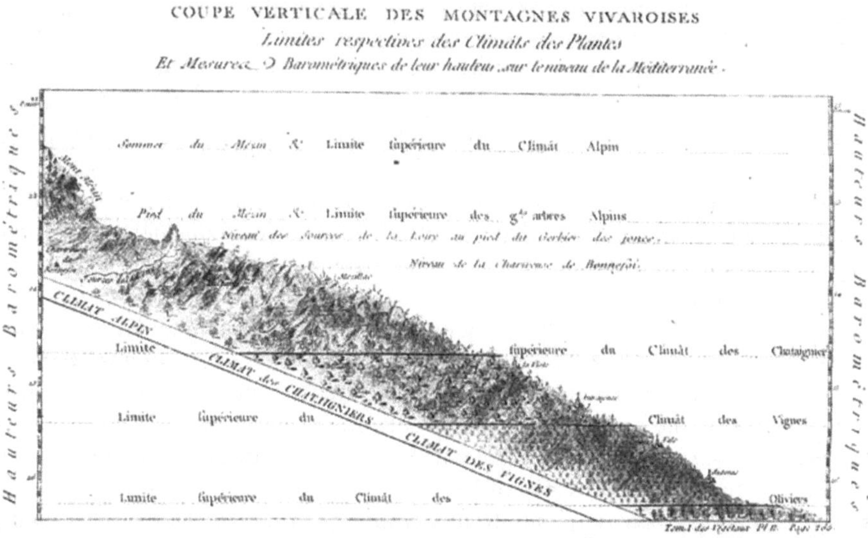

Fig. 3.4 Cross sections before Humboldt. (Giraud-Soulavie 1780–1784 "Coupe verticale des montagnes vivaroises," vol. 2/1)

verticale des montagnes vivaroises, Giraud-Soulavie depicted regional plant distribution by using a cross section (Fig. 3.4). This form of presentation was unusual (I will later return to this point)[34] and it allowed the reader to get a general view of the region and its elevation above sea level. Parallel lines that crossed the mountain slope indicated distinct climatic zones. Each zone was inhabited by particular plants in an ascending order: orange, vine, chestnut, fir tree, and alpine plants.

In Giraud-Soulavie's eyes, the *Coupe verticale* referred to both regional geography and more general principles of plant geography. It is important to note, however, that Giraud-Soulavie was primarily interested in the economic potential of his observations. For him, the cross section made visible principles of a "physical botany" that would enable naturalists to leave the constraints of pure research and make profitable contributions to society instead. This framework was precisely the one that the botanical garden in Paris would explicitly adopt a few years later. Instead of wasting time with classification and the naming of plants, Giraud-Soulavie claimed that botanists should instead concentrate on the "agricultural, economical, and industrial" usage of botanical knowledge.[35] In his eyes, this meant a heightened focus on the "atmospheric" dimension of plant distribution, a topic that was highly relevant for the acclimatization of exotic plants on French soils.[36]

[34] Before this period, cross-sections had only been used in geology and mining. See Rudwick (1976).

[35] Giraud-Soulavie (1780–1784, vol. 2/1: p. 43).

[36] Giraud-Soulavie (1780–1784, vol. 2/1: pp. 35–38).

The distribution borders of crops played a fundamental role in Giraud-Soulavie's concept of acclimatization, since they were supposed to signify the natural dispositions of a territory. Looking at the "natural garden"[37] of the province of Vivarais, Giraud-Soulavie was convinced that his homeland could serve as a model that could be used to reorganize the material culture of French botany. Because the boundaries of crops signified the limits of plant circulation, they also indicated opportunities for acclimatization. For instance, when discussing the regional distribution of the orange, Giraud-Soulavie found that this plant had not been indigenous to Vivarais but had been imported.[38] Thus, for him, its distribution areas made visible the regional determinants of plant distribution.

By applying his regional observations to the nation as a whole, Giraud-Soulavie sketched a plan of how to successfully manipulate the circulation of plants in France.[39] In particular, he recommended that the French state restructure the institutions involved in the acclimatization of plants. In the future, at least three botanical gardens should administer the botanical richness of the nation and its colonies. First, his homeland, the province of Vivarais, could be converted into a garden in order to acclimatize alpine plants from Europe and America. This project would work out best if the newly introduced species were planted in parallel belts around the mountain slopes. Second, the botanical garden in Montpellier could serve as an entrance point for plants from Africa and the tropics. Finally, the botanical garden in Paris should be responsible for plants coming from "temperate" climates.

In many ways, Giraud-Soulavie's suggestions concerning the rearrangement of the botanical gardens in France were implemented during and after the revolution. Although the province of Vivarais was not converted into a garden, over the course of the nineteenth century, many gardens became concerned with exactly the same functions that Giraud-Soulavie had imagined in the 1780s. Not only in Paris but also all over France and abroad, smaller gardens specialized in the acclimatization of plants from the colonies. They became hubs for plant geographical research, contributing in particular to the academic study of tropical agriculture.[40]

Candolle's *Carte botanique* referenced large numbers of botanical practitioners all over the country and displayed them as essential participants in the Paris garden's observational collective. By integrating the observations of naturalists like Young and Giraud-Soulavie in his map, Candolle left no doubt that a main goal of the *Flore francaise* was to link this botanical collective to "useful" applications. In this time period, the medium of a map promoted this message in a particularly clear way, because mapping was associated with the production of useful and applied knowledge. As Josef Konvitz has shown in detail, the rise of distribution mapping in France was closely connected to the state and to the reform of the educational

[37] Giraud-Soulavie (1780–1784, vol. 2/1: p. 188).

[38] Giraud-Soulavie (1780–1784, vol. 2/1: p. 194).

[39] Giraud-Soulavie (1780–1784, vol. 2/1: pp. 198–205).

[40] Matagne (1999). See also the article by Juhé-Beaulaton (1999) in the same volume "Du jardin des plantes médicinales de Paris aux jardins coloniaux."

system.[41] In this regard, Candolle's map is a specific example of a more general trend. At institutions like the Ecole Polytechnique, engineers and military officers were trained to cartographically depict all sorts of cultural and natural distribution phenomena (geology, population, etc.). With the *Carte botanique,* Candolle transferred this kind of cartographical framework to botany. He indicated that, in his view, a useful botany would only be possible if it contributed to the wealth of the state.

Coup d'oeil: Maps and the Inventories of States

Giraud-Soulavie had called his botanical distribution map a *coupe verticale,* a vertical view. Both Candolle and Young used similar but more generalized vocabulary in order to describe the epistemic value of their maps: *coup d'oeil* (translated into English as "at a glance").[42] This term, which originally emerged out of military and alpine observational practices, characterized the specific ability of maps to synthesize heterogeneous observations on a piece of paper and to make them visible all at once.[43] Interestingly, it was this synthesizing feature that made Arthur Young's writing popular among European audiences. Young was a prototype of the traveler with a "comparative gaze"[44]—someone who paid close attention to local details while simultaneously collecting his observations in a way that could later be processed via statistics and maps. Young often called his cartographically inspired approach a "general view."[45]

Due to his attentiveness to geographical conditions, his writings were widely read throughout Europe, and, as shown in the section before, his work encouraged several mapping enterprises, especially in France. Already by the early nineteenth century, however, German audiences exhibited a specific fondness for botanical distribution maps, and this preference was later amplified by the emerging cartographical market. In this section, I will show that a particular discourse had had a huge impact on early plant geography in the decades around 1800. That discourse was cameralism, the science of state and administration.[46] Here, the idea of inventories significantly merged with the visual culture of philanthropy, a broad humanistic reform movement of the enlightenment. Cameralists were particularly strong advocates of a specific feature of philanthropic pedagogy: the *coup d'oeil of* cartographic visualization. With Alexander von Humboldt among others, the philanthropy-inspired general view of cameralism would circulate among European readers and soon reach the Jardin des Plantes. Humboldt, probably today the most popular plant

[41] Konvitz (1987, chap. 6).

[42] Young (1792, p. 294). Candolle (1862) used this term later in *Mémoires et souvenirs*, p. 209.

[43] Bigg (2007) and Daston (2008, esp. 107–110).

[44] Green (2002).

[45] For instance, Young (1794).

[46] On the history of cameralism, see Wakefield (2009).

geographer of this generation, visited Paris immediately before and after his South American journey—exactly at the time when Candolle was designing his *Carte botanique.*

In the last two decades of the eighteenth century, botany was a new topic on the agenda of German cameralism. While the first generations of cameralists had concentrated on the classificatory fundaments of state science, by the turn of the century, a popular lay movement—consisting of farmers and agriculturists who were mostly organized in regional economic societies—had pushed for an academic discourse that focused more on the practical and environmental conditions of farming.[47] In the early 1800s, cameralists devoted a great deal of attention to "economic botany."[48] From the very beginning, this focus on economically relevant botany was flanked with maps.

In 1782, August Friedrich Wilhelm Crome, professor of cameralism at the University of Giessen, published his *New Chart of Europe,* one of the first maps to ever systematically deal with economic geography.[49] Crome used a plain map of Europe as a base on which he marked the major political borders, rivers, and mountains. In a second step, he inscribed the dominant economy of a region, ranging from farming to mining or sea trade. On the left and right side, he listed the most important economic features of each state. While looking at this map a reader could easily compare the economy of his own region to that of neighboring regions as well as to far distant parts of Europe.

For the next several decades, maps on economic geography, often displaying the distribution of crops, were published all over Europe.[50] For instance, in 1806, geographer Carl Ritter published a map on cultivated plants in his series *Six Maps of Europe* (*Sechs Karten von Europa*). The new cartographical genre matched with a more general trend among European economists to measure and to visualize the strength of states, most elaborately done by William Playfair in his *Political Atlas* (of 1786).[51] German cameralists would soon go so far to enlarge the scale of their maps to depict particular regions and woods.[52] Even public authorities systematically applied this new cartographical technique. Prussia was the first state ever that issued an *Administrative-Statistical Atlas,* a compendium of thematic maps displaying several features of the state's territory: from geology and population statistic to wine culture and the distribution of woods.[53] As mentioned in the beginning of this

[47] Schindler and Bonss (1980), Lowood (1991), Sandl (1999, pp. 91–92). For the broader context of this lay movement within discourses in *Oekonomie,* see the article by Denise Phillips, Chap. 2 in this volume.

[48] An early example was Suckow (1777). Short historical sketches concerning the history of economic botany are in Heiser (1986), Wickens (1990), with a broader focus on Victorian Britain: Endersby (2008).

[49] Nikolow (2001).

[50] Fick (1971) and Robinson (1982, pp. 140–147).

[51] See Nikolow (2001).

[52] See, for instance, the "illuminated forest map" in Hartig (1805). For the broader context of German forestry, see Lowood (1990).

[53] Scharfe and Neugebauer (1991).

chapter, by mid-century, the genre of crop and wood maps was so common that it was included in popular physical atlases as a matter of course (Fig. 3.2).

The first cameralists' maps on the distribution of cultivated plants appeared in the pedagogical setting of the philanthropic movement.[54] Crome developed his *New Chart of Europe* especially for the philanthropic school in Dessau; Ritter, too, was employed at a philanthropic school in Schnepfenthal when he published his *Six Maps of Europe*. Within the philanthropic movement, maps were systematically employed in order to mediate knowledge to students and Crome explicitly aimed to provide a new generation of "economists, cameralists, and scholars in general" with a more sophisticated understanding of geographical settings.[55] He also described how to use his map. Starting with a regional extract, the students should slowly zoom out before finally looking "at the whole at a glance."[56] *At a glance*—it was this particular idea of a cartographically mediated, synthesized view that was later invoked by Candolle and Young with the term *coup d'oeil*.

By 1800, Humboldt was the most prominent plant geographer who promoted the *coup d'oeil* as a basic scientific research technique.[57] He can also be seen as a key figure who transferred this philanthropic fondness of the general view to cameralism and, more generally, to all branches of physical science. Humboldt himself had received a philanthropic education, and he continued his academic career as a student of cameralism in Frankfurt (Oder). Later, he attended lectures on this topic at the University of Göttingen. Before he went on his South American journey, he was also employed as an officer of mining, where he learned the practice of mapping tunnels in a cross-sectional manner.[58] This practical expertise, combined with his philanthropic background, is an underemphasized part of Humboldt's biography, despite the fact that one can argue that it affected his research practices—and that of other plant geographers—in many ways. As historian of cameralism Andre Wakefield has argued for Humboldt and other cameralist-trained members of his generation the "earth sciences became fiscal sciences."[59]

Indeed, cameralism for Humboldt was first and foremost a practice for the hands, of making notes in a particular way and later cartographically drawing them together. One of the key features of cameralist practice was the employment of double-entry notebooks in order to administer the incomes and outputs of a state's bureaucracy and its natural resources.[60] Once having learned this technique in the lectures of Johann Beckmann at the University of Göttingen, Humboldt applied it systematically to South American landscapes on his 5-year journey (1799–1804).[61] When he wrote his diary during the passage to Venezuela, he recalled how his teacher

54 Nikolow (1999).

55 Crome (1782, pp. XII–XIII).

56 Crome, (1782), p. XV: "… das Ganze mit einem Blick zu übersehen…."

57 See especially Daston (2010).

58 Biermann (1990, pp. 149–168).

59 Wakefield (2009, p. 28).

60 Heesen (2005).

61 Lack (2004).

Beckmann would take notes on all of the phenomena surrounding him even under the most challenging of circumstances. Humboldt called it "the art of noting down everything in the pocket."[62]

Humboldt's individual style of note-taking mirrored a more general contemporaneous shift in European traveling practices. The "Italian way" of double-entry note-taking was also promoted in Leopold Berchtold's highly popular instructional text *The Patriotic Traveler*. Berchtold recommended that his readers "commit" everything "on paper" directly "upon the spot."[63] To this end, Berchtold formulated ca. 2400 questions on phenomena to which travelers should attend, ranging from observations on society, population, and bureaucracy to the composition of forests, the annual harvests, and the dominant cultivated plants.[64] There was probably not a single observation in Humboldt's notebooks that one could not link to one of Berchtold's questions.

When skimming through the notebooks of Humboldt's South American journey, one is struck by how seldom he observed the distribution of wild plants. Crops and other cultivated plants dominate the pages.[65] This pattern had to do with the particular division of labor in the expedition. His companion Aimé Bonpland had received a professional training at Paris' Jardin des Plantes and was primarily responsible for carrying out the botanical research. Humboldt, instead, concentrated on more general "physical" phenomena and measurements.[66] Thus, he mostly kept note of those plants that were most visible in the landscapes: either wild plants that dominated the site or crops.

Profiting from his cameralistic background, Humboldt also studied the agriculture of the places that he visited. For instance, while staying in Havana, he painstakingly examined the possibilities of acclimatizing and exporting sugarcane.[67] In the port of Guayaquil (today's Ecuador), he compiled a table that listed the export of particular cultivated plants like cocoa and coffee.[68]

As Jorge Cañizares-Esguerra has shown, Humboldt's interest in the economic features of South American plant geography was stimulated by an already existing discourse among local naturalists and officials on the environmental richness of the region's landscapes and climate.[69] Both regional elites and the colonial administration considered the provinces of Peru and Ecuador—which Humboldt mentioned as nearly ideal spots in which to study plant distribution[70]—as potential granaries of the world economy by this time. Humboldt designed his first cartographical draft

62 Humboldt (2000), p. 93: "… die Kunst, in der Tasche zu schreiben…."

63 Berchtold (1789, vol. 1, p. 43).

64 Berchtold (1789, pp. 195–515).

65 See Faak (2002, pp. 61–62).

66 Lack (2009).

67 Humboldt (1815–1832, vol. 6/2: pp. 279–305).

68 Today this document is part of his "Nachlass," Staatsbibliothek Berlin, manuscript department, "kleiner Kasten 7b," No. 71.

69 Cañizares-Esguerra (2005).

70 Humboldt (2003, vol. 2: p. 77).

of South America's plant distribution while in these particular areas. After he had crossed the Andes and arrived at the port of Guayaquil, he drew a cross section of the continent that served as a base for his famous *Tableau physique des Andes.*[71] He published it in 1807 as part of his *Essay on the Geography of Plants,* which today is regarded as one of the foundational documents of the discipline of plant geography.[72]

At first sight, this cross section is dominated by wild landscapes. In order to depict the distribution of wild plants, Humboldt had asked his companion Bonpland to report the distribution of particular species that he had collected throughout their journey. Humboldt marked their distribution area with letters at the inner side of the mountain. However, the closer one looks, the more obvious it becomes that agriculture played an important role in this map too. Not only did Humboldt display particular agricultural zones at the slope of the mountain but also labeled one of the columns at the side of the map as the "culture of the soil." Here, Humboldt listed the main agricultural areas in relation to the height above sea level. In the text, he explained in detail how the ground level affects the particular style of farming in this region. At up to 1000 m, farmers would mainly grow "indigenous" plants like maize; imported crops such as sugarcane, indigo, or coffee; or fruits like the orange and the pineapple. The higher a traveler got, the more the landscapes would be dominated by cotton plantations and, from 3000 to 4000 m, by the potato. The "scale of agriculture," Humboldt emphasized, "presents to the reader the whole image of human industry, ranging from within the mines to the snow-covered summits of the Andes."[73] When he published his *Essay on the Geography of Plants,* Humboldt pointed out the need to look at the wide range of physical factors "all at once." The synthetic medium of the map allowed him to integrate "at a glance" phenomena that reached far beyond the realm of wild, untouched nature. Instead, from the very beginning, agriculture was an important dimension of botanists' interest in the distribution of plants.

When Humboldt returned to Europe, he was closely allied with an institution that had already become famous for combining research on the distribution of wild plants and crops: the Jardin des Plantes in Paris. Directly after his return from South America, he stayed at the botanical garden for several months in order to analyze his field observations. Later, he sold his South American herbarium and his field notes to the garden. In Paris, he also presented the first outline of his *Essay on the Geography of Plants* in an 1806 talk at the Academy of Science. It is fair to assume that one of the listeners was a botanist who was employed at the Jardin de Plantes and who had already spent much time on charting plant distribution: Augustin-Pyramus de Candolle.

[71] For this map, which has become an icon of "Humboldtian science," see Dettelbach (1999).

[72] The "essai" is available in a new English translation: Humboldt (2009).

[73] Cited and translated from a German edition: Humboldt (1989), quoted on 153: "So bietet die Skala des Ackerbaus das Bild menschlicher Industrie von dem Innern der Bergwerke bis zu dem verschneiten Gipfel der Anden dar."

Lines of Cultivation, Lines of Nature: Botanical Cross Sections

When Candolle discussed Arthur Young's lines of cultivated plants in his *Carte botanique* (Fig. 3.3), he wondered about their connectedness with the boundary lines of "botanical regions." When comparing the shape of their respective distribution patterns, he found few similarities. Instead, he observed that the directionality of the crop lines—reaching form the Southeast to the Northwest—was "exactly the reverse of what we observe for wild plants."[74] With his observation, Candolle articulated a fundamental problem that puzzled many botanists when they cartographically displayed plant distribution in the early nineteenth century. Especially, when they used "plain" topographical maps as a base, patterns of plant distribution seemed quite irregular and in many cases disconnected. Many local factors (climate, soil conditions, etc.) affected the distribution of species and vegetational units, so that the overall outlook was often unconvincing and analytically confusing.

One advantage of cross sections—Humboldt called them "vertical projections"—was that distribution patterns became visible and were thus more striking. Especially because such cross sections seemed to make apparent more general laws of plant distribution, they became a popular genre of botanical distribution maps up to the mid-nineteenth century. As I will show in this section, they shaped the theoretical development of early plant geography in a particular way: The more botanists charted plant distribution on mountain slopes, the more they were convinced that the extensions of farming could serve as a general indicator in order to reveal the underlying laws of nature and even make visible a more fundamental pattern of global plant distribution.

Early plant geographers such as Giraud-Soulavie had mostly been interested in the boundary lines of cultivated plants due to their practical significance, but later plant geographers more systematically used these boundaries as theoretical guidelines. Humboldt in particular was concerned with the connections between the distribution patterns of wild and cultivated plants throughout his career. Not only did he start his *Essay on the Geography of Plants* with a long paragraph on the impact of cultivated plants on human life, in his later writing, he would return to this topic, especially when he wrote on the biogeography and climate of particular regions. While analyzing the vegetable production of New Spain, for instance, Humboldt discussed the relationship between the distribution patterns of wild and cultivated plants. He was struck by their similarity. "It would be possible," he remarked, "to treat the agriculture of New Spain according to the great divisions" which he had drawn in his cross sections by following the lines of cultivation, i.e., the boundary lines of crops.[75] As Humboldt noted, wild and cultivated plants were associating in

[74] Lamarck and Candolle1805, vol. 2: pp. VIII–IX. Quoted from: Ebach and Goujet (2006, p. 767).

[75] Humboldt (2008), pp. 344–345: "Ich könnte daher den Ackerbau von Neu-Spanien nach den grossen Abteilungen behandeln, welche ich oben bei meinem Entwurf des physischen Abrisses des mexicanischen Bodens auseinandergesetzt habe und könnte den Kultur-Linien folgen, die auf meinen geologischen Profilen gezogen…sind."

similar ranges on a mountain slope—a phenomenon that became visible in his cross sections, too (Fig. 3.1).

Humboldt's interest in lines of cultivation was caused by a distinctive meteorological feature of his travel destination—tropical America. After having studied climate and plant life throughout his journey, he developed the theory that close to the equator—and especially in mountainous regions like Peru—the temperature was less dependent on local factors and was essentially determined by the height above the sea level, which simultaneously and directly affected plant distribution.[76] Thus, the reduction of complexity in mountainous areas of the tropics made it easier to directly observe general principles of plant distribution as well as to make them visible in maps: "Near the equator meteorological phenomena as they affect the geography of plants and animals are under constant and easily recognizable laws."[77] Thus, the reader would be able to envision more general principles of plant distribution in a cross section of South America because nature was so to speak more concentrated in this region. The same was true for culturally modified plants. For Humboldt, lines of crops could signify more general dispositions of nature.

Throughout his career, Humboldt held the view that this regularity of plant distribution at the equator could also be transferred to other parts of the globe. In some ways, a mountain on the equator was a map itself because it potentially assembled all plant geographical zones on its slopes. If one travelled from the equator to one of the poles, Humboldt argued, the regular principles of distribution would generally remain the same; only the diversity of plant life would diminish. One of his followers, the botanist and traveler Franz Meyen, put it this way: "Judging from recent experiences it is not hard to recognize that the parallelism of plant zones on a mountain slope is equal to that found on the hemispheres between the equator and the poles. This is where one realizes the advantages and the impact which plant geography can have on agriculture and, more generally, on the culture of a country."[78]

Nevertheless, the regularity of plant distribution obviously got more complicated as soon as one left the tropics. Especially in the northern hemisphere, Humboldt admitted, the influence of localities was so high that the distribution of wild plant was highly fragmented and irregular, making it extremely difficult to make generalizations.[79] In order to restore some degree of regularity, his main source

[76] Humboldt (2008, pp. 344–347).

[77] Humboldt (2008, p. 345): "Die meteorologischen Phänomene wie die in der Geographie der Pflanzen und Tiere stehen unterm Äquator unter unveränderlichen und leicht kenntlichen Gesetzen."

[78] Meyen (1836, p. 31): "Auch ist es nach den gegenwärtigen Erfahrungen nicht mehr schwer zu erkennen, dass dieser Parallelismus genau mit jenem übereinstimmt, welcher sich, in Hinsicht der Wärme-Abnahme, zwischen den Entfernungen vom Äquator zum Pole und von der Ebene bis zur Schneegrenze zeigt. Hier wird man die Vortheile, welche die Geographie der Pflanzen auf den Ackerbau und überhaupt auf die Cultur des Landes ausüben könnte, zuerst recht deutlich erkennen lernen."

[79] Humboldt (2008, p. 345).

of comparison became, once again, the distribution of crops. Because the influence of the local climate was of "particular interest to farmers,"[80] more reliable observation had been made on this topic. And again, when dealing with this topic, Humboldt referred to the observations of Arthur Young and Augustin-Pyramus de Candolle.[81] He painstakingly compared their lines with lines of average temperature—so-called isotherms—in order to cartographically explore a system behind their distribution.[82]

Although such evidence remained highly controversial, an increasing number of botanists acted on the assumption that the zones of crops on mountain slopes gave more general insights into the principles of plant distribution. Humboldt had already claimed that crop distribution's significance resulted from the fact that farmers would always push the environmental limits of crops as far as their natural disposition would allow. "Cultivated plants," he wrote, "are so flexible in their organization that human care has forced them beyond the borders which the naturalist has determined for them."[83] Some botanists explained this observation with the "high degree of flexibility" (*hohen Grad an Biegsamkeit*) of cultivated plants.[84]

Whereas Humboldt did not clarify whether such borders were determined by "naturalists" or by "nature," some of his colleagues were more explicit. The Danish botanist Joakim Frederik Schouw, for instance, who published a frequently cited handbook on plant geography in 1823, argued that nature itself sets limits on how far culture could expand the distribution area of plants:

> Whether a cultivated plant is to be found at a particular spot, does not only depend on the climate, but also on the degree of cultural development, the activity of nations, and their interactions; sometimes it has to do with their manners and religious beliefs, etc. Nevertheless, climatic boundaries remain that cannot be exceeded by industry and art. Practically as well as theoretically, it is important for us to determine such boundaries.[85]

In an atlas that supplemented his handbook, Schouw displayed the worldwide distribution of wine, for him a particularly telling example of a plant that farmers have pushed to its limits.[86]

---

[80] Humboldt (2008, p. 345).

[81] Humboldt (2008, p. 346).

[82] See also: Meyen (1836, pp. 25–29).

[83] Humboldt (2008, p. 345): "[D]ie angebauten Pflanzen sind in ihrer Organisation so beweglich, dass die menschliche Sorgfalt sie äufig über die Grenzen hinaustreibt, die der Naturforscher ihnen zu bestimmen geruht hat."

[84] Meyen (1836, p. 108).

[85] Schouw (1823, p. 205): "Ob eine angebaute Pflanze an einem gewissen Orte vorkommt, hängt nicht nur von klimatischen Verhältnissen ab, sondern auch von dem Culturgrade, und der Betriebsamkeit der Völker, von ihrem gegenseitigen Verkehre, manchmal von ihren Sitten, ihren religiösen Vorstellungen u.s.w. Es bleiben indess doch noch immer Grenzen, welche, von den klimatischen Ursachen bestimmt, der Fleiss und die Kunst nicht übersteigen können. Diese zu bestimmen muss uns in theoretischer wie in praktischer Hinsicht wichtig seyn…."

[86] Schouw (1823, p. 206). Again, while discussing the distribution areas of this plant, Schouw referred to Young and Candolle's treatment of cultivated plants in France.

By the mid-nineteenth century, many botanists took it as a given that the boundaries of crop distribution had a high level of significance for the study of wild plants in particular areas. Up to a regional level, plant geographers would look for the main distribution areas of crops before integrating particular wild species into their grids. One of the most telling examples is the work of Hewett C. Watson on the plant geography of the British islands.[87] For decades, Watson had painstakingly accumulated observations on British plant distribution, while simultaneously searching for a reliable division of the country into sections or zones. At first sight, the result was frustrating. When Watson published his *Cybele Britannica* (the most elaborate work on regional plant geography in Victorian Britain) in 1847, he explained to the reader that any division of wild plants must be artificial: "In truth, however, the natural changes of vegetation being elsewhere gradual, any *line* will inevitably sever and divide that which is nearly alike; the vegetation being more similar on the contrary sides of any one dividing line, than it is on the two sides of the broad zone between two lines. This disadvantage attends all our arrangements and groupings of nature's realities."[88]

Despite this pessimistic note, Watson made a pragmatic decision. If any division of plant life is inevitably artificial, naturalists could rely on the most artificial line of plant distribution to be seen in the landscape, cultivated plants: "The primary division which is here to be proposed as one best applicable in Britain," Watson explained, "is ostensibly founded upon an artificial character; namely the presence or absence of cultivation. It is by this character that we may distinguish the lower from the upper zones of plants…."[89] By generally dividing Britain into "agrarian" and "arctic" regions, Watson was able to subdivide the islands into particular subdivisions.

The argument for the significance of cultivated plants was exactly the same as the one that Humboldt, Schouw, and others had brought forward. Especially when dealing with grain and corn, Watson argued, farmers had been extremely inventive in order to go as far as possible: "The interest of mankind are so intimately connected with the production of corn, that we shall everywhere find cultivated fields as far up the valleys and acclivities of the mountains, as their climate will allow."[90] Watson's argument was very pragmatic indeed. But practicality was precisely what attracted botanists in the mid-nineteenth century to the distribution patterns of cultivated plants. Despite being artificial by definition, crops became popular objects that could be used to depict nature in a state that would have remained invisible otherwise.

[87] Allen (1994, pp. 94–103)).

[88] Watson (1847–1859, vol. 1, pp. 30–31).

[89] Watson (1847–1859, vol. 1, p. 32).

[90] Watson (1847–1859, vol. 1, p. 33).

Audiences: The Cartographical Market in Germany and the Role of Humboldtianism

This chapter has shown that mapping crops was a pan-European technique that circulated between different milieus of knowledge production well into the 1840s. There were numerous particular centers for this activity—for instance, Candolle and Giraud-Soulavie in France, Young and Watson in England, Shouw in Denmark, and the cameralists in Germany. Many naturalists like Humboldt moved between these centers. During the first decades of the nineteenth century, botanical mapmakers developed very few iconic standards, but from the mid-nineteenth century onwards, the German states became a hub for botanical distribution mapping in general and a stronghold for crop maps in particular. Here, maps were most systematically applied as tools of theory.[91] In the late nineteenth century, agricultural geography with a strong emphasis on map-use even appeared as an academic subject.[92]

The strong German interest in distribution mapping has often been explained by the influence of Humboldtianism, supposedly a particularly important tradition in Central Europe.[93] Given the geographic diversity and Pan-European dimension of botanical distribution mapping during his lifetime (Humboldt died in 1859), one has to be careful not to overemphasize his actual impact. In this last section, I argue instead that maps became a widely used tool of theory because there existed a particularly vivid material and visual culture in Central Europe: a booming private cartographical market through which new audiences for these kinds for the maps emerged. When seen from an academic point of view, most readers of crop maps had a lay background.

As we have already seen, the first "cameralist" crop maps from the late eighteenth century were primarily designed to be used in education. August Crome and Carl Ritter had created their botanical maps for geography teaching in philanthropic schools. Whereas originally the use of such maps had been strongly related to philanthropic pedagogy, it quickly diffused to other contexts. Maps and atlases soon became a regular part of geography teaching in the German states, playing a particularly prominently role in more locally orientated lessons in "*Heimatkunde*."[94] School atlases were therefore sold in large numbers all over the German states in the first half of the nineteenth century.[95] By mid-century, a director at a gymnasium for young girls in Berlin, Ludwig Rudolph, even edited a whole atlas that dealt

[91] Güttler (2011).

[92] A good overview is to be found in Troll (1925).

[93] Especially Nicolson (1996).

[94] A history of the material culture of nineteenth-century geography teaching in Germany has not been written yet, but encyclopedias from the 1860 provide us with good impression of the map use in schools, see especially the articles "Geschichte und Geographie in der Volksschule" and "Geographie in höheren Schulen" in Schmid (1859–1875), vol. 2, pp. 704–715 and 806–820.

[95] Brogiato (1996).

exclusively with plant geography.[96] It goes without saying that the use of maps in schools was not restricted to the German states, but here the cartographical culture flourished with particular vigor. In 1885, Scott Keltie, Inspector of Education at the Royal Geographical Society, would enviously remark that due to the "wealth of good maps, reliefs, geographical pictures, and other apparatus," German schoolchildren had a better "chance of leaving school with a substantial knowledge of the subject" than students elsewhere in Europe.[97]

Besides the particularly strong educational demand for cartographical products, the rise of the private cartographical market in the German states was fueled by a specific social development. During the Napoleonic Wars, the military had employed hundreds of skilled cartographers all over the German states. Once the war was over, Prussia reassigned some of its military cartographers to carry out more civilian tasks. For several years, military officers collected data from statistical bureaus, and in the 1820s, they presented a cartographical *coup d'oeil* to the wider public: the *Administrative-Statistical Atlas of Prussia* [*Adminstrativ-statistischer Atlas von Preussen*].[98] Whereas this atlas emerged from a state-sponsored infrastructure, many former officers left service and sold their cartographical expertise to the private mapmaking companies that emerged all over the German states. The most prominent example was Heinrich Berghaus, the editor of the first *Physical Atlas*.[99] This brings me back to the map with which I started this chapter: Berghaus' crop map (Fig. 3.2), which would later be regarded as a role model for most botanical distribution maps of the late nineteenth century.

After leaving the military, Berghaus founded a private cartographical school in the town of Potsdam where he and his students designed maps that were later sold to cartographical companies.[100] The publishing house of Justus Perthes in the town of Gotha became his most sustained sponsor. Already in 1830 (2 years after Prussian officers had published the *Administrative-Statistical Atlas*), Berghaus wrote to Perthes that he planned to edit a similar atlas on a global scale, a thematic *coup d'oeil* of the whole world. It should be designed for the "geographical–physical amusement" of the wider public.[101] This *Physical Atlas* appeared from the mid-1840s onwards and was soon reissued by the Edinburgh cartographer Keith Johnston for the British market. Today, many historians regard Berghaus' atlas as a visual compendium of Humboldt's *Cosmos*.[102] But the closer one looks at the production process, it becomes clear that it was basically an undertaking that was pushed

---

[96] Rudolph (1852).

[97] Keltie (1885, quote 501). On this report, see Wise (1986).

[98] Scharfe and Neugebauer (1991).

[99] On this atlas, see note 2.

[100] On Berghaus, see Engelmann (1977).

[101] Berghaus called it a "Miscellen-Atlas für geographisch-physikalische Belustigungen" in a letter to his publisher Perthes from November 18, 1830. Archive of the Perthes Publishing Company, Research Library Gotha, University of Erfurt, Section "Mitarbeiter und Freunde des Verlages", Nr. 19A/1, 18.

[102] See note 2.

forward by Berghaus and the publisher Justus Perthes.[103] Both tried to establish a new and profitable map genre for Central Europe's booming cartographical market.

The botanical section of Berghaus' *Physical Atlas* shows the diversity of early botanical distribution mapping. It combined, for instance, a cross section in Humboldt's manner with statistical maps designed by Schouw. The global crop map (Fig. 3.2), however, would most strongly resemble botanical distribution maps of the second half of the nineteenth century. This resemblance did not just exist at the level of iconography. After the publication of this map, Perthes continuously invested in the genre of crop maps and published them with various scales, all the way down to a regional level. For the next decades, Perthes published dozens of botanical distribution maps.[104] They became an origin point of later botanical distribution mapping. When Berghaus' *Physical Atlas* was reissued in the 1880s, the plant geographical section became a visual standard within the international botanical community.

By this time, crop maps had already been overshadowed by maps that were devoted to "pure" botany. The editor of the section, the Dresden botanist Oscar Drude, spoke snidely of the atlas-map on cultivated plants, calling this branch of research "kitchen- and distilling flask-botany."[105] Such statements show that by the late nineteenth century, botanists had already begun to forget how much the rise of botanical distribution mapping had originally relied on the cartographical observation of crops.

References

Allen, David Elliston. 1994. *The naturalist in Britain: A social history,* 2nd ed. Princeton: University Press.

Anonymous. 1865. *Inhaltsverzeichnis von Petermann's 'Geographischen Mitteilungen' 1855–1864*. Gotha: Justus Perthes.

Berchtold, Leopold. 1789. *Essay to direct and extend the inquiries of patriotic travellers*, 2 vols. London: Robinson.

Berghaus, Heinrich. 1845–1848. *Physikalischer Atlas oder Sammlung von Karten, auf denen die hauptsächlichsten Erscheinungen der anorganischen und organischen Natur nach ihrer geographischen Verbreitung und Vertheilung bildlich dargestellt sind.* Gotha: Perthes.

Biermann, Kurt-R. 1990. *Miscellanea Humboldtiana*. Berlin: Akademie.

Bigg, Charlotte. 2007. The panorama; or, la nature á coup d'oeil. In *Observing nature-representing experience: The osmotic dynamics of Romanticism, 1800–1850,* ed. Erna Fiorentini, 73–95. Berlin: Reimer.

Bourguet, Marie-Noëlle. 2002. Landscape with numbers: Natural history, travel and instruments in the late eighteenth and early nineteenth centuries. In *Instruments, travel and science:Itineraries*

[103] See the letters of Berghaus in the Archive of the Perthes Publishing Company, Research Library Gotha, University of Erfurt, Section "Mitarbeiter und Freunde des Verlages," Nr. 19A.

[104] See also note 8.

[105] Drude made this statement in a letter that he wrote to his atlas editor on September 30, 1882: Archive of the Perthes Publishing Company, University of Erfurt, Section "Mitarbeiter und Freunde des Verlages," Nr. 19B/6 (Correspondence of Hermann Berghaus), pp. 12–13.

of precision from seventeenth to the twentieth century, ed. Marie-Noëlle Bourguet, Christian Licoppe, and H. Otto Sibum, 96–125. London: Routledge.
Brockway, Lucile H. 1979. *Science and colonial expansion: The role of the British Royal Botanical Gardens*. New York: Academic.
Brogiato, Heinz Peter. 1996. 'An dem Knochen wir von vielen genagt': Zur Entwicklung der geographischen Schulatlanten im 19. Jahrhundert. *Internationale Schulbuchforschung* 18:35–66.
Browne, Janet. 1983. *The secular arc: Studies in the history of biogeography*. New Haven: Yale University Press.
Browne, Janet. 1992. A science of empire: British biogeography before Darwin. *Revue d'histoire des sciences* 45:453–475.
Browne, Janet. 1996. Biogeography and empire. In *Cultures of national history,* ed. N. Jardine, A. Secord, and E. C. Spary, 305–321. Cambridge: Cambridge University Press.
Camerini, Jane R. 1993. The physical atlas of Heinrich Berghaus: Distribution maps as scientific knowledge. In *Non-verbal communication in science prior to 1900,* ed. R. G. Mazzolini, 479–512. Florence: Olschki.
Candolle, Augustin Pyramus de. 1862. *Mémoires et souvenirs de Augustin-Pyramus de Candolle, écrits par lui-même et publiês par son fils*. Geneva: Joél Cherbuliez.
Cañizares-Esguerra, Jorge. 2005. How derivative was Humboldt? Microcosmic nature narratives in early modern Spanish America and the (other) origins of Humboldt's ecological sensibilities. In *Colonial botany: Science, commerce, and politics in the early modern world,* ed. L. Schiebinger and C. Swan, 148–165. Philadelphia: University of Pennsylvania Press.
Crome, August Friedrich Wilhelm. 1782. *Europens Produkte. Zum Gebrauch der Neuen Produkten-Karten von Europa*. Dessau: Self-published.
Daston, Lorraine. 2008. On scientific observation. *Isis* 99:97–110.
Daston, Lorraine. 2010. The Humboldtian gaze. In *Science as cultural practice. 1 Vol. Cultures and politics of research from the early modern period to the age of extremes,* ed. Moritz Epple and Claus Zittel, 45–60. Berlin: Akademie.
Dettelbach, Michael. 1996. Humboldtian science. In *Cultures of national history,* ed. N. Jardine, A. Secord, and E.C. Spary, 287–304. Cambridge: Cambridge University Press.
Dettelbach, Michael. 1999. The face of nature: Precise measurement, mapping and sensibility in the work of Alexander von Humboldt. *Studies in the History and Philosophy of Biology and Biomedical Sciences* 30:473–504.
Drayton, Richard. 2000. *Nature's government: Science, imperial Britain, and the 'improvement' of the world.* New Haven: Yale University Press.
Ebach, Malte C., and D. F. Goujet. 2006. The first biogeographical map. *Journal of Biogeography* 33:761–769.
Eckert, Max. 1921–1925. *Die Kartenwissenschaft. Forschungen und Grundlagen zu einer Kartographie als Wissenschaft,* 2 vols. Berlin: Walter de Gruyter.
Endersby, Jim. 2008. *Imperial nature: Joseph Hooker and the practices of Victorian science*. Chicago: Chicago University Press.
Engelmann, Gerhard. 1977. *Heinrich Berghaus: Der Kartograph von Potsdam*. Halle: Deutsche Akademie der Naturforscher Leopoldin.
Espenhorst, Jürgen. 2003–2008. *Petermann's planet*. Trans. G. C. Grossmann, 2 vols. Schwerte: Pangaea Verlag.
Faak, Margot. 2002. *Alexander von Humboldts amerikanische Reisejournale: Eine Übersicht*. Berlin: Akademie der Wissenschaften.
Fick, Karl E. 1971. Die kartographische Darstellung wirtschaftsgeographischer Sachverhalte im 18. Jahrhundert. *Geographische Zeitschrift* 59:131–139.
Giraud-Soulavie, Jean Louis. 1780–1784. *Histoire naturelle de la France méridionale,* 2 vols. Paris: Coste Belle.
Green, Nancy L. 2002. The comparative gaze: Travelers in France before the age of mass tourism. *French Historical Studies* 25:423–440.
Güttler, Nils. 2011. Scaling the period eye: Oscar Drude and the cartographical practice of plant geography, 1870–1910s. *Science in Context* 24:1–41.

Hartig, Georg Ludwig. 1805. *Anweisung zur Taxation und Beschreibung der Forste, zweiter oder practischer Teil,* 2nd ed. Giessen: Heyer.

Heesen, Anke te. 2005. Accounting for the natural world: Double-entry bookkeeping in the field. In *Colonial botany: Science, commerce, and politics in the early modern world,* ed. L. Schiebinger and C. Swan, 237–251. Philadelphia: University of Pennsylvania Press.

Heiser, Charles B. 1986. Economic botany: Past and future. *Economic Botany* 40:261–266.

Hinds, Richard Brinsley. 1835. Observations on the construction of maps in geographical botany. *Magazine of Natural History* 8:498–501.

Humboldt, Alexander von. 1815–1832. *Reise in die Äquinoctial-Gegenden des neuen Kontinents in den Jahren 1799, 1800, 1801, 1802, 1803 und 1804,* 6 vols. Stuttgart: Cotta.

Humboldt, Alexander von. 1989. Ideen zu einer Geographie der Pflanzen nebst einem Naturgemälde der Tropenländer, *Studienausgabe,* 1 vol, ed. Hanno Beck, 42–161. Darmstadt: Wissenschaftliche Buchgesellschaft.

Humboldt, Alexander von. 2000. *Reise durch Venezuela. Auswahl aus den amerikanischen Reisetagebüchern,* ed. Margot Faak. Berlin: Akademie.

Humboldt, Alexander von. 2003. *Reise auf dem Rio Magdalena, durch die Anden und Mexiko,* ed. Margot Faak, 2nd ed. 2 vols. Berlin: Akademie.

Humboldt, Alexander von. 2008. Versuch über den politischen Zustand des Königreichs Neu-Spanien. In *Studienausgabe,* vol. 4, ed. Hanno Beck. Darmstadt: Wissenschaftliche Buchgesellschaft.

Humboldt, Alexander von. 2009. *Essay on the geography of plants,* ed. Stephen Jackson. Chicago: University of Chicago Press.

Johnston, Alexander Keith. 1848. *The physical atlas: A series of maps and notes illustrating the geographical distribution of natural phenomena.* Edinburgh: Blackwood.

Juhé-Beaulaton, Dominique. 1999. Du jardin des plantes médicinales de Paris aux jardins coloniaux: Développement de l'agronomie tropicale française. In *Le jardin entre science et représentation,* ed. Jean-Louis Fischer, 267–283. Paris: CTHS.

Keltie, Scott J. 1885. Geographical education. *Scottish Geographical Journal* 1 (10): 497–505.

Konvitz, Joseph W. 1987. *Cartography in France, 1660–1848: Science, engineering, and statecraft.* Chicago: University of Chicago Press.

Lack, Hans Walter. 2004. The botanical field notes prepared by Humboldt and Bonpland in tropical America. *Taxon* 53 (2): 501–510.

Lack, Hans Walter. 2009. *Alexander von Humboldt and the botanical exploration of the Americas.* Munich: Prestel.

Lamarck, Jean-Baptiste, and Augustin Pyramus de Candolle. 1805. *Flore Française: ou descriptions succinctes de toutes les plantes qui croissent naturellement en France,* 3rd ed., 5 vols. Paris: H. Agasse.

Lowood, Henry E. 1990. The calculating forester: Quantification, cameral science, and the emergence of scientific forestry managment in Germany. In *The quantifying spirit in the 18th Century,* ed. T. Frängsmyr, J. L. Heilbron, and R. E. Rider, 315–342. Berkeley: University of California Press.

Lowood, Henry E. 1991. *Patriotism, profit, and the promotion of science in the German Enlightenment: The economic and scientific societies, 1760–1815.* New York: Garland.

Matagne, Patrick. 1999. Des jardins écoles aux jardin écologiques. In *Le jardin entre science et représentation,* ed. Jean-Louis Fischer, 307–315. Paris: CTHS.

McClellan, James E. III, and François Regourd. 2000. The colonial machine: French science and colonization in the Ancien Régime. *Osiris* 15: 31–50.

Meyen, Franz Julius Ferdinand. 1836. *Grundriss der Pflanzengeographie mit ausführlichen Untersuchungen über das Vaterland, den Anbau und den Nutzen der vorzüglichsten Culturpflanzen, welche den Wohlstand der Völker begründen.* Berlin: Haude und Spenersche Buchhandlung.

Nelson, Gareth. 1978. From Candolle to Croizat: Comments on the history of biogeography. *Journal of the History of Biology* 11:269–305.

Nicolson, Malcolm. 1996. Humboldtian plant geography after Alexander von Humboldt: The link to ecology. *British Journal for the History of Science* 29:289–310.

Nikolow, Sybilla. 1999. 'Die Versinnlichung von Staatskräften': Statistische Karten um 1800. *Traverse* 6 (3): 63–82.
Nikolow, Sybilla. 2001. A.F.W. Crome's measurement of the 'strength of the state': Statistical representations in Central Europe around 1800. In *The age of economic measurement,* ed. Judy L. Klein and Mary S. Morgan, 23–56. Durham: Duke University Press.
Robinson, Arthur Howard. 1982. *Early thematic mapping in the history of cartography*. Chicago: University of Chicago Press.
Rudolph, Ludwig. 1852. *Atlas der Pflanzengeographie über alle Theile der Erde für Freunde und Lehrer der Botanik und Geographie nach den neuesten und besten Quellen entworfen und gezeichnet*. Berlin: Nicolai.
Rudwick, Martin J. S. 1976. The emergence of a visual language for geological science, 1760–1840. *History of Science* 14:149–195.
Sandl, Marcus. 1999. *Ökonomie des Raumes. Der kameralwissenschaftliche Entwurf der Staatswirtschaft im 18. Jahrhundert*. Köln: Böhlau Verlag.
Scharfe, Wolfgang, and Wolfgang Neugebauer. 1991. Administrativ-statistischer Atlas vom Preussischen Staate: Vorgeschichte, Entstehung und Umfeld des Preussischen Nationalatlases von 1827/28. *Forschungen zur Brandenburgischen und Preussischen Geschichte* 1:242–284.
Schiebinger, Londa, and Claudia Swan, eds. 2005. *Colonial botany: Science, commerce, and politics in the early modern world*. Philadelphia: University of Pennsylvania Press.
Schindler, Norbert, and Wolfgang Bonss, 1980. 'Praktische Aufklärung' – Ökonomische Sozietäten in Süddeutschland und Österreich im 18. Jahrhundert. In *Deutsche patriotische und gemeinnützige Gesellschaften,* ed. Rudolf Vierhaus, 255–353. München: Kraus.
Schmid, Karl Adolf. ed. 1859–1875. *Encyklopädie des gesammten Erziehungs- und Unterrichtswesens,* 10 vols. Gotha: Besser.
Schouw, Frederik Joakim. 1823. *Grundzüge einer allgemeinen Pflanzengeographie. Textband und Atlas*. Berlin: Reimer.
Spary, Emma. 2000. *Utopia's garden: French natural history from Old Regime to revolution*. Chicago: University of Chicago Press.
Stafleu, Frans A., and Richard S. Cowan. 1976–1988. *Taxonomic literature: A selective guide to botanical publications and collections with dates, commentaries and types,* 2nd ed. Utrecht: Bohn, Scheltema & Holkema.
Suckow, Georg Adolf. 1777. *Ökonomische Botanik: zum Gebrauch der Vorlesungen auf den hohen Kameralschulen zu Lautern*. Mannheim: Schwan.
Troll, Karl. 1925. Die Landbauzonen Europas in ihrer Beziehung zur natürlichen Vegetation. *Geographische Zeitschrift* 31:265–280.
Vetter, Jeremy. 2011. Introduction: Lay participation in the history of scientific observation. *Science in Context* 24 (2): 127–141.
Wakefield, Andre. 2009. *The disordered police state: German cameralism as science and practice*. Chicago: University of Chicago Press.
Watson, Hewett C. 1836. Observations on the construction of maps for illustrating the distribution of plants, with reference to the communication of Mr. Hinds on the same subject. *Magazine of Natural History* 9:17–21.
Watson, Hewett C. 1847–1859. *Cybele Britannica; or British plants and their geographical relations,* 4 vols. London: Longham.
Wickens, G. E. 1990. What is economic botany? *Economic Botany* 44:12–28.
Williams, Roger L. 2001. *Botanophilia in eighteenth-century France: The spirit of the enlightenment*. Dordrecht: Kluwer Academic Publishers.
Wise, Michael J. 1986. The Scott Keltie Report 1885 and the teaching of geography in Great Britain. *The Geographical Journal* 152:367–382.
Young, Arthur. 1792. *Travels, during the years 1787, 1788, and 1789, undertaken more particularly with a view of ascertaining the cultivation, wealth resources, and national prosperity, of the Kingdom of France*. London: J. Rackham.
Young, Arthur. 1794. *General view of the agriculture on the County of Suffolk*. London: C. Macrae.

Nils Güttler Nils Güttler is a postdoc at the Chair for Science Studies at ETH, Zürich, Switzerland. He holds a PhD in history of science from Humboldt University of Berlin and the Max Planck Institute for the History of Science. His research interests include the history of ecology, especially biogeography and environmental history, the role of images in science, as well as the history of popular science and scientific publishing.

Chapter 4
Rose and Pear Breeding in Nineteenth-Century France: The Practice and Science of Diversity

Cristiana Oghina-Pavie

Introduction

During the first half of the nineteenth century, horticulturists were concerned with obtaining a wide diversity of plants. In their desire to produce this diversity, to understand its causes and to master its mechanisms, the horticulturists were confronted with some fundamental problems of plant science. One of these problems was the issue of variation. For horticulturists, this question was a practical one: How did one generate the greatest variation, in order to then be able to select the most appropriate plants to meet market demands? In response to this dilemma, horticulturists established selection strategies in which the choice of seeds or artificial fertilization was coupled with actions that affected the development of seedlings.

Theorists, meanwhile, were more concerned with the causes of variation. Was variation the result of environmental influences or an effect of generation? The process of selection raised both theoretical and practical issues. How did one define, describe, and classify plant varieties? For practitioners, hybrids were mixtures of two or more "kinds" of plants; they were often ranked as "intermediate" categories. Theorists interpreted the hybrid issue in relation to the difficult question of the boundaries and possible transformation of species. Plant systematics included both horticultural classifications, based on criteria of usage, and botanical classifications, based on morphological criteria. But the two types of classifications posed the problem of the relationship between the origin and the polymorphism of cultivated plants. Both horticulture and botany were deeply concerned with the conceptual and evidentiary construction of the notion of heredity.

C. Oghina-Pavie (✉)
CERHIO – Maison des sciences humaines, Université d'Angers, 5bix Bd Lavoisier, 49045 Angers, France
e-mail: cristiana.pavie@univ-angers.fr

D. Phillips, S. Kingsland (eds.), *New Perspectives on the History of Life Sciences and Agriculture,* Archimedes 40, DOI 10.1007/978-3-319-12185-7_4

Past historical work on the formation of the "epistemic space"[1] of heredity has examined the role of pre-Mendelian breeders. These studies have shown particular interest in the breeding of annual plants. In the French case, the best-known examples are the writings of Louis Vilmorin on beets and those of Charles Naudin on cucurbits,[2] both from the middle of the nineteenth century. In this chapter, I am focusing on the breeding of two perennial, woody plants: the rose and the pear. This choice is motivated by the particularities of these two plants: They are woody plants, vegetatively propagated by grafting or cuttings, with a lifespan of several years or decades. The issues of character variation and transmission did not arise in the same ways as for annual plants. Roses and pears are also plants for which the individual differences were, and still are, very marked and very popular. As a luxury product, the commercial value of a new rose lies in the details of form, color, perfume, or habit. Breeders were thus very sensitive to minute variations. Finally, the rose and the pear were subject to intense breeding activity in France in the nineteenth century.

The selection of new varieties was part of the development of commercial horticulture, which responded to consumer demand for new and varied plants. From the fifteenth century, the pear was the emblematic fruit tree in the French aristocratic garden. At the end of the eighteenth century, the catalogue of Chartreux de Paris' nursery described 102 varieties of pears. In the 1860s and 1870s, pomological dictionaries and nursery catalogues contain lists of 600–1000 varieties of pears. The rose, meanwhile, became a true "French passion" at the end of the eighteenth century, under the influence of Empress Josephine's collection at Malmaison and the celebrated books of roses illustrated by Redouté.[3] The beginnings of rose selection were contemporary with the first collections that gathered all cultivated roses already known in France and new varieties from England, Asia, the USA, and Bourbon Island. About 120 varieties were present in the largest French collections in the early nineteenth century. Fifty years later, about 3000 varieties were in trade.

The people involved in the breeding of pears and, even more, of roses, belonged to various occupational categories. Amateurs, professional horticulturists, and learned societies obtained new varieties that were presented in competitions and horticultural exhibitions, and were named, observed, exchanged between breeders, multiplied, collected, or sold. "A noble emulation seized lovers of Flora"[4] said Vibert, a rose grower, to describe the French *rosomania*.

Objects of passion and trade, rose and pear breeds were described in the practical literature; popular publications such as the *Almanach du Bon Jardinier* and gardening handbooks spread tips for sowing and selecting new varieties. Selection techniques and hybrid classifications were discussed in learned societies. Commercial catalogues of nurseries and rose growers also described new varieties. The authors of these writings formed a milieu integrated through the circulation of knowledge.

---

[1] Müller-Wille and Rheinberger (2007).

[2] Naudin (1862); Vilmorin (1859).

[3] Joyaux (2001, pp. 80–81).

[4] Vibert (1824, p. 20).

They were active in central and local societies in which they were in contact with the works of scholars who belonged to public scientific institutions, either because these scholars were themselves members of those societies or because their publications were read and discussed there.[5] Breeders, who may also be called "practitioners," were not ordinary gardeners. They were nurserymen or educated amateurs who had access to scholarly botanical knowledge. They were the French equivalent of the British *scientific horticulturists*.[6] They were concerned about scholarly publications on physiology and plant systematics.[7] But they were not mere consumers of scientific knowledge. Seeking explanations and intellectual tools to understand and manage horticultural techniques with greater efficiency, they were also suppliers of observations in greenhouses, nurseries, and gardens. The practitioners were purveyors of facts from observation and experience.

In this chapter, I wish to describe how horticultural breeding became an area of intersection between concepts and practices, how it participated in the construction of a common space of problematization in the nineteenth century, and how confrontations and mutual influence provoked rethinking and re-formulation of scientific and practical problems. Sowing, crossbreeding, selections, and classifications of new varieties were activities that required simple technical know-how but complex conceptual references. I will analyze them in the order of complexity, from practical to theoretical issues, focusing on the period from the early nineteenth century until the 1860s.

From Sowing to Crossbreeding

In the early nineteenth century, the method of sowing was generally accepted as the most natural way to propagate plants. André Thouin (1747–1824), a gardener at the Jardin du Roi and, after the French Revolution, the first professor of *[Plant] Cultivation* at the National Museum of Natural Sciences, exhorted cultivators to sow in order to obtain strong seedling rootstocks, and also to improve varieties or races that would acclimatize more easily to the ground in which they were born.[8] André Thouin was a practitioner who belonged to the scientific world, working closely with botanists in the Museum and enjoying great prestige among gardeners and nurserymen.

Pear and rose sowings were carried out in order to reach two goals: propagation or improvement. Cultivators observed that seedling roses or pears could not "retain" or "reproduce" the variety that produced the seeds. Sowing did not preserve the variety.[9] Thus, horticulturists used seedlings for propagating plants intended to serve

[5] Fox (2012, pp. 52–94).

[6] Loudon (1822, p. 837).

[7] Oghina-Pavie (2011).

[8] Thouin (1805, p. 441).

[9] Falk (2009, p. 15).

as rootstocks. In this case, they were looking for qualities such as vigor, well-developed roots, and hardiness. Flowers and fruits had no importance in these seedlings; their variation was not relevant. Horticulturists sowed seeds of wild pears or seeds collected from pear crushes in order to obtain subjects that would then receive the grafted selected varieties. For roses, wildings produced by rose hips or *Rosier des chiens* were similarly designated to become wild stocks.

The approach was different when it came to sowing carried out so as to obtain new varieties. Eighteenth-century authors (Henri-Louis Duhamel du Monceau on trees,[10] Antoine-Nicolas Duchesne on strawberries[11]) had argued for the importance of seedlings for obtaining new varieties. That became the usual practice among amateurs and nurserymen looking to obtain new varieties of roses or pears. Sowing was a simple operation. It required no special skills except the constant care given to seedlings for several years, the time necessary for the first flowers on roses and the first fruits on pear trees. Breeders then had to choose among individuals that presented particular interest and, after that, multiply them by graft. If the technique was simple, it required time, patience, and planting space for a risky outcome. Pear breeders usually sowed 3000 or 4000 seeds at once[12] and selected one or two trees after 5 or 6 years. There were *semis du hasard* (random seedlings, seeds collected without any concern for their origin) or *semis ordonnés* (methodical seedlings, seeds collected on chosen varieties).

It is interesting to note that until the 1850s, breeders only paid attention to the "mother" plant that produced the seeds. They used the term "mother" in a broad sense. It could refer to the plant that provided the seeds as well as the plant which provided scions for grafts, or cuttings, or the root divided to be replanted elsewhere. The vegetative propagation by separation of the parts of plants might be conceived as a mode of generation.[13] The mother plant was the one that gave rise to an organized individual, distinct and separated from the other, either by seed or by cuttings or layering. The seed was only one of the forms of continuity between mother and offspring.

Since the eighteenth century, variation had been attributed to the mixing of garden varieties. For Duhamel du Monceau, in 1758, plant diversity could not be due to chance because organized beings could not leave a fortuitous arrangement of material parts. According to him, the high number of varieties grown together in gardens caused mixtures by "fertilizing dust," giving rise to *mongrels* and thus explaining the prodigious multitude of varieties.[14]

A change occurred in breeding in the 1830s and 1840s when practitioners were no longer content to wait for variation in seedlings but were advised to use artificial crossing. In practice, artificial crossing meant more or less controlled fertilization, most often through shaking the flowers of one variety over another. Examples of

---

[10] Duhamel du Monceau (1758, pp. 292–299).

[11] Duchesne (1766).

[12] Petit-Thouars (1828, p. 92).

[13] Massey (1828).

[14] Duhamel du Monceau (1758, p. 3).

successful artificial fertilization multiplied and attracted the attention of practitioners. In France, the writings of Augustin Sageret (1763–1851) marked this stage. Sageret was one of the most original French connoisseurs. He studied law and was a magistrate for a few years. After 1791, he left his office and was passionate about the most current topics of agronomy and gardening. Inspired by the work of Thomas Andrew Knight and Joseph Gottlieb Kölreuter, Sageret began using artificial crossbreeding on cucurbits and tobacco in order to show that crossing facility depended on the affinity between plants. He observed that the mother and the father have a major influence on some characters, and described an unequal distribution of "dominant" characters in offspring. Then he turned to the cultivation of fruit trees. He described his practice in a book entitled *Physiological Pomology* in 1830 and in an article that promoted hybridization in the improvement of fruits in 1840.[15] He suggested ways to "give rise to new species and varieties and direct their creation" by transferring pollen from one variety to another. I will draw on his views on hybridization in several parts of this chapter.

Another book highly acclaimed by practitioners was a work published by Henri Lecoq (1802–1871) in 1845. Lecoq was a pharmacist, professor of botany and director of the botanical garden in Clermont Ferrand. He was not, like Thouin or Sageret, a gardener. He was a provincial scholar, somewhat removed from Parisian academic networks, and a popularizer of botanical science. Nevertheless, his writings, composed in a highly metaphorical style,[16] enjoyed great popularity among home gardeners. In a book on fecundation and hybridization, he explained in detail how to carry out artificial fertilization for various crops, including the pear and the rose.[17] He presented crossbreeding as an imitation of nature that reduced the element of chance: "We must not rely any more on the inconsistency of a more or less favourable fortune, or on the flight of an insect, to create new roses; the brush must produce it, the taste, aided by the experience and the intelligence to direct them."[18]

Authors who recommended artificial fertilization, such as Sageret and Lecoq, did not pretend to give guarantees to ensure the certain success of sowings. Mastering the mixture is not the same as controlling the variation, which they attributed to fertilization as well as to "external circumstances," i.e., the conditions of cultivation.

Causal Factors of Variation

Practitioners had made two sensible observations: Plants in gardens varied more than plants in wild, and seedlings did not always resemble the mother plant. They concluded that seedlings were *modified* in gardens from one generation to the other. What were the causes of this variation when passing from mother to offspring

[15] Sageret (1826, 1830, 1840).

[16] Drouin and Fox (1899).

[17] Lecoq (1845).

[18] Lecoq (1845, p. 115).

through seeds? They believed that generation by seed had a disorganizing role: Sowing altered, disrupted, perverted, or denatured the organization of individuals.[19] Instead of reproducing an identical being, they come out altered. As breeders, they were interested in alteration only as an improvement. What seed should they sow to obtain improved flowers and fruits?

For most breeders, sowing the seeds produced by the best varieties offered the best chance for success. "Best" was defined by a set of horticultural criteria: color, flowering time, floriferous, double flowers, hardiness, fragrance for roses, early or late maturity, fruit size, fineness, sweetness, and softness of fruits for pears. But the opposite opinion also existed: Good fruits always have degenerate offspring.[20] For Pierre-Antoine Poiteau (1766–1854), the reasoning that encouraged farmers to sow the seeds of the best pears was modeled on the fact that a well-developed man and woman usually gave birth to well-developed children, with the children inheriting the physical and moral qualities of the parents. Poiteau was a gardener first in market gardens near Paris, then in the Jardin des Plantes of the National Museum of Natural Sciences, and later in several botanical gardens in San Domingo, Haiti, and Guiana, and in the royal nurseries of Versailles and Fontainbleau. He was also a professor of horticulture. He wrote a natural history of orange trees and numerous botanical descriptions and comments on physiology. As editor-in-chief of the *Revue Horticole* and *Almanach du Bon Jardinier,* he combined horticultural practice with extensive but not systemic knowledge of plant morphology and physiology. About the artificial fertilization of pear trees, he wrote, "[It] is not perhaps bad in itself, but we probably execute it in a wrong or incomplete manner," since the cultivators' seedlings rarely or never gave excellent new pears.[21] From observations he made in France and the USA, he deduced that the improvement of plant characteristics was a process that took place over several generations. Therefore, he exhorted cultivators to sow, sow, and sow again, even if the first trees did not produce suitable fruits. Nature "does not jump in its concessions, and it is only gradually and slowly that it gives us what we ask."[22] This method was the same one used by Van Mons, a Belgian horticulturist, who predicted that the best fruit would come only in the sixth generation of a pear tree.[23] Poiteau and Van Mons based this idea on the observation that the offspring did not fully reproduce the qualities of the parents. By repeating sowing, good pears appeared. They inferred from this that repeated sowing produced cumulative effects on the generation.

These three different views illustrate different conceptions of the offspring and inheritance. In the first view, the most common characters of the parents could be reproduced by seed and might even improve. In the second, varieties tended to return to their original type spontaneously much like wild pears and roses, losing the qualities that cultivation had inculcated in them. And finally, Van Mons and

[19] Vibert (1824, p. 28).

[20] Brébisson (1809).

[21] Poiteau (1828, p. 289).

[22] Poiteau (1828, p. 294).

[23] Mons (1835).

Poiteau supposed that there was progress in generations (understood here as successive descents) but that it took place over a long period of time. The diachronic view of variation was related to the antiquity of cultivation. As pears and roses had been grown since ancient times, they had become prodigiously capable of producing varieties because they had lost, as a result of domestication, the ability to maintain the characters of the species. For pears, practitioners found that old varieties, already described in the seventeenth and eighteenth century, were the best. They then concluded that those varieties had improved over time and were gradually moving away from their original type, the wild pear. The process of improving and accelerating change by sowing seeds had already travelled part of the road that led to perfection, and this process must consequently be identified. In this view, generation by seed was viewed historically. In contrast, if vegetative propagation was understood, not as generation, but as a continuation of the same individual (as in T. A. Knight's writings), antiquity was synonymous with aging. Varieties propagated by grafting were only an extension of the same individual and they were the age of the first selected tree.[24] After a period of adolescence and maturity, they got old, lost their qualities and could even approach natural death. The old varieties became unable to produce strong and healthy offspring; as a result there are no good varieties in seedling varieties that were excellent, but old.

Variation was considered to be the result of many casual factors, all united by the word "culture." Horticulturists compared wild roses and pears with crops and they observed that certain characters were found only in varieties "submitted to the yoke of our industry."[25] Wild roses were once flowering (non remontant) and did not produce double flowers. Wild pear fruits were small and bitter. An increase in flowering, double flowers, and improved size and sweetness of fruits were thus the result of horticultural techniques: graft, pruning, transplantation, nurture, etc. Roses became double under the action of an overabundance of manure in gardens[26], colors might vary as a result of the "atmosphere"[27] and, from time to time, "accidents" or "monstrosities" like variegation or color changes might be produced under the effect of one or all of these cultural factors.

The movement of plants could also change their nature. For example, shipping fruit from France to America would affect its very nature; this fact held true both for individual displacement (transplanting the tree) and generational displacement (sowing). The American soil imprinted a more primitive and a wilder character on fruits. Roses from America did not resemble roses from Asia because the climate and soil imprinted different characters on them.[28]

For Augustin Sageret, powerful factors could create variation before the formation of the seed (climate, soil, manure, and all artificial interventions on the mother plant: transplantation, root cutting, grafting, incision, ligation, layering) or during

[24] Knight (1841).

[25] Vibert (1824, p. 55).

[26] Boitard (1836, p. 152).

[27] Vibert (1824, p. 42).

[28] Boitard (1836, p. 92).

seed formation (fruit maturity, exposure to the sun or wind, desiccation of seeds). Other causes of variation occurred after seed formation. Sageret considered this issue crucial: "This is, for horticulture and plant physiology, of extreme importance: it is, in fact, a question of considering whether the seeds, once formed and harvested, have definitely received, and forever, the imprint, the character that nature had imprinted on them at birth or even at their creation, or, if the circumstances in which they could eventually sooner or later find themselves, would affect their characters, could modify or change them, and what the limits of these changes are."[29]

His response consisted of a series of examples that illustrated the link between variation due to crossing and variation induced by external factors. Sowing carried out at a different time than that assigned by nature to a plant in its native climate was more likely to produce variations. All the other changes in natural conditions (temperature, humidity, even electricity of the atmosphere) were factors that had a significant influence on variation. Some places were more favorable than others to causing changes, and they gave a particular character to the fruit through the combination of these factors. How else could one explain that the seeds of the same pear, harvested in Paris, gave different fruit when they were sown in Flanders, Holland, or America? As for the seeds of the same pear being sown in one place, Sageret explained the diversity of the offspring by the double or triple paternity exerted on the same seed "by mixing fertilizing dust of the same family."[30]

Was variation due to external conditions experienced by the parents and transmitted to the offspring? This issue was not already formulated as a question of the "heredity of acquired characters."[31] Sageret did not explicitly raise the issue of inheritance. He did not describe a hierarchy or a succession of causal factors on variation. Lecoq, more influenced by the scientific literature, saw in variation the effect of a shaken stability that he described in terms borrowed from Lamarck.[32] To cause plant variation, the practitioner needed to make the plant lose its habits. Changing climate conditions, temperature, or soil caused morphological changes in seedlings.[33] These changes in external form, even if small, were a sign of disturbed stability. Cultivators must sow the seeds of those individuals that were more susceptible to producing new variations.[34] A recently obtained variety would produce seeds that would vary more easily than an older one. The permanent care of gardeners would then give the variety new habits and create favorable circumstances to its stability, to the propagation by seed of the morphological variations. The permanence of the causes that determined the variation thus leads to its constancy and preservation, also in cases of generation by seeds.

This analysis of the causes that created variations reflected the close link between the transmission of characters and plant development. The diachronic variation

[29] Sageret (1830, p. 140).

[30] Sageret (1830, p. 126).

[31] Gayon (2006).

[32] Lamarck (1809, pp. 221–227).

[33] Lecoq (1845, p. xv).

[34] Lecoq (1845, p. 21).

from one generation to another and the synchronic variation between individuals from the same seed were not dissociated. They were conceived as joint effects of physiological processes (fertilization, growth, nutrition), influenced by culture-related techniques and conditions.

Horticultural Hybrids

To describe the nature of variation, botanists took species as a conceptual framework. They classified variations depending on their intensity, degree of generality, and permanence. For de Candolle, these different degrees of variation were characteristic of varieties, races, variants, and monstrosities.[35]

Horticulturists used the word "species" in the trivial sense of "kind,"[36] or they used the phrase "species and varieties" as a merger of the two classificatory categories. The conservation of the species by seed and hybrid (between species) sterility was of no particular importance for rose and pear breeders, who "fixed" the new varieties by vegetative propagation. However, they were confronted with the distinction between species and varieties when they are needed to classify new plants.[37]

What did the word *hybrid* in practitioners' writings mean? Botanists and zoologists reserved it for the (most often) sterile result of sexual fecundation between two individuals belonging to two different species. Horticulturists, in contrast, designated the result of a crossing between two different plants as *hybrids*. "Different" might refer either to species, varieties, races, groups, tribes, or other classification division. Therefore, any new plant grown from seeds was not a hybrid. Practitioners indeed used this term for plants that had the appearance of a composite being and which they could not classify in already established categories. This horticultural sense of hybridity was ambiguous. On the one hand, it expressed the idea of character transmission, because the presence of composite characters indicated the mixed origin of the hybrid. On the other hand, it also expressed the idea of a continuum, because hybrids were intermediate categories in the classification. According to Duhamel du Monceau, older fruits were combined in new seedling fruits. The Colmar Pear was a combination of Bon Chrétien and Autumn Bergamot. The Colmar Pear *partook* of the qualities of the other two pears that produced it. The Colmar Pear *shared* its characters with that mixed varieties that had generated it. The terms "partake" and "share" were not synonymous with "transmit," but the idea of filiations was present here. It was sufficient to carefully observe and taste new fruit varieties to identify the ones that have produced them. However, pears were rarely qualified as *hybrids* for the simple reason that the classification of pears referred to old classifications by La Quintinye and Duhamel du Monceau. Pear "species and varieties" were classified according to the date of maturity of the fruit, which was

[35] Candolle (1832, pp. 720–721). See also Brisseau de Mirbel (1828, pp. 470–472).

[36] Duhamel du Monceau (1758, p. xxxviii).

[37] Müller-Wille and Orel (2007).

a horticultural criterion. This simplification did not prevent the emergence of difficulties when it came to the regularization of pears in commercial nomenclature.

The classification of roses was a more complex issue. In the nineteenth century, any attempt at classification proved a serious challenge. Botanical classifications of roses (such as the one found in Linnaeus, done according to the shape of the fruit) were incompatible with the increasing diversity of cultivated roses. Beginning in the early nineteenth century, it became impossible to reconcile botanical criteria with horticultural ones. However, known botanists had made attempts in this regard, such as John Lindley,[38] who added an appendix on garden varieties to his 1820 monograph on roses. Auguste de Pronville, who translated Lindley's book into French 4 years after the English edition appeared, completely changed Lindley's divisions with the aim of making the original classification more useful to cultivators and better suited to French roses.[39] In these various classifications, the botanists and horticulturists added to traditional categories like species and varieties some ad hoc divisions; these included, depending on the author, families, groups, subvarieties, subgroups, tribes, series, etc. In these classifications, a *hybrid* was a rose that an author could not classify in a category because it presented a variety of characters that, for him, were dichotomous between the classes he had already established. At the same time, the novel combination of characters allowed the breeder to assume the origin of the new rose.

In practice, a plant was described as *hybrid* on morphological criteria. This was an *a posteriori* qualification, which confronted a classification by analogy and one by origin. Thus, for practitioners, some roses are a mixture of several varieties of characters. Poiteau deduced that one particular rose was a hybrid between several varieties: *Rosa perpetuosissima* was a hybrid of *damask, Ile Bourbon, Noisette, majalis, bengal, tea,* and of *cent-feuilles*. The commercial name of this rose was quite telling: "Le Désespoir des Amateurs" (The Despair of Amateurs).[40]

Other authors were more concerned with clarifying the limits imposed by species on hybridization. One such author was Pierre Boitard (1789–1859), an officer, and later an amateur naturalist and novelist. He considered this an abuse of the word hybrid in horticulture:

> They sow the seeds of *cent-feuilles*, they obtain roses that have some analogy with *damask, alba* etc. At once they decide that these roses are hybrids of *cent-feuilles* and *damask, alba,* etc. This is moving a little to fast! Others, however, are even more expeditious: they sow seeds collected at random, then, when seedling roses are in bloom, they study them and classify them arbitrarily among the hybrids of a particular species, because they believe that they would recognize the specific characters of these two species; or it may happen, and it happens very frequently for that matter, that these so-called hybrids originated from a seed that belonged to neither one of the other two species the characters of which they bear.[41]

[38] Lindley (1820).

[39] Lindley (1824). Pronville (1818) published also his own classification of roses.

[40] Poiteau (1833, p. 325).

[41] Boitard (1836, p. 75).

Sometimes, the characters of different varieties were "melted," mixed in a manner so that the observer could not distinguish what variety they come from. According to the rose grower and breeder Jean-Pierre Vibert (1777–1886), this resulted from the reciprocal fertilization between roses that had opposite characteristics, such as contrasting colors. Alterations could thus be gradual: The seeds of a red rose fertilized by a white one could produce white and red flowers, or also all the shades of pink.

Classifying roses in the nineteenth century was a difficult task; the rules of botanical taxonomy were hard to reconcile with the intentionally fostered diversity found in cultivated plants. Numerous groups were created for new varieties (*Noisette, Bengali, Portland,* etc.). Groups might be related to species (*Rosa gallica*), or lineages (the *Noisette*, all descended from a rosebush that Louis Noisette has received from his brother, Philippe Noisette, a nurseryman in the USA). The groups entitled "hybrid of…" or "uncertain hybrid" were becoming more numerous and larger. In each group, the horticulturists identified a *group leader* that they saw as both the type (which might mean species in some classifications) and the origin of all varieties. The description and classification of hybrids focused on linguistic ambiguities that proved to be conceptual problems as well. For example, Louis-Augustin Bosc D'Antic (1759–1828), naturalist and agronomist, professor at the National Museum of Natural Sciences,[42] observing a rose whose color, smell, and leaf shape attached it to one group (*bipinné* rose), but whose flower shape belonged to another group (Provins), called it *Bosc rosa intermedia.* It was, he said, probably a variety of the other groups, but the different roses' characteristics were different enough that one was forced to separate them "at least in the practice of gardening."[43] Nevertheless, designated as a species, with a Latin name, this *intermedia* hybrid was conceptualized in horticultural writings as "an intermediary link in the chain" between the two divisions. The hybrids that appeared every day in seedlings filled the "distance" between the two groups. They therefore raised another fundamental issue: that of the continuity of species and their transformation.

Hybrids and the Limits of Species

From the 1830s until the1860s, the debate about the nature of species that animated the French scientific community also included discussion of the diversity of horticultural crops. In 1835, Pierre Boitard presented a memoir on the nomenclature of roses at the Academy of Sciences. The following year, he published a longer version of his memoir, entitled "A Complete Manual of Roses for Amateurs"[44] in which he openly argued with horticulturists like Vibert who suggested fanciful classifications of roses. Boitard saw species as natural entities. Hybrids between species were rare

[42] Bosc D'Antic (1809).

[43] Bosc D'Antic (1809).

[44] Boitard (1836).

in nature, unlike the hybrids between varieties that were often found in gardens. If fecundation was possible between two groups, this meant that they belonged to a unique species or, in rare cases, two species that had a strong analogy. He proposed only three species of roses (*simplicifolia* rosa, *lutea,* and *centifolia*), between which crossings were never observed. All the other groups, divisions or tribes were only varieties.

Boitard's reasoning was that *facies,* i.e., all forms that give the appearance of a rose, were a set of traits that were all dependent on the living conditions of the plant. Cultivated roses were classified by botanists according to the form of fruits, styles, stamens, petals, sepals, etc. All these organs were affected by cultivation. The care provided by the gardeners was the cause of accidental changes. But these changes were not invariable. Their reproduction by seed over several generations depended on external conditions too. Their inheritance existed only if the variety were placed in the same conditions as those of the initial accident. "Let's leave it to nature and it will quickly return to its type, or it will perish." The same causes rigorously produced the same effects: Variety was thus the result of a set of constant causes. In this deterministic vision, he explained the differences between Asian and American roses through the effects of a different climate. These different kinds of roses were local varieties, not distinct species. According to Boitard, the only specific character that was invariable and constant, and that escaped the influence of climate and culture, was the color, "considered from a physiological viewpoint." More precisely, he identified the yellow color as a constant character in roses. He explained that the yellow color was permanent because it was given by the presence of an alkali. It depended on the chemical composition of the plant, unlike *facies,* which was the result of development. The yellow color was therefore a character that distinguished the members of a species because its presence and inheritance were independent of the environment.

Double flowers, for example, appeared because of an excess of food, which would explain the lack of double roses in spontaneous, wild flora. The character was reproduced irregularly in the offspring: Only a part of the seeds from double roses would produce double flowers in seedlings. However, this irregular inheritance depended on the constancy of the conditions that provoked it. In the absence of overabundant food, there would be no double roses in seedlings. For other characters, Boitard did not explain the mechanisms by which culture affected the shape of the organs, or why the yellow color would be free from any external influence. His classification into three species failed to convince botanists and practitioners. However, his manual attempted to find a coherent explanation of the determinism of generations, one that could reconcile Buffon's definition of species with the diversity of cultivated roses. He focused on the distinction between development (morphology, influenced by climate variables and culture) and chemical composition (stable essence, independent of external conditions).

The French analytical school of botany in the mid-nineteenth century was also interested in these debates. This school believed that species were not polymorphic or unstable. Unlike the school of synthetical botany (George Bentham and Joseph Dalton Hooker), which operated with distinct Linnaean species, the botanists of the analytical school defended the idea that all distinctions between organisms that did not hybridize were signs that they belonged to different species. Therefore, their taxonomy took into account all the differences, even tenuous, for qualifying species. The most radical French "species maker" was the naturalist Alexis Jordan (1814–1879) from Lyon. Jordan and his followers raised questions about the limitations of botanical species[45] that included a discussion of crops, especially those that showed a greater propensity to vary, such as pears and roses. According to them, almost all varieties of cultivated plants should be considered as fixed species. Variations in a natural state and variation in a cultivated state both needed to be understood as consistent with the fixity of species. Jordan relied on the example of pear trees to demonstrate the origin of cultivated varieties: All new morphological forms of unknown origin that were emerging in cultivation could not be hybrids. They could not be anything other than formerly cultivated species that had been forgotten and had degenerated. Now that these species were being cultivated and improved again, they had resumed the rank of species that they had once lost.[46]Alexandre Boreau (1803–1875), director of the botanical garden of the city of Angers and disciple of Jordan, protested against the sharp reduction of the number of species of roses in botanical and horticultural taxonomy. For him, each dichotomous character was minor when taken separately: shape and direction of thorns, leaflet's shape, leaf serrations, glabrous condition, etc. The extreme ease with which the rose "plays in our gardens" showed that fixed characters were uncommon, but did not prove the nonexistence of species. According to Boreau it would be wrong to give these forms "the vain title of varieties."[47] Following Jordan and Boreau, botanist Alfred Déséglise (1823–1883) made several classifications of the genus *Rosa,* also multiplying the number of species.[48] He protested against the *hybridolaters* who saw hybridization as a creation of intermediate forms between species and therefore a transformation. If these forms were constant, i.e., if they reproduced through seed; they must be recognized as species. Otherwise, they were only random accidents. This stationary school[49] sought to identify species in the spontaneous flora that corresponded to kinds of cultivated roses. Analytical botanists were opposed to the "progressive school" according to which garden plants came from a few original types that were successively transformed. According to Jordan, this was "the key of the genus"; it would be very easy to understand the diversity of cultivated roses by observing the criteria that distinguished wild roses.[50]

Scholars on both sides of this controversy regarded the practice of breeders as a source of examples that could be employed when constructing theories of natural or physiological heredity.[51] Empirical observations performed at the level of spe-

[45] Matagne (2011).

[46] Jordan (1853).

[47] Boreau (1844, p. 53).

[48] Déséglise (1861, p. 8).

[49] Déséglise (1877).

[50] Jordan (1847, p. 77).

[51] Gayon (1995).

cies, variety or horticultural group were confronted with another level of generalization, that of the living beings in their present state and their history.[52] The theorists of transformation borrowed from the horticulturists whatever facts and conceptual representations they need. Reciprocally, they attempted to explore horticultural plants with experimental methods in order to test their theoretical hypotheses with empirical data. Experiments conducted on pears by Joseph Decaisne (1807–1882), professor of culture at the Museum of Natural History, joined the search for links between "experimental biology, natural history and improved domestic species in order to unravel the mysteries of heredity."[53]

Experimentation and Evolution

In 1863, Joseph Decaisne published two very similar texts on the variability of pears in the *Annales de sciences naturelles* and *Comptes rendus de l'Académie des Sciences.*[54] He presented an experiment that he had done at the Museum in order to clarify, with scientific methods, the nomenclature of pears. Decaisne explained the absence of a standard method in the description of pear trees and the great number of local names and synonyms. To name a variety or choose a name among available options, it was necessary to define the criteria that distinguished a given pear from others. Even more crucially, one needed to define the characteristics of the entire *genus* that contained all of the relevant varieties.

Were there several natural species in the group of the cultivated pear trees, or were there only subdivisions of a primitive type modified in different ways by cultivation areas and methods? In other words, the first practical purpose—that of establishing a unique nomenclature—was connected to an issue that "presents the highest interest and which we have to consider as one of the bases of the Science"[55]: the relevance of the concept of species in the particular case of the pear tree.

Decaisne conceived an experimental approach to the variability of the pear tree as a part of his taxonomic research, as well as an empirical ground for the observation of the limits of the species he studied. He began his experiments in 1853. He sowed many pear seeds taken from four varieties clearly distinguished by horticulturists: the old pear of *England,* the *Bosc* pear, the *Belle-Alliance* pear, and the *Cirole* pear. Only a few seedlings produced fruits. However, Decaisne studied the differences from the previous generation and he noted a lot of variation among these four varieties: changes in shape, size, and color of the fruit, leaf shape, general appearance of the trees, sap, and, finally, form and flower size.

[52] Müller-Wille and Rheinberger (2012, pp. 15–25).

[53] Gayon (1997, p. 390).

[54] Decaisne (1863). The text was also published, with some modifications, in Decaisne (1871–1872).

[55] Decaisne (1871–1872, p. 5).

He concluded that there was wide variability and no stability between generations. His results took on complex significance in the debates that took place during the second part of the nineteenth century. Indeed, the interpretation of variation was an important part of the debate on the limits and origins of species. He wanted to find experimental observations to invalidate both fixism, especially Jordan's hypothesis, and transformationism, especially Darwin's theory.

Decaisne described the process of variation; he thought that in a lot of species of plants, especially in cultivated ones (including, among others, the pear tree), varieties were products of cultivation and came from a unique ancestral form. For Decaisne, variation within a species was founded on the following principle: In the beginning, a main species was divided into secondary species, which, still belonging in part to the primordial plasticity, were subject to the action of cultivation that produced our current races and varieties. Over centuries, races and varieties continued to multiply by themselves, but were never able to change from one kind to another. However, for Descaine, there was a strict boundary between species. Variations never led to other species. He claimed that if we transported a French variety of pear tree into all the regions of the world wherever it could live, it would tend to be in harmony with the environment. We could be sure that after a few generations, it would give new and numerous varieties. Decaisne noticed that this fact had been observed for all the economic plants disseminated in the entire world. This fact explained the origin of polymorphism in species, so problematic for botanists; it was the result of dissemination. In other words, Decaisne linked variation both to the conditions of the environment but also to the practices of cultivation. It is very important to note, however, that he did not accept Lamarck's thesis about the influence of the environment[56] and the transformation of species.

Moreover, he explicitly said that he did not agree with Darwin about the stability of variations across generations. Here, we can see two different concepts of variation in play, and also two different theoretical contexts. As is well known, Darwin included variation as a basis for the theory of natural selection. He regarded variation as a constant process that led to the explanation of the origin of new species by means of selection. For Decaisne, variation was also a fact. But he did not speak of variation in nature, and he studied only the case of cultivated plants and trees. For him, variations existed within the boundaries of the species, but could not produce new species. Darwin used Decaisne's results[57], which were the most complete available on the variation of pear trees, and seemed satisfied by the fact that Decaisne claimed that every variety derived from one ancestral form. Their theoretical positions were, however, antagonistic.

Decaisne's and Darwin's conceptions centered on the same empirical data about variation, a process that takes place in nature nowadays. The two naturalists had two opposite interpretations of the role of this process. On the one hand, Decaisne conceived of a history limited to each species. Darwin, on the other hand, had a broader perspective of a history unlimited to species. The arguments over the variation in

[56] Gayon (1997, p. 342).

[57] Darwin (1868, p. 372).

the Museum's pear trees were characteristic of the multiplicity of concrete debates in which fixism and transformationism were involved during the first decades of Darwinism.

Decaisne's position was indebted to earlier discussions of the effects of generation and culture on the diversity of horticultural plants. However, this corpus could not be reduced to a simple set of data. Indeed, it was closely linked to the theoretical interpretations of variation that were part of fundamental debates in the life sciences. It was precisely during Decaisne's experiment that this context changed. Decaisne had probably sown with the intention of responding to Jordan; he interpreted the results of his observation with the intention of responding to Darwin. In addition to Darwin's book, in the intervening years between the start and conclusion of his experiments, many works on heredity had been published. In France, between 1859 and 1863 Vilmorin,[58] Duchartre,[59] Godron,[60] Lecoq[61] and Naudin[62] completely rewrote the conceptual relationship between variation, generation, and heredity.

Decaisne adapted the interpretation of his observations to this moving intellectual context. Taking the pear as an object of experimentation, he was in a deadlock. While he waited for his pear trees to grow and bear fruit, Naudin observed several generations of squash and drew conclusions on the changes per generation that were comparable to those of hybridizers like Vilmorin or Gartner. Decaisne remained, meanwhile, confined to an experimental practice modeled on the breeding methods of gardeners: He combined variation and nomenclature issues, he did not know the origin of the pollen he used, and he did not proceed to successive sowings to study variation over several generations. There was a mismatch between the theoretical questions and the horticultural species subjected to the experiment.[63]

Conclusion

Decaisne's experiment on the pear reveals one of the peculiarities of roses and pears, woody plants with a long life cycle: the difficulty of understanding the relationship between variation and time. Those perennial plants acquired a particular epistemological status,[64] simultaneously bearing the mark of their living nature and their ancient culture.

Both savants and practitioners conceived that roses and pears were bred within these limitations. Breeders connected selection to a longer history of the accumulated influence that cultivation techniques had exercised over variation. They used

[58] Vilmorin (1859).

[59] Duchartre (1862).

[60] Godron (1862).

[61] Lecoq (1845).

[62] Naudin (1861, pp. 396–399). See also Naudin (1862, 1865).

[63] Burian (2005, p. 11).

[64] Tirard (2012, pp. 16–17).

the past to explain the present state of plants. Their view of hybridization showed a tension between the temporal vision of the plant's origin and the static view of its place in systems of classification. This equivocal concept of hybridization fed on scientific controversies. In turn, it provided arguments to these more theoretical debates.

The debates on evolution in the second half of the nineteenth century placed plant breeding in a new context. The variation observed and analyzed by horticulturists in the first decades of the nineteenth century became a crucial issue in theories of evolution. In particular, the works and debates on woody plants were appropriated to reveal the complexity of the interactions between practice and theoretical issues.

Acknowledgments I am grateful to Denise Phillips and Sharon Kingsland for helpful comments and to Oana Babîi for translation. My recent research on nineteenth-century hybrids has been conducted within the project FLORHIGE (GDO–IRHS, CERHIO, CFV and partners) with support from Region des Pays de la Loire and project HYBRIDES with support from the University of Angers.

References

Boitard, Pierre. 1836. *Manuel complet de l'amateur de roses, leur monographie, leur histoire et leur culture*. Paris: Librairie encyclopédique Roret.

Boreau, Alexandre. 1844. Notes sur quelques espèces de plantes françaises. In *Bulletin de la société industrielle d'Angers et du département de Maine-et-Loire*, ed. Cosnier et Lachèse, 15e Année, 48–69. Angers: Cosnier et Lachèse.

Bosc D'Antic, Louis-Augustin. 1809. Rosier, Rosa. In *Nouveau cours complet d'agriculture théorique et pratique, contentant la grande et la petite culture, l'économie rurale et domestique, la médecine vétérinaire, etc. ou Dictionnaire raisonné et universel d'agriculture, etc.*, ed. F. Rozier, vol. 11, 237–266. Paris: Deterville.

Brébisson. 1809. Poirier. In *Nouveau cours complet d'agriculture théorique et pratique, contentant la grande et la petite culture, l'économie rurale et domestique, la médecine vétérinaire ou Dictionnaire raisonné et universel d'agriculture, etc.*, ed. F. Rozier, vol. 10, 234–269. Paris: Deterville.

Brisseau de Mirbel, Charles-François. 1828. Théorie fondamentale de la botanique. In *Dictionnaire des sciences naturelles dans lequel on traite méthodiquement des êtres de la nature, soit considérés en eux-mêmes, d'après l'état actuel de nos connoissances, soit relativement Ã l'utilité qu'en peuvent tirer la médecine, l'agriculture, le commerce et les arts, suivi d'une biographie des plus celebres naturalistes,* ed. F. Cuvier, vol. 53, 470–508. Paris: Le Normant.

Burian, Richard. 2005. *The epistemology of development, evolution and genetics: Selected essays.* Cambridge: Cambridge University Press.

Candolle, Augustin Pyramus de. 1832. *Physiologie végétale, ou exposition des forces et des fonctions vitales des végétaux, pour servir de suite Ã l'organographie végétale, et d'introduction Ã la botanique géographique et agricole*. Paris: Béchet.

Darwin, Charles. 1868. *The variation of animals and plants under domestication*. London: Murray.

Decaisne, Joseph. 1863. De la variabilité de l'espèce de poirier: Résultats d'expériences faites au Muséum d'histoire naturelle de 1853 Ã 1863 inclusivement. *Comptes rendus hebdomadaires des séances de l'Académie des Sciences* 57:6–17. (Reprinted in Annales des Sciences Naturelles (Botanique) 4th ser. 20:188–200).

Decaisne, Joseph. 1871–1872. *Le jardin fruitier du Muséum ou iconographie de toutes les espèces et variétés d'arbres fruitiers cultivés dans cet établissement avec leur description, leur histoire, leur synonymie etc.*, vol. I. Paris: Didot frères.

Déséglise, Alfred. 1861. *Essai monographique sur cent cinq espèces de rosiers appartenant Ã la flore de la France*. Angers: Cosnier et Lachèse.

Déséglise, Alfred. 1877. *Catalogue raisonné ou énumération méthodique du genre rosier pour l'Europe, l'Asie et l'Afrique, spécialement des rosiers de la France et de l'Angleterre*. Genéve: Ch. Menz.

Drouin, Jean-Marc, and Robert Fox. 1899. Corolles et crinolines: Le mélange de genres dans l'œuvre de Henri Lecoq (1802–1871). *Revue de synthèse* 120 (4): 581–599.

Duchartre, Pierre. 1862. Rapport sur la question de l'hybridité dans les végétaux. *Annales des sciences naturelles. Botanique*, ser. 4, 19:125–134.

Duchesne, Antoine Nicolas. 1766. *Histoire naturelle des fraisiers contenant les vues d'économie réunies Ã la botanique, suivie de remarques particulières sur plusieurs points qui ont un rapport Ã l'histoire naturelle ou générale*. Paris: Didot jeune.

Duhamel du Monceau, Henri-Louis. 1758. La physique des arbres, où il est traité de l'anatomie des plantes et de l'économie végétale, Première partie. Paris: Chez H.L. Guérin et L.F. Delatour.

Falk, Raphael. 2009. *Genetic analysis: A history of genetic thinking*. Cambridge: Cambridge University Press.

Fox, Robert. 2012. *The savant and the state: Science and cultural politics in nineteenth-century France*. Baltimore: John Hopkins University Press.

Gayon, Jean. 1995. Entre force et structure: Genèse du concept naturaliste de l'hérédité. In *Le paradigme de la filiation,* ed. Jean Gayon and Jean-Jacques Wunenburger, 61–65. Paris: L'Harmattan.

Gayon, Jean. 1997. Le Muséum national d'histoire naturelle et l'amélioration des plantes au XIXe siècle. In *Le Museum au premier siècle de son histoire,* ed. Claude Blanckaert, et al., 375–401. Paris: Muséum national d'histoire naturelle, Coll. Archives.

Gayon, Jean. 2006. Hérédité des caractères acquis. In *Lamarck, philosophe de la nature,* ed. Pietro Corsi, Jean Gayon, Gabriel Gohau, and Stéphane Tirard, 105–163. Paris: Presses Universitaires de France.

Godron, Dominique-Alexandre. 1862. Des hybrides végétaux considérés au point de vue de leur fécondité et de la perpétuité ou de la non-perpétuité de leurs caractères. *Annales des sciences naturelles. Botanique*, ser. 4, 19:135–179.

Jordan, Alexis. 1847. Observations sur plusieurs plantes nouvelles, rares ou critiques de la France, Lues Ã la Société Linnéenne de Lyon le 12 avril 1847, Sixième Fragment. Paris: Ballière.

Jordan, Alexis. 1853. *De l'origine des diverses variétés ou espèces d'arbres fruitiers et autres végétaux cultivés pour le besoin de l'homme. Mémoire lu en séance publique Ã l'Académie des sciences, belles-lettres et arts de Lyon le 14 décembre 1852*. Paris: Ballière.

Joyaux, François. 2001. *La rose, une passion française: Histoire de la rose en France 1778–1914*. Paris: Editions Complexe.

Knight, Thomas Andrew. 1841. On the culture of pear-tree (read May 13, 1813). In A selection of physiological and horticultural papers published in the transactions of the royal and horticultural societies, ed. Longman, Orme, Brown, Green, and Longmans, 201–204. London: Longman.

Lamarck, Jean-Baptiste. 1809. *Philosophie zoologique*. Paris: Dentu.

Lecoq, Henri. 1845. *De la fécondation naturelle et artificielle des végétaux et de l'hybridation, considérée dans ses rapports avec l'horticulture, l'agriculture et la sylviculture*. Paris: Librairie horticole de H. Cousin.

Lindley, John. 1820. *Rosarum Monographia; or a botanical history of roses, to wich is added an appendix for the use of cultivators, in which the most remarkable garden varieties are systematically arranged*. London: James Ridgeway.

Lindley, John. 1824. *Monographie du genre rosier, traduit de l'anglais de M. J. Lindley, avec des notes de M. L. Joffrin, et des changements importants, suivie d'un appendice sur les roses cultivées dans les jardins de Paris et environs, trans. Auguste Pronville*. Paris: Audot.

Loudon, John Claudius. 1822. *An encyclopaedia of gardening, comprising the theory and practice of horticulture, floriculture, arboriculture and landscape-gardening*. London: Longman.

Massey, Placide. 1828. Végétaux. In *Dictionnaire des sciences naturelles dans lequel on traite méthodiquement des êtres de la nature, soit considérés en eux-mêmes, d'après l'état actuel de nos connoissances, soit relativement Ã l'utilité qu'en peuvent tirer la médecine, l'agriculture, le commerce et les arts, suivi d'une biographie des plus celebres naturalistes, etc.*, ed. F. Cuvier, vol. 67, 3–16. Strasbourg/Paris: Levrault/Le Normant.

Matagne, Patrick. 2011. Les botanistes de l'Ouest et les faiseurs d'espèces. *Bulletin d'épistémologie et d'histoire des sciences de la vie* 18 (2): 157–168.

Mons, Jean-Baptiste Van. 1835. *Arbres fruitiers. Leur culture en Belgique et leur propagation par la graine ou Pomonimie belge, expérimentale et raisonnée, on y a joint le catalogue descriptif abrégé de bons fruits nouveaux procréés et cultivés Ã la pépinière d'expérience de l'auteur Ã Louvain,* vol. 1. Louvain: Dusart et Vandenbroeck.

Müller-Wille, Staffan, and Vitezslav Orel. 2007. From Linnaean species to Mendelian factors: Elements of hybridism, 1751–1870. *Annals of Science* 64 (2): 171–215.

Müller-Wille, Staffan, and Hans-Jörg Rheinberger. 2007. Introduction. In *Heredity produced: At the crossroad of biology, politics, and culture, 1500–1870,* ed. Staffan Müller-Wille and Hans-Jörg Rheinberger, 5–9. Cambridge: MIT Press.

Müller-Wille, Staffan, and Hans-Jörg Rheinberger. 2012. *A cultural history of heredity*. Chicago: University of Chicago Press.

Naudin, Charles. 1861. Sur les plantes hybrides. *Revue horticole* 10:396–399.

Naudin, Charles. 1862. Nouvelles recherches sur l'hybridité dans les végétaux. *Annales des sciences naturelles. Botanique*, ser. 4, 19:180–203.

Naudin, Charles. 1865. Nouvelles recherches sur l'hybridité dans les végétaux: Mémoire présenté Ã l'Académie en décembre 1861. *Archives du Muséum* I:25–176.

Oghina-Pavie, Cristiana. 2011. Horticulture et physiologie végétale au début du XIXe siècle: Un espace de savoir partagé. *Bulletin d'histoire et d'épistémologie des sciences de la vie* 18 (2): 113–119.

Petit-Thouars, Louis-Aubert du. 1828. Rapport fait au nom du comité des arbres fruitiers, sur trois poires soumises Ã leur examen. *Annales de la société d'horticulture de Paris et journal spécial de l'état et des progrès du jardinage* 2:86–94.

Poiteau, Pierre-Antoine. 1828. Considérations sur le procédé qu'emploient les pépiniéristes pour obtenir de nouveaux fruits améliorés et sur celui que paraît employer la nature pour arriver au même résultat. *Annales de la société d'horticulture de Paris et journal spécial de l'état et des progrès du jardinage* 2:288–298.

Poiteau, Pierre-Antoine. 1833. Le désespoir des amateurs. *Annales de la société d'horticulture de Paris et journal spécial de l'état et des progrès du jardinage* 13:325–326.

Pronville, Auguste. 1818. *Nomenclature raisonnée des espèces, des variétés, et sous variétés du genre rosier, observées au Jardin royal des plantes, dans ceux de Trianon, de Malmaison, et dans les Pépinières des environs de Paris*. Paris: Huzard.

Sageret, Augustin. 1826. *Mémoire sur le cucurbitacées, principalement sur le melon, avec des considérations sur la production des hybrides, des variétés, etc*. Paris: Huzard.

Sageret, Augustin. 1830. *Pomologie physiologique ou Traité du perfectionnement de la fructification, avec recherches et expériences sur les moyens d'améliorer les fruits domestiques et sauvages, d'augmenter et d'assurer leur produit, de faire naître des espèces et variétés nouvelles*. Paris: Huzard.

Sageret, Augustin. 1840. *Des fécondations étrangères spontanées et artificielles et de la production des hybrides considérées dans leurs rapports avec l'amélioration des fruits. Extrait du Journal des connaissances usuelles et pratiques*. Paris: Delanchy.

Thouin, André. 1805. *Cours complet d'agriculture théorique, pratique, économique, et de médecine rurale et vétérinaire,* vol. 12. Paris: Marchant.

Tirard, Stéphane. 2012. La sélection variétale: Maîtriser l'évolution biologique? In *La protection juridique du végétal et ses enjeux économiques,* ed. Serge Blondel, Sophie Lambert-Wiber, and Camille Maréchal, 15–24. Paris: Economica.

Vibert, Jean-Pierre. 1824. *Essai sur les Roses*. Paris: Huzard.
Vilmorin, Louis. 1859. *Notices sur l'amélioration des plantes par semis et considérations sur l'hérédité dans les végétaux, précédées d'un mémoire sur l'amélioration de la carotte sauvage*. Paris: Librairie Agricole.

Cristiana Oghina-Pavie is an assistant professor in the Department of History of the University of Angers, member of Centre de Recherches Historiques de l'Ouest (CERHIO UMR CNRS 6258). Her research deals with the relationship between horticulture and plant sciences in the nineteenth and twentieth centuries with focus on physiology and heredity. She develops interdisciplinary approaches with epistemologists and biologists on woody plants. They currently study the effects of breeding, trade, and collection on the genetic diversity of cultivated roses.

Chapter 5
Napoleonic Cotton Cultivation: A Case Study in Scientific Expertise and Agricultural Innovation in France and Italy, 1806–1814

Joseph Horan

Research for this article was made possible by a grant from the Fondation Napoléon.

Introduction

In January of 1807, Napoleon Bonaparte instructed his interior minister to initiate an extensive project designed to introduce the cultivation of cotton to southern France and to areas of Italy that had recently fallen under French control. The undertaking was announced in the major newspapers at the end of March, and explained as an effort on the part of the regime to "stimulate usefully the efforts of rural industry."[1] Behind this rather bland declaration was a daring geopolitical calculation. Napoleon's decision to encourage the acclimatization of cotton to France and Italy was a key element of the Continental System, his grand strategy for victory in the ongoing struggle against Great Britain. The Continental System had been initiated in November of 1806 with the Berlin Decrees, which effectively banned British commerce from the European Continent in a bid to provoke the economic ruin of Napoleon's most persistent foe.[2] Napoleon and his officials recognized the commercial advantages that Great Britain derived from advanced textile manufacturing and from control of the global cotton trade. They calculated that if French manufacturers could not match the productivity of their counterparts across the channel, British textiles would continue to enter Europe through the smuggling networks that spanned the Continent, thus providing the British with a commercial lifeline and effectively neutralizing the Continental System. By introducing cotton cultivation

[1] (1807) "Le Ministre de l'Interieur, a Monsieur le Prefet du Department d'…." All translations by author unless otherwise noted.

[2] For a recent analysis of the Continental System, see Marzagali (2005).

J. Horan (✉)
Division of Liberal Arts & International Studies,
Colorado School of Mines, 1500 Illinois St., 80401 Golden, CO, USA
e-mail: jwhoran@mines.edu

D. Phillips, S. Kingsland (eds.), *New Perspectives on the History of Life Sciences and Agriculture,* Archimedes 40, DOI 10.1007/978-3-319-12185-7_5

to the areas he controlled, Napoleon hoped to create a secure and convenient source of raw cotton for the French textile industry. If French manufacturers could provide cotton textiles competitive with their British counterparts, Napoleon believed, the rest of Europe would be more inclined to support his measures against British commerce. Deprived of its income from European markets and without allies on the Continent, Great Britain would be forced to sue for peace on Napoleon's terms. With these goals in mind, in early 1807, the French Ministry of the Interior launched an ambitious project to alter the economy and environment of a large section of the Napoleon's empire.

The Napoleonic project for the acclimatization of cotton to France and Italy constitutes an important case study in the history of science, providing valuable insight into the interaction between botanical knowledge, agricultural innovation, state power, and biological exchange at the start of the modern era. Yet this episode has largely passed into obscurity, receiving only a handful of references in more general studies of the Napoleonic era.[3] The only substantial scholarship on the subject can be found in two articles by the French historian Alain Blondy, who is primarily interested in the activities of Mikel Anton Vassalli, a Maltese native charged by the French government to direct experiments with cotton cultivation in the Bouches du Rhône department. As for the parallel efforts that were taking place across much of France and Italy, Blondy notes only that they "had no great importance for French agriculture or industry."[4] The fact that France and Italy did not become major centers of cotton production has meant that the Napoleonic experiments with cotton cultivation have passed into obscurity following the termination of the project with the fall of Napoleon's regime in the spring of 1814.

While the history of biological exchange has in recent decades been the subject of a growing body of innovative scholarship, this research has focused almost exclusively on the most successful examples of plant acclimatization. Indeed, in a recent essay on the subject, historians William Beinart and Karen Middleton have argued that studies of plant transfers in history should focus primarily on cases of successful acclimatization, suggesting that "most transfers become important, historically and ecologically, if they spread."[5] Yet, a closer analysis of the Napoleonic experiments with cotton cultivation reveals that even an unsuccessful acclimatization project can offer insight into a range of important topics, including the role of the state in promoting environmental change, the links between the development of botanical science and agricultural innovation, and the wide range of human and natural factors that can facilitate or inhibit the introduction of plants to new regions of the globe. If scholarship on the history of plant transfers continues to concentrate only on the most dramatic success stories, our understanding of the role of human agency in the history of biological exchange will remain incomplete.

In this chapter, I analyze the Napoleonic experiments with cotton cultivation in order to make several interrelated arguments concerning the history of botanical and agricultural science in the early modern world. The first section describes

[3] Lefebvre (1953, p. 253), Godechot (1979, p. 39), Tulard (1992, p. 263), and Donati (2008, vol. 1, p. 42).

[4] Blondy (1993, 2006).

[5] Beinart and Middleton (2004).

the antecedents of Napoleon's acclimatization project. An assessment of French interaction with the cotton plant prior to the start of the nineteenth century reveals a gradually expanding familiarity, made possible by scientific specialization and colonial expansion. However, while these trends produced a basic corpus of knowledge that made possible the vision of large-scale acclimatization to France by the middle of the eighteenth century, it is important to recognize that political and economic considerations played a central role in shaping the outcome of such projects. Indeed, geopolitical rather than strictly scientific or agricultural factors account for the sharp distinction between the scattered experiments with cotton in France in the seventeenth and eighteenth centuries and the more extensive project initiated by Napoleon in 1807.

As I show in the second section, this undertaking was made possible above all by the initiative of the Napoleonic regime and its centralized state apparatus, which accomplished an unprecedented mobilization of intellectual and biological resources to achieve its objective. At their most effective, the Napoleonic experiments with cotton cultivation were able to translate the botanical knowledge which had accumulated over the previous centuries into a concerted effort at agricultural innovation, producing results which offered a tantalizing promise of a "cotton kingdom" stretching from the Atlantic coast of France to central Italy. In a closer analysis of *savant* participation in this undertaking, it is possible to discern a basic specialization in which the regime relied upon botanical experts to provide specific advice on the cotton plant, while various agricultural institutions accepted the task of publicizing the project in order to extend the experiments across the fields of France and Italy. This dynamic suggests the extent to which a vision of large-scale acclimatization was made possible at the start of the modern era by a distinct convergence of state resources, botanical expertise, and agricultural innovation.

Ultimately, however, the effective merger of scientific knowledge and tangible agricultural practice required for the creation of Napoleon's "cotton kingdom" proved impossible to achieve, and despite a handful of successful efforts in various parts of France and Italy the undertaking was ultimately stymied by a combination of human and natural obstacles. If the scale of the Napoleonic acclimatization project suggests a growing confidence in the capacity for states and *savants* to successfully manipulate natural resources at the start of the modern era, the ultimate fate of this undertaking serves as an important reminder of the difficulties inherent in moving from an expanded botanical knowledge of a plant to its large-scale introduction into the fields.

Antecedents of the Napoleonic Acclimatization Project

Napoleon's decision to gamble on the acclimatization of cotton did not occur in an intellectual vacuum. This project was carried out in the midst of a rising European interest in plant acclimatization and biological exchange, a product of the expanding reach of European commerce and colonization during the early modern era.

While much of the recent scholarship on this subject has focused on the colonies as a setting for acclimatization experiments, Europeans were also interested in the advantages that could be obtained from the introduction of foreign crops to the soil of their own continent. Experiments with exotic crops were carried out in Spain by the mid-sixteenth century and the other major powers of Western Europe were quick to follow this example, eager to exploit every advantage offered by their expanding knowledge of the world outside Europe. The influential Swedish botanist Linnaeus spent much of his time promoting such acclimatization schemes and in 1746, he even managed to obtain cottonseeds, trying without success to cultivate them in a Stockholm garden. French *savants* were no less interested in the potential benefits to be obtained through acclimatization. By the second half of the eighteenth century, hundreds of exotic plants were being raised at the *Jardin du Roi* in Paris and various other gardens throughout France.[6] The influence of this trend was evident in the naturalist Alexander von Humboldt's *Essai sur la Géographies des Plantes,* which was published 2 years before Napoleon initiated his effort to introduce cotton to France. In this treatise, which is considered a foundational text of modern biogeography, Humboldt marveled that "unquiet and laborious man, in traveling to the diverse corners of the world, has forced a certain number of vegetables to inhabit all the climates and all the elevations."[7] This expanding European interest in acclimatization was part of a broader movement of agricultural innovation that developed in the seventeenth and eighteenth centuries.

Historical scholarship has called into question the idea of a sweeping "agricultural revolution" in eighteenth-century France, noting that grandiose projects for agricultural improvement did little to alter basic conditions in the countryside, where established practices and routines maintained a strong hold. Nonetheless, the "agronomanie" expressed by many leading *savants* encouraged a culture of experimentation and innovation, an important precondition for the more widespread transformation of French agriculture in the nineteenth century.[8] For Europeans of the early modern era, the myriad possibilities offered by plant acclimatization were on display in gardens, fields, and plantations on their own continent and across the globe, a fact which helps to explain the broad support Napoleon's acclimatization project received from leading figures in French and Italian scientific circles.

On a more specific level, the Napoleonic acclimatization project was linked directly to an expanding French familiarity with the cotton plant and its fibers that developed over the early modern era. Information about cotton had been circulating in Europe since antiquity, long before the plant itself arrived on the shores of the Mediterranean. While descriptions of cotton in this early literature were often

[6] On early Spanish acclimatization projects, see De Vos (2006). On Linnaeus's experiment with cotton in Sweden, see Koerner (1999, p. 33). On acclimatization and the early history of the *Jardin du Roi,* see Mukerji (2005). On acclimatization at the *Jardin du Roi* during the eighteenth century, see Spary (2000).

[7] Humboldt (1805, pp. 24–30).

[8] For a detailed account of French agronomy during the early modern era, see Bourde (1967). For an analysis of the links between agricultural science and practical innovation, see Moriceau (2002, pp. 236–272).

speculative and imprecise, over the course of the early modern era European scholars collected and codified a wide range of information about cotton.[9] This familiarity was a product of the increasingly formalized study of plants that emerged in Western Europe during the early modern era, a development which historian Brian W. Ogilvie has identified as a key step in the emergence of modern botanical science. As Ogilvie has shown, it was during this period that the formerly distinct fields of medicinal botany, natural philosophy, and agriculture began to converge into a common discipline pursued by an expanding community of *savants* scattered across Western Europe.[10] The emergence of "Renaissance natural history," to use Ogilvie's term for the new field, was an important precursor to acclimatization experiments. Indeed, the first clearly documented cultivation of cotton in France was carried out in the early seventeenth century by Guy de La Brosse, the personal physician to Louis XIII and a leading figure in herbalist circles. In 1633, La Brosse was given permission to start a botanical garden in Paris for the purpose of raising plants with potentially useful medicinal properties, forming the basis for what would become the *Jardin du Roi*. Three years later, La Brosse published an extensive catalog of the plants he managed to cultivate, including cotton. Cotton also appeared in a catalog of plants cultivated at the botanical garden of Montpellier, another crucial center of herbalist studies, which was compiled in 1697 by the director of the garden, Pierre Magnol.[11] While Magnol and La Brosse were primarily interested in medicinal uses of the cotton plant and did not envision a more extensive cultivation beyond their protected gardens, the appearance of *gossypium* at the leading French botanical institutions during the seventeenth century was a clear sign of expanding French familiarity with the crop.

The cultivation of cotton in the gardens of Paris and Montpellier was followed in the middle of the eighteenth century by the first serious effort to incorporate the crop into French agriculture. The initiative for this project was provided by Johannis Althen, an Armenian Christian from Persia who arrived in France during the 1740s after being held for 16 years as a slave in Ottoman Turkey. In the fall of 1743, Althen dictated a letter to the powerful Director General of Finances, explaining that he had acquired experience with cotton cultivation during his captivity and proposing an experiment with the crop in southern France. This proposal was rewarded with a small stipend and a promise of support from local officials, and the following year Althen was able to raise several plants on a plot of land that he had leased on the outskirts of the town of Castres in Languedoc. The Director General justified his support for the project with the vague suggestion that the introduction of cotton cultivation to the region "might be advantageous."[12] While the administration never articulated a more extensive explanation for its decision to offer limited support to the Armenian, the episode was shaped fundamentally by political and economic

[9] Deham (1919).

[10] Ogilvie (2006). See also Davy de Virville (1954) and Morton (1981, pp. 115–164).

[11] La Brosse (1636, p. 101) and Magnol (1697, p. 90).

[12] Archives Departementales de l'Hérault (hereafter ADH) C 2629, "Memoire pour le Sieur Baptiste Joannis Althen" (undated); Orry to le Nain (14 October 1743).

considerations directly related to the early phases of European industrialization. Demand for cotton textiles had been rising steadily across Europe since the fifteenth century. While the French government initially sought to ban the produce in an effort to protect domestic silk and woolen manufactories, by the mid-eighteenth century it seemed increasingly clear that prohibiting the use of a highly popular product among a population of millions was beyond the capacity of the royal government.

By the 1740s, there was a growing consensus in administrative and intellectual circles that France would need to develop its own cotton industry.[13] The French government's interest in cotton cultivation can only be understood as a part of its effort to address the evident failure of the protectionist legislation. In this context, we can see the influence of a body of eighteenth-century economic thought recently identified by historian Fredrik Albritton Jonsson. In contrast to the classical liberal emphasis on international trade, the ultimate goal of this "naturalist" approach to political economy was the achievement of autarchy through acclimatization and similar schemes for the "improvement" of nature.[14] This was the solution to the "cotton question" proposed by Althen, taking the logic of import substitution to its natural conclusion with the introduction of cotton to French soil.

Unfortunately for this early effort to create a "cotton kingdom" in the fields of Languedoc, Althen's plants never achieved full maturity and produced only a small quantity of inferior cotton, causing the local Intendant to report to his superiors that "the cotton plants cultivated by Johannis do not inspire great hopes." While the government terminated its support for the undertaking at this stage, Althen relocated his efforts to nearby Montpellier and in the autumn of 1750, he once again petitioned the administration for support. Significantly, Althen was also able to produce a strong recommendation from the Royal Scientific Society of Montpellier, which had charged several of its members with investigating his efforts. This commission, which was headed by the noted naturalist François Boissier des Sauvages, confirmed that Althen's cotton plants "have always appeared to be in a good state," and recommended that the government should "put Althen in a state to execute his project." Based on this recommendation the administration once again extended its support to Althen, initiating two more years of experiments with acclimatization. Yet this second stage in the experiment was no more successful than the first, and after 1753, Althen abandoned the effort.

Nonetheless, Althen's activities in the mid-eighteenth century reveal several themes that are important for understanding the more extensive experiments later carried out under Napoleon. A report on the experiment produced in late 1744 by Jean Hellot, an influential *savant* and advisor to the government, described a fundamental element of the relationship between botanical expertise and agricultural innovation. Whatever success Althen might obtain through his efforts, Hellot noted, "one should not conclude anything from experiments which are conducted in the garden of a single individual, where foreign and curious plants will always be better cared for than all others." Hellot concluded that in order for such efforts to acquire

[13] Depitre (1912) and Cole (1943, pp. 165–177).

[14] Jonsson (2010).

real success, "it is necessary to have experiments in the field, which require nothing more than the routine plowing and weeding." As Hellot recognized, the knowledge of cotton that Althen possessed would not necessarily translate into widespread cultivation of the crop, an observation confirmed by the experience of the Armenian in the fields of Languedoc. Althen's efforts had suffered, for example, from the refusal of the proprietor of the land on which he planted his cotton to allow him to draw a sufficient quantity of water from the well, from the damage caused by local livestock which were allowed to graze in the area, and from the theft of cotton plants by the local population. At one point, the frustrated Armenian lamented that "the entire world" was against him.[15] This problem was further compounded by the hesitance of central and local authorities to extend their support for the project to the degree envisioned by Althen. While Althen did receive some funding from the administration, the limited nature of this support allowed for only limited experiments, and no concerted effort was made to extend interest in cotton to the general population or facilitate the integration of the crop into local agricultural practice.

Although the French government lifted the bans on cotton textiles in 1759 and thereafter sought to promote the development of a domestic cotton industry, interest in promoting cotton cultivation in France was not revived until the turn of the nineteenth century. Because the expanding textile mills could be adequately supplied with cotton fibers from French colonies overseas, there was little reason for the state to encourage experiments with the crop in France. Indeed, the growth of the European textile industry stimulated interest in the crop among French colonists, and by 1770, the colony of Saint Domingue alone was producing more cotton than all the British and Dutch colonies of the Caribbean combined. Altogether, the French colonies produced more than half the raw cotton shipped across the Atlantic in the late eighteenth century.[16] Significantly, the growth of cotton planting in the colonies played an important role in promoting French familiarity with the crop, greatly facilitating the spread of information on cotton and its cultivation among French *savants* during the final decades of the *ancien régime*. In 1788, for example, the published proceedings of the Royal Society of Agriculture included a series of articles on cotton presented by colonists from the Caribbean. One of these articles, produced by a colonist from Guadeloupe named Badier, described 19 distinct varieties of cotton that Badier had cultivated in the colony over the previous decade, likely the most systematic investigation of the crop conducted by a European colonist in the Americas to that point. A second article was written by the influential creole *savant* Médéric Louis Moreau de Saint-Méry and described a variety of cotton that had been recently introduced to Saint Domingue. At the request of Moreau de Saint-Méry a commission of experts (which included André Thouin, the long-serving director of the *Jardin du Roi*) investigated a sample of this cotton. In their report, the experts not only praised the quality of the cotton but also suggested that such systematic examination of colonial cotton by metropolitan *savants* offered "a

[15] Archives Nationales, Paris, F/12/655, "Culture du coton dans le Languedoc"; ADH, C 2629, Althen to Le Nain (5 July 1744) Le Nain to Orry (August 15 1744).

[16] Statistics from Eltis (1997, pp. 114–115).

simple and equally sure method of recognizing and appreciating exactly the different qualities of cotton which the cultivators can present to us."[17] As this document suggests, the expansion of cotton cultivation in the French colonies helped to stimulate interest in the plant among French scientific circles. Yet while the rise of cotton planting in the French Caribbean played a central role in expanding French familiarity with the crop, the success of colonial cotton cultivation also meant that the French state had little incentive to back acclimatization experiments like the one proposed by Althen.

This situation changed dramatically in the course of the 1790s, when the French overseas empire entered into a period of intense social and political upheaval, while also becoming a major theater in the global struggle between France and Great Britain. While war and revolution in the Caribbean did not entirely disrupt the commerce in plantation products between France and the colonies, these events did produce significant anxiety on the part of many contemporaries concerning the future of the colonial enterprise.[18] As a result, the 1790s witnessed a revival of interest in the possibility of acclimatizing colonial crops to southern France. In the most notable of these projects, a refugee colonist from Saint Domingue named Jean-François Bermond experimented with a wide range of exotic crops on a property he owned outside of Nice, informing the minister of the interior in September of 1796 that he had obtained "three pounds of superb cotton." Recognizing the potential advantages of this project, the Interior Ministry agreed to furnish Bermond with a limited stipend intended to offset the cost of his project. Bermond, however, insisted that the funds he had received were insufficient, and in 1798, he abandoned his project in order to return to Saint Domingue in the hope of contributing to the revival of the plantation system.[19] Nonetheless, the renewal of official interest in acclimatization was a sign that the shifting geopolitical situation had produced the conditions necessary for a more extended effort to introduce cotton cultivation to France.

Indeed, the "cotton crisis" which had started in the 1790s was made even more acute in the early years of the nineteenth century by the growing intensity of the struggle between France and Britain for commercial supremacy. Since the mid-eighteenth century, French textile mills had been gradually improving their capacity to produce finished textiles from raw cotton, a trend Napoleon was determined to extend as far as possible in order to advance his efforts to overturn the economic and commercial preeminence of Great Britain. This policy culminated in a decree issued in February of 1806, which significantly increased the tariff on foreign yarn and strictly banned the introduction of finished fabrics into France and the areas it controlled.[20] Napoleon initiated the acclimatization project less than a year later,

---

[17] Badier (1788), Moreau de Saint-Méry (1788), and Nicholas Desmarets et al. (1788).

[18] The revolutionary crisis in the Caribbean at the end of the eighteenth century is the subject of a substantial and growing body of research. For a recent overview of these events and critical analysis of the historical literature, see Geggus (2010).

[19] Archives Nationales, Paris, F/10/433–434, Bermond to Bénézech (16 Fructidor An IV).

[20] On the development of the French cotton industry during this period, see particularly Chassagne (1991). For a recent analysis of Napoleon's commercial and economic policies, see Horn (2006, pp. 216–240).

once success on the battlefield had secured his dominance of Western Europe. By this point, knowledge of the cotton plant in France had expanded from imprecise and speculative entries in the herbalist texts to the more systematic investigation of the *savants* and the direct experience of the colonists. Yet while this growing familiarity with cotton played an important role in the project that unfolded after 1807, the undertaking itself was a product not only of the enhanced scientific understanding of cotton cultivation but also of a distinct convergence of political and economic developments.

Acclimatization and Centralization in Napoleonic France and Italy

In its execution as well as its origins, Napoleon's effort to promote cotton cultivation in the areas under his control was fundamentally shaped by political conditions. The centralized administrative apparatus that had been created in the early years of Napoleonic rule played a particularly important role in the project. For the first time in French history, the central government was both motivated to provide extensive support to experiments with cotton cultivation and capable of carrying out such an undertaking. This outcome reflects a broader trend in world history at the start of the modern era. As historian John F. Richards has shown in an influential study, the scale of human intervention in the natural environment increased notably between the sixteenth and nineteenth centuries, a trend Richards explains in part by emphasizing the growing pace of state centralization.[21] Viewed from this perspective, Napoleon's acclimatization project presents an opportunity to examine in detail the ways in which the expansion of state power shaped human interaction with the environment, and in particular the ongoing process of biological exchange between different regions of the globe. In a recent article, historians Christopher M. Parsons and Kathleen S. Murphy have emphasized the extent to which European ships of the eighteenth century became "ecosystems under sail," facilitating the transport of myriad plants and animals across the world's oceans. Indeed, as Parsons and Murphy show, the French government in particular sought to encourage these transfers through a favorable maritime policy.

The Napoleonic experiments with cotton cultivation represent not only a continuation of this trend but also a considerable expansion of direct state involvement in the circulation of plant species across the globe. Earlier examples of such exchange generally involved small quantities of a myriad of biological specimens, which were most often destined for cultivation in greenhouses or study in museums and private collections.[22] Between 1807 and 1814, the Napoleonic regime systematically collected information related to the cotton plant and its cultivation from

[21] Richards (2003).

[22] Parsons and Murphy (2012). On biological exchange in early modern Europe and the European colonies, see also Schiebinger (2004) and Schiebinger and Swan (2005).

leading figures in French scientific circles as well as from less prominent individuals who had direct experience with the crop. In addition, central and local authorities acquired over 5000 kg of cottonseeds and distributed them to aspiring cotton farmers across France and Italy, offering a financial reward for those who succeeded in producing raw cotton. While Althen's experiments in the mid-eighteenth century suggested the possibility of introducing cotton cultivation to French fields, a sustained and extensive effort at acclimatization was only made possible by the distinct convergence of economic and political developments that had emerged by 1807.

The intellectual mobilization accomplished by Napoleon's administration in its effort to naturalize cotton to France was notable both in its extent and in the diversity of expertise involved. Leading scientific institutions played a central role in the project, both as advisors to the officials directing the project and as promoters of acclimatization in the public sphere. This collaboration began early in the project, when the Interior Ministry consulted the noted agricultural expert Henri-Alexandre Tessier, a long-standing member of the Society of Agriculture in Paris and the author of a substantial number of books and articles on agricultural subjects. Asked to assess the feasibility of the project, Tessier described the growing familiarity with cotton among French *savants* as well as the scattered acclimatization projects that had taken place over the previous decades, concluding that "what was previously seen as a hazardous enterprise has become at this moment an endeavor likely to succeed." Tessier maintained a leading role in the acclimatization project as it unfolded in the spring of 1807, composing a short instruction pamphlet on cotton cultivation that was published at the expense of the administration and distributed in conjunction with the shipments of cottonseeds that took place in the following years. In this document, Tessier argued strongly for the feasibility of the project, noting that "so many plants have been acclimatized among us against all appearance of success, it is to be believed that cotton will have an equally happy fate." While Tessier admitted that cotton was typically associated with "the hottest climates," he also emphasized that "little by little it has been brought to the temperate zones, both on the old and the new continents."[23] By situating Napoleon's acclimatization project in the broader history of global cotton cultivation, Tessier's instruction lent credibility to the undertaking and provided a powerful impetus for participation in the effort on the part of the general population. Such support from leading *savants* was a vital tool for the Napoleonic regime as it sought to promote its geopolitical objectives via acclimatization.

Prominent scientists and intellectual institutions played specialized roles in the acclimatization project, assisting the regime in ways that reflected their particular capacities and expertise. Botanical experts typically provided specialist advice on the cotton plant and conducted experiments in the protected conditions of greenhouses and gardens, while agricultural specialists publicized the project and stimulated interest among the general population. The Society of Agriculture in Paris, which was the leading forum for *savants* interested in agriculture from across the

[23] Archives Nationales, Paris F/10/420, *Mémoire sur l'introduction en France de la culture du coton* (undated); *Instruction sur la maniere de cultiver le coton en France* (6 March 1807).

territories controlled by Napoleon, announced in late 1807 its intention to support the project by offering monetary prizes to be awarded in 1809 for the "best treatises in which, after providing a description of different cotton plants, one determines by the results of exact and well-proven experiments, *what are the species and varieties of cotton which can be cultivated with the most advantage in France, in regard to the quantity and quality of product*." In this way, the society sought to harness the knowledge and experience of partisans of "agromanie," facilitating the spread of information to the more general population in a manner that was not seen in earlier acclimatization experiments. In 1810, the society announced that it had received only a single entry for the contests, a treatise written by J. Paris, the subprefect of Tarascon. Although the society determined that Paris had not merited the full prize it had offered 3 years earlier, it did award him a prize of 300 francs and announced that his treatise would be "printed and distributed to the cultivators of cotton in the southern departments."

Significantly, the society also used this opportunity to advance a more general assessment of the acclimatization project. The announcement explained that Paris had encountered numerous setbacks in the course of experiments conducted since 1808, while emphasizing that "far from being discouraged, he used them to enlighten himself and to adopt more effective measures." Such difficulties, the article suggested, "will not surprise those who know how much effort was required to introduce to France the vine, which originated in the Levant and the tropics, and so many other useful and agreeable plants which now prosper here."[24] In this way, the society sought both to promote the spread of useful information and to encourage those participants in the experiment who had obtained disappointing results in previous years.

The director of the Interior Ministry's Bureau of Agriculture, Augustin-François Silvestre, often discussed the project in his reports to the Society of Agriculture, reports that were included in the proceedings of the society and thus helped to publicize the experiment. In 1808, Silvestre explained that over the previous year cotton had been cultivated "in more than twenty of our southern departments; in almost all it succeeded." Silvestre also explained the considerations that had motivated the regime to undertake the acclimatization project, noting that the ultimate goal was to obtain "in our own soil a material of primary necessity" and expressing his view that "it is not difficult to be persuaded that, within just a few years, we will have undoubtedly achieved the desired result." In 1810, Silvestre explained that unseasonable weather during the previous year had "not permitted the obtainment of results as advantageous as might be desired." Nonetheless, Silvestre insisted, the experiments undertaken had demonstrated "the certainty that in several parts of the empire the cotton plant can be grown." Silvestre continued to promote the acclimatization project to the end, and in his report to the society in 1813 he praised experiments with cotton in the department of Pyrénées Orientales, which had produced a "very satisfying" harvest.[25] Local societies of agriculture throughout France and

[24] Neaufchateau and Silvestre 1807, Silvestre (1808, pp. xiv–xlviii, lxxix, 1810b).

[25] Silvestre (1808, pp. xiv–xlviii, lxxix, 1810a, 1811, 1813).

Italy played a similar role as publicists for the acclimatization experiments. In 1808, for example, the Society of Agriculture in recently annexed Turin published information on cotton cultivation compiled by one of its members, in order to "assist with instructions the many people to whom seeds have been distributed."[26] This type of support was important not only as a means of circulating information about cotton cultivation but also as a way of promoting the experiment and publicizing the cooperation between the regime and the *savant* networks of France and Italy.

While the societies of agriculture focused their efforts on stimulating interest in the countryside, the administration turned to leading experts in the increasingly specialized French botanical sciences to provide more precise advice on the cotton plant. Early in the project the Interior Ministry recruited the Swiss botanist Augustin Pyramus de Candolle, an ideal choice not only because of his recognized expertise in botany but also because in 1806 he had been charged by the government to undertake a series of voyages across France with the purpose, as Candolle later explained in his memoirs, of "studying botany in its links with geography and agriculture."[27] Tasked with gathering information related to "the means necessary to multiply a crop which it is important for the government to encourage as much as possible," Candolle delivered a detailed report on the subject in October of 1807. Drawing on his extensive knowledge of the region's flora, Candolle emphasized that "one sees cultivated here and there in the fields plants indigenous to countries as hot as those where the cotton plant grows," citing species from as far afield as India, Syria, and Ethiopia. Yet Candolle also calculated that it would be difficult to convince the farmers and landowners of Languedoc to take up cotton cultivation, because "it is doubtful that this harvest alone can equal the numerous products of this privileged soil." Indeed, reflecting on the prospect of acclimatizing cotton to the south of France, Candolle concluded that "the principal difficulties are found in the character of the inhabitants of these provinces," because "after having embraced this project with ardor, they will abandon it at the first difficulty." Success could be achieved, Candolle concluded, through more systematic experiments backed by a detailed knowledge of cotton cultivation, which would "diminish the repugnance which the cultivators have toward attempting new experiments."[28] Candolle's report provided a model of the ways in which the French intellectual elite could help produce such an outcome. In addition to applying his detailed understanding of botany to the problem of acclimatization, Candolle also provided the administration with a careful analysis of the ways in which local agricultural practice and economic considerations might play a role in shaping the project.

In its efforts to promote the introduction of cotton to French soil, the Napoleonic regime also turned for advice to the botanical experts at Museum of Natural History in Paris. The professors at the museum were asked by the Interior Ministry to assist

---

[26] Nuvolone Pergamo (1808, p. 3).

[27] Candolle (1862, p. 169).

[28] Archives Nationales, Paris, F/10/420, Champagny to Candolle (6 March 1807); F/10/202, *Copie du Rapport à Son Excellence le Ministre de l'Interieur sur la Naturalisation du cotonnier dans les Départements du sud ouest* (7 October 1807).

the project by providing precise information on the areas "where it would be convenient to acquire cotton seeds," as well as the "species whose cultivation should be preferred in France." In addition, the ministry requested that the professors report on the possibility of using the museum's protected gardens to cultivate "seed-bearing cotton plants [*cotonniers porte-graines*]," which would be used "to disseminate the seeds to the south." In May of 1809, André Thouin and René Desfontaines, leading professors at the museum whose prominence in French botanical science dated back to the *ancien régime,* delivered a response to this inquiry. In order to identify the regions that were most appropriate for seed acquisition, the museum professors began with the basic observation that the areas "most analogous in temperature, soil, and method of cultivation are those for which the plant products will acclimatize with most ease from one to the other." Based on this straightforward assessment of biological exchange, Thouin and Desfontaines identified a number of specific locations that would be appropriate to furnish seeds for experiments in France.

Some areas recommended by the professors, such as Italy, Spain, and the USA, had already been identified by the administration as a potential source of seeds. Others, including Egypt, the Barbary Coast, and the Aegean islands, had not previously been considered and were consequently the target of successful acquisition efforts. Thouin and Desfontaines also suggested that seeds already harvested in France would present the best prospects for future cultivation, and suggested that the administration should "invite the French landowners who already cultivate cotton to harvest the seeds and indicate to their respective prefects the precise quantities that they have harvested in excess of their needs." In a circular letter distributed to the prefects in July of 1809, the Interior Minister Crétet cited this recommendation as the basis for his order that the prefects should "collect all the good seeds and utilize them," an instruction which, significantly, had not formed part of the directives which Crétet addressed to the prefects the previous year.[29] In this manner, the expertise of the professors was translated into experimental reality thanks to the activity and influence of the central government.

The support which the museum was able to provide to the project was, however, subject to limits, as Thouin and Desfontaines noted in their report. On the crucial question of which types of cotton were most suitable for cultivation in France, the professors admitted that "we possess only a vague understanding of these objects" and that more precise information could be obtained only through experiments carried out "in a climate less variable and warmer than that of Paris." As for the possibility of using cotton plants raised in the gardens of the museum to obtain seeds for distribution, Thouin and Desfontaines explained that while they had "sown all the cotton seeds found in the Museum," the end result would produce only enough seeds for three *arpents* of land (roughly 600 square feet), a quantity which was

[29] Archives Nationales, Paris, F/10/420, "Rapport a l'administration du Museum par les Professeurs Thouin et Desfontaines" (20 May 1809); "Rapport Présenté au Ministre de l'Interieur" (24 June 1809); "A Monsieur le Prefet du…" (8 July 1809).

"far from sufficient for the requirements of cultivation in the south of France."[30] While the botanists at the museum possessed extensive experience in the field of acclimatization, the location and nature of the institution constrained the professors to a purely advisory role in the Napoleonic experiments with cotton cultivation. As the *savant* Hellot first noted in the midst of Althen's experiment during the 1740s, it was relatively easy for experts to cultivate cotton under favorable conditions, but truly effective acclimatization could only be accomplished once the crop had been introduced to French fields, a more daunting proposition.

Yet while earlier governments had been reluctant to commit to such an undertaking, the Napoleonic regime was determined to promote a more extensive acclimatization project that was intended to attract the support not only of the leading scientific institutions but also of farmers and landowners across southern France and northern Italy. Between 1807 and 1814, hundreds of individuals accepted the cottonseeds distributed by the administration, motivated by a combination of support for the strategic aims of the project and desire to obtain the reward that was offered for those who managed to obtain raw cotton from their fields. A closer look at this aspect of the project sheds light on the intersection of scientific expertise and agricultural practice at the start of the modern era, demonstrating both the sheer ambition of the Napoleonic experiments with cotton cultivation, and the challenges inherent in such an undertaking.

The most successful of the participants in the Napoleonic acclimatization experiments were able to combine botanical expertise with an understanding of local agricultural practice and conditions. One particularly notable example is Casimir Freycinet in the department of Drôme, described by the local prefect as "a distinguished, educated, and hardworking agriculturalist." Freycinet had the advantage of possessing a garden and fields which he could devote to cotton cultivation, and was able to produce detailed accounts of his experiments in which he not only reported the results of his efforts but also provided extensive reflection on the most effective means to achieve the objectives articulated by the regime. Early in the project, Freycinet came to the conclusion that it was too soon to hope for the cultivation of cotton on a large scale in Drôme, explaining to the local prefect in 1809 that "I do not believe planting in the open field, the only means used for major crops, can at this point present the least success in this department." For the moment, Freycinet advocated "preliminary experiments," which would be "absolutely indispensable to the success of large-scale cultivation." The goal of these efforts, Freycinet explained in another letter, would be to "obtain seeds of perfect maturity" by cultivating cotton plants in protected conditions early in the year and then transplanting them to the fields in late spring. Freycinet viewed this time-consuming and costly exercise as the best means of "causing this plant to follow the laws of naturalization," and his activities over the course of the experiment were devoted to achieving this result as the first step in the acclimatization process. In early 1813, Freycinet informed

[30] Archives Nationales, Paris, F/10/420, "Rapport a l'administration du Museum par les Professeurs Thouin et Desfontaines" (20 May 1809); "Rapport Présenté au Ministre de l'Interieur" (24 June 1809).

the prefect that he was determined to "engage with perseverance in small experiments, with the object of obtaining good seeds," gradually creating new generations of plants that would be "successively more acclimatized."[31] While Freycinet envisioned a gradual process that could not satisfy the short-term objectives of the regime, he was firmly convinced of the long-term feasibility of acclimatization and possessed a familiarity with botany and agriculture that made him an ideal collaborator in Napoleon's effort to promote the introduction of cotton to France.

Many of the individuals who participated in the experiments as well as the officials charged with overseeing the project shared Freycinet's dedication to the undertaking, and many also came to appreciate his more long-term vision of acclimatization. This was the view expressed, for example, by Camille Philippe Casimir de Tournon, who was the prefect of Rome following its annexation to France in 1809. As the southernmost territory directly administered by Paris during the Napoleonic era, this region seemed to offer the best hopes for the successful introduction of cotton cultivation. Yet in late 1813, Tournon reported to the interior minister that many of the participants in the effort had been discouraged by several years of unsuccessful experiments. He lamented that "a crop which began on a large scale is now confined only to small gardens." Nonetheless, Tournon assured the government of his certainty that "cotton can be cultivated with success in the department of Rome so long as it is given the necessary care," and he promised to use his influence to stimulate further interest in the project.[32]

The administration could also look hopefully at the success of Paolo Savonatti, a Maltese immigrant charged by the administration with directing a large-scale cotton plantation in the department of Pyrénées Orientales on the French border with Spain. Savonatti, who had acquired direct experience with cotton cultivation on his native island, obtained successively more promising results between 1808 and 1813, harvesting over 200 kg of raw cotton in the final year of the project. Significantly, Savonatti's success also provided valuable publicity for the experiment, helping to encourage those less familiar with the crop that acclimatization was indeed possible. In November of 1810, for example, Savonatti reported that his cotton field had been visited by "up to four hundred people," who "have come to see the beautiful view which the plantation offers at this moment; the whiteness of the cotton and the great quantity of open capsules are the object of their admiration."[33] To be sure, the regime could not hope to find many individuals in France with the level of knowledge related to cotton cultivation that Savonatti possessed. Yet it could be hoped that his success would provide a stimulus to more widespread participation in the project, serving as the starting point for a process of acclimatization that might

---

[31] Archives Nationales, Paris, F/10/416, d'Escourches to Crétet (4 January 1808); F/10/420, "Observations sur la culture du cotonnier dans la commune de Miremande" (30 December 1808), Freycinet to D'Escourches (undated); F/10/425–426, Freycinet to d'Escourches (13 February 1813).

[32] Archives Nationales, Paris, F/10/424, Tournon to Montalivet (20 January 1811); Montalivet to Tournon (13 March 1811); Tournon to Montalivet (6 March 1812); Tournon to Montalivet (30 September 1813).

[33] Archives Nationales, Paris, F/10/417–419, Savonatti to Montalivet (29 November 1810).

not have obtained the quick results desired by Napoleon, but nonetheless offered the promise of success for those patient enough to see the experiment through to the end.

Any hopes of promoting a long-term acclimatization process came to an end with the fall of the regime in early 1814 and the subsequent revival of French overseas trade, which could now supply cotton from the expanding plantations of the antebellum USA. Just as the Napoleonic acclimatization experiments were made possible by a distinct convergence of political and economic developments, the project was ultimately terminated as a response to a changing geopolitical environment. Nonetheless, the difficulties inherent in translating scientific knowledge of cotton into substantial agricultural innovation became apparent long before the end of the experiment in 1814. In contrast to the dedication of Freycinet, Savonatti, and a handful of others, most of the amateur cotton farmers involved in the project lost interest following initial setbacks. The natural obstacles to successful acclimatization were indeed daunting. The prefect of Herault in the Rhône valley, for example, reported that out of the 12 kg of cottonseeds distributed to landowners in his department, "the majority did not germinate; those which did germinate vegetated very slowly, as a result of extreme drought which was felt here during the spring and a great part of the summer." Nature, however, was far from the only impediment faced by the French state and its collaborators in this undertaking. Ignorance, apathy, and outright hostility combined with droughts, frosts, and aphids to threaten the success of Napoleon's incipient cotton kingdom. In northern Italy, for example, the prefect of Genoa reported wearily that years of wasted effort had thoroughly discouraged the people of his department, "and it is only by force that one can oblige the landowners and farmers to take a chance with new sacrifices."[34] Such sentiments were common by the final years of the project, as the cumulative effect of disappointing results dampened the initial confidence of the participants.

Conclusion

Although the Napoleonic experiments with cotton cultivation did not create a "cotton kingdom" in France and Italy, this episode reveals several important themes in the intertwined histories of botany and agriculture. By time that Napoleon launched his effort to introduce cotton cultivation to France, interest in plant acclimatization was common among European governments and intellectual circles. Indeed, the growing specialization of European plant science during this era paved the way for a process of biological exchange which caused profound change in the global distribution of plant species. Yet the expanded understanding of acclimatization that accompanied the rise of modern botanical science offers only a partial explanation for this episode. Equally important were the geopolitical trends that converged in

[34] Archives Nationales, Paris, F/10/425–426, Nogares to Montalivet (15 September 1813); Bourdon to Montalivet (4 September 1813).

the early nineteenth century to pave the way for an undertaking more extensive than any seen before. Once the Napoleonic regime was committed to promoting the introduction of cotton cultivation to France, it was able to draw on an impressive array of intellectual resources for the project. Significantly, this mobilization of science in the interest of the state developed in a manner that emphasized the distinct yet complementary roles of agricultural and botanical science. While the agricultural societies of France and Italy concentrated on publicizing the experiment, botanical experts supplied the administration with expert advice based on a substantial body of scientific knowledge. As the project unfolded, a model of successful acclimatization began to emerge where botanical expertise and familiarity with local agricultural practices converged, yet these success stories were the exception rather than the rule. While the difficulties posed by climate and environment played an important role in frustrating Napoleon's effort to bring cotton cultivation to the territory under his control, human factors must also be recognized as a central element in the outcome. The challenges of bridging the gap between botanical science and agricultural routine must take a central place in explaining the process of biological exchange in the early modern world.

References

Champagny, Jean-Baptiste Nompère de. 1807. Le Ministre de l'Interieur, a Monsieur le Prefet du Department d'.... *Gazette nationale ou le Moniteur Universel*, April 1.

Badier. 1788. Observations sur differentes espèces de cotonniers cultivèes a la Guadeloupe. *Mémoires d'Agriculture, d'Économie Rurale et Domestique publiés par la Société Royale d'Agriculture* (no volume) Autumn: 118–131.

Beinart, William, and Karen Middleton. 2004. Plant transfers in historical perspective. *Environment and History* 10 (1): 3–29.

Blondy, Alain. 1993. Mikel-Antoine Vassalli et les réfugiés maltais, principaux agents de la culture du coton en France (1807–1814). *Journal of Maltese Studies* 23–24:201–214.

Blondy, Alain. 2006. La culture du coton en Provence sous le 1ere Empire. *Provence Historique* 56:251–279.

Bourde, André. 1967. *Agronomie et agronomes en France au XVIIIe siècle*. Paris: S.E.V.P.E.N.

Candolle, Augustin Pyramus de. 1862. *Mémoires et souvenirs de Augustin-Pyramus de Candolle, écrits par lui-même et publiés par son fils*. Geneva: Joël Cherbuliez.

Chassagne, Serge. 1991. *Le coton et ses patrons: France, 1760–1840*. Paris: Éditions de l'École des hautes études en sciences sociales.

Cole, Charles Woolsey. 1943. *French mercantilism, 1683–1700*. New York: Columbia University Press.

Davy de Virville, Adrien. 1954. *Histoire de la botanique en France*. Paris: Société d'Édition d'Enseignment Supérieur.

Deham, H. J. 1919. *Gossypium in pre-Linnaean literature*. London: Oxford University Press.

Depitre, Edgard. 1912. *La toile peinte en France au XVIIe et au XVIIIe siècles: Industrie, commerce, prohibitions*. Paris: M. Rivière et cie.

Desmarets, Nicholas, et al. 1788. Rapport du mémoire de M. Moreau de Saint-Méry, sur le coton de soie. *Mémoires d'Agriculture, d'Économie Rurale et Domestique publiés par la Société Royale d'Agriculture* (no volume) Autumn: 145–150.

De Vos, Paula Susan. 2006. The science of spices: Empiricism and economic botany in the early Spanish Empire. *Journal of World History* 17 (4): 399–427.

Donati, Edgardo. 2008. *La Toscana nell'Impero Napoleonico*. 2 vols. Florence: Edizioni Polistampa.

Eltis, David. 1997. The slave economies of the Caribbean: Structure, performance, evolution and significance. In *General history of the Caribbean,* ed. Franklin W. Knight, vol. III, 105–137. London: UNESCO.

Geggus, David. 2010. The Caribbean in the age of revolution. In *The age of revolutions in global context, c. 1760–1840*, eds. David Armitage and Sanjay Subrahmanyam, 83–100. New York: Palgrave Macmillan.

Godechot, Jacques. 1979. Le dirigisme agricole sous le Premier Empire. *Annuario del Istituto Storico italiano per l'età moderna e contemporanea* 31:35–58.

Horn, Jeff. 2006. *The path not taken: French industrialization in the age of revolution, 1750–1830*. Cambridge: MIT Press.

Humboldt, Alexander von. 1805. *Essai sur la géographie des plantes*. Paris: Fr. Schoell.

Jonsson, Fredrik Albritton. 2010. Rival ecologies of global commerce: Adam Smith and the natural historians. *American Historical Review* 115 (5): 1342–1363.

Koerner, Lisbet. 1999. *Linnaeus: Nature and nation*. Cambridge: Harvard University Press.

La Brosse, Guy de. 1636. *Description du Jardin Royale des Plantes Medicinales etably par le Roi Louis le Juste a Paris*. Paris: Johannis d'Éstrees.

Lefebvre, Georges. 1953. *Napoléon*. Paris: Presses Universitaires de la France.

Magnol, Pierre. 1697. *Hortus Regius Monspeliensis, sive catalogum plantarum quæ in Horto Regio Monspeliensi demonstrantur*. Montpellier: Honoratum Pech.

Marzagali, Sylvia. 2005. Le Blocus continental pouvait-il réussir? In *Napoléon et l'Europe: Regards sur une politique: actes du colloque, 18 et 19 novembre 2004,* ed. Thierry Lentz, 103–114. Paris: le Grand livre du mois.

Moreau de Saint-Méry, Médéric Louis. 1788. Mémoire sur une espèce de Coton nommé a Saint-Domingue, Coton de Soie, ou Coton de Sainte-Marthe. *Mémoires d'Agriculture, d'Économie Rurale et Domestique publiés par la Société Royale d'Agriculture* (no volume) Autumn: 132–144.

Moriceau, Jean-Marc. 2002. *Terres mouvantes: Les campagnes françaises du féodalisme à la mondialisation, 1150–1850*. Paris: Fayard.

Morton, A. G. 1981. *History of botanical science: An account of the development of botany from ancient times to the present day*. London: Academic.

Mukerji, Chandra. 2005. Dominion, demonstration, and domination: Religious doctrine, territorial politics, and French plant collection. In *Colonial botany: Science, commerce, and politics in the early modern world,* ed. Londa Schiebinger and Claudia Swan, 19–33. Philadelphia: University of Pennsylvania Press.

Neaufchateau, N. L. François de, and A. F. Silvestre. 1807. Sur l'introduction en France de la culture du cotonnier. *Mémoires publiés par la Société d'Agriculture du Département de la Seine* 10:lii–lvi.

Nuvolone Pergamo, G. 1808. *Della coltivazione del cotone*. Turin: Stamperia della Societa' di Agricoltura.

Ogilvie, Brian W. 2006. *The science of describing: Natural history in Renaissance Europe*. Chicago: University of Chicago Press.

Parsons, Christopher M., and Kathleen S. Murphy. 2012. Ecosystems under sail: Specimen transfer in the eighteenth-century French and British Atlantics. *Early American Studies* 10 (3): 503–539.

Richards, John F. 2003. *The unending frontier: An environmental history of the early modern world*. Berkeley: University of California Press.

Schiebinger, Londa. 2004. *Plants and empire: Colonial bioprospecting in the Atlantic world*. Cambridge: Harvard University Press.

Schiebinger, Londa, and Claudia Swan, eds. 2005. *Colonial botany: Science commerce, and politics in the early modern world*. Philadelphia: University of Pennsylvania Press.

Silvestre, A. F. 1808. Rapport sur les travaux de la Société d'Agriculture du département de la Seine, pendant l'année 1807. *Mémoires publiés par la Société d'Agriculture du Département de la Seine* 11:xiv–xlviii.

Silvestre, A. F. 1810a. Compte Rendu des Travaux de la Société d'Agriculture du département de la Seine, pendant l'année 1809, à la Séance générale du 15 juillet 1810. *Mémoires publiés par la Société d'Agriculture du Département de la Seine* 13:10–31.

Silvestre, A. F. 1810b. Rapport sur le concours pour la culture comparative des diverses espèces de cotonniers. *Mémoires publiés par la Société d'Agriculture du Département de la Seine* 13:68–72.

Silvestre, A. F. 1811. Rapport sur les travaux de la Société, pendant l'année 1810, lu à la Séance générale du 21 juillet 1810. *Mémoires publiés par la Société d'Agriculture du Département de la Seine* 14:5–31.

Silvestre, A. F. 1813. Rapport sur les travaux de la Société, pendant l'année 1812, lu à la Séance générale du 25 avril 1813. *Mémoires publiés par la Société d'Agriculture du Département de la Seine* 16:19–44.

Spary, E. C. 2000. *Utopia's garden: French natural history from Old Regime to Revolution.* Chicago: University of Chicago Press.

Tulard, Jean. 1992. *Napoléon.* Paris: Livre de Poche.

Joseph Horan is an assistant professor in the Division of Liberal Arts and International Studies at the Colorado School of Mines, where he teaches courses in environmental history and ethics. His essay is based on part of his dissertation entitled "Fibers of Empire: Cotton Cultivation in France and Italy During the Age of Napoleon," which was defended in the Department of History at Florida State University in October 2013.

[illegible]. 1806. Rapport sur les travaux de la Société [illegible] depuis [illegible] 1807. Mémoires publiés par la Société [illegible]
[illegible]. 1808. [illegible]
[illegible]. 1810. Rapport sur les travaux [illegible]
[illegible]
[illegible]
[illegible]
[illegible]

Chapter 6
Whale Oil Pesticide: Natural History, Animal Resources, and Agriculture in Early Modern Japan

Jakobina Arch

Introduction

Ōkura Nagatsune, a prolific agricultural writer in nineteenth-century Japan, wrote 28 books promoting more effective farming techniques.[1] In the midst of one of his texts, *Jokōroku* (*Record of Abolishing Locusts*, first published in 1826), there are six pages of drawings showing different types of whales and dolphins. He also included a short description of basic species classification for the five species of baleen whales and six types of toothed whales and dolphins that could be found in Japan. The images are like those that would appear in natural history texts, with important parts such as the teeth (or baleen) and different fins labeled.[2] Why would a writer of agricultural improvement manuals include such a digression into whale description and classification? In fact, this section was not a digression at all. The central argument of *Jokōroku* was that whale oil was the most effective insecticide to use against the insects he refers to as locusts (*inago* or *kō*) and which most likely were brown planthoppers, *Nilaparvata lugens,* still known today for their damaging outbreaks in the rice-dependent areas of the world.[3] The fact that he devoted an entire volume to the efficacy of whale oil is interesting. Also, his basic images of whale species and image of whalers from the Gotō Islands capturing whales must have been copied from circulating descriptions of whales and whaling. The inclusion of these images indicates the close ties between the expansion of natural

[1] All names in the main text are provided in the Japanese order, family name first, but the reference list uses English-language name order. For a good description of Ōkura's work and his role in nineteenth-century agricultural improvement, see Smith (1998, Chap. 8, pp. 173–198).

[2] Ōkura ([1826]/1977, pp. 33–38).

[3] Okuta et al. (2012), researched the prediction of migrations of planthoppers because these are related to the kinds of outbreaks in Japan that tended to cause famines before the modern era.

J. Arch (✉)
Whitman College, Maxey Hall, 345 Boyer Ave., Walla Walla, WA 99362, USA
e-mail: archjk@whitman.edu

D. Phillips, S. Kingsland (eds.), *New Perspectives on the History of Life Sciences and Agriculture,* Archimedes 40, DOI 10.1007/978-3-319-12185-7_6

history knowledge and the development of agricultural science in early modern Japan. Furthermore, it is a good example of shifts in resource use during this period.

In this chapter, I will discuss how Ōkura's promotion of whale oil as an insecticide for rice crops depended on the practical understanding of animals and plants developed by scholars of natural history (in Japan, known as *honzōgakusha*). Early modern Japanese scholars, while they recognized particular disciplinary areas of study such as *honzōgaku* (natural history or materia medica) or *jitsugaku* (practical or applied learning), tended to combine knowledge from different fields or specialties in their studies. Farm manuals such as Ōkura's were aimed at the ordinary farmer, intended for practical use, and yet this does not mean that they avoided referencing other forms of scholarship. The difficult scholarly Chinese prefaces in works by Ōkura and other popular agricultural writers show that these texts were intended partly for an audience of village administrators or samurai officials, not just for the people doing the actual farming. Farmers' literacy would have been more limited, and the bodies of the same texts have pronunciation guides for characters and abundant illustrations for less-educated readers.[4] From the complex prefaces, it is clear that at least some of the readership was expected to be well-versed in fields of study beyond practical farming. In fact, as Jennifer Robertson's study of plant-gender categorization in farm manuals indicates, scholars like nativist intellectual Hirata Atsutane wrote agricultural treatises in part to spread their philosophies into the countryside, so some manuals went far beyond the basic concerns of farmers.[5] In the case of Ōkura's *Jokōroku*, a philosophical agenda was less strongly developed than in Hirata's nativist treatises, but the work was still written with reference to philosophical investigations like natural history. In this period, someone with any kind of interest in learning was likely to study a variety of topics. The intellectual world of Tokugawa Japan (1603–1868) was highly interconnected and without strict disciplinary boundaries, even in cases where there were terms for different areas of study. For example, the physician and wealthy farmer Takano Chōei wrote an agricultural treatise in response to a major famine in the 1830s where he drew on many types of sources, including both Japanese natural history (*honzōgaku*) and the Western sources that were available through trade with the Dutch.[6] Therefore, Ōkura's text can be seen as both an instructional manual intended to promote rationalized agriculture, and also an example of how the less apparently practical areas of natural history were important in discussions of agricultural improvement.

To fully exploit their environment and produce the food necessary to support their population and protect them from scarcity and famine, agricultural writers like Ōkura considered more than just the potential of the local farm environment. Many of the authors of these manuals had a variety of interests and came from diverse

[4] Rubinger (2007, pp. 88–91).

[5] Robertson (1984).

[6] For more on Takano and how different types of knowledge were brought together by people looking for practical solutions to famines and crop failures, see Nakamura (2005), including her translation of *Kyūkō nibutsukō* (*Treatise on Two Things for the Relief of Famine*) in Appendix A, pp. 183–198.

backgrounds, and the authors did not always agree with each other or with the most popular opinions held by farmers.[7] These manuals had a varied but extensive audience, and some individual volumes went through a large number of reprints. For example, the first systematic manual, *Nōgyō zensho* or *General Treatise on Agriculture,* went through at least seven printings.[8] While the majority of these printings may well have been read by higher-level village administration and scholars less directly involved in farming, Ōkura's own experience shows the possibility that farmers would also have been interested in the contents. It also highlights the role of more affluent farmers, ones who were able to afford time for study and travel, in the development of this kind of literature.

Ōkura was born into a farming family in northern Kyushu. Early in his life, he grew, processed, and sold cotton, and then later worked for a relative who manufactured wax from lacquer trees. He had to balance his ambitions as a scholar with his father's desire to focus on farming, and did not settle into writing until he moved to Osaka, where he sold lacquer trees imported from his native Kyushu. Certainly he thought that at least some of his audience would be farmers and not just scholars or village administrators, as he addressed his writings directly to farmers in the text. In the back of some of his books he also advertised his business selling lacquer seedlings to readers.[9] Another clue that these manuals could be read by farmers for practical advice is in the second edition of *Jokōroku,* where Ōkura responded to the difficulties that people in some regions could have in following his original advice. This 1844 reprint includes text focusing on the more affordable substitutions that could be made for whale oil by farmers in mountain villages unable to acquire or afford the expense of whale oil. In a new preface for this edition, Ōkura also noted that there were areas where whale oil was limited or scarce, particularly the northeastern region of Tōhoku, which was farthest from the whaling areas of Japan. He therefore looked for other methods of killing insects. The alternatives he presented included not only other types of oils but also brine and materials that could be gathered in local areas, like the leaves of Japanese andromeda (*Pieris japonica*). He included these new methods in the text as cheaper options that might be effective in small fields if used quickly enough.[10] This period was one where all kinds of agricultural products were becoming commodities, and thus attempts to boost productivity in individual farms could arise from the purely monetary desire to increase profits.[11] Thus, Ōkura's suggestions of more inexpensive substitutes for whale oil or rapeseed oil could also have been used by frugal readers even where the more costly but effective oils were available.

---

[7] For example, Robertson (1984, pp. 246–248) notes that Ōkura's publication *Saishuhō* (*The Method of Double-Cropping Rice*) dismissed categorization of seed quality based on assigned gender even though he had written earlier texts relying on this categorization.

[8] Rubinger (2007, p. 91).

[9] Smith (1998, pp. 174–176).

[10] Ōkura ([1844]/1977, p. 62). (The entire 1844 version of *Jokōroku* is on pp. 57–118.)

[11] Some examples of work discussing the commercialization of agriculture and cash crops in the period include Smith (1969); Saito (1986); Howell (1989); and Hauser (1974).

Even though profit was certainly a motive in increasing the efficiency of farming methods, agricultural writers like Ōkura tended to focus on the more noble motive of social benefit rather than directly addressing the issue of money. In *Jokōroku,* he recognized that any knowledge about living creatures could potentially be applied to one of the biggest problems of a peaceful Tokugawa state, famine. The danger of famine and scarce resources drove scholars to find ways to help the state keep people fed, requiring extensive knowledge of the natural world's resources, including resources like whale oil from areas far outside the fields.

Early Modern Agriculture and Whaling in Japan

In the Tokugawa period, rice was Japan's main agricultural product. It was also the center of the economy, as taxes and the stipends paid to samurai were generally paid in units of rice, and only later converted into cash by selling the rice to merchant bankers. Variable rice production thus affected not just the supply of a food staple but also the amount that merchants were willing to pay for rice, and thus the exchange rate for samurai stipends. However, not all areas were equally capable of producing enough rice to both feed the local population and pay taxes. Problems in marginal rice-growing areas led to the rise of cash crops and diversification of farming inputs to increase productivity in this same period. Measures to increase agricultural productivity, including new techniques, tools, and fertilizers, all were important as assorted agricultural products (including rice) became commodities in the early modern markets.[12] Some of these new resources to increase productivity included marine inputs into agriculture. Conrad Totman argues that the restriction on foreign travel under the Tokugawa rule, along with shifts in land use and coastal shipping, promoted the development of specialized fisheries. Such fisheries directly contributed new foods for human consumption, but also provided important fertilizers for crops.[13] These specialized fisheries, including whaling, could provide not just replacements for nutrients no longer available in the soil but also products with whole new effects. The use of whale oil as an insecticide is one of the more dramatic examples of this trend.

Whaling became a major commercial enterprise during the Tokugawa period. The most easterly whaling area was near the merchant city of Osaka. This whaling area stretched along the coast of today's Wakayama Prefecture, in the Kumano region in what was then Kii or Kishū domain. Whalers also operated to the west and south: on Shikoku's Pacific coast in Tosa domain, around the many islands off the coast of northern Kyushu (the Saikai area, including the Gotō Islands, Iki, and Tsushima Islands), and on the Japan Sea side of Honshu in Chōshū domain. Because whales tended to follow major currents, all four of these areas are situated along coastal whale migration routes: Tosa and Kumano lie along the Kuroshiō current,

[12] Howell (1989, p. 351) in particular. See also Totman (1995); Saito (1986) and Howell (1995).

[13] Totman (1995, pp. 273–274).

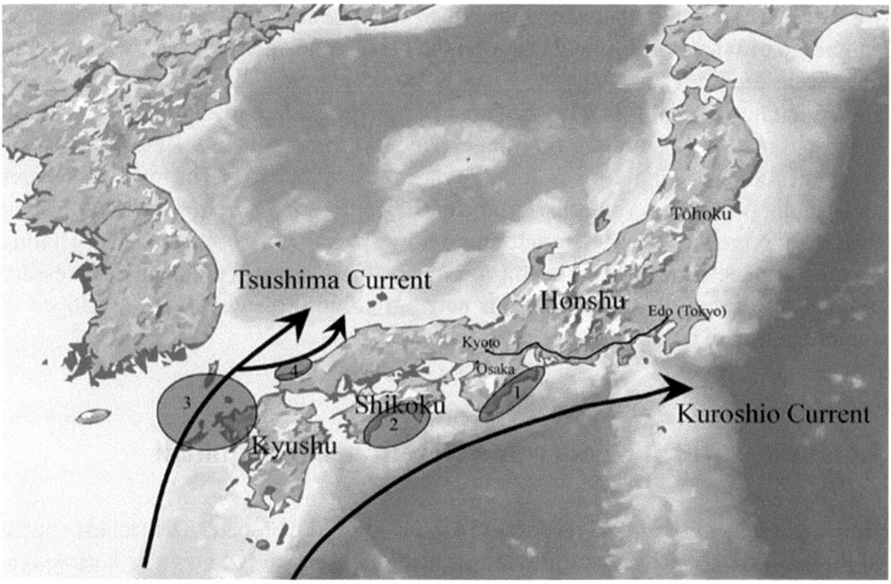

Fig. 6.1 Whaling areas in early modern Japan are shown in numbered circles: *1* The Kumano coast in Kii or Kishū domain, *2* Tosa Bay off the island of Shikoku, *3* the Saikai area of northern Kyushu, and *4* the area around the village of Kayoi in Chōshū. Whales migrated along the Kuroshio and Tsushima Currents (*arrows*), bringing them close to shore in these areas. Also shown are the major cities in early modern Japan and the Tokaidō road connecting Kyoto and Edo (modern Tokyo)

and Saikai and Chōshū along the Tsushima current (Fig. 6.1). Although whale meat was salted for preservation and transportation inland, whale oil was a more easily transportable and storable commodity. Oil from the Hirado whaling group in Kyushu was shipped as far as Edo (modern Tokyo), over 750 miles away.[14] At first, it was sold primarily for use in oil lamps. As methods for producing plant-based oils were developed, however, the smellier whale oil fell out of favor with higher-ranking people who could afford to be more selective with their lighting.[15]

While the growing city of Edo accounted for much of the whale oil sales for use in lamps, whale oil pesticide was at first used mostly in Kyushu, where the greatest number of whales were caught. At the start of the Tenmei famine in 1786, however, wholesalers and whaling groups reportedly provided nearly 1700 barrels of whale oil for use in combating insect damage all over the country. At least some of the domains in Kyushu had begun collecting emergency stores of whale oil in the 1760s, and by 1820 every county in Fukuoka domain was supposed to have 1500 barrels of emergency whale oil on hand.[16] Once the Tokugawa shogunate heard of this method, they sent out orders in 1787 and again in 1796 for all domains to use

[14] Referenced in Nakazono and Yasunaga (2009, pp. 145–146).

[15] Nakazono and Yasunaga (2009, p. 146), from the *Honchō shokkan* which commented that whale oil was better than fish oil but was not used when people could get flaxseed oil instead.

[16] Nakazono and Yasunaga (2009, p. 148).

whale oil when insect damage appeared in rice paddies, thus spreading the word and usage of whale oil on fields beyond the areas in northern Kyushu where it was most well-known.[17]

While Ōkura's book is now the most famous reference to the use of whale oil in agriculture, he was not the first to describe or promote its use. Before *Jokōroku,* other mentions of whale oil pesticide were most likely to be in texts on whales and whaling, such as Yamase Harumasa's *Geishi* (*Whale Essay*) in 1760, rather than texts on agriculture.[18] To determine how specialized descriptions of whaling became connected with scholarship on agricultural improvement, it is necessary to consider the process by which this new agricultural resource was developed in conjunction with natural history explorations.

Japanese Natural History and Resource Development

How were new agricultural resources like whale oil developed? The richest source of information about the natural world was the field of study known as *honzōgaku*. *Honzōgaku* was originally based on Chinese classification of natural substances in reference to their effectiveness in medical preparations (materia medica). Even though the term *honzōgaku* can also simply mean the study of materia medica, the substances it described were not limited to a small subset of the natural world. In traditional Chinese medicine, all natural substances had potential medical uses and were possible ingredients in pills and tonics used to rebalance internal flows of vitality.[19] Thus, there were few boundaries between practical medical knowledge and the scholarly description and classification of the natural world that we generally refer to as natural history. In fact, one historian of science, Ueno Masuzō, dates the start of scientific natural history in Japan to 1613, with the arrival of a specific Chinese materia medica known as the *Bencao gangmu* (in Japanese, *Honzō kōmoku, Compendium of Materia Medica*).[20] This book was first imported seventeen years after it was published in China, and was part of a long line of materia medica books dating back to at least the eighth century. It was more influential for natural history than its precursors because it did not focus solely on medical effects. The *Bencao gangmu* was an encyclopedic text that listed the properties of many objects and species in the natural world. It organized things under the categories for the five phases: water, fire, earth, metal (including stones), and wood (including all plants). These basic categories served a similar purpose to European categorization by the four elements, indicating the essential nature of a substance and how it might affect

[17] Matsubara (1984, pp. 23–24). Although he notes that petroleum was first used against planthoppers in 1869, whale oil's effectiveness was still being considered in 1872 in Aichi prefecture.

[18] Yamase ([1760]/1944).

[19] For a comprehensive view of the history of traditional Chinese medicine, see Unschuld (1985).

[20] For a discussion of *Bencao gangmu,* its contents and its place in the development of natural history in early modern China, see Nappi (2009).

the balance of forces in the body. However, the author also included three other categories: implements made of natural materials, beasts of all kinds, and different types of people. These sections offered a categorization more like descriptive natural history. The animals were ordered by outer characteristics such as scales, armor, feathers, or fur. Within each of these categories, they were further grouped by similarity, sometimes with consideration of their habitat, and from the most prototypical example of the category to the most unfamiliar or dangerous.[21] In many entries, natural history information was provided "because the qualities of a creature in life help[ed] determine its use in death."[22]

In translating the *Bencao gangmu,* scholars looked for useful qualities of Japanese natural resources to correspond to those listed for Chinese entries. One of the most influential developments in natural history appeared in 1709, when Kaibara Ekiken published a Japanese equivalent of the *Bencao gangmu* called *Yamato honzō* (*Japanese Herbal/Materia Medica*). His version focused more narrowly on practical usability of natural substances. It listed only those species found in Japan and provided only information of direct relevance to pharmacology or agronomy.[23] Kaibara used a similar general organization of water, fire, metal, plants, fish, shells, birds, beasts, and people, leaving out the section on tools. His work inspired the collection of information about new plants and animals found in Japan and not in China, a process that helped make *honzōgaku* into a form of natural history research rather than just materia medica. However, Kaibara himself was focused on the promotion of individual health as the first step towards a moral life and as an important component of the health and welfare of the state.[24] The development of agriculture was an important part of both of these types of health, and thus became closely tied to investigations of natural resources. Especially in the case of medically useful plants, such as ginseng, which were not native to Japan, agricultural science was also employed in an effort to grow these plants or find native substitutes for them.[25]

Even though the field of study known as *honzōgaku* was conceived of as a practical search for medically useful products, historian of science Federico Marcon explains that "by the first half of the eighteenth century [*honzōgaku*] had developed into an eclectic discipline of nature study consonant with what was known in early modern Europe as 'natural history'."[26] Scholars of *honzōgaku* in Japan provided useful information not just for doctors, but also for people concerned with agricultural improvement and those simply curious about the natural world and the possible uses of its resources. This led to practical explorations of the possibilities of

[21] Nappi (2009, especially pp. 71 and 113–114).

[22] Nappi (2009, p. 57).

[23] Kaibara ([1709]/1911).

[24] See Chap. 2, especially p. 79 in Marcon (2007).

[25] For example, Kasaya (2001) looks at Yoshimune's "rediscovery" of Dodonaeus' botanical text in the form of currency reforms, rationalization and promotion of domestic products, and most especially expensive imported medicines. He notes that there was a very trial-and-error experimental production of live ginseng in Japan.

[26] Marcon (2013, p. 191). See also Marcon (2007, Chap. 4, pp. 210–306).

natural substances like whale oil, which garnered a more general curiosity due to the unusual nature of its animal source.

Because scholars often combined different types of knowledge in their studies, there were few boundaries between what we would think of as the distinct fields of natural history or agricultural science. For example, the process of transplanting and acculturating foreign medicinal plants like ginseng involved both a natural-historical identification of possible native replacements, and also the testing of different environmental conditions for growing live specimens smuggled into the country from Korea. A major focus of this area of scholarship, however, was the desire to improve farmers' abilities to prevent or mitigate famines in the face of crop failures.[27] This was the motivating factor for the development of whale oil pesticide. As environmental historian Conrad Totman has pointed out, the near doubling of the Japanese population in the seventeenth century led to resource depletion and to associated searches for "ways to maximize the biosystem's immediate utility."[28] As a result of the population pressure that pushed the system to its environmental limits, there were increasingly severe crop failures and famines in the latter half of the Tokugawa period. During this time, there were at least three major famines following disastrous failures of rice crops which are believed to have led to tremendous death tolls: the Kyōhō famine of 1732–1733, the Tenmei famine of 1782–1787, and the Tenpō famine of 1833–1839. The suffering experienced during these major famines was a strong motivating factor in attempts at agricultural reform and improvement to prevent further famines. The first of these, the Kyōhō famine, was most severely felt in central and western Japan. After a series of poor harvests, then heavy rains in early 1732 which ruined the winter wheat and barley crops, a disastrous outbreak of *unka* (planthoppers or ricehoppers) destroyed possibly as much as 90 % of the following season's rice crop and pushed conditions into outright famine.[29] In response, in 1735 shogun Tokugawa Yoshimune instituted the Kyōhō reforms.[30]

This famine had enough of an impact that it was still being noted as part of the inspiration for Ōkura to publish texts such as *Jokōroku* in the following century. But I focus on this first famine not just because Ōkura referenced it as inspiration but also because Yoshimune's Kyōhō reforms heavily influenced future plans for agricultural improvement. As part of the reforms, he commanded that each domain conduct surveys to describe all the agricultural products, plants, and animals within its boundaries, including marine products. Yoshimune directed a *honzōgaku* scholar, Niwa Shōhaku, to organize these surveys.[31] Most of the records produced were

[27] As Ōkura notes in his justification for writing, *Jokōroku,* 1826 ([1826]/1977, p. 12).

[28] Totman (1995, p. 234).

[29] Kalland and Pedersen (1984, p. 40). Both the Japanese name *unka* and the English name planthopper cover a wide variety of species of insect in the order Hemiptera which feed on the phloem of rice plants, killing them. Walker (2010), in his discussion of the Kyōhō famine identifies these planthoppers as from three different species: the brown planthopper *Sogatella furcifera*, the white-backed planthopper *Nilaparvata lugens*, and the six-spotted leafhopper *Cicadula sexnotata*.

[30] Henceforth, referred to by his given name, Yoshimune, to avoid confusion.

[31] These surveys have been collected and reproduced in 21 volumes in Yasuda (1996–2005). For a description of the pivotal influence these surveys had on the development of natural history in Japan, see Marcon (2013), pp. 189–206.

simple lists of available products rather than detailed notes on the natural history of each domain; however, they followed the organizing and classificatory principles of natural history scholarship familiar to Niwa. His natural history scholarship extended to more than just classification. By the start of the seventeenth century, *honzōgaku* included the theory that food in general should be treated as medicinal, and care should be taken in everyone's diet to promote health. Thus, scholars began to consider the classification of Japanese foods in ways that may not have matched Chinese views of medicine, including many shellfish, fish, and whales that do not appear in the *Bencao gangmu*.[32] Meat was one of the major products of the early modern Japanese whaling industry, but whales were a food not eaten in China. Therefore, whales were one of the animals for which Japanese *honzōgaku* scholars needed more information.

The highly influential encyclopedia published by the physician Terajima Ryōan in 1712, known as the *Wakan sansai zue* (*Illustrated Sino-Japanese Encyclopedia*), includes entries for 125 different fish (broadly defined, including whales, jellyfish, and shrimp). While Terajima did rely on the *Bencao gangmu* for some of his basic descriptions of species, there are only 59 fish classified in *Bencao gangmu* and only 52 given in Terajima's other major inspiration, the Chinese encyclopedia *Sancai tuhui* (*Illustrated Compendium of the Three Powers [Heaven, Earth and Man]*, Jp. *Sansai zue*). For example, while Terajima preferred to cite the *Bencao gangmu* at the start of entries where there was an equivalent in China, he was forced to reference the Chinese encyclopedia's vague legends of whales to start his entry. Most of his information on whales is from other Japanese sources, including Kaibara Ekiken's *Yamato honzō*, which includes a reference to whaling and whales' edible parts, as well as describing the oil that was extracted for lamps.[33] Terajima's encyclopedia entry is longer than Kaibara's, relying on sources outside of natural history, but includes details on the major parts of whales' bodies, descriptions of their seasonal migration patterns, and individual descriptions of each of the six species commonly found near the Japanese coast. The rest of his entry on whales does mention other products like meat and baleen derived from whales. However, his first species description, for the favored right whale (*Eubalaena japonica*), only mentions how much oil could be extracted from them—one indication of oil's preeminence as a product of whaling.[34]

Popular works like Terajima's and Kaibara's show the process of consolidating information about the natural world that became even more comprehensive with the product surveys instituted by Yoshimune's Kyōhō reforms. These surveys were ostensibly carried out as part of Niwa Shōhaku's revision of a work for the shogunate called *Shobutsu ruisan*—an encyclopedia by Ino Jakusui intended to be "an all-inclusive and rational arrangement of all *honzōgaku* sources and a definitive

[32] Morita (1994), discusses the links between *honzō* and whales/whaling in Chap. 4.7 "Honzōgaku to kujira," pp. 210–216.

[33] Kaibara ([1709]/1911, p. 333). This entry is in the 13th maki (volume or scroll) of *Yamato honzō*, devoted to ocean fish.

[34] Terajima ([1712]/1985, vol. 7, pp. 198–203). This entry is in the start of maki (volume) 51, on bay and ocean fish without scales.

classificatory system."[35] Unfortunately, most of the survey data never appeared in Niwa's revision, which was furthermore never published, so it could seem as if the surveys had little to no influence on natural history in Japan. However, Marcon argues that these surveys were central to Yoshimune's agricultural reform policy expressed in the Kyōhō reforms. They were an effort to "acquire precise data on land productivity and exploitable resources," so that agricultural reforms to increase productivity could be enacted using those resources.[36] Some of the exploitable resources they were considering were marine resources. The survey from Chikuzen, for example, provided a list of various types of fish and also described the fishing methods through which they were caught.[37] In Yoshimune's home domain of Kii (modern Wakayama Prefecture), where he was daimyo until he became shogun in 1716, there was a longstanding interest in marine resources expressed in picture scrolls with different whale and fish species. These scrolls are similar to the simple product lists of Niwa's surveys, in that they generally include a series of images of different species of whales and sometimes exotic fish without any descriptive text or sometimes even labels designating their species names. But sometimes they include details such as labels for body parts like fins, blowholes, and what type of teeth they have. Also, there can be notes written at the end of these works discussing the sources from which they were copied, which demonstrate the conceptual and practical context for such simple lists or series of illustrations, whether in Niwa's surveys or from other sources. In one example, *Kozaura hogei emaki (Koza-village Whaling Picture Scroll*), the note explains merely that it is a scroll illustrating the whales that have been caught in Koza, a village in Kii domain that often hosted domainal officials passing through on the well-travelled coastal pilgrimage route.[38] However, from references on other scrolls, it seems that some original version of *Kozaura hogei emaki* was under control of government officials in Koza at least as early as 1726, and copies made from that original were sent to the domain's chief retainer, with known dates of copying in 1751 and 1798.[39] This set of scrolls from Kii and their link both to the government and to *honzōgaku* scholarship is more than just an interesting offshoot of the same sort of thinking about natural resources that Yoshimune implemented in his Kyōhō reforms. These scrolls show local scholars' and officials' strong interest in describing and classifying whales. Whales were such important resources to the domain that the earliest known work solely about whales in Japan, *Geishi,* was written by a *honzōgaku* scholar based on his personal experience of trips to the whaling villages of Taiji and Koza along the Kumano coast of Kii domain.[40]

---

[35] Marcon (2013, p. 198).

[36] Marcon (2013, p. 199).

[37] Yasuda (1996–2005, vol. 12).

[38] Both the author and date of this set of scrolls are unknown but the scroll was reprinted in 2008. See Anonymous ([n.d.]/2008).

[39] Harima (2008, p. 630).

[40] Ueno (1987, p. 247).

So how could these surveys of potential domainal resources like whales lead to agricultural improvements? As noted earlier, one driving force behind the desire to classify Japanese species and natural resources was their practical contribution to the welfare of the state. In this case, the state could mean individual domains, but sometimes could refer to the larger shogunal government. For Tokugawa Yoshimune, concern for the welfare of his domain of Kii transferred to concern for a larger state when he became shogun. Under the political sponsorship of Yoshimune, *honzōgaku* as a field of independent scholarship really began to flourish, in part because of his support for *honzōgaku* scholarship and data collection in the Kyōhō reforms. These reforms had three goals: a comprehensive survey of all Japanese plants and animals, the development of agricultural technologies that could prevent or mitigate the effects of famines, and the establishment of a shogunal medicinal garden which could supply pharmacological substances (like ginseng) which were at that time solely available through Korean or Chinese imports.[41] By combining all of these things, he hoped to improve the economy of Japan by no longer bleeding out silver to pay for foreign medicinal substances that had native counterparts, and also to improve the use of native natural resources in ways that would alleviate the harm of future weather events or pest attacks that in the past had led to major famines.[42] Even the basic lists of what animals and plants were found in domainal surveys contributed to this project, as they also contributed to domainal ambitions to be able to support themselves economically in competition with other domains. The tight interrelationship between these three goals is apparent in the illustrated whale scrolls like *Kozaura hogei emaki,* which show that the domainal government under Yoshimune's tenure as *daimyo* had a keen interest in the animal resources in Kii. This interest continued within the domain even after Yoshimune left to become shogun, probably because of the importance of the whaling enterprise in villages such as Koza.

The influence of diverse interests intersecting in widely curious scholars should not be ignored. One Kii-domain scroll dated to 1764 comments on the source from which the author copied his images of whales, noting that they were done from life on the order of the shogunal government in 1721.[43] It was in this year that Niwa Shōhaku wrote a preface that appears on at least two different scrolls (probably from some other original copy) noting that his friend had visited Taiji and Koza in that year and asked whalers about the whales there, including making drawings of them to bring back to Niwa.[44] Niwa, the same natural history scholar who planned the domainal product surveys to improve the output of rice crops and avoid future famines in the Kyōhō reforms, was thus closely linked to the men who went down to find out more about whales caught in Koza and Taiji.

[41] For details of this project, see Marcon (2013) and also Chap. 3 of Marcon (2007).

[42] For an example of how the prosperity of domains and the shogunate was also linked to non-medicinal cash crop development, see Kō (2010).

[43] Ueno (1987, pp. 246–247).

[44] Isono (1994, p. 27).

New uses for animal products like whale oil could have developed out of known agricultural practices with other animal parts. Even though there was relatively little livestock farming in most of Japan, the ties between agricultural development and biological knowledge were not limited to plants. Probably the most common use for animal products was as a fertilizer. Ōkura's various treatises on agricultural improvement do not focus specifically on fertilizers, but fertilizers are the focus of another agricultural treatise about cultivation entitled *Baiyō hiroku* (*Secret Notes on Cultivation*). This book by Satō Nobuhiro describes the various types of fertilizers important in successful agriculture in nineteenth-century Japan. Satō was employed by the shogunal government and is best known for his promotion of theories of political economy and Westernization. He was especially concerned with economic growth and with building up military power to defend against the increasing pressure from Western ships trying to open trade with Japan in the mid-nineteenth century. His focus on the development and optimal use of natural resources to strengthen Japan came in part from family experience in agriculture, horticulture, forestry, and mining.[45] In *Baiyō hiroku,* written in 1840, he discusses the effectiveness of human waste (extensively used in farming in Japan), the waste of domestic animals, and also the use of animal products such as fish or oils and fats.

These discussions are not accompanied by images of the associated animals, so unlike in Ōkura's case there is not a direct link here between natural history-type descriptions of organisms and the use of those organisms in agriculture. Nevertheless, in the chapter specifically devoted to fish fertilizers, he notes that the best fertilizer is dried sardines and sardine oil, followed by whale products as the next-best option. Thus, the interest in listing off the whales that were available in Koza and Taiji may be traced partly to their efficacy as a fertilizer. The method he describes of rendering down whale fat is similar to one provided for the extraction of oil from wild boar, but the ash from burning whale bones is also noted as the most effective type of animal-ash fertilizer.[46] His description of whale oil also notes that "it has the mysterious benefit of killing insects," and includes instructions on how much and when to apply it to rice fields to suppress outbreaks.[47] Satō points out the importance of these fertilizers not just for growing staple crops like rice, but also for growing the increasingly popular cash crops such as sugar and tobacco (particularly important in Kyushu), a point which shows that one of the driving forces behind the spread of agricultural techniques was not merely the desire to prevent famine but also the desire to expand agricultural output into non-food areas.[48] The use of ash from plant products such as nut husks and from shells was apparently cheap enough to be quite popular, but Satō warns farmers that they must supplement this treatment with oil-based fertilizers or the ground will lose its fertility. This point is not necessarily limited to a scientific understanding of nutrients, since his theory of why the treatment is effective is based on a conception of balancing essential characteristics

[45] De Bary et al. (2005, vol. 2, pp. 601–602).

[46] Satō ([1840]/1977, p. 308 and 314, respectively).

[47] Satō ([1840]/1977, p. 308).

[48] Satō ([1840]/1977, p. 317).

of the ground such that the earth's *ki* (in Chinese, *qi*) or vital essence does not become diminished and make the ground ill or let it starve.[49] As useful as whale parts seem to have been in the role of fertilizers, Satō's brief aside about how to apply oil during pest outbreaks became the focus of Ōkura Nagatsune's book describing the agricultural benefits of their oil.

Ōkura Nagatsune and Whale Oil in Agriculture

The spread of whale oil in agriculture beyond the whaling areas where it was first deployed is a fascinating demonstration of the interconnections between interests in natural history and practical applications of that knowledge to new agricultural techniques. This development process is similar to Western experimental science but was not necessarily modeled on the scientific method, although by this time some Western scientific books were available in Japan and did influence scholars. Whale oil was most likely developed as a pesticide more than once, in different places at different times. In his timeline of Japanese pesticide development, Matsubara Hiromichi lists three or four separate occasions where this method was discovered or rediscovered, from a secret family recipe for insecticide (which does not specify the ingredients) in 1641, to the sprinkling of whale oil on rice paddies by unrelated individuals in different counties in Hizen domain in 1670, 1720, and 1732.[50] According to the neighboring Fukuoka domainal records, in 1670 a farmer named Kuratomi Kichiemon discovered that whale oil spread on his fields was an effective protection from insects, and a local shrine's priest passed this information along to the county officials during the Kyōhō famine in an effort to mitigate the damage. A different explanation traces the method back to 1720, when a farmer named Ōmaru Hikoshirō of Kasuya county began using whale oil on his rice crop (Fig. 6.2).[51] However, it is the final date on this list, 1732, which is linked to the most well-known (re)discovery of the effectiveness of whale oil on planthoppers, as detailed by Ōkura Nagatsune.

The important thing to note is not where exactly in Kyushu the technique was first invented. Instead, it is noteworthy that this is the area credited with the discovery, in part because it was from farmers here that Ōkura first heard of the technique and described it to the wide audience for his agricultural manuals. Ōkura says that a certain Mr. Yahiro in Chikuzen saw planthoppers falling into the oil of shrine lamps while he was praying for protection from insects during the massive infestation of

[49] Satō ([1840]/1977, p. 317). Balancing this vital force or *qi* was a particular concern in traditional Chinese medicine. See Nappi (2009, pp. 62–63). Satō reference here to the earth becoming ill shows how the concern for balancing essential forces and qualities was both central to medical theory and applied to more than just human health.

[50] Matsubara (1984, p. 23).

[51] Fukuoka domain's "Gaichū kujo hakkensha chō [Investigation of harmful insect extermination discovers]" as cited in Nakazono and Yasunaga (2009, p. 147). These incidents are also listed with less detail in Matsubara (1984, p. 23).

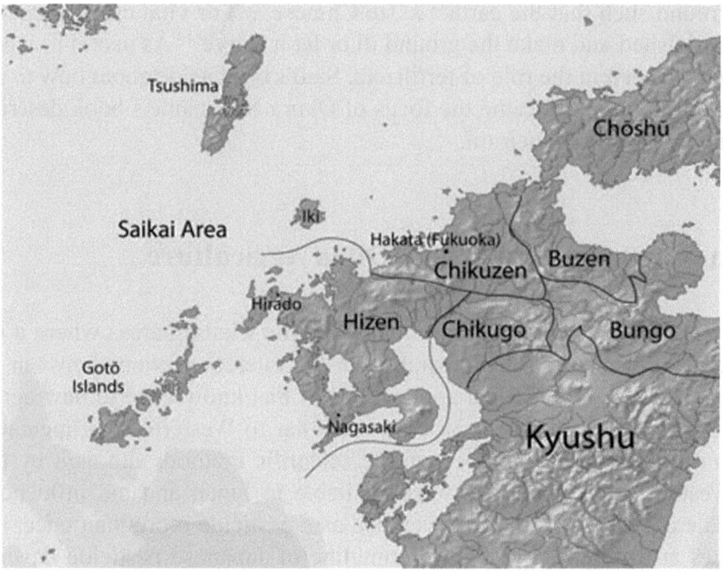

Fig. 6.2 Domains in northern Kyushu and their major cities during the Tokugawa period

1732. Yahiro observed the whale oil's ability to kill insects attracted to the light, and saw this as the answer to his prayers. Because whale oil lamps were relatively common, he was certainly not the first person to recognize this effect, but Yahiro did take the next important step and test out the effectiveness of whale oil sprinkled on the water on one of his fields. "After this, day and night he exhausted himself bringing oil, and the rice revived again and that field could be harvested."[52] Furthermore, he wrote down the effect to show his thanks to the gods for whom he was lighting the lamp, and this presumably is why Ōkura was able to recount the story nearly a hundred years later. According to Ōkura, before the end of the seventeenth or the beginning of the eighteenth century no one knew of this technique, instead depending on a traditional twilight ritual intended to drive insects off with torches and gongs.

In describing his desire to write the book *Jokōroku,* Ōkura includes an anecdote from when he was travelling along the Tōkaidō road, watching a farmer who clearly did not know about the effectiveness of whale oil. This happened during an insect outbreak in 1825. He describes meeting a farmer named Sanzaemon, who explained that he "saw the withering of the [rice] stalks and noticed quickly that insects came forth afterwards, so [he] divided the fields into three and within those, into the field that had many insects he put a lot of rapeseed oil five times, in the field that had fewer insects he put in half the oil three times, and in the field with the fewest insects he tried putting in not even a little. The first field produced seven or eight parts [out of 10], four parts for the next, and the last were totally withered, and just

[52] Ōkura ([1826]/1977, p. 13).

as this person told me, I also went to see and it was just as he said. Ah! If at that time he had provided whale oil, I sighed sadly thinking there might not have been insects at all."[53] This anecdote demonstrates the setting in which agricultural improvement manuals such as Ōkura's were produced: The farmer Ōkura watched battle an insect outbreak, without any apparent input from the travelling scholar, was carefully experimenting with different concentrations of rapeseed oil treatments to discover the most effective one. The writers of agricultural manuals were able to circulate many copies of their treatises because there was an avid audience for a kind of improvement that relied on this kind of experimentation and adaptation to local conditions of techniques developed elsewhere.

While Ōkura did not know why whale oil was more effective than plant-based oils in driving off and killing insects, he did note that pure whale oil had proved to be the most effective treatment. Therefore, he told farmers they should try to use it whenever they could. Apart from possible difficulties in supply, whale oil was far more expensive than the plant-based oils such as rapeseed oil or tung oil that were used to combat insects, even before adding in the transportation costs introduced when the use of whale oil moved from the whaling region of northern Kyushu to other parts of the country.[54] He cautioned readers that whale oil could be difficult to distinguish from other kinds of fish oil, and yet "if you put in one field one gō [0.2 liters] of whale oil and in another field five gō [1.0 liters] of assorted fish oil, the effect of the true whale oil will not spread to the other. If you do not know how to separate out the types, then no matter how much effort you put in, you will not get the result you are looking for."[55] His desire to determine the difference between the effective whale oil and the less effective generic fish oil seems to have led him to look for natural history information about whales. He provides a quick summary of the categories of whales seen in Japan and notes that, "there are some whose oil is useless."[56] The problem with assorted or generic fish oil was that its main components were sardine, shark, and tuna oil, and also that it was an unrefined oil. Possibly, in the interests of showing the source of proper whale oil instead of this mixed fish oil, or possibly simply because he thought farmers would be curious about whales if they had never used their oil before, Ōkura included six pages of diagrams of whales and dolphins with some minor labels of external anatomical features.[57] These diagrams are quite similar to the ones that appear in the lists of whale species developed by *honzōgaku* scholars visiting the whaling groups in Kii domain. It was certainly not necessary to know these details about different whale species in order to buy their oil and spread it on one's rice field, but the inclusion of such details shows the ties between this type of knowledge and the development of agricultural improvements. Furthermore, he also included a section discussing the

[53] Ōkura ([1826]/1977, p. 15).

[54] Ōkura notes that farmers using rapeseed oil instead of whale oil should use twice as much to get the same effect ([1826]/1977, p. 47).

[55] Ōkura ([1826]/1977, p. 31).

[56] Ōkura ([1826]/1977, p. 32).

[57] Ōkura ([1826]/1977, pp. 33–35).

types of insects that would attack rice crops, because people needed to know some basic biological details of the pests in order to block their outbreaks. At the start of this section he cited Kaibara Ekiken's *Yamato honzō* as a major source, so Ōkura's interest in and use of natural history information was not confined just to descriptions of whales.[58]

Ōkura's example serves to highlight the availability of many different types of information in print during this period. It also shows how observations made in one particular area were brought to bear on larger problems. Ōkura's own observations while travelling brought home to him the importance of making local knowledge more widespread, particularly in cases where someone had already come up with a solution to a problem, but other people had not yet heard of the technique. His work was also informed by work on natural history, whether in encyclopedias like Terajima's or in reports on whales from people interested in describing the whaling industry. Local information circulated within a broader reading public that flourished with the vibrant print culture that developed during the Tokugawa period. Such widely circulated information about whaling groups and their targets came from visitors to villages in Kii or other whaling areas, such as the prosperous Gōtō islands' whaling group in Kyushu whose illustrations Ōkura copied in *Jokōroku.*

The same commercialization that led to the proliferation of printing houses and a variety of products—including agricultural manuals that could be borrowed, rented, or purchased by interested farmers or scholars—also drove the search for new, profitable resources. Thus, whaling groups were interested in circulating descriptions of their work and their products, including details about the different species they caught, which might draw the curiosity of an even wider public to buy their oil. Works like *Jokōroku,* with its description of a novel use for whale oil, promoted the sales of a product that, before its widespread use in agriculture, was mostly used on a much smaller scale for lighting. Once it was being promoted as an insecticide, some merchants would increase the price of whale oil in years where there were insect outbreaks, so the importance of commercial interests and profit driving the use of new resources should not be discounted.[59]

Conclusion

The use of whale oil in agriculture provokes curiosity by its paradoxical nature. Closer investigation of this unique process shows the ways in which natural history knowledge worked together with agricultural improvement. As far as Ōkura's practical suggestions for farmers go, the section describing different whale species is not particularly relevant, but such apparently unconnected information slakes just the sort of curiosity about the workings of the natural world that would also tend to promote tinkering and experimentation within traditional farming practices.

---

[58] Ōkura ([1826]/1977, pp. 26–30).

[59] Ōkura ([1826]/1977, p. 31).

The origin story for the use of whale oil pesticides that Ōkura gives is not the only way in which this practice could have begun, as other sources seem to record independent inspirations by different farmers in northern Kyushu. Given the ease with which ideas circulated in this period, many possible influences on the practice might not have been recorded. There are records in Chinese books about using oil as a treatment for rice pests, although it is not whale oil.[60] Nagasaki was the main port for trade with China during the early modern period, and it is possible that people in Kyushu had access to some of the books and ideas brought over from China through contact with merchants and scholars in Nagasaki. Knowing about the reference to oil in general as a treatment for pests could have led farmers to try the oil they had on hand, which happened to be whale oil for their lamps. In other words, information from many different places was intersecting during this period, and the richness of contact with ideas from outside one's very local area was a characteristic of early modern Japan that led to agricultural manuals like Ōkura's.

A complex array of intersections appears in the use of whale oil in agriculture: between knowledge of animals, use of animals, and agricultural development. Ōkura's publication of *Jokōroku* included seemingly irrelevant images of different whale species, showing that a focus on the useful parts of animals did not preclude examination of their characteristics as a whole. The culture of curiosity and the desire for practical knowledge also combined to develop the system in the Chinese *Bencao gangmu* into a Japanese scholarship more closely resembling natural history than simple materia medica. Agriculture was one of the areas in which the interconnection of theoretical and practical knowledge from many sources is quite apparent, thanks to the production of agricultural manuals from writers like Ōkura. The success of agriculture in this period of population growth depended on finding ways to expand both general knowledge and practical applications for all parts of the living world. Agricultural manuals of the nineteenth century show how the study of the natural world—what we would classify today as natural history, as agricultural science, or as basic environmental understanding—was harnessed to support an increasing population and stave off scarcity and famine. As the promotion of supplementary materials from beyond the shores of Japan shows, agriculture during this period required more than just the use and knowledge of the conditions of a particular field or farm, but rather a wider environmental connection to new sources of nutrients, pesticides, and other resources, including knowledge from a variety of scholarly endeavors.

References

Anonymous. [n.d.]/2008. Kozaura hogei emaki. In *Kozachō shiryō – Hogei hen kaiga shiryō*, ed. Kushimoto chōshi hen-san iinkai, 41–51. Kushimoto-chō: Nishioka Sōgō Insatsu KK.

De Bary, William Theodore, et al. 2005. *Sources of Japanese tradition*. 2 vols. New York: Columbia University Press.

[60] Itō (2001).

Harima, Yoshinori. 2008. Kaiga shiryō ni tsuite. In *Kozachō shiryō – Hogei hen,* ed. Kushimoto chōshi hen-san iinkai, 623–631. Kushimoto-chō: Nishioka Sōgō Insatsu KK.

Hauser, William B. 1974. The diffusion of cotton processing and trade in the Kinai region in Tokugawa Japan. *Journal of Asian Studies* 33 (4): 633–649.

Howell, David L. 1989. Hard times in the Kantō: Economic change and village life in late Tokugawa Japan. *Modern Asian Studies* 23 (2): 349–371.

Howell, David L. 1995. *Capitalism from within: Economy, society and the state in a Japanese fishery.* Berkeley: University of California Press.

Isono, Naohide. 1994. Edo jidai kujirarui zusetsu kō (Old illustrations of whales). *Hiyoshi Review of Natural Science Keio University* 16:22–36.

Itō, Seiji. 2001. *Sanemori kigen kō: Nitchū hikaku minzokushi.* Tokyo: Seidosha.

Kaibara, Ekiken. [1709]/1911. Yamato honzō. In *Ekiken zenshū,* ed. Ekikenkai, 1–426. Tokyo: Ekiken Zenshū Kankōbu.

Kalland, Arne, and Jon Pedersen. 1984. Famine and population in Fukuoka Domain during the Tokugawa Period. *Journal of Japanese Studies* 10 (1): 31–72.

Kasaya, Kazuhiko. 2001. The Tokugawa Bakufu's policies for the national production of medicines and Dodonaeus' Cruijdeboeck. In *Dodonaeus in Japan: Translation and the scientific mind in the Tokugawa Period,* ed. W. F. Vande Walle and Kazuhiko Kasaya, 167–189. Leuven: Leuven University Press.

Kō, Ochiai. 2010. The shift to domestic sugar and the ideology of 'the national interest.' In *Economic thought in early modern Japan,* ed. Bettina Gramlich-Oka and Gregory Smits, 89–110. Boston: Brill.

Marcon, Federico. 2007. The names of nature: The development of natural history in Japan, 1600–1900. Ph.D. dissertation, Columbia University.

Marcon, Federico. 2013. Inventorying nature: Tokugawa Yoshimune and the sponsorship of Honzōgaku in eighteenth-century Japan. In *Japan at nature's edge: The environmental context of a global power,* ed. Ian J. Miller, J. A. Thomas, and B. L. Walker, 189–206. Honolulu: University of Hawai'i Press.

Matsubara, Hiromichi. 1984. *Nihon nōgakushi nenpyō.* Tokyo: Gakkai Shuppan.

Morita, Katsuaki. 1994. *Kujira to hogei no bunkashi.* Nagoya-shi: Nagoya Daigaku Shuppankai.

Nakamura, Ellen Gardner. 2005. *Practical pursuits: Takanō Chōei, Takahashi Keisaku, and western medicine in nineteenth-century Japan.* Cambridge: Harvard University Asia Center.

Nakazono, Shigeo, and Hiroshi Yasunaga. 2009. *Kujiratori emonogatari.* Fukuoka-shi: Gen Shobō.

Nappi, Carla. 2009. *The monkey and the inkpot: Natural history and its transformations in early modern China.* Cambridge: Harvard University Press.

Ōkura, Nagatsune. [1826]/1977. Jokōroku, 1826. In *Nihon nōsho zenshū,* ed. Tatsuo Yamada, et al., 33–38. Tokyo: Nōsangyoson Bunka Kyōkai.

Ōkura, Nagatsune. [1844]/1977. Jokōroku, 1844. In *Nihon nōsho zenshū,* ed. Tatsuo Yamada, et al., 57–118. Tokyo: Nōsangyoson Bunka Kyōkai.

Okuta, Akira, et al. 2012. Prediction of overseas migration of the small brown planthopper, *Laodelphax striatellus* (Hemiptera: Delphacidae) in East Asia. *Applied Entomology and Zoology* 47 (4): 379–388.

Robertson, Jennifer. 1984. Sexy rice: Plant gender, farm manuals, and grass-roots nativism. *Monumenta Nipponica* 39 (3): 233–260.

Rubinger, Richard. 2007. *Popular literacy in early modern Japan.* Honolulu: University of Hawai'i Press.

Saito, Osamu. 1986. The rural economy: Commercial agriculture, by-employment, and wage work. In *Japan in transition: From Tokugawa to Meiji,* ed. Marius B. Jansen and Gilbert Rozman, 400–420. Princeton: Princeton University Press.

Satō, Nobuhiro. [1840]/1977. Baiyō hiroku. In *Nihon nōsho zenshū,* ed. Tatsuo Yamada, et al., 153–406. Tokyo: Nōsangyoson Bunka Kyōkai.

Smith, Thomas C. 1969. Farm family by-employments in preindustrial Japan. *Journal of Economic History* 29 (4): 687–715.

Smith, Thomas C. 1998. *Native sources of Japanese industrialization, 1750–1920*. Berkeley: University of California Press.
Terajima, Ryōan. [1712]/1985. Wakan sansai zue. In *Tōyō Bunko*, ed. Isao Shimada, Atsuo Takeshima, and Motomi Higuchi, no. 471, 198–203. Tokyo: Heibonsha.
Totman, Conrad D. 1995. *Early modern Japan*. Berkeley: University of California Press.
Ueno, Masuzō. 1987. *Nihon dōbutsugakushi*. Tokyo: Yasaka Shobō.
Unschuld, Paul U. 1985. *Medicine in China: A history of ideas*. Berkeley: University of California Press.
Walker, Brett L. 2010. *Toxic archipelago: A history of industrial disease in Japan*. Seattle: University of Washington Press.
Yamase, Harumasa. [1760]/1944. Geishi. In *Sangyō gijutsu hens. Nōgyō, seizōgyō, gyogyō, in series Nihon kagaku koten zensho*, ed. Kōkichi Kanō and Hiroto Saigusa, vol. 11, 455–518. Tokyo: Asahi Shinbunsha.
Yasuda, Ken. 1996–2005. *Edo kōki shokoku sanbutsuchō shūsei*. Tokyo: Kagaku Shoin.

Jakobina Arch is an assistant professor in the History Department at Whitman College. She holds a PhD in History and East Asian languages from Harvard University. Her dissertation is a marine environmental history of whales in early modern Japan, focused on how whales and whaling knit land and sea together.

Chapter 7
Forests, Climate, and the Rise of Scientific Forestry in Russia: From Local Knowledge and Natural History to Modern Experiments (1840s–early 1890s)

Anastasia A. Fedotova and Marina V. Loskutova

Introduction

The question of the environmental impact of forests has been the subject of a lively debate over the past two or three centuries. Early ideas about causal links between deforestation and deterioration of climate developed first in the seventeenth and eighteenth centuries in the context of French and British colonial expansion to tropical islands. As some scholars have recently argued, it was these ideas that first paved the way for the rise of environmental consciousness among Europeans.[1] By the early nineteenth century, concerns about climate and deforestation began to be voiced about Europe itself. Many famous scientists of that age, such as Alexander von Humboldt in Germany and François Arago in France, contributed to the debates by referring to their own travel observations or by providing general theories on the mechanisms of the climatic impact of forests. Their writings considerably enhanced the credibility of such arguments; however, the academic community and the wider public did not universally accept these claims, since factual proof remained slim and evidentiary standards were not yet well defined.

From their inception, debates about the climatic impact of forests were inextricably intertwined with the politics and practices of territorial governance. They also had implications for the competing ways that different social or ethno-cultural

[1] Grove (1995).

A. A. Fedotova (✉)
Institute for the History of Science and Technology, RAS, Universiteskaya naberezhnaya 5/2, 199034 St. Petersburg, Russia
e-mail: f.anastasia.spb@gmail.com

M. V. Loskutova
History Department, National Research University - Higher School of Economics, Promyshlennaia ul. 14a, 198099 St. Petersburg, Russia
e-mail: mvlosk@yandex.ru

D. Phillips, S. Kingsland (eds.), *New Perspectives on the History of Life Sciences and Agriculture,* Archimedes 40, DOI 10.1007/978-3-319-12185-7_7

groups utilized nature. But these debates were simultaneously ones in which participants referred to "science" as their principal source of credibility. Furthermore, the type of evidence used to support or invalidate an opinion in this discussion changed considerably over the course of the nineteenth century, as the study of nature evolved from natural history and natural philosophy to the modern life sciences. A number of recent studies have looked at debates about forests and climate; most studies, however, have focused primarily on the ideas and concepts articulated by the debates' principal protagonists.[2] Much less attention has been paid, in contrast, to changes in observational practices or in forms of empirical evidence. Past work on this topic also has not fully considered the fact that these observations and experiments were carried out by people whose social and epistemic identities changed considerably over time. In this chapter, we are going to examine precisely this issue—the gradual transition from a natural historical set of evidentiary standards and observational practices to a very different regime of modern scientific experimentation.

Russia offers an interesting place to examine these issues for several reasons. Its academic community and general public were very much familiar with the debates in Western and Central Europe and frequently borrowed arguments, research programs, and practices from these regions. Yet at the same time, the geography and history of Russian imperial expansion provided a setting that was in many ways similar to the tropical context of the early Western European debates. From the eighteenth century onwards, large numbers of Russian and Ukrainian peasant settlers migrated from the forest zone to the southern steppes, moving into land that had been previously occupied by nomadic pastoralists. This development prompted the imperial academic community, the state administration, and a larger public to reflect on this unfamiliar environment and to speculate about how it would be transformed under the impact of peasant colonization.[3] In this way, the debates about the climatic impact of forests in nineteenth-century Russia involved both Western and Central European ideas and ideas drawn from domestic experience.

The arid grasslands of Russia's southern frontier had exceptionally fertile black soils and seemed very promising for agricultural development. Crop production in the region could be both a lucrative and fragile business, however, as these provinces periodically suffered from acute droughts, bad harvests, and resulting famines.[4] As a result, steppe climate and its malleability emerged very early as one of the central themes in debates about the relationship between humans and the environment in

[2] For recent research on the nineteenth century debates about the environmental impact of forests, see Grove (1995, pp. 309–379), Rajan (2006), Andreassian (2004), and Weigl (2004).

[3] Moon (2010, pp. 251–275).

[4] One of the major impediments for agricultural production in the steppe region is a dramatic fluctuation in the amount of rainfall from one year to the next. As a result, an average annual rainfall is a figure of a little practical importance in this region. For details, see, e.g., Kovda and Samoĭlova (1983), Mordkovich et al. (1997), and Moon (2013).

Russia. This issue became all the more important over the course of the nineteenth century, as the region became a major center of agricultural production that was of strategic importance for domestic food security and Russia's export trade.

An almost complete lack of wood in the steppes was another problem that was noted very early as an impediment to colonization. Wood was a principal source of fuel and construction material for Russian peasants, whose whole material culture was shaped by the forests of northern and central Russia. In the early nineteenth-century, many naturalists, foresters, and writers in Russia assumed that the southern Russian steppes had an arid climate precisely because they were devoid of forests.[5] They argued that an enlargement of forested areas would make the climate more humid and moderate, which would in turn make yields more stable. A few enlightened landlords and German colonists in the region repeatedly tried to plant forests on their steppe estates and farmsteads. In the 1840s, the Russian state administration joined these efforts by establishing several forestry districts in the region.

In our chapter, we will focus on observations and experiments that were carried out at one of these state forestry districts, the Velikii Anadol' forestry district in the Ekaterinoslav province (now a nature reserve of the same name in the Eastern Ukraine). The Velikii Anadol' forestry district was created in 1843. In 1892, it became one of the areas explored by the Special Expedition of the Forestry Department led by Vasilii Dokuchaev, a pioneer of soil science on a global scale and the founder of this discipline in Russia. Six years later, in 1898, it was transformed into one of the first experimental forestry districts in the Russian empire. In this chapter, we will not provide a detailed account of the Special Expedition of the 1890s, since historians of Russian forestry and science have paid considerable attention to it already. They unanimously consider it to be the first scientific research project concerned with the environmental impact of forests in Russia. Therefore, the Special Expedition usually functions as a starting point for a further discussion of late imperial and Soviet environmentalism, with a predictable emphasis on the novel features it introduced to forestry's concepts and practices. Our work, in contrast, traces the "prehistory" of experimental research on forests. In what follows, we examine the changing nature of observations and experiments and analyze how foresters' evolving research practices related to the transformation of broader conceptual frameworks and evidentiary standards. From this perspective, the Special Expedition can also be seen as part of a longer, more complicated history of changing evidentiary standards in nineteenth-century life sciences.

Our choice of the Velikii Anadol' forestry district as a privileged object of analysis can be explained by several factors. Its history is exceptionally well document-

[5] For the history of academic debate on the absence of forests in the steppe zone of European Russia and the role of this debate for the making of plant geography and plant ecology in Russia, see Fedotova (2012).

ed.[6] As the earliest and most successful steppe forestry district, the Velikii Anadol' district has always attracted the attention of historians of Russian forestry. At the same time, its place in Dokuchaev's Special Expedition undoubtedly contributed to its fame. Dokuchaev has always been considered an iconic figure in the history of science in Russia. The issue of steppe afforestation enjoyed a similar status, especially during the Soviet era, with its ambitious plans to transform nature.[7] Yet the history of the Velikii Anadol' has so far been written from a particular standpoint: Historians of forestry have been predominantly interested in documenting its administrative transformations and in demonstrating its remarkable success in afforestation (the enlargement of forest plantations, adopted technologies, afforestation costs, etc.). We, however, would like to also emphasize the importance of the district as an early research site.

Early Debates over Forests and Climate in Russia and the Foundation of the Velikii Anadol' Forestry District (the 1840s–the early 1860s)

Beginning in the 1770s and 1780s, when the Russian empire annexed the Black Sea coastal areas under Catherine II, much hope was placed on these vast uncultivated territories with their exceptionally fertile black soils and warm climate. Yet the most striking feature of the steppe landscape—the absence of forests or even small groves—constituted a serious drawback, since wood was the principal material used by Russian and Ukrainian peasants in construction and as a fuel. In the early nineteenth century, attempts were made to promote the planting of forests in the region, and the afforestation efforts of Mennonite immigrants from Central Europe enjoyed marked success. In return for substantial land grants and exemption from military conscription, they were obliged to plant and grow fruit and forest trees in the steppe.[8] By the 1830s, small groves that surrounded Mennonite settlements became an attractive feature of the local landscape and were much admired by travelers.[9]

---

[6] Russian State Historical Archive (Rossiiskii Gosudarstvennyi Istoricheskii Arkhiv, hereafter—RGIA), f. 387 (Forestry Department) for the years 1843–1917. All further references to documents from Russian archives are given here in compliance with established academic practice: the name of the archive is followed by collection (*fond* or f.), inventory (*opis'* or op.), file (*delo* or d.), and folio (*list* or l., ll. in plural form). See also Red'ko (1994). The book narrates the early history of the Velikii Anadol' forestry district, from its establishment to the last years of Graff administration. It is based on an extensive range of archival documents; its author is a forestry specialist who spent many years researching the history of forestry in Russia. See also Filonenko (2000), Bark (1872), Polianskii (1888), and Tsvetkov (1957).

[7] Brain (2011).

[8] On peasant colonization of the steppe zone in the Russian empire, see, e.g., Sunderland (2004).

[9] On afforestation of the southern Russian provinces in the 1800s–1840s, as it was carried out by Mennonites and landlords, see Red'ko and Red'ko (2003) and Tsvetkov (1957).

In the 1830s, the issue of steppe afforestation attracted considerable governmental and public attention in Russia. Unlike in the previous decades, when a few occasional pieces of legislation were adopted to encourage afforestation, this time the drive was more consistent, and it was explicitly linked to the debates over climate change caused by the destruction of forests. It well might be that these fears were provoked or exacerbated by the serious drought of 1832–1834,[10] yet at the same time, they were quite clearly promoted and exploited by the Ministry of Finance, chaired by Georg Cancrin—a man who corresponded extensively with Alexander Humboldt and was well informed about other German and French publications that linked deforestation with desiccation and climate change. These theories became his personal conviction, which he apparently shared with a few of his immediate subordinates.[11] Cancrin was also alarmed by the pace of industrialization in the provinces around Moscow. Factories and mills mushrooming in the area consumed vast quantities of wood, pushing up fuel prices and causing the rapid destruction of forests along major waterways.

As a result of these concerns, Cancrin became the first nineteenth-century Russian statesman to actively promote forestry education and public interest in forestry. He considerably expanded and upgraded the Forestry Institute in St. Petersburg, and he founded and sponsored the first Forestry Society in the empire, which among other things encouraged steppe afforestation by running competitions among interested landowners and assisting them with seeds and plants.[12] It was Cancrin who sponsored a stream of articles in the Russian press that linked the destruction of forests with droughts and the depletion of rivers. While advocating tight governmental control of forests along the waterways in central Russia, these publications principally referred to French legislation and public debates over this issue in France. At the same time, however, they also proposed the idea that the Russian steppes had once been forested areas, with their forests destroyed by later invasions of nomadic tribes. To support these claims, some authors of a more academic leaning started looking for evidence in ancient Greek and medieval sources.[13]

When in 1837 the forestry administration was transferred from the Ministry of Finance to the newly established Ministry of State Domains (MSD), the latter ministry immediately withdrew its support for these types of publications, and even arranged an inquiry into the matter by a commission set up by the St. Petersburg Academy of Sciences. Its report, which was signed by several academicians, was careful to avoid a serious engagement with the general issue of the climatic impact of forests, yet it downplayed fears of the rapid destruction of Russian forests and explicitly argued against the notion that the Russian steppes had ever been covered with trees.[14] Unlike Cancrin, the new minister of state domains Pavel Kiselev and his closest associates were not particularly convinced that deforestation generally

[10] Moon (2010, p. 257).

[11] For details, see Loskutova (2012b).

[12] Bozherianov (1897).

[13] For details, see Loskutova (2012b).

[14] Köppen (1841).

led to the depletion of rivers and to droughts; they were also not interested in tightening laws for private forest owners. They were even less keen on raising public concern over deforestation. Unlike Cancrin, who did much to encourage civil initiative, Kiselev and his subordinates preferred to rely on the bureaucratic chain of command.

Yet the MSD was no less committed to the cause of steppe afforestation than the Ministry of Finance had been. The new ministry had been created with the explicit aim of providing a more efficient management of the state land domains and a more active guardianship with regard to the state peasants. While the tsarist government was still unwilling to commit itself to the abolition of serfdom, it hoped that the new ministry could gradually navigate the way towards emancipation by modernizing the local administration and improving the agricultural productivity of state peasants. Indeed, in the 1840s and early 1850s, the MSD pursued actively interventionist policies. It supervised large-scale peasant resettlement from densely populated heartland provinces to the southern and eastern frontiers of European Russia; it also carried out land and forest cadasters; encouraged the cultivation of potatoes, tobacco, sugar beet, and fruit trees; supported livestock husbandry; disseminated agronomical knowledge; and promoted elementary education.[15] The new ministry considered the improvement of forestry a high priority, particularly when it concerned the afforestation of southern steppe provinces—the destination region for peasant resettlement from the overpopulated central parts of Russia.

As early as 1840, the MSD began devising plans for promoting afforestation in the Black Sea coastal provinces of the empire. The MSD officials who drafted these plans were well versed in the contemporary French literature that linked deforestation and desiccation, and they often referred to the beneficial impact of forests on climate and soil fertility as a rationale behind the proposed schemes.[16] Yet, given earlier criticisms expressed by the MSD leadership and its experts with regard to Cancrin's propaganda, it remains unclear to what extent these claims could be taken as an expression of a genuine belief shared by leading forestry officials. We can only say that by the 1840s the argument about the climatic impact of forests was well known to Russian forestry officials in St. Petersburg and they often employed it when it suited ministerial policies.

The Velikii Anadol' forestry district was established by the MSD in 1843 as a part of these afforestation schemes. It was meant to be a model forestry, one among several that would be set up at the same time in each of the southern coastal provinces. The ministerial experts assumed that poor progress in steppe afforestation could be explained entirely by the local serfs and landowners' lack of commitment and the poor quality of their seeds. It was hoped, therefore, that the future success of Velikii Anadol' and a few similar model plantations would eventually prompt local peasants to grow forests on their own initiative.

[15] For details, see Mironov (2012, pp. 200–202).

[16] See, e.g., RGIA, f. 387 (Forestry Department), op. 1, d. 465, ll. 2–21 ("On the measures to promote afforestation in southern Russia," a draft report of the MSD third department, December 27, 1840).

In 1843, the MSD sent one of its leading forestry specialists to Ekaterinoslav province in order to choose the location of the planned forest plantation, and a few months later, it appointed Viktor Graff—a recent graduate of the Forestry Institute in St. Petersburg—as its head forester. When Graff first arrived in the southeastern part of the province, it was still sparsely populated, with a few recently founded Russian, Ukrainian, German, and Greek villages. The region's inhabitants lived on the production of grain and the raising of sheep. A few miles away, there was a road that led to the town of Mariupol' on the Black Sea coast. In accordance with the objectives set up by the ministry, Graff and his senior colleague from the MSD headquarters chose the most inhospitable location for the future plantation—a plot of a so-called high steppe with dry heavy clay soils that was a watershed between two small creeks.[17]

When the area had been chosen, Graff paid a visit to Johann Cornies, an informal leader of Mennonite colonists and an acknowledged expert in steppe afforestation.[18] With Cornies, Graff studied Mennonite methods of planting trees in the steppe—their "local knowledge" acquired by years of trial and error. Another visit he paid was to the Crimea, where he met with the respected naturalist Christian Steven, who served there as a senior inspector of agriculture, and with Nikolai Hartwiss, the head of the botanical garden in Nikita. These two men provided him with academic advice on the tree species that would be most appropriate for his purpose and furnished him with some seeds.[19]

Acting upon their recommendations, Graff started his model plantation. His objective was primarily to ensure a stable growth of trees and the expansion of the plantation territory by perfecting techniques of forest cultivation and by identifying appropriate species and varieties of plants. The forestry district was also meant to serve as a ground for acclimatizing—by means of trial and error—some foreign forest species to the Russian steppe environment.[20] The Velikii Anadol' was also a center for disseminating practical knowledge: Graff opened a forester school for peasant boys who could learn the trade by assisting his personnel in their daily work on the forest plantation.[21]

Neither senior officials of the MSD nor Graff questioned the basic assumption underlying the project—that forests could be grown anywhere in the steppes, if only sufficient efforts were applied to the purpose. Also, none of them, apparently, considered the economic feasibility of afforestation. In the pre-emancipation

[17] RGIA, f. 387, op. 1, d. 10415, ll. 107–09 (Graff's report, September 1, 1843); 143–51 (Graff's report, October 25, 1843).

[18] From 1845, Johann (Ivan Ivanovich) Cornies (1789–1848) was the first head of Staroberdiansk forestry district. On Cornies, see Brandes (1993) and Epp (1946), *Johann Cornies*.

[19] RGIA, f. 387, op. 1, d. 10415, ll. 147–51 (Graff's report, October 25, 1843).

[20] For the period when Graff was the head of the forestry district (1843–1866), it acclimatized more than 30 tree species and 40 shrub species in the arboretum. A smaller number was used for afforestation, however. For this purpose, the personnel of the forestry district still used not only local species but also a few introduced species, including some American trees. The forestry district objectives were outlined in a number of sources, e.g., Bark (1873).

[21] Red'ko (1994).

period, labor could be requested from state peasants, and additional labor duties were invariably discussed in terms of their effect on social stability rather than their cost.[22] At the same time, the MSD implicitly assumed that, in the long term, a more hospitable environment in the southern provinces would compensate for the initial expenses of afforestation.

Meteorological Observations in Velikii Anadol' in the 1840s

For all these reasons, the MSD did not order Graff to carry out meteorological observations. The forest in Velikii Anadol' was not meant to be a testing ground for exploring correlations between afforestation and meteorological phenomena. Indeed, such experiments were not considered anywhere in Europe during this period. When in the 1850s the debates in France first moved away from a general discussion to the search for quantifiable evidence, the impetus came not from foresters, but from hydraulic engineers who focused on hydrometric measurements.[23] Yet Graff had a taste for natural history, and he wished to start meteorological and phenological observations of a kind that were pursued at that time in a few model farms, schools of horticulture, and plant nurseries subordinated to the MSD. Indeed, from the early 1840s, the ministry began promoting these observations by providing instruments, such as minimum and maximum thermometers, barometers, psychrometers, and rain gauges, and distributing instructions, which were written by the leading Russian meteorologist of the period, academician Adolf Kupffer (sometimes Kupfer). In the 1830s and 1840s, Kupffer cooperated closely with Carl Friedrich Gauss and Wilhelm Weber in Germany; Edward Sabine, Humphrey Lloyd, and John Herschel in Britain; and Adolphe Quetelet in Belgium in their attempts to create pan-European networks to observe meteorological phenomena and the earth's magnetism, and he was ultimately interested in expanding the number of observation stations in the Russian empire.[24] The MSD assisted Kupffer in his project for its own reasons: Its leadership was keen on demonstrating its reliance on the most modern, enlightened means of territorial governance, which included the compilation of all sorts of statistical data and maps. In 1843, the ministry committed itself to the production of the first climatic map of European Russia (published eventually in 1851).[25]

It is not clear how well Graff was informed about this project, yet his own pursuits were certainly in line with the MSD's recent interest in meteorological observations. As early as 1844, Graff requested that the MSD send him some instruments

---

[22] See, e.g., RGIA, f. 387, op. 1, d. 465, ll. 56–203 ("On measures to promote the afforestation of the southern Russia", March 15, 1841, see esp. ll. 174–75).

[23] Andreassian (2004).

[24] Pasetskii (1984).

[25] For details, see Loskutova (2012a).

for meteorological observations.[26] However, it is by no means clear exactly what instruments he received from the ministry in response to his request, and when he received them. From other cases, we know that the making of instruments and their subsequent delivery to remote provincial places on famously poor roads were a very demanding and time-consuming task that could take more than a year. By 1845, Graff had definitely received at least a maximum thermometer, yet it took almost 10 years before he would finally obtain a barometer and a rain gauge. In 1847, the question of meteorological observations in Velikii Anadol' emerged once again.[27] This time the MSD emphatically endorsed the idea and authorized the construction of a meteorological observatory in the forestry district. In order to have a trained observer at Velikii Anadol', academician Kupffer, who supervised the making of the instruments, was also requested to mentor a graduate of the forestry school. And indeed, some sources suggest that Viktor Graff was sending meteorological data from Velikii Anadol' to the Main Physical Observatory in St. Petersburg as early as in 1847.[28] Unfortunately, this information is not supported by the *Meteorological Review of Russia,* a periodical that Kupffer published from 1850 that listed data from his observation stations across the empire. This discrepancy might easily be explained, however, by different observational standards employed by Kupffer and Graff.

From his informants, Kupffer requested precise instrumental measurements of air temperature and atmospheric pressure, which were to be taken several times a day at specified hours. Many local observers fell short of his rigorous standards: They sent in descriptive everyday accounts of the weather that highlighted extraordinary occurrences at particular places, while often experiencing problems with providing exact statistical averages that fit in with a contemporaneous scientific discourse aimed at exploring global regularities. As a few other cases clearly indicate, Kupffer's instructions were not sufficient to produce standardized observations. His observers, who were mostly civil servants, physicians, and secondary schoolteachers, had not only to learn how to handle their instruments but also to understand the rationale behind his project. Indeed, numerous articles published by the MSD in its official journal from the 1840s through the early 1850s served precisely this purpose: They explained basic concepts of Humboldtian meteorology and the ways in which local data could contribute to a better understanding of Russia's climatic place within Europe.

Graff's meteorological observations from Velikii Anadol' were fairly typical of those generated in the 1840s by the MSD's network of provincial observers. Despite his vast responsibilities as the head of the forestry district, he managed to find time to produce a few papers on local nature and on his efforts at steppe afforestation.[29]

---

[26] RGIA, f. 387, op. 1, d. 10415, l. 224 (a letter from the head of Ekaterinoslav chamber of state domains to his superiors at the MSD in St. Petersburg, May 1, 1844).

[27] RGIA, f. 387, op. 2, d. 22244.

[28] Red'ko and Red'ko (2002).

[29] Graff (1850, 1855). RGIA archival collections also contain comprehensive reports written by Graff.

These papers contained information about "periodical phenomena" in living nature and a few observations on local climate and weather. He recorded extreme air temperatures, average monthly temperatures and wind direction, as well as the number of clear and overcast days in a month. Concerning rainfall, wind strength, humidity, and groundwater, Graff provided only the most general descriptions. "The spring of 1848 was disastrously dry," "there were two small and one decent rain and a hail in April, 1848," and autumn rains "soaked the soil pretty well."[30]

Velikii Anadol' in the Great Reforms Era and "Cultivation Experiments" of the 1870s

By the mid-1850s, the Velikii Anadol forestry district, which already encompassed more than 50 *desiatina* (about 0.21 square miles) of forested land, could finally boast a meteorological observatory housed in a special building. Yet the next decade did not show the substantial progress in meteorological observations that one might have expected; it was in fact the most difficult period in the nineteenth-century history of the forestry district. With the defeat in the Crimean War and the death of Nicholas I, Russia entered a new era known as the "Great Reforms" period, with the abolition of serfdom (1861) as its most important achievement. In the 1860s, the MSD dramatically changed its policies by adopting a laissez-faire approach to forestry, as well as to most other areas in which it had previously been active. Historians of forestry in Russia usually assume that afforestation work in Velikii Anadol' was considerably reduced in scope because the station lost its principal source of labor with the abolition of serfdom. However, a detailed analysis of Graff's reports suggests that even in the pre-reform period, it relied mostly on students of its forestry school to carry out most of the work. Indeed, their number was on increase in the late 1850s and early 1860s, reaching its peak in the early 1860s when the school had about 120 boys. It was the school closure and not the abolition of serfdom that dealt the hardest blow to the forestry district. Yet the MSD had its own rationale for making this move.

First of all, by the 1860s, the forestry department of the MSD assumed that the Velikii Anadol' plantation had achieved its objectives. It had demonstrated that even Russians, and not just German colonists, could grow forests in the steppe zone. Secondly, after the abolition of serfdom, the MSD was relieved from guardianship over state peasants. As a result, peasant education was no longer an issue on the ministry's agenda. The MSD was also discharged from "promoting forest culture in the country in general and assisting private forest owners."[31] Because of this change, the Forestry Department refused to continue supporting the forestry school in Velikii Anadol'. Instead, the ministry concentrated its efforts on promot-

[30] Graff (1850, 1855). Apart from meteorological data, Graff published the results of his phenological observations on cultivated and wild plants.

[31] RGIA, f. 387, op. 3, d. 24744, l. 19.

ing higher education. Attempts were made to transfer the school to local elected authorities (*zemstvo*), but local society failed to raise money for this purpose. In 1863, the school admitted its last group of students, and in 1866, the last graduates left the school.[32]

The MSD cut spending on the Velikii Anadol' forestry district to one third of previous levels; personnel was reduced from 38 men to 12. Yet a new head of the district, Ludwig Bark (1835–1882),[33] was too stubborn to give up afforestation and, given the reduction in funding, started experimenting with cheaper methods. He was quite successful, and the areas under afforestation experienced no reduction in size. Bark arrived at the station in 1862, 4 years prior to Graff's departure from Velikii Anadol'.[34] Therefore, he had enough time to learn Graff's methods of afforestation and to get to know the local environment.

In 1872, Bark made his first test felling at the station. Most likely, he acted upon the request of the Forest Society in St. Petersburg. The results were supposed to be presented at the Polytechnic Exhibition in Moscow.[35] From these results, Bark concluded that steppe afforestation could be a lucrative business. In fact, his calculations of expected revenues may have been misleading. He was felling trees that had been grown by Graff, who used highly complex and expensive techniques of afforestation, but he based his financial calculations on his own much simpler and cheaper methods. He assumed that his own plantations would be no less successful than those created by his predecessor, yet the future demonstrated that this assumption was mistaken.

In the early 1870s, however, the MSD was impressed by Bark's arguments. At that time, ministerial policies changed again, following a broader pattern of Russian domestic politics. The MSD resumed its interventionist approach, but was now most interested in maximizing profits. The Forestry Department increased Velikii Anadol's budget and enabled Bark to expand the territory of new plantations. Already in 1873, he planted 70 *desiatina* (about 0.3 square miles) of land instead of the 10 planted in the previous year. Subsequently, the figures increased up to 100 or even 150 *desiatina* per year. At the same time, the officials at the Forestry Department, after a decade of indifference and neglect, realized that Bark was one of very few specialists who knew how to plant forests in the steppe zone.[36] As a result, dissemination of his knowledge and skills became a part of Bark's job. Young graduates of the Forestry Institute in St. Petersburg and Petrovskaia Agricultural Academy in Moscow were sent to Velikii Anadol' for training. A newly established

---

32 RGIA, f. 387, op. 3, dd. 24743–24745.

33 For his biography, see Red'ko (1994).

34 In 1866, he became the professor of forestry in Moscow Agricultural Academy.

35 The Forestry Society in St. Petersburg was established just before these events. It was very active in arranging the forestry section of the exhibition: see [1871–1878] "Izvestiia o deiatel'nosti Lesnogo obshchestva," 1871. The 1872 exhibition was an important public event; its exhibits formed the basis for the Polytechnic and Historical Museums in Moscow. See [1872] *Obshchee obozrenie*; [1922] *Piatidesiatiletie Politekhnicheskogo muzeia v Moskve*. For the history of the Forestry Society, see Beilin (1962).

36 RGIA, f. 387, op. 3, d. 24744, l. 334–343.

Lesnoi Zhurnal (*Forestry Journal*) and other periodicals began to publish papers, in which they described the success of the forestry district. Afforestation again appeared on the agenda of various institutions. Other state forestry districts in the steppe zone became engaged in afforestation.[37] By the late 1870s, in Ekaterinoslav province alone, five forestry districts expanded their afforested territories by 400 *desiatina* within a year. In the Kherson province, 11 forestry districts planted about 350 *desiatina*.[38] Apart from the MSD, the Don and Ural Cossack Military Administrations started promoting afforestation in their territories.[39]

In the autumn of 1874, Bark submitted to the Forestry Department his "Program of experiments that are proposed on afforestation and exploitation of planted forests in the Velikii Anadol' Model Steppe Forestry District." His program can best be described as a series of "cultivation experiments"; this expression was employed by some Russian agronomists in the 1870s, 1880s, and 1890s in order to distinguish trial-and-error tests like those of Bark from scientific experiments that involved a clearly defined initial hypothesis, control plots, and mandatory record-keeping. Bark, furthermore, did not conceive of his experiments as a scientific endeavor. He was merely looking for ways to improve techniques of afforestation; he wanted to make them cheaper, less labor intensive, and more productive at the same time. Like landowners in the region and elsewhere, he proceeded by trial and error. At the end of a "cultivation experiment," a farmer or a forester would work out guidelines to follow in the future, but he would not be able to explain the causes of his success or to predict if the same methods would work in a different environment.[40]

In the summer of 1876, the Forestry Department supported Bark's proposals; it only stipulated that the experiments should take place "on sites that are representative of different soil types and different locations" and recommended the keeping of detailed records.[41] We have so far failed to discover the exact guidelines for the selection of sites and the recording of experiments. Perhaps these guidelines never existed in the first place. It seems that in a few cases, Bark had something like control plots, yet it is by no means clear how careful he was in meeting this essential requirement.

We can safely say, therefore, that in the 1870s, no experiments that would match the standards of the late nineteenth-century life and earth sciences were conducted at the Velikii Anadol' forestry district. Indeed, it was not an objective of the station and Bark never positioned himself as belonging to the community of academic naturalists. Unlike Graff, who was a member of several naturalists' societies in Russia and Europe, Bark never joined such groups, and identified himself only with

[37] See, e.g., Poletaev (1878) and Kvest (1878).

[38] Bark (1880).

[39] For afforestation in the forestry districts of the Don Cossack Military Region, see Turskii (1884) and Reviako et al. (2004). On the project to establish a model forestry district in the Ural Cossack Military Region, see RGIA, f. 387, op. 25, d. 77, ll. 128–133.

[40] On "cultivation experiments" and the need to start full-scale scientific experiments, see, e.g., Izmail'skii (1893).

[41] RGIA, f. 387, op. 3, no. 24744, ll. 228–245, 246, 334–343, 351–365.

the foresters' corps. Significantly, publications from this period that concerned the station provided virtually no botanical or entomological data, and presented very cursory descriptions of local soils and climate. They were principally focused on afforestation techniques and the expenses involved.[42]

Given the considerable progress made by academic research in Russian universities in the 1860s and 1870s, the epistemic and social distance between forestry and the life sciences substantially increased in cases where the practice of forestry remained tied to its earlier foundations in the *Kameralwissenschaft*. Yet from the late 1860s and early 1870s, both in Russia and in other countries of continental Europe, we begin to discern an opposite trend within forestry, as leading practitioners began to champion the new experimental methods of the life sciences. In the next section, we will see how this trend affected the Velikii Anadol' forestry district and its research.

The "Forest Question" in Russian Public Opinion and the Idea of Observation Networks in the 1870s

In the liberal climate of the Great Reforms era, the "forest question"—large-scale destruction of forests in the country and its detrimental impact upon climate and human welfare—emerged in the pages of newspapers and nonspecialized journals.[43] Most authors were convinced that forests did improve climate, regardless of any experiential evidence to the contrary. This view certainly became a dogma in the foresters' corps, although the most committed and outspoken dissenting voices also came from this group.

In this respect, Bark was no different from the majority of his colleagues when in 1872 he reflected on his forestry district and its achievements. He acknowledged that so far the Velikii Anadol' plantation had failed to meet one of its principal objectives: the improvement of climate in the region. Yet he ascribed this failure entirely to the modest size of the forest plantation—161 *desiatina* of land.[44] He believed that scientists had proven in principle that forests impacted the environment. Forests created barriers for dry winds and wind erosion, they diminished sharp oscillations of air temperatures, slowed down surface runoff, and therefore facilitated soil's absorption of melted snow in spring and summer rains. Bark assumed that forests were also beneficial for human and animal health; they provided shade, ozonized fresh air, and purged air "from excessive harmful carbonic acid."[45] For Bark, the only problem was that all of these factors were poorly researched. As a result, the arguments in favor of afforestation did not sound convincing to those who were scared by the expenses and technical difficulties of planting trees in the steppe.

[42] Permskii (1876) and Orlukovich (1880).

[43] Moon (2010) and Costlow (2003).

[44] Bark (1872).

[45] Bark (1872, p. 86).

When Bark was expressing these ideas in the early 1870s, the Forestry Society, which had just been established in St. Petersburg, started discussing the idea of improving experimental research in this field. A new stimulus came from an International Congress of Agriculture and Forestry in Vienna in 1873, which called for international cooperation in experimental research on the climatic impact of forests and recommended the establishment of permanent observation stations.[46]

In the late 1860s and 1870s, German specialists in scientific forestry began redefining their discipline's relationship to the natural sciences; this redefinition altered the field's epistemic foundations and its institutional infrastructure. Meteorology provided German foresters with a role model for experimental research, and they began establishing a network of experimental forestry stations, veritable scientific laboratories set up in the woods. In 1868, the Bavarian Forestry Department created eight stations for experimental forestry. Ernst Ebermayer (1829–1908), professor of forestry in Aschaffenburg, supervised the network and analyzed the data collected by forest officers daily at these stations.[47] This network carried out the first long-term series of observations that focused on the interrelations between forests and their natural environment: air and soil temperatures, the moisture and ozone concentration in the air, snowfall and precipitation, and wind force and direction. It also conducted one of the first experiments that measured these variables simultaneously both in the woods and in open fields.[48]

Russian foresters were well informed about these and earlier experiments in Western Europe. In the 1870s, the *Lesnoi Zhurnal* (*Forestry Journal*) published a whole series of reviews devoted to experiments made by German and French specialists.[49] The Forestry Society enthusiastically embraced the idea of instrumental observations and experiments carried out by a network of research stations based in the woods. It began discussing research programs, organizational issues, and funding for these stations, and the Forestry Department gave its official blessing to this initiative.[50]

In those years, the idea of a network of experiment stations was very much in the air in Russia: The congresses of Russian agriculturalists debated the establishment of experiment stations in agronomy and the Russian Entomological Society considered setting up entomological experiment stations.[51] These debates arguably

---

[46] Österreich (1873) and Krauze (1878)

[47] Hölzl (2010, pp. 431–460, esp. 452–453).

[48] The results obtained by Ebermayer suggested that forests increased the humidity of air and soil, thus giving evidential support to earlier assumptions. Later on, this assumption was proven to be wrong: Ebermayer's experiments were focused on comparing relative humidity of air under forest canopy and in an open space, but they did not take into account the amount of water evaporated by plants.

[49] Rudzkii (1873) and Kravchinskii (1876) and [1877] "Muettrich Jahresbericht ueber die Beobachtungs."

[50] Shafranov (1876a) and Sobichevskii (1876) and Shafranov (1876b) and Red'ko and Red'ko (2002).

[51] Morozov (1875); St. Petersburg Branch of the Archive of the Russian Academy of Sciences (SPF ARAN), f. 724, op. 1, d. 80 (K. Gernet , "On the Establishment of Entomological Stations in Russia," 1873).

reflected a general trend visible in many applied disciplines that were linked to the life sciences. It was certainly related to an increasing fascination with the laboratory research methods that had come to prominence through the advancement of physiology.[52] Yet at the same time, it undoubtedly reflected a growing distance between professionally trained naturalists—university faculty, forestry specialists, etc.—and lay observers from various walks of life, who in earlier periods had supplied interested scholars with local data.

Experiment station networks required considerable organizational, technical, and human resources, and as a result, building them in Russia took several decades. In the 1870s, the Forestry Society was only able to send 14 sets of rain gauges to people who expressed their interest in carrying out observations on "the quantity of falling waters."[53] The logic behind this step is easily discernible: Organizing "pluviometric observations" was seen as a relatively easy task, and the influence of forests upon precipitation was perceived as the key factor in the debates about the environmental impact of forests. Iurii I. Morozov, a well-known professor of physics and meteorology at Khar'kov University, volunteered to process incoming data from the provinces. Yet after repeatedly making this promise in the pages of the *Forestry Journal,* he never published any results.[54] For a few years, a meteorological station functioned in the Lisinskii forestry district in St. Petersburg province. However, the observations conducted there had the opposite objective: They were aimed at exploring "the impact exercised by climatic phenomena on the growth rates of tree species and the results produced by forest improvement."[55]

Soil Science Enters the Debate

In the 1870s and 1880s, steppe areas under afforestation were expanding. The Forestry Department produced special guidelines for forestry districts in the steppe zone, but planting forests in the steppes was still a hard business that often failed to bring success. Not all foresters approved of the Forestry Department's guidelines, and one observer reported that trainee-foresters recruited from local populations achieved much better results than graduates of the St. Petersburg Forestry Institute. Yet it would be misleading to take such views as an argument for the superiority of "local knowledge." These criticisms came from a professional forester who experienced considerable difficulties in his early career in adapting knowledge he had acquired in St. Petersburg to the steppe environment, but in criticizing the For-

[52] Benson (2009, pp. 76–89, esp. 84–86).

[53] [1871–78] "Izvestiia o deiatel'nosti Lesnogo obshchestva" 1875 no. 1: 112–144; 1876 no. 1: 99–100. A few foresters, university, and agronomical schools' faculty members volunteered to carry out observations.

[54] [1871–1878] "Izvestiia o deiatel'nosti Lesnogo obshchestva" 1878 no. 2.

[55] Turskii and Shafranov (1876, p. 121 fn).

estry Institute, he was not suggesting that foresters should "go native." Instead, he emphasized the need to improve the teaching of natural science in the institute so that its graduates would be able to analyze the physiology of various tree species in order to see how they were affected by climate and soils. That would facilitate the development of standard guidelines for tending forest trees in ways appropriate to the specific local environment.[56]

In 1878, Bark himself realized that his program of "cultivation experiments" was insufficient and asked the Forestry Department to provide additional support for proper scientific observations at Velikii Anadol'. Being preoccupied with many other commitments, Bark selected just one variable to explore. He requested funds for ordering analytical balances and some other devices "for carrying out scientific research on humidity of different soils."[57] In other words, he intended to focus on the key factor that mattered for afforestation in the southern steppes, and his request definitely anticipated the future direction of research. In his letter to the Forestry Department, Bark explained that he had already contacted the Agricultural Academy in Moscow and the Forestry Institute in St. Petersburg for advice, but so far, he had received no reply to his questions.[58] The Forestry Department approved the expenses, but Bark soon left Velikii Anadol' and his plans probably remained unrealized.

As we see, by the late 1870s, many foresters, including Bark, were keen to establish closer contacts with the rapidly developing life sciences—an alliance that would transform forestry into a more scientific endeavor. By that time, meteorology was not the only discipline to which they could turn for a model. At the end of the 1870s, the new discipline of soil science was making rapid progress in Russia.[59] Soil science opened up new opportunities for research on the environmental impact of forest vegetation. The problems related to erecting expensive, well-equipped meteorological stations prompted naturalists, agronomists, and foresters to focus on the cumulative effect of meteorological phenomena instead of monitoring a whole range of factors. Scientists began to measure the humidity of soils and subsoils and their annual dynamics. This research program appeared spontaneously; many people came to it independently. Among them were scientists who are now considered part of the Dokuchaev school, but there were also others who were not directly related to Dokuchaev and his projects. Research in this new field emerged at the intersection of forestry and the natural sciences, and these developments were directly related to the history of Velikii Anadol' forestry district.

In the 1870s, Vasilii Dokuchaev, one of the founders of the discipline, had just begun his research on the black soils of southern Russia. His first seminal work, crucial to the formation of soil science, was *Russian Chernozem,* released in 1883.

[56] Poletaev (1878).

[57] Analytical balances were needed for establishing soil weight in its normal and in an absolutely dry state. This method was a standard way of measuring soil moisture.

[58] RGIA, f. 387, op. 3, d. 24745, l. 319.

[59] For a very general overview of the making of this discipline in Russia and the role of Dokuchaev, see Evtuhov (2006).

Another significant contribution, Pavel Kostychev's *Soils of the Chernozem Region of Russia,* appeared in 1886. In 1882, Alexander A. Izmail'skii (1851–1914),an agronomist who collaborated with Dokuchaev and his team, published the results of his own research, *Soil Moisture in Relation to Its Cultural Condition.* Later, in 1893–1894, he produced a more elaborate treatment of this problem.[60]

Izmail'skii's experience clearly demonstrated a number of problems that Bark would have encountered if he had conducted his own experiments measuring the annual dynamics of soil humidity.[61] Measuring soil humidity proved to be a very demanding task. Izmail'skii made a breakthrough in this field by managing to work out a consistent research program that precisely defined all the technical issues involved, from the choice of sites where soil samples were to be collected up to the final analysis of data. With the support of a few students, he was able to analyze about 6000 samples.[62] True, he carried out his observations on a remote steppe estate where he served as a manager. Yet he knew in advance that his research program was very time consuming, and he was careful to make a provision in his job contract that left him sufficient time for scientific research. Bark, as a head of the forestry station, would have had no spare time for regular observations of this kind.

Izmail'skii demonstrated that soil covered with vegetation kept snow cover longer and therefore in the spring was more humid than bare soil. In summer, soils covered with vegetation had higher evaporation rates than bare soils, but the latter ones eroded faster. Therefore, vegetation turned out to be the decisive factor in producing the desired net effect. Dokuchaev's Special Expedition adopted Izmail'skii's approach to observations on the annual dynamics of soil humidity in different natural and cultivated environments. Several other scientists used these methods as well, among them students of Dokuchaev like Georgii N. Vysotskii and Georgii F. Morozov. Dokuchaev incorporated Izmail'skii's results into his own concept of soil science, and his experiments also exercised a profound influence upon Russian agronomy, plant geography,[63] and experimental forestry.

Izmail'skii, however, carried out his observations in ploughed fields and steppe pastures. Fairly similar research on the environmental impact of forest plant communities in the steppe zone was carried out by Pavel Kostychev in Velikii Anadol'.[64] Kostychev's impact on the life sciences in general did not equal that of Dokuchaev, but in the field of soil science proper, his contribution was almost equally important. Unlike Dokuchaev, who was a university professor and an experienced field geologist, Kostychev graduated from the St. Petersburg Forestry Institute and was best known for his laboratory research.[65] Some of his works explored the rela-

[60] Izmail'skii (1894).

[61] See correspondence between Dokuchaev and Izmail'skii in Dokuchaev (1961).

[62] Izmail'skii (1894, p. 90).

[63] Similar observations were carried out by Andrei N. Krasnov, who was a plant geographer and a student of Dokuchaev, and who made important theoretical contribution to the debates on the "steppe question." See Krasnov (1892); see also Fedotova (2012).

[64] See Krasnov (1892).

[65] On him, see Krupenikov (1987).

tions between soils and vegetation. Already in the 1870s Kostychev consistently argued against a prevailing opinion that forests had covered the southern Russian steppes in antiquity[66]; his understanding of the correlations between soil types and vegetation convinced him that forests could not grow naturally on the steppe soils.[67] In the 1880s, in order to prove his ideas, Kostychev made a few research trips to the southern provinces and carried out a series of chemical and microbiological experiments that confirmed his hypothesis.[68]

In the spring of 1889, Kostychev, at that time an assistant professor at the Forestry Institute in St. Petersburg and a member of the Academic Committee at the MSD, suggested that the Forestry Department carry out systematic observations in Velikii Anadol' that would focus on soil humidity in forests and open spaces.[69] As he explained in the proposal, he was convinced that black soils were unfavorable environments for forest vegetation, yet he wished to explore the impact of artificially planted forests on the soils in this zone, evidently hoping that afforestation might have created a friendlier environment for forest vegetation. Kostychev predicted that every new generation of trees would grow better than a preceding one, since, with afforestation, soil humidity would increase.[70]

Kostychev's project was quite modest. He applied for 225 roubles to be spent on equipment (for a drying cabinet and analytical balances) and 25 roubles a year on fuel for the drying cabinet. As for the personnel on the ground, he hoped that one of his students could be sent to Velikii Anadol' to carry out observations. His proposal was approved by the Forestry Department, and in early 1891, Sergei Khramov, a graduate of the St. Petersburg Forestry Institute, went to Velikii Anadol'. Khramov proved to be a diligent researcher, and in the next year, he and Kostychev suggested that they expand the research program. They intended to carry out more detailed forest measurements and meteorological and hydrological observations in order to track environmental changes on the plantation.[71] Unfortunately for Kostychev, by summer 1892, Dokuchaev had chosen Velikii Anadol' as one of the experimental sites for his Special Expedition. Kostychev was on very strained terms with Dokuchaev, and therefore, in 1893, he decided to transfer his observations to the eastern parts of European Russia. Nevertheless, Kostychev's project demonstrates

[66] These views were primarily substantiated by general theories borrowed from natural sciences and on citations from ancient and medieval authors; see, e.g., Schleiden (1870) and Palimpsestov (1890).

[67] Kostychev (1876).

[68] Kostychev (1886).

[69] RGIA, f. 387, op. 5, d. 31378.

[70] Later, soil scientists came to a conclusion that the principal difference between forest and steppe soils was in the type of litter. In steppe soils, litter consists for the most part of rapidly degradable roots of herbaceous plants. In forest soils, litter is formed by those parts of plants that grow above the earth's surface.

[71] Khramov (1893) published his results in "O vlazhnosti pochvy v Veliko-Anadol'skom lesnichestve." His aim was to find out if soils were always more humid under any vegetation cover. He concluded from his observations that forest soils retained more snow in winter and water in spring, yet they evaporated more in summer, as compared to tilled fields.

that Dokuchaev's Special Expedition was not unique in its research design. Its objectives and program reflected ideas that were expressed by many of the naturalists and foresters who were increasingly adopting research methods developed by soil scientists.

Velikii Anadol' as an Observation Site for Dokuchaev's Special Expedition

In 1891–1892, a catastrophic drought followed by a famine prompted the MSD to support the "Special Expedition of the Forestry Department on testing and inventorying various methods and techniques of forestry and water management in the Russian steppes" led by Vasilii Dokuchaev. In 1892, the forestry district in Velikii Anadol' became one of the sites for the Special Expedition. The very name of the project emphasized its applied objectives. However, its organizational structure resembled an academic research institute.

Observations carried out by the Special Expedition had all the essential features of a scientific experiment. Before the observations started, its members studied in detail all of the major environmental parameters—local geology, soils, vegetation, etc.[72] From the very beginning, the Special Expedition set up six observation sites equipped with meteorological instruments: these sites were meant to be representatives of the three forestry districts in different provinces. For each forestry district, two observation sites were established: one was located under the forest canopy, the other on open ground. The Special Expedition employed not only university naturalists but also a sufficient number of highly qualified technical staff. Research output was promptly published in the *Proceedings* of the expedition.[73] The Special Expedition was very generously funded by the Forestry Department, which also provided support with logistics.

To some extent, the Special Expedition differed from Kostychev's and other projects not in its conceptual framework but in scale. However, the scale of the project had a decisive influence on its output. An ample budget (more than 40,000 roubles a year) enabled Dokuchaev to recruit highly qualified specialists and trained technical personnel. Some of these people had already had experience of working with Dokuchaev and some had just started to establish themselves in the academic world. Among them were the plant geographer Georgii N. Tanfil'ev, the meteorologist Nikolai P. Adamov, the soil scientist Nikolai M. Sibirtsev, the hydrologist Pavel V. Ototskii, and the forestry scientist Georgii N. Vysotskii, all of whom made substantial contributions to the advancement of their disciplines in Russia.

In 1898, the Special Expedition was dissolved and its experimental sites were converted into experimental forestry districts, the first such experimental forestry

---

[72] Sibirtsev and Dokuchaev (1893).

[73] [1894–1898] *Trudy Osoboi ekspeditsii*. This became the precursor for the first specialist journal in experimental forestry, *Trudy po opytnomu lesnomu delu*.

districts in Russia. Their network gradually expanded with the course of time, as did their research programs, so that by the eve of the First World War they published a substantial amount of data analysis. Both within the framework of the Special Expedition and in later years, these experimental sites were involved not only in applied forestry research but also in research that would today be considered part of forest ecology. In this earlier period, such work took place under the label of "a biological approach to forestry." Experimental forestry districts carried out research in ornithology, entomology, plant geography, plant physiology, soil science, hydrology, and meteorology.

The results produced by the Special Expedition in Velikii Anadol' summarized earlier research conducted by foresters, meteorologists, and soil scientists and, at the same time, dramatically changed the perspective on steppe afforestation. The Special Expedition demonstrated that "stable" artificial forest plantations in the steppe zone could be created only in certain localities (habitats), while the typical steppe environment was not humid enough for forests. It also established that forests affected only microclimates; they had no impact on the regional climate. Foresters had to abandon the idea of planting vast forest plantations in the steppe zone, opting instead for protective forest belts and small woods on those territories that were not suitable for agriculture. Members of the Dokuchaev team demonstrated—both within the framework of the Special Expedition and in later years—that forest vegetation, with its massive leaf area, evaporates more than any other type of vegetation, and much more than bare soil. Forests affect "climate" (or environment) not by exercising any influence upon rainfall but predominantly by reducing surface runoff and soil erosion.[74] Ideas and methods developed by the Special Expedition laid the foundation for later Soviet forest melioration projects.[75]

Conclusion

The history of the observations carried out at the Velikii Anadol' forestry district from the 1840s till the 1890s clearly illustrates a gradual transformation in the debates on the environmental impact of forests in Russia. Over these decades, the early nineteenth century's mode of natural historical research gave way to systematic experiments modeled on advances in the life and soil sciences. In the process, forestry as a discipline increasingly adopted sophisticated instrumental measurements and a research design borrowed from the exact sciences. Simultaneously, the very concept of "climate" underwent a profound change from a broad philosophical concept, which in the early nineteenth century encompassed a variety of natural and even moral phenomena, to a much more specific set of variables. If Viktor Graff's observations and experiments from the 1840s were treated as a remote precursor of modern environmental studies in Russia, they would certainly lack a number of

[74] Ototskii (1905) and Vysotskii (1938).

[75] See, e.g., Brain (2011) and Maslov (1999).

essential features required by later evidentiary standards. Yet we must remember that a totally different perspective informed his activities. His meteorological observations had little to do with testing correlations between air temperature, precipitation, and the growth of his forest plantation. Instead, he was trying to contribute to the MSD's effort to graphically represent the climatic diversity of Russia as a part of broader European patterns. Yet even as he took part in this ministerial endeavor, Graff evidently did not feel himself quite familiar with its research agenda: He was more inclined to stress the local and particular phenomena that mightily interfered with his daily labor on the plantation.

His immediate successor, Ludwig Bark, was even more focused on afforestation as a commercial enterprise. His "cultivation experiments" would have been familiar to any nineteenth-century agriculturist who wished to rationalize his business by adopting the most labor- and cost-efficient techniques of farming. For this reason, Bark was relatively isolated from the naturalists' community in Russia, which in the same decades came to be dominated by university faculty.

In the 1870s, however, Russian forestry specialists, following German professionals, began reorienting their discipline towards the life sciences by trying to adopt and adapt laboratory methods of experimental research. By the late 1870s and 1880s, the rise of a new discipline—genetic soil science—offered promising strategies for researching the environmental impact of forests and also dramatically changed the setting in which this research took place. Soil science helped to identify a set of clearly defined variables for instrumental measurements. Bark was evidently affected by these developments; by the end of his service in Velikii Anadol', he began contemplating experiments focused on measuring soil humidity.

The history of Dokuchaev's Special Expedition of the early 1890s can thus be seen not as a "foundation narrative" for Soviet environmentalism but as an episode in a long series of transformations in the practices and evidentiary standards of research on the climatic impact of forests. In many ways, the Special Expedition was no different from experiments and observations that had been carried out in Velikii Anadol' just a few years before its commencement. Yet a dramatic difference in the scale of these observations and experiments enabled Dokuchaev and his team to change the very understanding of relations between forests and climate.

Acknowledgments We would like to thank Mark Finlay, Kari Korhonen, Grigorii Isachenko, David Moon, and the editors of this volume, Sharon Kingsland and Denise Phillips, for their very helpful comments and suggestions.

References

1871–1878. Izvestiia o deiatel'nosti Lesnogo obshchestva [News about the activities of the Forestry Society], *Lesnoi Zhurnal* [Forestry journal] 1871, no. 2: 65–66, 1875, no.1: 112–144; 1876, no. 1: 99–100; 1878, no. 2: 59–76.

1872. Obshchee obozrenie Moskovskoi politekhnicheskoi vystavki [General overview of the Moscow Polytechnic Exhibition]. Moscow: Moskovskii Universitet.

1877. [Book review] Muettrich Jahresbericht ueber die Beobachtungs. *Lesnoi Zhurnal* (2): 120–122.

1894–98. *Trudy Osoboi ekspeditsii snariazhennoi Lesnym Departamentom pod rukovodstvom V. V. Dokuchaeva* [Proceedings of the Special Expedition dispatched by the Forestry Department under the supervision of V. V. Dokuchaev], vols. 1–4. St. Petersburg: Forestry Department.

1922. *Piatidesiatiletie Politekhnicheskogo muzeia v Moskve. 1872–1922. Kratkii istoricheskii ocherk* [The fiftieth anniversary of the Polytechnic Museum in Moscow. 1872–1922. A brief historical overview] Moscow: Tip. MKKh.

Andreassian, Vazken. 2004. Waters and forests: From historical controversy to scientific debate. *Journal of Hydrology* 291:1–27.

Bark, Ludwig G. 1872. Veliko-Anadol'skoe stepnoe lesnichestvo [Velikii Anadol Steppe Forestry District]. *Zapiski obshchestva sel'skogo khoziaistva Iuzhnoi Rossii* [Notes of the Agricultural Society of Southern Russia] 1 (2): 115–125; 2 (4): 78–87; 2 (5): 5–18.

Bark, Ludwig G. 1873. Stepnoe lesorazvedenie v Ekaterinoslavskoi gubernii [Steppe afforestation in the Ekaterinoslav Province]. *Lesnoi Zhurnal* [Forestry Journal] 5: 1–26.

Bark, Ludwig G. 1880. Lesorazvedenie na iuge Rossii [Afforestation in Southern Russia]. *Sel'skoe khoziaistvo i lesovodstvo* [Agriculture and Forestry] 133: 313–329.

Beilin, Isaak G. 1962. *Ocherki po istorii lesnykh obshchestv dorevoliutsionnoi Rossii* [Essays on the history of forest societies of pre-revolutionary Russia]. Moscow: Goslesbumizdat.

Benson, Keith R. 2009. Field stations and surveys. In *The Cambridge history of science,* ed. Peter J. Bowler, John V. Pickstone, vol. 6, 76–89. Cambridge: Cambridge University Press.

Bozherianov, Ivan N. 1897. *Graf Egor Frantsevich Kankrin, ego zhizn', literaturnye trudy i dvadtsatiletniaia deiatel'nost' upravleniia Ministerstvom finansov* [Count Egor Frantsevich Cancrin, his life, literary works and the twenty years of his administration of the Ministry of Finance]. St. Petersburg: Izdatel'stvo grafa I. V. Kankrina.

Brain, Stephen. 2011. *Song of the forest: Russian forestry and Stalinist environmentalism, 1905–1953*. Pittsburgh: University of Pittsburgh Press.

Brandes, Detlef. 1993. *Von den Zaren adoptiert. Die deutschen Kolonisten und die Balkansiedler in Neurussland und Bessarabien 1751–1914*. Muenchen: R. Oldenbourg.

Costlow, Jane. 2003. Imaginations of destruction: The 'forest question' in nineteenth-century Russian culture. *Russian Review* 62:91–118.

Dokuchaev, Vasilii V. 1961. *Sochineniia* [Works]. 8 vols. Moscow: Izd-vo AN SSSR.

Epp, David H. 1946. *Johann Cornies. Züge aus seinem Leben und Wirken.* Steinbach: Echo.

Evtuhov, Catherine. 2006. The roots of Dokuchaev's scientific contributions. In *Footprints in the soil: People and ideas in soil history*, ed. Benno P. Warkentin, 125–148. Amsterdam: Elsevier.

Fedotova, Anastasia A. 2012. The study of steppe vegetation in European Russia (1850–1917): The making of geobotany. Ph.D. dissertation: St. Petersburg.

Filonenko, Ivan E. 2000. *Osobaia Ekspeditsiia* [The Special Expedition]. Moscow: Prima-Press-M.

Graff, Viktor E. 1850. Nabliudeniia, proizvedennye na Veliko-Anadol'skoi plantatsii (v Ekaterinoslavskoi gubernii) nad vliianiem klimaticheskikh peremen na rastitel'nost' voobshche i na zhivotnykh v 1848 g. [Observations made at Velikii Anadol' plantations (in Ekaterinoslav Province) about the effect of climate change on vegetation in general and animals in 1848]. *Lesnoi Zhurnal* [Forestry Journal] (36): 283–287; (37): 290–295; (38): 297–299.

Graff, Viktor E. 1855. Neblagopriiatnye vliianiia 1853 g. na Veliko-Anadol'skuiu plantatsiiu i na rastitel'nost' voobshche [Unfavorable impacts during 1853 on Velikii Anadol' plantations and vegetation in general]. *Trudy Vol'nogo ekonomicheskogo obshchestva* [Proceedings of the Free Economic Society] 3 (8): 33–69; 2nd pagination.

Grove, Richard H. 1995. *Green imperialism: Colonial expansion, tropical island Edens and the origins of environmentalism, 1600-1860.* Cambridge: Cambridge University Press.

Hölzl, Richard. 2010. Historicizing sustainability: German scientific forestry in the eighteenth and nineteenth centuries. *Science as Culture* 19:431–460.

Izmail'skii, Alexander A. 1893. Znachenie kul'turnykh opytov v sel'skom khoziaistve [The value of cultivation experimentation in agriculture]. *Vestnik russkogo sel'skogo khoziaistva* [Bulletin of Russian agriculture] 30:22–35.

Izmail'skii, Alexander A. 1894. *Vlazhnost' pochvy i gruntovaia voda v sviazi s rel'efom mestnosti i kul'turnym sostoianiem poverkhnosti pochvy. Rezul'taty issledovanii vlazhnosti pochvy v Poltavskom uezde s 1886 po 1893 g.* [Soil moisture and ground water relative to the topography and cultivation state of the surface of the soil. The results of the study of soil moisture in the Poltava County from 1886 to 1893]. Poltava: Tip. L. Frishberga.

Khramov, Sergei. 1893. O vlazhnosti pochvy v Veliko-Anadol'skom lesnichestve [About the soil moisture in Velikii Anadol' Forestry District]. *Lesnoi Zhurnal* (2): 140–146.

Köppen, P. von. 1841. Über den Wald- und Wasser-Vorrath im Gebiete der obern und mittlern Wolga. Ein Bericht an die Commission zur Untersuchung der Frage über den Einfluss der Verminderung der Wälder auf die Verminderung des Wassers in der obern Wolga von P. v. Köppen, Mitgliede der Akademie der Wissenschaften. Mit einem Vorworte des Herausgebers. *Beiträge zur Kenntnis des Russischen Reiches* 4:165–253.

Kostychev, Pavel A. 1876. Sposobstvuet li razvedenie lesov unichtozheniiu zasukh? [Could afforestation reduce droughts?]. *Otechestvennye Zapiski* 3:1–33; 2nd pagination.

Kostychev, Pavel A. 1886. *Pochvy chernozemnoi oblasti Rossii, ikh proiskhozhdenie, sostav i svoistva* [Black earth soils of Russia, their origin, composition and properties]. St. Petersburg: A. F. Devrien.

Kovda, Viktor A., and E. M. Samoĭlova, eds. 1983. *Russkii chernozem. 100 let posle Dokuchaeva* [Russian black soils. 100 years after Dokuchaev]. Moscow: Nauka.

Krasnov, Andrei N. 1892. Geo-biologicheskie nabliudeniia nad pochvami Khar'kovskoi gubernii [Geo-biological observations of the soil in Kharkov Province]. *Zhurnal Khar'kovskogo obshchestva sel'skogo khoziaistva za 1891 g.* [Journal of Kharkov Agricultural Society for 1891] 2:57–64 (supplement).

Krauze, E. 1878. Napravlenie i organizatsiia lesnogo opytnogo dela [The aims and organization of forestry experimentation]. *Zapiski Obshchestva sel'skogo khoziaistva Iuzhnoi Rossii* [Proceedings of the Agricultural Society of Southern Russia] 2:88–106, 128–143.

Kravchinskii, Dmitrii. 1876. Vopros o vliianii lesa na klimat [The influence of forests on climate]. *Lesnoi Zhurnal* (6): 25–54.

Krupenikov, Igor' A. 1987. *Pavel Andreevich Kostychev. 1845–1895.* Moscow: Nauka.

Kvest, G. 1878. Lesorazvedenie v Chernolesskoi dache [Afforestation in Chernolesskii Forestry District]. *Lesnoi Zhurnal* (4): 179–192.

Loskutova, Marina. 2012a. 'Svedeniia o klimate, pochve, obraze khoziaistva i gospodstvuiushchikh rasteniiakh dolzhny byt' sobrany…': prosveshchennaia biurokratiia, gumbol'dtovskaia nauka i mestnoe znanie v Rossiiskoi imperii vtoroi chetverti XIX v. ['Information about climate, soils, ways of economy, and dominant plants must be collected…': Enlightened bureaucrats, Humboldtian science and local knowledge in the Russian empire ca. 1830s-1850s]. *Ab Imperio* 4:111–156.

Loskutova, Marina. 2012b. 'Vliianie lesov na obmelenie rek est' tol'ko nedokazannaia ipoteza': prikladnaia nauka i gosudarstvennaia politika po upravleniiu prirodnymi resursami v Rossiiskoi imperii vtoroi chetverti XIX v. ['The influence of forests on river shallowing is just an unsubstantiated hypothesis': Applied research and governmental policies in forest management in the Russian empire in the 1830s–1840s]. *Studies in the History of Biology* 4 (1): 9–32.

Maslov, Boris S. 1999. *Ocherki po istorii melioratsii v Rossii.* Moscow: Meliovodinform.

Mironov, Boris. 2012. *The standard of living and revolutions in Russia, 1700–1917.* ed. Gregory L. Freeze. Abingdon: Routledge.

Moon, David. 2010. The debate over climate change in the steppe region in nineteenth-century Russia. *Russian Review* 69:251–75.

Moon, David. 2013. *The plough that broke the steppes: Agriculture and environment on Russia's grasslands, 1700–1914.* Oxford: Oxford University Press.

Mordkovich, V. G., A. M. Giliarov, A. A. Tishkov, and S. A. Balandin. 1997. *Sud'ba stepei* [The fate of the Steppes]. Novosibirsk: Mangazeia.

Morozov, Iurii I. 1875. O neobkhodimosti ustroistva sel'skokhoziaistvennykh meteorologicheskikh stantsii v Rossii [The need for agricultural meteorological stations in Russia]. In *Trudy IV*

s'ezda russkikh sel'skikh khoziaev v Khar'kove v dekabre 1874 g [Proceedings of the IV Congress of Russian landowners in Kharkov in December 1874]. pp. 117–139. Odessa.
Österreich Ackerbau-Ministerium. 1873. *Verhandlungen des Agrarischen Congresses. Nach den stenographischen Protokollen zusammengestellt im k. k. Ackerbau-Ministerium*. Wien: Hof- und Staatsdruckerei.
Orlukovich, R. 1880. O lesorazvedenii v Ekaterinoslavskoi gubernii [About afforestation in the Ekaterinoslav Province]. *Lesnoi Zhurnal* (5): 287–323.
Ototskii, Pavel V. 1905. *Gruntovye vody, ikh proiskhozhdenie, zhizn' i rasprostranenie* [Ground waters, their origin, life and distribution]. St. Petersburg: Gerold.
Palimpsestov, Ivan U. 1890. Stepi iuga Rossii byli li iskone vekov stepiami i vozmozhno li oblesit' ikh? [Were the steppes of Southern Russia always steppes, and is it possible to afforest them?]. Odessa.
Pasetskii, Vasilii M. 1984. *Adol'f Iakovlevich Kupfer. 1799–1865*. Moscow: Nauka.
Permskii, A. 1876. Nyneshnie kul'tury v Veliko-Anadol'skom stepnom lesnichestve [The current plantations in the Velikii Anadol' Steppe Forestry District]. *Lesnoi Zhurnal* (1): 1–14, (2): 1–22.
Poletaev, I.S. 1878. O rezul'tatakh iskusstvennogo lesorazvedeniia v Novorossiiskikh stepiakh posevom [The results of afforestation by sowing in the Novorossia steppes]. *Lesnoi Zhurnal* (2): 68–75.
Polianskii, Khristofor F. 1888. Opisanie Veliko-Anadol'skogo obraztsovogo lesnichestva Ekaterinoslavskoi gubernii [Description of Velikii Anadol' model forestry district in Ekaterinoslav province]. *Lesnoi Zhurnal* [Forestry Journal] (5): 728–740.
Rajan, Ravi S. 2006. *Modernizing nature: Forestry and imperial eco-development*. Oxford: Clarendon Press.
Red'ko, Georgii I. 1994. *Polkovnik Korpusa lesnichikh* [A colonel of the Forestry Corps]. Kiev: Ukrainian Ministry of Forestry.
Red'ko, Georgii I., and N. G. Red'ko. 2002. *Istoriia lesnogo khoziaistva Rossii* [The history of forestry in Russia]. St. Petersburg: MGUL.
Red'ko, Georgii I., and N. G. Red'ko. 2003. *Lesnoe khoziaistvo Rossii v zhizneopisanii ego vydaiushchikhsia deiatelei* [Russian forestry in the biographies of its eminent personalities]. Moscow: MGUL.
Reviako, I. V., A. A. Kulygin, I. I. Reviako, and A. M. Lepshov. 2004. K istorii stepnogo lesorazvedeniia na Donu [On the history of steppe afforestation in the Don Region]. *Lesnoi Zhurnal* (2): 121–122.
Rudzkii, Alexander. 1873. O vliianii lesov na klimat: po povodu sochineniia d-ra Ebermaiera. Bibliograficheskoe obozrenie [On the influence of forests on climate: About Dr. Ebermayer's work. Bibliographical review]. *Lesnoi Zhurnal* (2): 81–113.
Schleiden, Matthias J. 1870. *Für Baum und Wald. Eine Schutzschrift an Fachmaenner und Laien gericht*. Leipzig: Engelmann.
Shafranov, Nikolai S. 1876a. K voprosu ob ustroistve v Rossii opytnykh lesnykh stantsii [On the organization of experimental forestry stations in Russia]. *Lesnoi Zhurnal* (2): 86–94.
Shafranov, Nikolai S. 1876b. Eshche k voprosu ob ustroistve v Rossii opytnykh lesnykh stantsii [More on the organization of experimental forestry stations in Russia]. *Lesnoi Zhurnal* (6): 55–57.
Sibirtsev, Nikolai M., and Vasilii V. Dokuchaev. 1893. *Osobaia ekspeditsiia Lesnogo departamenta po ispytaniiu i uchetu razlichnykh sposobov i priemov lesnogo i vodnogo khoziaistva v stepiakh Rossii. Predvaritel'nyi otchet o deiatel'nosti ekspeditsii s iiunia po noiabr' 1892 g. i obshchii proekt opytnykh rabot ee* [Special expedition of the Forestry Department on testing and inventorying various methods and techniques of forestry and water management in the Russian Steppes. A preliminary report on the activities of the expedition in June-November, 1892, and the overall project of experimentation]. St. Petersburg.
Sobichevskii, Vasilii T. 1876. K voprosu ob ustroistve v Rossii opytnykh lesnykh stantsii [On the question of the organization of experimental forestry stations in Russia]. *Lesnoi Zhurnal* (5): 47–56.
Sunderland, Willard. 2004. *Taming the wild steppe: Colonization and empire on the Russian steppe*. Ithaca: Cornell University Press.

Tsvetkov, Mikhail A. 1957. *Izmenenie lesistosti Evropeiskoi Rossii s kontsa XVII stoletiia po 1914 g.* [Change in the forest cover of European Russia since the end of the 17th century until 1914]. Moscow: Izdatel'stvo AN SSSR.

Turskii, Mitrofan K. 1884. Iz poezdki v lesa Srednei i Iuzhnoi Rossii [About a trip to the forests of Central and Southern Russia]. *Lesnoi Zhurnal* (5–6): 289–325.

Turskii, Mitrofan, and Nikolai Shafranov. 1876. K tablitsam meteorologicheskikh nabliudenii, proizvodimykh na Lisinskoi uchebnoi dache [On the tables of meteorological observations taken in Lisinskii Training Forestry District]. *Lesnoi Zhurnal* (3): 121–125.

Vysotskii, Georgii N. 1938. *O gidrologicheskom i meteorologicheskom vliianii lesov* [On hydrological and meteorological effects of forests]. Moscow: Gosudarstvennoe lesotekhnicheskoe izdatel'stvo.

Weigl, Engelhard. 2004. Wald und Klima: Ein Mythos aus dem 19. Jahrhundert. *Humboldt im Netz. International Review for Humboldtian Studies* 9:2–20.

Anastasia A. Fedotova is a senior researcher at the St. Petersburg Branch of the Institute for the History of Science and Technology, Russian Academy of Sciences. She received her Ph.D. in the history of plant ecology in Russia. Her research interests include the role of scientific knowledge in the rationalization of agriculture and forestry and history of applied biology from mid-nineteenth to mid-twentieth century in the Russian Empire and the USSR.

Marina V. Loskutova is an associate professor at the History Department at the National Research University-Higher School of Economics, St. Petersburg campus, Russia. She works on the history of exploration of natural resources and the interaction between academic communities, state administration, and the wider public in the nineteenth-century Russian Empire.

Chapter 8
The Rise of Applied Entomology in the Russian Empire: Governmental, Public, and Academic Responses to Insect Pest Outbreaks from 1840 to 1894

Marina V. Loskutova and Anastasia A. Fedotova

Introduction

The early history of applied entomology, prior to the widespread use of synthetic insecticides, has conventionally been written through the prism of its institutionalization and professionalization. The creation of special offices for applied entomologists within the state administration and the establishment of positions for them in colleges and universities signified a major shift from self-supported amateurs to professional investigators. This shift was also of profound significance for the discipline's conceptual concerns and its increasing emphasis on practical applications. The new personnel who came to staff these positions were strongly motivated to reorient the research from the collection of rare, exotic species and the description of their geographical distribution to the study of the physiological and adaptive responses of common species, the structure of their communities, and their life cycles.[1]

North America, and the USA in particular have been justly perceived as the region that set the trend for other countries to follow. American predominance in agricultural entomology in the nineteenth and early twentieth centuries has been explained by the early commercialization and mechanization of agriculture and horticulture in a moving frontier zone. American conditions encouraged the practice of monoculture, which in turn created a favorable environment for the multiplication of insect

[1] See Palladino (1996, Chaps. 1–2), Sorensen (1995, Chap. 4), Castonguay (2004b, Chap. 1), and McWilliams (2008, Chaps. 2–4).

M. V. Loskutova (✉)
History Department, National Research University - Higher School of Economics, Promyshlennaia ul. 14a, 198099 St. Petersburg, Russia
e-mail: mvlosk@yandex.ru

A. A. Fedotova
Institute for the History of Science and Technology, RAS, Universiteskaya naberezhnaya 5/2, 199034 St. Petersburg, Russia
e-mail: f.anastasia.spb@gmail.com

D. Phillips, S. Kingsland (eds.), *New Perspectives on the History of Life Sciences and Agriculture,* Archimedes 40, DOI 10.1007/978-3-319-12185-7_8

pests. The large-scale expansion of cultivated areas also made it nearly impossible to use traditional methods of controlling insect pests by handpicking. In contrast, these methods remained in use much longer in many European countries where farms were smaller and labor resources more abundant. On the "demand" side, these were the factors that accounted for the emergence of applied entomology as a distinctive field of research. On the "supply" side was a growing community of college-educated naturalists who were keen to establish their authority and who were looking for employment opportunities. A decentralized, democratic political system that was flexible and provided ample room for public initiative initially served as a favorable milieu. Later, in the early decades of the twentieth century, the emergence of a new post-Darwinian conceptual perspective on nature that emphasized dynamic interrelations between organisms and their environment provided justifications for a more interventionist vision of the federal state and the centralization of applied entomological services.[2]

It is quite instructive that the history of agricultural entomology in Russia involves similar themes, even though the established account of the developments in this country was produced independently from Western scholarship. Indeed, in her book written half a century ago, Larisa Chesnova explicitly linked the rise of agricultural entomology to the spread of large-scale commercial farming in the southern steppe provinces, a rapidly developing "frontier" region in the nineteenth-century Russian empire. In the southern steppes of Russia, she also found a propensity for monoculture, which in turn made the region a hotbed for insect pest outbreaks.[3] Furthermore, she emphasized the role of public initiative, and in particular the activities of local elected authorities (*zemstvos*), in supporting the institutionalization and professionalization of agricultural entomology, a development that first took place in this peripheral area.

The striking parallels between the Russian and the Western scholarship on this topic become explicable when we recollect that early practitioners of applied entomology were themselves the first historians of their discipline.[4] They themselves constructed a narrative about the establishment of professional entomological services within their respective national frameworks. They were also well aware of the impetus given to the institutionalization and professionalization of their discipline by large-scale commercial farming in frontier areas where monoculture and insect pest outbreaks were widespread. Given that public support provided a growing portion of their income, early agricultural entomologists stressed its role in promoting professionalization; they also advertised their own civic commitments. Moreover, in the Russian case the argument about a strong connection between applied entomology and civic initiative acquired an additional dimension by becoming a part of a broader liberal discourse that explicitly contrasted such public activism with the inertia and ineptitude of the tsarist government.[5]

---

[2] Sorensen (1988) and Castonguay (2004a).

[3] Chesnova (1962).

[4] For example, Marchal (1896), Mokrzhetskii (1900), and Howard (1930).

[5] Blunt statements of this type can be found in both prerevolutionary and Soviet literature, e.g., Mokrzhetskii (1900) and Chesnova (1962).

The purpose of our essay is not to revise the conventional account but to alert scholars to a few important aspects that so far have remained on the margins of the story. The making of the discipline inextricably involved not just the emergence of a professional community equipped with specialized knowledge and practices. It required a stable flow of mass-produced field data that would match the standards of scientific research. The very number of salaried positions in the early days of the discipline's history suggests its strong reliance on field data supplied by outsiders to the profession, or "lay" collectors and observers, as they are called in the current historiography on the production and circulation of knowledge in the modern and early modern world.[6] Historians have only recently come to appreciate the importance of lay participation in science, and, more specifically, the importance of the scientific observations that lay participants produce. Scholarly interest in this topic arose from an increasing dissatisfaction with the rigid dichotomy between "amateurs" and "professionals" offered in straightforward accounts of professionalization.[7]

Moreover, while research on the circulation of knowledge initially suggested a vision of highly uneven, hierarchical relations in which leading scholars located in "centers of calculation" accumulated specimens and observations from distant locales, and then analyzed and codified these specimens and observations into authoritative knowledge,[8] some recent works have emphasized much more complex and reciprocal relations among the participants of these networks. Knowledge, it has been argued, flows in all directions, and certain local centers might sometimes be no less important than the more obvious focal points located in a few world capitals.[9] This observation may be particularly applicable to the history of the applied sciences, which presupposes a much broader array of protagonists involved in the production of knowledge. Indeed, the making of agricultural entomology should properly be understood not as an application of a preexisting "pure" form of the discipline to pragmatic concerns but as a joint endeavor of naturalists and agriculturalists aimed at devising efficient means of pest control—a task that required the integration of knowledge about insects (their taxonomy, life cycles, geography, and ultimately their ecology) with agricultural practices.

In this chapter, we examine the role of networks that connected naturalists and agriculturalists, academic scholars, state officials, and local public activists in the nineteenth-century Russian empire. These networks enabled the circulation of specimens, observations, research guidelines, and farming recommendations, thus ultimately leading to the advancement of knowledge about insect pests. Our analysis is limited to the early history of insect pest research in Russia: We start from the period in which certain branches of the Russian state administration began collecting information on insect pest outbreaks with an explicit aim of providing local

---

[6] For the recent discussion of the role of lay observers in the production of scientific knowledge, see, e.g., Naylor (2006), Güttler (2011), Brenna (2011), and Vetter (2011).

[7] See, e.g., Secord (1994), Alberti (2001), Desmond (2001), Allen (2009), and Lucier (2009).

[8] Latour (1987).

[9] For example, Terrall (2010).

authorities and landowners with academically sanctioned advice on pest control. Our overview stops in 1894, the year when the first professional entomological service (the Bureau for Applied Entomology at the Academic Committee of the Ministry of Agriculture and State Domains) was created in St. Petersburg to supervise insect pest control in the whole territory of the Russian empire.

Within this period, we trace several stages in the development of these networks, each with different principal agents and different agendas. In the 1840s and 1850s, when the central government in St. Petersburg first attempted a systematic collection of scientific data on insect pest outbreaks, the impetus came from the upper echelons of the Russian civil service, which was principally interested in demonstrating its own efficiency and commitment to modern, rational means of governance. Although the civil servants involved in this initiative were not lacking in academic expertise, they encountered serious problems when communicating with the provincial public that was expected to provide the necessary field observations.

In the late 1850s and 1860s, the Russian government withdrew from this field, ceding initiative to learned societies in St. Petersburg, namely to the recently created Russian Entomological Society (RES). In this chapter, we examine in detail why this group, which certainly did not lack academic credibility, failed to establish itself as the true center of insect pest research, with the result that in the 1870s and 1880s major events took place in the southern provinces of the country.

Finally, we examine a complex interaction between the capital of the empire and the provinces in the 1880s and 1890s. In this period, entomologists began to appreciate the problems involved in adjusting their knowledge of insect pests to the agricultural practices of a particular place, while agriculturalists gradually learned how to produce field data that would be compatible with the language of the life sciences.

Insect Pest Research in the Ministry of State Domains in the 1840s and Early 1850s

The first systematic collection of data on insect pest outbreaks began in the Russian empire in the late 1830s and 1840s when a newly established Ministry of State Domains (MSD) committed itself to this task. The ministry was founded in 1837 with the explicit aim of providing a more efficient management of the state land domains and a more active guardianship of the state peasants. The decision to create a special agency for this purpose came after several secret committees failed to recommend ways to modernize Russian agriculture and local administration in the countryside. As the government was still unprepared to risk the abolition of serfdom, the new ministry was meant to set an example for the nobility to follow. The MSD was expected to improve the economic position of peasants living on state land by systematizing and codifying the obligations between them and the state treasury, and administering their affairs more efficiently. The leadership of the new ministry, which was chaired by Count Pavel D. Kiselev, was firmly convinced that a detailed assessment of the country's natural, social, and economic conditions was a prerequisite for drafting workable administrative and legislative measures. In this way, systematic data collection on insect outbreaks was a part of a much broader interventionist policy pursued by the MSD in the 1840s and early 1850s.[10]

Within less than a year from its establishment, the ministry issued a circular that requested its regional chambers to report on insect outbreaks, providing descriptions of insects and, if possible, sending specimens to St. Petersburg.[11] The chambers were also instructed to report on adopted measures of pest control if these means had proven to be effective. It was hoped that information assembled in this way would form the basis for a reference book that could be used by local administrators, foresters, and landowners. The circular generated a steady stream of reports from various localities across the European part of the empire (with the exception of Finland and Russian Poland, which were exempt from the ministry's authority).[12]

Subsequent historians of nineteenth-century Russian science have tended to dismiss this and similar initiatives on the grounds that they were executed by the state bureaucracy instead of by "scientists," i.e., the university faculty. However, a detailed examination of research on insect pests in the 1840s and 1850s demonstrates that the MSD relied on the best academic advisors available in those days both in Russia and abroad. Admittedly, academician Peter Koeppen (1793–1864)—the man who in the early 1840s emerged as the principal "expert" of the MSD in this area—is largely known as a statistician and ethnographer.[13] Yet he was also an amateur naturalist and was on friendly terms with entomologists of international standing such as Christian von Steven (1781–1863), Arvid Hummel (1778–1836), or Johann Gotthelf Fischer von Waldheim (1771–1853). Crucially, Koeppen relied on professional consultations provided by Édouard Ménétries (1802–1861), a curator at the Zoological Museum in St. Petersburg and a former student of Cuvier and Latreille at the Museum of Natural History in Paris.[14]

The MSD was also able to capitalize on the close contacts that existed between the Forestry Institute in St. Petersburg (an institution directly subordinated to the ministry) and the Neustadt-Eberswalde Forst-Institut in Prussia. In the 1830s and 1840s, the Forestry Institute at Neustadt-Eberswalde emerged as the leading center of forest entomology in Europe, thanks to the efforts of its professor of natural history, Julius Theodor Ratzeburg, the founder of this discipline and the author of a pioneering book on parasitic insect species.[15] Forest entomology was indeed the first area of research on insect pests to institutionalize and achieve academic recognition, as forestry itself became a profession very early in continental Europe,

[10] For details, see Lincoln (1982, pp. 30–34) and Mironov (2012, pp. 200–202).

[11] Russia. Ministry of State Domains (1841, pp. lix–lx).

[12] See, e.g., Rossiiskii Gosudarstvennyi Istoricheskii Arkhiv [The Russian State Historical Archive—hereafter RGIA], St. Petersburg, fond (collection) 398, opis' (inventory) 6, delo (file) 1583 ("A common file on insects that are harmful for agriculture for the year 1842"); f. 398, op. 8, d. 2493 ("On insect pests that appeared in fields in various provinces in 1844").

[13] On Koeppen, see Lincoln (1982), Koeppen (1912), Sukhova (1993), and Sukhova and Krasnikova (2002).

[14] Anonymous (1863).

[15] Schwerdtfeger (1972), Aguilar (2006, p. 88), and Smith et al., ed. (1973, pp. 368–369).

and education in this field involved the study of natural history already in the early decades of the nineteenth century.

Consequently, the scientific credibility of the MSD's experts and advisors cannot be considered as the cause that prevented their initiatives from having a substantial impact on insect pest control in Russia. The real problem lay elsewhere: it was the type of evidence provided by observers "on the ground." Indeed, the descriptive language employed by the provincial chambers of state domains fell short of scientific standards. Russian provinces swarmed with "greenish worms," "hairy beetles," or "shell-skinned flying bugs."[16] The request to send specimens also generated little useful information: Insects arrived to St. Petersburg in such a poor state that no entomologist could identify them.

Koeppen was quick to appreciate the difficulties inherent in the situation. Upon his recommendations, the MSD substantially modified its approach. Potential local observers had first to familiarize themselves with normative scientific descriptions, visual images of insect pests, and standard Latin and Russian names before they could supply the ministry with reliable data. Instead of compiling a reference book on insect pests from local data, the ministry decided to produce the intended volume by translating Ratzeburg's popular treatise *Die Waldverderber und ihre Feinde* (*Forest Pests and Their Enemies*) and adding information from a few other seminal books by French and German authors. While drafting the book proposal, Koeppen emphasized the target audience for the book: "Its principal aim is to serve as a manual for landowners, foresters and local authorities when insect pests appear. We should keep in mind not experienced naturalists but people who are more or less educated and who have some knowledge of nature."[17] The reference book was published in two volumes in 1845–1851, and it was lavishly illustrated.[18] The MSD took special care disseminating it to its personnel in the provinces.

At the same time, the MSD worked to improve its guidelines on insect pest specimens. The 1838 circular had merely requested that insect pests be sent to St. Petersburg; it had not even specified if the insects in question should be shipped alive or as preserved specimens. Instructions printed in the first volume of the reference book *On Insect Pests* in 1845 were more informative on this matter. It provided introductory questions to aid in describing an insect, along with its transformations and the environment in which it was observed. The instructions further suggested sending not just an imago but also eggs, larvae, nymphs, and pupae. However, the recommendations concerning the shipment of live insects still fell far short of scientific standards of the day: "When dispatching worms and caterpillars they should be placed into a small box filled with turf and soil taken from ploughed field.... You should try to put into the box a few plants of the crops that have been damaged by these insects and have not yet withered, and quite a number of live caterpillars. The soil should be sprinkled with water and slightly pressed; however, you should take

[16] Probably shield bugs (family Scutellariidae).

[17] RGIA, f. 398, op. 5, d. 1106, l. 51ob.

[18] Russia. Ministry of State Domains (1841).

care not to crush caterpillars. The box should be completely filled with soil; otherwise the caterpillars will be damaged in shipment."[19]

Already when Koeppen was working on the reference book manuscript, he was fully aware of the inadequacy of these instructions. More than once in his replies to provincial chambers he clarified the basic techniques of preparing specimens for shipment: pinning adult insects and preserving immature ones in spirit.[20] A few years later, in 1850, the MSD asked Édouard Ménétries to compose new guidelines, which were then published in the ministry's official journal.[21] Ménétries' document explained insect life cycles and transformations and advised local observers on keeping detailed notes on insect pests and the environment in which they were found. He provided a list of guiding questions and offered exact instructions on how to kill insects, transform them into specimens, and package them for shipment.

It is hard to judge whether or not these labors produced the desired effect. The MSD's announcements about its reference book did indeed generate some response from its provincial personnel: Requests came from about 20 provinces in the European part of Russia, principally from its central, southern, and northeastern regions. Most of the book subscribers were local officials or professional foresters, but there were also a few landowners, a merchant and even a peasant.[22] By the late 1850s, we can perhaps discern a minor improvement in a few of the reports "on the ground" that arrived at the ministry's headquarters: Insect descriptions were more detailed and informative, while occasionally local observers (mostly those who were foresters or who had a medical background) were able to identify some taxonomic groups of insects and in a few cases even pest species.

Insect Pest Research in the Russian Entomological Society in the 1860s–1870s

By the late 1850s, however, the MSD had abruptly withdrawn from this field, leaving the initiative to newly created local authorities, voluntary associations, and learned societies. The Crimean War debacle (1853–1856) and the death of Nicholas I (1855) signified the dawn of a new era. The country entered the period of "the Great Reforms," with the abolition of serfdom (1861) as its most important achievement. The MSD concentrated on acute social and economic issues; at the same time, it followed a broader pattern within the Russian government, which in the late

---

[19] Russia. Ministry of State Domains (1841, pp. 265–266).

[20] RGIA, f. 398, op. 6, d. 1583, l. 9; for similar recommendations, see also: ibid, ll. 21, 33, 53ob., 66ob.-67; op. 8, d. 2493, l. 36ob.; d. 2493, ll. 40ob., 36ob.

[21] Russia. Ministry of State Domains (1850). The instructions contain no reference as to the author of the document. Our claim that they were written by Ménétries is based on the minutes of the MSD Academic Committee dated 13 October 1858. At the meeting, Ménétries stated that "a few years ago I wrote guidelines on shipping insects." See RGIA, f. 398, op. 21, d. 7163, l. 66ob.

[22] RGIA, f. 398, op. 5, d. 1106, ll. 242, 262, 362.

1850s and1860s adopted a far more liberal, less interventionist approach. The *zemstvo* reform of 1864 introduced a new system of elected assemblies at the provincial and county levels that were empowered to levy taxes and in this way to support primary and secondary education, public health, roads, and other social services.[23] Insect pest control was thus assumed to fall into the *zemstvos*' sphere of responsibility, although the MSD continued to amass data from the provinces.

At the same time, in late 1859 and early 1860 independent initiative led to the foundation of the RES in St. Petersburg, one of the first learned societies in the country that specialized in the life sciences, second only to the Moscow Naturalists' Society (1805). From its early days, the RES demonstrated its commitment not only to "pure" science but also to the public good by establishing its insect pest commission.[24] This step was quite predictable, given that the society's leadership was initially formed by people who in one way or another had been involved in or at least had been familiar with the MSD projects of the 1840s and early 1850s. Indeed, the first steps to founding the society were made as early as 1846–1847, and, at that stage, the link between the MSD and the society's potential founders was even more obvious. Academicians Karl Ernst von Baer and Johann Brandt, who chaired the RES in 1860–1864, had earlier provided consultations to Koeppen and collaborated with the MSD on many other projects that involved academic expertise, while the next president, a major-general of the Forester Corps named Viktor Semenov, was the ministry's employee and a member of its academic committee. He was also the author of the second volume of the MSD reference book *On Insect Pests,* published in 1851.

Like many other Russian scientific associations, the RES strongly depended on the government for financial support. Consequently, the establishment of the insect pest commission might have also been a well-calculated move to demonstrate the utility of the society's pursuits. The MSD for its part was quite happy to delegate some of its unfinished projects to the RES. When the provincial chambers submitted reports and inquiries that referred to insect pests, the MSD began handing these over to the Entomological Society.[25] It did the same with the manuscript and materials that had been collected for the third volume of the reference book *On Insect Pests,* evidently hoping that the society would bring the undertaking to completion.[26]

By the late 1860s and 1870s, however, the RES had apparently lost much of its initial enthusiasm for insect pest research. To some extent, this decline in interest might have been due to a generational change in its leadership. With Semenov's

---

[23] For an introduction to the *zemstvo* reform, see Eklof et al., ed. (1994).

[24] Jacobson (1910). See also minutes of the society's meetings for 1860–1869 in RES, Simashko (1861, pp. 17–24, 28–29, 37–41, 43–44), Koeppen (1865–66, pp. 4, 17–24, 26–27, 62–64), *Trudy Russkogo entomologicheskogo obshchestva* 4 (1867–69), pp. 22–26; St. Petersburg Branch of the Archive of the Russian Academy of Sciences (hereafter SPF ARAN), f. 724 (the RES collection), op. 1, d. 71, 75, 77.

[25] See, e.g., the minutes of the society's meeting on October 2, 1861, and December 4, 1861 in RES, *Trudy Russkogo entomologicheskogo obshchestva* 2 (1863), pp. xix, xxii–xxiv.

[26] RES, *Trudy Russkogo entomologicheskogo obshchestva* 2 (1863), pp. xxvii, xxxi–xxxvi.

death in 1868, the presidency passed to a new younger leader, Oktavii Radoshkovskii. An officer of horse artillery, Radoshkovskii belonged to a group of similar-minded military men from the elite guard corps who were visible among the RES members from the start, but who began to shape the society's outlook only in the 1870s. These amateur entomologists were primarily interested in collecting exotic, rare species, and their service commissions made them familiar with the southern and southeastern frontiers of the Russian empire—regions that were of a considerable interest for taxonomists and animal geographers. Although Radoshkovskii was never explicitly hostile to agricultural entomology, and even took part in the insect pest commission in the 1860s, the commission was nevertheless dissolved in the first year of his presidency. In the 1870s and 1880s, the RES channeled most of its energy into amassing collections of insects from the Caucasus, Central Asia, and Siberia.[27] By the 1880s, the RES had begun to acquire a reputation as an exclusive aristocratic club by courting its connections with the Grand Duke Nikolai Mikhailovich, an amateur entomologist and an enthusiastic sponsor of expeditions to the Near East and Central Asia.[28]

The reasons for the society's declining contribution to applied entomology might have been even more complex, however. Indeed, the RES inherited a few important predilections from the MSD's earlier efforts. First of all, the RES was manifestly committed to accumulating, verifying, analyzing, and publicizing data from the country as a whole. Yet this desire to create an exhaustive, authoritative coverage of insect pests in Russia was not matched by the society's network of contacts. In 1871, St. Petersburg residents composed about 71 % of the RES membership, and this proportion remained largely unchanged over time. In 1892, it had risen to 73 %. Most of the remaining members were disproportionally concentrated in Moscow, Warsaw, or the Baltic provinces.[29] The RES had virtually no contact with the emerging milieu of provincial public activists, the deputies of provincial and county *zemstvo* assemblies, and the executive officers of their boards. It was these people, however, who were the first to confront insect pest outbreaks and who had the right and duty to devise control measures.

Indeed, in the early 1870s, when local elected authorities sought advice on insect pests from men of science, they were more likely to address their inquiries to the Free Economic Society (FES) in St. Petersburg. The FES, established in 1765, was the oldest learned society in Russia. Its public visibility was much greater than that of the RES, and it had an extensive network of provincial correspondents.[30] By the

[27] Our analysis is based on the RES membership lists and the lists of expeditions and publications. See Jacobson (1910), pp. 4–6, 11–13, Semenov-Tian-Shanskii (1910), and RES (1873), pp. 67–118.

[28] The Grand Duke became the RES's president in 1881. In 1884–1901, he edited and published nine volumes of *Memoirs sur les Lepidopteres*. For details, see Semenov-Tian-Shanskii (1910), p. 31.

[29] Our calculations are based on the society's membership lists: see RES (1872), pp. 53–57 and RES (1893).

[30] See RGIA, f. 91, op. 1, d. 347. On the Free Economic Society, see Leckey (2011) and Bradley (2009), pp. 38–84.

1870s, the FES had positioned itself as an informal advisory board or coordinating center for *zemstvo* activists; at the same time, it had been able to recruit a number of prominent professors from St. Petersburg University as its leading members. In this way, the FES was certainly one of the leading centers of civic science in Russia. Entomology, however, was not the strongest area of FES expertise. For this reason, the FES did not venture to assume leadership in this field, and insect pests remained a low priority on its expanding agenda. It was on fairly friendly terms with the RES, and occasionally they even cooperated.[31] Yet there were no signs that the RES ever used FES contacts in the provinces for enhancing its own presence outside of the imperial capital. Most likely, the RES leadership was not particularly interested in cultivating its links with *zemstvo* activists, as it was skeptical about the quality of the field observations that such people might provide.

In 1873, this skepticism was expressed unequivocally in a project to set up a network of permanent entomological observation stations in Russia. This project was drafted by one of the RES's founding members, Karl Gernet, an amateur entomologist who served in one of the branches of the Russian state administration.[32] His proposals can be seen as summarizing the RES's vision of research on insect pests. It had to be comprehensive and concerned with the Russian empire as a whole, and it also required a stable flow of field observations. Such observations could not be generated by isolated field trips, and therefore needed to come from permanent residents in the provinces. Yet this mission could not be entrusted to amateur entomologists or local elected authorities, as these people either were not able to carry out proper scientific observations, or, if they did have some background in entomology, were simply not interested in meticulously recording data concerning the life cycles of common species and their adaptive interaction with environment.

There was a certain irony in Gernet's self-directed criticism of amateur entomologists, but the ideas he expressed should not be taken just for his personal prejudices. In the 1870s, the idea of setting up a network of experimental stations was very much in the air. It certainly reflected increasing professionalization of research in the life sciences, along with a fascination with new laboratory research methods. Characteristically, when discussing the organizational and financial support required for the proposed network, Gernet assumed that the MSD ought to take up responsibility for all such matters: Salaried local observers should be its employees. Characteristically, Gernet was very elusive about the issue of pest control. He insisted that the staff of entomological stations should not be held responsible for devising pest control measures, yet he did not suggest anyone else who might take up this task. In other words, he was concerned with creating an efficient centralized system for the collection of field observations that could be processed and analyzed by scientists in the capital, but he did not design any feedback from this system to agriculturalists in the provinces.

---

31 For example, in 1871–1872 the two societies established a joint commission for bird protection. See RGIA, f. 91, op. 1, d. 347, ll. 40–43.

32 SPF ARAN, f. 724, op. 1, d. 80 ("About the establishment of entomological stations in Russia," 1873).

Another reason why the RES failed to take up leadership in applied entomological research could be related to its major areas of expertise, which were in turn determined by the legacy of the MSD's earlier initiatives and by the society's social contacts. From its early days, the RES concentrated its efforts either on forest and garden pests or on locusts.[33] The society's interest in forest entomology is understandable if we remember that this field had been at the forefront of MSD initiatives in the 1840s. Professional gardeners (particularly those who looked after botanical gardens, plant nurseries, or aristocratic estates in the capital and its vicinity) also had close ties to the group of amateur entomologists and scientists from the Imperial Academy of Sciences who established the RES.[34] The society's interest in locusts was obviously due to the fact that these insects had traditionally been considered the principal enemy of southern agriculture and horticulture. Stories abounded about locusts destroying vast areas of crops in the Black Sea coastal provinces in just a matter of days, and newspapers and journals reported on the desperate attempts of local administration and landlords to organize resistance to their invincible columns.[35] Before the 1870s, no one could have predicted that in the next few years thousands of roubles would be spent (or rather wasted) on control measures against a new enemy—*Anisoplia austriaca,* a medium-sized beetle of the Scarabaeidae family that would cause enormous damage to grain crops in the southern steppe provinces.

It might seem at a first glance that the choice of certain biological species as a privileged object of analysis is of no significance for the social history of the life sciences. Yet it is not quite so. Locusts migrate over large distances and therefore are difficult pests to control; moreover, in this period, the measures used to control locusts stood in little or no relation to farming practices in a particular region affected by these insects. Before locusts became airborne, landlords and peasants could try to destroy them with harrows and rollers or with fire; people might also herd cattle into an infested area in order to trample the locusts. All these measures were very expensive, however, and required coordination of efforts across large areas. As a result, from the 1840s forward research on locusts tended to focus on their migration habits and the location of their eggs. In contrast, *A. austriaca* is a species indigenous to a particular area, and, as we shall see, the discussion of its control measures quickly evolved into a specific discussion of farming practices. Yet, this arena was precisely the one in which the RES had only a very limited interest and understanding.

The life of Fedor Koeppen (1834–1908), the leading authority on insect pests in Russia in the mid-nineteenth century and possibly even later, exemplifies the kind of insect pest research that was carried out by the MSD and the RES from the 1850s to

---

[33] Simashko (1861, pp. 17, 19, 37–38, 43–47), Anonymous (1863, pp. 23, 39–53), Koeppen (1865–1866, pp. 19–24, 26–27, 54, 62–63), *Turdy Russkogo entomologicheskogo obshchestva* 4 (1867–1869), pp. 22–26.

[34] Simashko (1861) and Jacobson (1910, pp. 4–5).

[35] For example, Pot'e (1846), Russia. Ministry of State Domains (1847), Baziner (1851), Demol' (1846), and Skarzhinskii (1846).

the 1870s.[36] A son of Peter Koeppen, Fedor seemingly tried to follow in his father's footsteps by choosing the study of *Kameralwissenschaft,* agriculture, and natural science at the St. Petersburg and Dorpat universities. In 1858, he defended his master's thesis on insect pests in Russia at Dorpat University and took up the position of a subinspector of agriculture at the MSD's branch in the Crimea. After 5 years, however, Koeppen junior left the MSD, perhaps because he was disappointed by its withdrawal from the research area that interested him most, or perhaps because he was attracted by the prospects of an academic career in the capital. For the next few years, he served at the Ministry of Education in St. Petersburg, apparently with the aspiration of receiving an appointment as a professor of agriculture at one of the Russian universities. Indeed, the 1860s were the period when universities began to dominate academic life in the country and to enjoy enormous respect and attention from the Russian public. By opting for a university career instead of civil service, Koeppen seemed to follow the *Zeitgeist*. When he returned to Russia in 1872 after a 2-year research trip to Western Europe, however, he discovered that the Ministry of Education had no vacant university chairs. As a result, Koeppen had to take up a position at the Public Library in St. Petersburg, where he served for most of his later life. As one might expect, he was also very active at the RES in the 1860s, being one of its founding members, its academic secretary, and a member of its commission on insect pests. It was in the early 1860s when Koeppen commenced his study of locusts and other insects of the Acridoidea family.[37] By the 1870s, he had already been acknowledged as the leading Russian expert in this field. Upon his return to Russia in 1872, however, his relations with the RES soured, and he left the society in 1875.[38] He did not abandon his project of compiling an authoritative guide to Russian insect pests, a task that had been commissioned to him in the early 1860s by the MSD. His three-volume monograph was eventually published in 1881–1908,[39] yet the lengthy time it took for him to produce his masterwork persuaded some of his critics in the provinces that the task would never be accomplished.[40]

Agricultural Entomology in Southern Russia in the 1870s and 1880s

Until the mid-nineteenth century, research on insect pests in Russia was carried out primarily in St. Petersburg. Even Moscow with its university and its Naturalists' Society could not rival the capital of the empire in this respect. In 1846, Moscow naturalists felt a duty to produce a popular booklet on "a grain worm" (cutworm, larvae of turnip moth, *Agrotis segetum*) when an outbreak of these insects occurred

[36] For his biography, see Adelung (1908).

[37] Koeppen (1865–1866, 1870).

[38] Trudy Russkogo entomologicheskogo obshchestva 9, 1875–1876, pp 57–59.

[39] Koeppen (1881–1908).

[40] [1878] "Protokoly zasedaniia komissii, sostoiashchei pri Odesskoi zemskoi uprave."

in many provinces of European Russia.[41] Yet it was probably the only contribution of some substance made by Moscow scientists before the mid-nineteenth century. If that was the state of affairs in the second major city of the empire, what could be expected of those very few provincial agricultural societies that had been established in the first half of the nineteenth century? Indeed, a careful examination of the *Notes of the Agricultural Society of Southern Russia*—one of the oldest and largest Russian provincial agricultural societies, founded in Odessa in 1828—shows that in the 25 years prior to 1855 the society published just nine papers on insect pests. Most of them merely reported on the experience of local landowners in devising improvised measures against locusts, while only one was written by a trained naturalist.[42]

The situation began to change somewhat in the 1860s and 1870s, however, at least in those few cities that had a university. The university statute of 1863 substantially enhanced the schools' autonomy, improved the quality of teaching, and, crucially, promoted research by authorizing the establishment of university-affiliated learned societies. In the 1860s, the state invested heavily in the university system by increasing the salaries of the university faculty, adding to the number of student bursaries, and sponsoring research trips to Western Europe.[43] By the mid-1870s, Russian universities could therefore provide provincial centers such as Odessa and Khar'kov with credible experts in the natural sciences. At the same time, the Great Reforms era left a profound imprint upon the university culture by forging its links with the wider public and promoting various forms of civic science.[44] University professors consciously assumed the role of speakers for various civic groups addressing all sorts of social concerns. Insect pest control, however, did not attract public attention before a large-scale outbreak of *A. austriaca* that hit the southern provinces in the 1870s.

A. austriaca was indigenous to the south, and prior to the 1870s it had been considered fairly harmless. Some locals even believed that *A. austriaca,* or the Kuz'ka-beetle, as it was commonly known, was a "useful" insect that pollinated corns.[45] By the 1870s, however, the situation had changed dramatically due to the large-scale expansion of commercial agricultural production in the area. The southern provinces of the Russian empire—the present-day territories of southern Ukraine and the Russian Federation—were once a boundless dry grassland stretching from the Black Sea coast almost up to Khar'kov, Kursk, and Voronezh. When they were annexed to the Russian empire in the late eighteenth century, these territories were relatively sparsely populated by nomadic and seminomadic pastoralists. But in the nineteenth century, Russian, Ukrainian, and to a much lesser extent German peasants migrated into the region, and the southern provinces became a major center of fast-growing

[41] Rul'e and Farenkol' (1847).

[42] Our conclusions are based on the contents of the journal published in Palimpsestov, *Otchet*, appendix.

[43] On the 1863 university reform in Russia, see Vucinich (1970, pp. 42–61) and Kassow (1994).

[44] See Bradley (2009) and Hachten (1991).

[45] RGIA, f. 1287 (collection of the Economic Department of the Ministry of the Interior), op. 4, d. 1364 ("On insect and animal pests that appeared in 1878"), ll. pp. 35–36.

large-scale commercial agricultural production, specializing in grain crops and sugar beets. By the 1870s, a continuous reduction of pasture and wildlife areas in the region evidently created the environment favorable for magnifying to disastrous proportions an occasional population increase of *A. austriaca*.

Faced with this problem, local elected authorities and provincial governors appealed to the Ministry of the Interior for professional advice. Their reports indicate that the state of entomological knowledge in the provinces was in most cases hardly any better than in the 1840s. Only in the two major southern centers, Odessa and Khar'kov, were local authorities able to identify the insect species. The Ministry of the Interior forwarded these local appeals for help to the MSD, and in the summer of 1878 the latter agency sent one of its employees, Iosif Porchinskii, to the southern provinces.[46]

Porchinskii's career demonstrates the very close ties that still existed between the MSD and the RES. In 1871, Prochinskii graduated from St. Petersburg University and took up the position of curator and librarian at the RES. In 1874, he was promoted to the position of its academic secretary (a post he filled until 1896). In 1875, he got a junior position at the MSD upon the society's recommendation. In the next few years, Porchinskii would emerge as the principal ministerial expert in applied entomology and one of the leading authorities in this field in Russia.[47] During this period, another MSD employee whose future research agenda would be shaped by the *A. austriaca* outbreak also visited the southern provinces. Before 1876, Karl Lindeman (1844–1928), a young zoology professor at the Moscow Agricultural Academy, had produced a few publications, of which only one book (his doctoral thesis) was focused on forest insect pests.[48] In the 1880s and 1890s, however, he would become one of the most respected Russian specialists in the field of agricultural entomology. In subsequent years, local authorities from various regions of European Russia repeatedly asked the MSD to send a specialist in applied entomology to their provinces. The MSD had to respond to these appeals, and Porchinskii made himself indispensable by specializing in this particular area. In 1883, the MSD hired another university graduate, Viktor Filip'ev (1857–1906), to act as a travelling consultant on agricultural entomology.[49] Both positions, however, were not yet officially designated as "entomologists."

These field trips never lasted more than a couple of summer months, during which these few ministerial specialists had to visit several provinces. They brought little consolation to the local authorities who had to cover their travel costs. From

[46] Porchinskii (1879).

[47] For details of Porchinskii's career at the MSD, see RGIA, f. 398, op. 59, d. 18445 ("About assigning a graduate of St. Petersburg University Iosif Porchinskii to the Department of Agriculture"). For Porchinskii's activities at the RES, see SPF ARAN, f. 724, op. 1, d. 82 ("About the society's engagement in insect pest control measures, correspondence with zemstvos, provincial and country administration and central administration," 1879–1887).

[48] Lindeman (1875). On Lindeman's and Porchinskii's trip to the southern provinces in 1878, see RGIA, f. 1287, op. 4, dd. 1376, 1377 ("On measures for the extermination of the grain beetle in the southern provinces of Russia, part. 1, 2").

[49] RGIA, f. 1287, op. 4, d. 1551, ll. 4–5; f. 398, op. 47, d. 15345, ll. 23–26.

the Great Reforms era of the 1860s the Russian public had been extremely enthusiastic about science, perceiving it as a powerful tool of social and economic modernization, and even expecting it to perform almost like a magic wand working in incomprehensible ways to produce an immediate impact. In practice, however, Porchinskii, Lindeman, and Filip'ev could only inspect infested fields, identify a pest, hand out a few booklets, and give recommendations that were far too general to be of much service. Often they arrived too late for any effective action to be taken.[50]

For these reasons, in the late 1870s southern *zemstvos* began to look elsewhere for academic advice. Public initiative was particularly strong in Kherson province. It was the first to experience an outbreak of *A. austriaca,*[51] and its biggest city, Odessa, could boast a university that was particularly strong in the life sciences. Among its professors at that time were Ilia Mechnikov (Elie Metchnikoff, 1845–1916) and Alexander Kovalevskii (1840–1901), known specialists in experimental zoology. By 1878, the local authorities in Odessa had established a special entomological commission staffed by members of their executive board, a few landowners, and university faculty.[52] The other university center in the region, Khar'kov, was much less prominent in this area of research, despite the fact that in 1879 the Khar'kov *zemstvo* followed Odessa's example and established its own entomological commission. The commissions debated at length various means of pest control. At first, they had high hopes for certain mechanical devices, or "beetle-traps," but enthusiasm for the traps faded rapidly. Mechnikov instead suggested looking for biological methods of pest control: He was experimenting with pathogenic fungi such as *Metarhizium anisopliae* (the cause of so-called green muscardine disease) that killed insect larvae.[53] In the 1880s, Mechnikov and his colleagues were even able to develop efficient methods to mass-produce this fungal pathogen, yet they failed to obtain necessary funding for field tests.[54]

Local authorities, however, had no time for research and experiments; the commissions had to propose emergency measures. In the mid-1870s, local authorities in the south decided to mobilize peasants to handpick beetles, and landowners had to pay a special "bug duty" to cover the costs. On May 11, 1879, Alexander II authorized the *zemstvos* of six southern provinces to carry out these plans, if they deemed

---

[50] RGIA, f. 398, op. 47, d. 15345, ll. 52–56, 163, 184–185, 281–282.

[51] RGIA, f. 1287, op. 4, d. 1008 ("About the specialist's trip to Kherson Province to study appearing pests and raising funds for their extermination"); [1879] *Pervyi opyt.*

[52] Local society, the *zemstvo,* and municipal authorities in Odessa were very active in promoting various forms of civic science. Thus, in the 1880s they also established and supported the first Russian bacteriological station. See Hachten (1991). For details on the Odessa entomological commission, see [1878] "Stenograficheskii otchet"; [1879] *Stenograficheskii otchet.* For the Khar'kov commission, see [1883] *Trudy oblastnogo (III) s'ezda.*

[53] [1879] *Stenograficheskii otchet*, pp. 4–5. See also Mechnikov (1879a, b).

[54] Krasil'shchik (1886, pp. 13–23). On similar experiments with the fungal pathogens of locusts in the 1890s, see RGIA, f. 398, op. 56, d. 18056 g ("On the study of grasshoppers in Tobolsk province"). Entomopathogenic fungal spores were produced in France in the 1890s; in some cases, Mechnikov personally supervised the production process (Marchal 1896).

them necessary, for the period of 2 years.[55] Mobilization for beetle handpicking, however, was met with hostility from the peasants. The Ministry of the Interior could not afford to risk social unrest in a country already stricken by the massive buildup of tax arrears, a notoriously inefficient system of provincial administration, a state budget destroyed by the war with the Ottoman Empire, and mounting terrorist attacks on the emperor.[56] In 1881, the emergency legislation expired. By then local authorities in Odessa had expressed their wish to convene a conference on insect pest control with the representatives of other southern *zemstvos*. The Ministry of the Interior faced a hard choice. In principle, it was manifestly hostile to any initiatives that were aimed at promoting inter-*zemstvo* cooperation among provinces; the government evidently feared that these steps might eventually lead to the formation of an all-Russia union of local elected authorities, which in its turn might become a proto-Parliament. This time, however, it supported the *zemstvo* petition and secured the emperor's consent. Fear of *A. austriaca* evidently overweighed all other considerations. Both the Ministry of the Interior and the MSD were flooded with reports from the south about substantial damage caused by the beetle, yet the information they contained on the efficacy and feasibility of improvised control measures was extremely contradictory.[57]

Local authorities, for their part, lost their hopes fairly quickly that machinery or handpicking could be effective pest control measures. Their disillusionment was manifest at the first regional conference on insect pest control that took place in Odessa in late February and early March of 1881. The conference was supposed to work out recommendations concerning a prolongation or a repeal of the emergency 1879 law, and it assembled *zemstvo* delegates from eight southern provinces, the leadership of the Imperial Agricultural Society of Southern Russia, a group of university faculty members from Khar'kov and Odessa universities (including Mechnikov), and a few agronomists.[58] It voted against the prolongation of the emergency law, yet its delegates were unsure about possible alternatives. The only recommendations they could give concerned professionalization of applied entomology. Some of the delegates suggested that the entomological commissions in Odessa and Khar'kov should become permanent institutions, while salaried positions of entomologists could be created in all other southern provinces. At that time, however, even some prominent university professors who attended the conference saw the latter step as redundant. They assumed that the task of identifying insects and

[55] Russia (1881).

[56] For a detailed discussion of the challenges faced by provincial administration in the late 1870s and early 1880s, see Pearson (1989).

[57] Reports from the southern provinces: see RGIA, f. 1287, op. 4, d. 1376, 1377, 1378 (On measures for the extermination of the grain beetle in the southern provinces of Russia, part 1, 2, 3). In autumn 1880, the Interior Ministry sent its representative to the southern provinces, who supported the idea of the regional zemstvo meeting. See RGIA, f. 1287, op. 4, d. 1500.

[58] [1881] *Khlebnyi zhuk*.

providing basic recommendations on their control could be performed for free by faculty and students during summer vacations.[59]

On May 3, 1881, acting upon the conference recommendations, the new emperor Alexander III signed a law that shifted all the responsibility for insect pest control in the south to local authorities. They could still impose the "bug duty" or mobilize peasants for handpicking beetles, but the choice of control techniques was left to their discretion.[60] In the next decade, these conferences or "regional congresses" on applied entomology became regular events; they convened every year either in Odessa or Khar'kov with *zemstvo* delegates from eight southern provinces and with invited specialists.[61] It is very telling that these congresses were authorized by the government in the years known as the period of conservative "counter-reforms." No governmental agency, including the MSD, was willing to assume responsibility for insect pest control, but the matter was evidently considered a high-priority issue that required extensive consultations with all available experts. For this reason, the influence of the congresses reached beyond the field of agricultural entomology. They were used as a forum for discussing various problems of southern agriculture and as a result became an important stimulus for the rise of agricultural science in Russia.[62]

While none of the MSD specialists from Moscow or St. Petersburg attended the first congress in Odessa, Fedor Koeppen, Iosif Porchinskii, and Karl Lindeman were later invited to these meetings. Their inclusion indicated a growing recognition of their research by members of the Khar'kov and Odessa entomological commissions. Indeed, in 1878–1879, the Odessa commission was against the idea of developing contacts with the RES, having been persuaded by Mechnikov that the society was just a bunch of "amateurs who were only collecting and pinning insects."[63]

All the congresses repeatedly advocated further professionalization of applied entomology. Initially they only proposed the creation of salaried positions and experimental agricultural stations in the provinces, with the expenses to be covered jointly by local authorities and the central government.[64] From 1884, however, the congresses substantially expanded their vision. They began petitioning for the transformation of the MSD into a Ministry of Agriculture and arguing that a Central Entomological Bureau should be created in St. Petersburg along with the new

[59] [1881] *Khlebnyi zhuk,* pp. 61–62.

[60] Russia (1885).

[61] RGIA, f. 382, op. 1, d. 570; f. 1287, op. 4, d. 1379–1382 (On measures for the extermination of the grain beetle in the southern provinces of Russia, part 4–7); [1882] *Trudy oblastnogo s'ezda predstavitelei zemstv*; [1883] *Trudy oblastnogo III s'ezda predstavitelei zemstv*; [1884] *Trudy IV entomologicheskogo s'ezda predstavitelei zemstv*; [1885] *Trudy oblastnogo s'ezda predstavitelei zemstv*; [1886] *Trudy VI entomologicheskogo s'ezda*; [1887] *Trudy VII oblastnogo entomologicheskogo s'ezda*; [1888] *Trudy VIII entomologicheskogo s'ezda*; [1890] *Trudy IX oblastnogo entomologicheskogo s'ezda.*

[62] See Minutes of the regional entomological congresses 1881–1884 (footnote 61); Elina and Savchuk (1998, pp. 96–110).

[63] [1878] "Stenograficheskii otchet," p. 40.

[64] RGIA, f. 382, op. 1, d. 570, l. 1; *Khlebnyi zhuk*, p. 78.

ministry.[65] This request also signaled southern entomologists and agriculturalists' expectation that the state would establish a research institution for applied entomology.[66] Both steps, however, were made only in 1894, once the government again assumed a much more interventionist position towards agriculture and the peasant economy. Until this change in policy, the MSD continued collecting data on insect pests in Russia and provided advice, but its human and financial resources remained rather meager in comparison to the rising demand for professional entomological consultations.[67] Meanwhile, the entomological commission in Odessa created a salaried position for a regional entomologist funded entirely by the local budget. From 1887 until 1893, this person provided entomological consultations in five southern provinces.[68]

In the course of the 1880s, the number of people professionally involved in pest control steadily increased from one annual congress to the next. At the same time, the congress discussions became progressively more focused. The early congresses debated a broad spectrum of agricultural issues and limited their discussion of insect pest species to those that bred on grain crops. By the end of the decade, however, applied entomology dominated the discussions, and speakers began to devote papers to many other types of insect pests. They gave presentations on insects that were dangerous for horticulture, vineyards, and tobacco plantations, and also discussed the fungal pathogens that attacked these insects. In the course of the 1880s, local authorities also improved their methods of collecting field data on insect pests, so that by the end of the decade this task was performed not only by the ordinary employees of *zemstvo* executive boards but also by hired specialists with backgrounds in the life sciences (most of them local university students or teachers of natural history).

In the course of the 1880s and the early 1890s, most people who discussed the issue of insect pests realized that their proliferation in southern Russia had been caused by monoculture. Most also thought that science could not offer an instant universal means of pest control. Solutions to the problem could be found only in modifying farming practices in ways that would obstruct insect pest reproductive cycles. Among other things, insect pest prevention required improved crop rotations, better weed control, and better tillage. Entomologists could not develop these recommendations alone; they required field experiments and feedback from farmers. Agriculture itself would have to become a scientific enterprise practiced by a growing body of professional agronomists. The correspondence exchanged between entomologists employed by the MSD and provincial local authorities, who in the

[65] *Trudy IV entomologicheskogo s'ezda,* pp. 82–84; *Trudy VII oblastnogo entomologicheskogo s'ezda*, p. 59; RGIA, f. 398. op. 51, d. 16586, l. 10–13; op. 54. d. 17529 etc.

[66] Elina (2008), RGIA, f. 398. op. 51, d. 16586, l. 10–13; op. 53, d. 17172, l. 65 etc.

[67] The archival documents testify that every year the MSD refused to send its entomologists to some provinces on the grounds that there were no available specialists. Those few who were employed by the ministry had been already sent elsewhere (RGIA, f. 398, op. 47, d. 15345; op. 51, d. 16586; op. 53. d. 17172; op. 56, d. 18056b, g; op. 57, d. 18116b, etc.).

[68] See, e.g., Zabarinskii (1887).

1880s and 1890s began creating salaried positions for "agricultural supervisors" or agronomists, demonstrates this changing perspective on insect pest control. The discussion became more technical, it increasingly concerned local farming practices, and it progressed from occasional observations to controlled field experiments.[69]

Conclusion

In this essay, we have tried to sketch an early history of insect pest research in nineteenth-century Russia. Rather than focusing on particular works produced by the most prominent practitioners of applied entomology, we explored the social and institutional environment in which academic scholars, state officials, and local public activists looked for means of effective pest control. When examined from this perspective, the history of nineteenth-century applied entomology can be understood as a series of attempts to establish a working network that would enable not only a stable flow of observations from localities to trained naturalists but also efficient ways of integrating knowledge produced by the latter into agricultural practices on a local level.

At the earliest stage of this process, all efforts were understandably concentrated on ensuring the quality of local observations, since in the first half of the nineteenth century natural history could be of little help beyond establishing the taxonomic identity of a pest. The identification of insects was no small achievement in this period, however, given that the language of scientific taxonomy remained unknown to the vast majority of the provincial population in Russia. Stumbling upon the problem of field data, the upper echelons of the imperial administration, guided by their academic advisors, had to devise some means of "calibrating" local informants—instructing them in basic entomology, its language, concepts, and practices—in order to secure the flow of proper specimens and meaningful descriptions from the provinces to the capital of the empire. In this respect, the history of the MSD initiatives in the 1840s serves as a particularly interesting case that highlights the interactive nature of entomological "popularization." The decision to publish the first Russian reference book *On Insect Pests* was prompted by an acutely perceived need to bring local observations up to current natural historical standards.

Yet establishing an insect's taxonomic identity in itself mattered little for provincial agriculturalists. Ultimately, they would become interested participants in these observational networks only if they obtained some feedback, a type of knowledge that could directly inform attempts at pest control. Producing this type of knowledge required data not only on insects but also on local agricultural practices, and local practices were also changing, increasingly affected by scientific research and the commercialization of agriculture. In the 1870s and 1880s, St. Petersburg might have lost its leading position in agricultural entomology for a variety of reasons

[69] RGIA, f. 398, op. 47, d. 15345; op. 51, d. 16586; op. 53, d. 17172; op. 56, d. 18056 b, g; op. 57, dd. 18116b, v; op. 59, d. 18445; f. 382, op. 9, d. 235 etc.

unrelated to this problem. Yet the rise of the southern Russian provinces to prominence in this field was undoubtedly due to the fact that the crucial process of mutual adjustment between entomologists and agriculturalists, with their differential languages, observations, and practices, took place in the localities where pest control measures were urgently required.

Acknowledgments We would like to thank Joseph Horan, Aleksei Kouprianov, and the editors of this volume, Sharon Kingsland and Denise Phillips, for their very helpful comments and suggestions.

References

4th Entomological Congress of Zemstvo Representatives of Southern Russia. 1884. *Trudy IV entomologicheskogo s'ezda predstavitelei zemstv Iuzhnoi Rossii s 10 po 21 fevralia 1884 v Odesse* [Proceedings of the 4th Entomological Congress of Zemstvo Representatives of Southern Russia, 10-21 February 1884 in Odessa]. Odessa.

6th Regional Entomological Congress. 1886. *Trudy VI oblastnogo entomologicheskogo s'ezda v Odesse* [Proceedings of the 6th Regional Entomological Congress in Odessa]. Odessa.

7th Regional Entomological Congress. 1887. *Trudy VII oblastnogo entomologicheskogo s'ezda, sostoiavshegosia v marte 1887 g. v Odesse* [Proceedings of the 7th Regional Entomological Congress, held in March 1887 in Odessa]. Odessa.

8th Regional Entomological Congress. 1888. *Trudy VIII oblastnogo entomologicheskogo s'ezda* [Proceedings of the 8th Entomological Congress]. Odessa.

9th Regional Entomological Congress. 1890. *Trudy IX oblastnogo entomologicheskogo s'ezda sozyvavshegosia v aprele 1889 g. v Odesse* [Proceedings of the 9th Regional Entomological Congress, held in April 1889 in Odessa]. Odessa.

Adelung, Nikolai. 1908. Fedor Petrovich Koeppen. *Russkoe entomologicheskoe obozrenie [Russian Entomological Review]* 8 (1): xv–xviii.

Aguilar, Jacques d'. 2006. *Histoire de l'entomologie.* Paris: Delachaux et Niestle.

Alberti, Samuel J. M. M. 2001. Amateurs and professionals in one county: Biology and natural history in late Victorian Yorkshire. *Journal of the History of Biology* 34:115–147.

Allen, David. 2009. Amateurs and Professionals. In *The Cambridge history of science*, vol. 6, ed. P. J. Bowler and J. Pickstone, 15–33. Cambridge: Cambridge University Press.

Anonymous. 1863. Notice biographique sur Mr Edouard Ménétries. *Trudy Russkogo entomologicheskogo obshchestva [Transactions of the Russian Entomological Society]* 2:1–7; 2nd pagination.

Baziner. 1851. O sostoianii raznykh otraslei sel'skogo khoziaistva v Iuzhnoi Rossii v 1850 godu [On the state of various branches of agriculture in Southern Russia in 1850]. *Zhurnal Ministerstva Gosudarstvennykh Imushchestv* 40:109–112.

Bradley, Joseph. 2009. Voluntary associations in tsarist Russia: Science, patriotism and civil society. Cambridge: Harvard University Press.

Brenna, Brita. 2011. Clergymen abiding in the fields: The making of the naturalist observer in eighteenth-century Norwegian natural history. *Science in Context* 24:143–166.

Castonguay, Stéphane. 2004a. Naturalizing federalism: Insect outbreaks and the centralization of entomological research in Canada, 1884–1914. *Canadian Historical Review* 85:1–34.

Castonguay, Stéphane. 2004b. *Protection des cultures, construction de la nature. Agriculture, foresterie et entomologie au Canada, 1884–1959.* Sillery: Septentrion.

Chesnova, Larisa V. 1962. *Ocherki iz istorii prikladnoi entomologii v Rossii [Essays on the history of applied entomology in Russia]* Moscow: AN SSSR. [Eng. trans.: *Essays from the history of applied entomology in Russia.* Jerusalem: Israel Program for Scientific Translation, 1968.]

Demol' I. 1846. Ob istreblenii saranchi. [On the destruction of locusts]. *Zapiski Imperatorskogo Obshchestva sel'kogo khoziaistva Iuzhnoi Rossii [Notes of the Imperial Agricultural Society of Southern Russia]* 1:68–72.

Desmond, Adrian. 2001. Redefining the X axis: 'Professionals,' 'amateurs' and the making of mid-Victorian biology: A progress report. *Journal of the History of Biology* 34:3–50.

Eklof, Ben, John Bushnell, Larissa Zakharova, eds. 1994. *Russia's great reforms, 1855–1881*. Bloomington: Indiana University Press.

Elina, Olga. 2008. *Ot tsarskikh sadov do sovetskikh polei: Istoriia sel'skokhoziaistvennykh opytnykh uchrezhdenii, XVIII v.—20-e gg. XX v.* 2 vols, Moscow: Egmont Russia [From Tsar's gardens to Soviet fields: A history of experimental agricultural institutions from the 18th century to the 1920s].

Elina, Olga, and Varfolomei Savchuk.1998. Opytnye sel'skokhoziaistvennye stantsii Rossiiskoi imperii: tendentsii razvitiia na rubezhe XIX–XX vv. [Experimental agronomic stations in the Russian Empire: Trends at the turn of the nineteenth and twentieth centuries]. *Vestnik Dnepropetrovskogo universiteta [Bulletin of Dnipropetrovsk University]* 4:96–110.

Güttler, Nils. 2011. Scaling the period eye: Oscar Drude and the cartographical practice of plant geography, 1870s–1910s. *Science in Context* 24:1–41.

Hachten, Elizabeth A. 1991. *Science in the service of society: Bacteriology, medicine, and hygiene in Russia, 1855–1907*. Madison: University of Wisconsin Press.

Howard, Leland O. 1930. *A history of applied entomology (somewhat anecdotal)*. Washington: Smithsonian Institute.

Imperial Society of Agriculture of Southern Russia. 1878a. Protokoly zasedaniia komissii, sostoiashchei pri Odesskoi zemskoi uprave dlia razrabotki voprosa o vrednykh v sel'skom khoziaistve nasekomykh ot 7 oktiabria i 25 noiabria 1878 goda [Minutes of the meetings of the commission affiliated with the Odessa Zemstvo Executive Board for solving the question of insects harmful for agriculture, from 7 October and 25 November 1878]. In *Zapiski Imperatorskogo Obshchestva sel'skogo khoziaistva Iuzhnoi Rossii*, November-December, appendix: 2–4.

Imperial Society of Agriculture of Southern Russia. 1878b. Stenograficheskii otchet o zasedaniiakh Odesskoi uezdnoi zemskoi upravy sovmestno so spetsialistami i khoziaevami-praktikami dlia izyskaniia mer k istrebleniiu vrednykh nasekomykh, osobenno kuz'ki *Anisoplia austriaca* [Stenographic report on the meetings of the Odessa County Zemstvo Executive Board together with specialists and farmers for discovering means of insect pest control, particularly the 'Kuz'ka Beetle', *Anisoplia austriaca*]. In *Zapiski Imperatorskogo Obshchestva sel'skogo khoziaistva Iuzhnoi Rossii*, Nov.–Dec, 1878, appendix: 3–4.

Jacobson, Georgii G. 1910. Kratkii ocherk deiatel'nosti Russkogo entomologicheskogo obshchestva za pervye 50 let ego sushchestvovaniia [A brief essay on the activities of the Russian Entomological Society for the first 50 years of its existence]. *Trudy Russkogo entomologicheskogo obshchestva [Transactions of the Russian Entomological Society]* 39:14–15.

Kassow, Samuel D. 1994. The University Statute of 1863. In *Russia's Great Reforms*, ed. Ben Eklof Bloomington: Indiana University Press. 247–263.

Koeppen, Fedor. 1865–66. Zametki o nasekomykh Tavricheskoi gubernii, preimushchestvenno o vrednykh [Notes on insects of Tavrida Province, mainly about insect pests]. *Trudy Russkogo entomologicheskogo obshchestva* 3:3–16; 2nd pagination.

Koeppen, Fedor. 1870. O saranche i drugikh vrednykh priamokrylykh iz semeistva Acridoidea, preimushchestvenno po otnosheniiu k Rossii [About locusts and other insect pests of the Acridoidea family, mainly in relation to Russia]. *Trudy Russkogo entomologicheskogo obshchestva* 5:1–352; 2nd pagination.

Koeppen, Fedor. 1881–1908. *Vrednye nasekomye v Rossii [Insect pests in Russia]*. 3 vols., St. Petersburg: Departament zemledeliia i sel'skoi promyshlennosti.

Koeppen, Fedor P. 1912. *Biografiia P. I. Koeppena [Biography of P. I. Koeppen]*. St. Petersburg: Tip. Imp. Akademii nauk.

Krasil'shchik, Issak M. 1886. O fabrichnom proizvodstve zaraznykh gribkov [On the factory production of infectious fungi]. In *Trudy VI oblastnogo entomologicheskogo s'ezda v Odesse*

10–17 fevralia 1886 g. [Proceedings of the 4th Regional Entomological Congress in Odessa, February 10-17, 1886], 13–23. Odessa.

Latour, Bruno. 1987. *Science in action: How to follow scientists and engineers through society*. Cambridge: Harvard University Press.

Leckey, Colum. 2011. *Patrons of enlightenment: The Free Economic Society in eighteenth-century Russia*. Newark: University of Delaware Press.

Lincoln, Bruce W. 1982. *In the vanguard of reform: Russia's enlightened bureaucrats, 1825–1861*. DeKalb: Northern Illinois University Press.

Lindeman, Karl E. 1875. *Monografiia koroedov v Rossii [A monograph on Scolytidae in Russia]*. Moscow: Izd. Obshchestva liubitelei estestvoznaniia.

Lucier, Paul. 2009. The professional and the scientist in nineteenth-century America. *Isis* 100:699–732.

Marchal, Paul. 1896. *L'entomologie appliquée en Europe*. Paris: 41 rue de Lille.

McWilliams, James E. 2008. *American pests: The losing war on insects from colonial times to DDT*. New York: Columbia University Press.

Mechnikov, Ilia I. 1879a. *Khlebnyi zhuk (bolezni lichinok khlebnogo zhuka) [The grain beetle (diseases of its larvae)]*. Odessa: Komissiia, sostoiashchaia pri Odesskoi zemskoi uprave dlia razrabotki voprosa o vrednykh dlia zemledeliia nasekomykh.

Mechnikov, Ilia I. 1879b. Miuskardina khlebnogo zhuka [The grain beetle Muscardin]. *Zemledel'cheskaia gazeta [Agricultural Gazette]* 25:185.

Mironov, Boris. 2012. *The standard of living and revolutions in Russia, 1700–1917*, ed. Gregory L. Freeze, Abingdon: Routledge.

Mokrzhetskii, S. A. 1900. *Opytnye stantsii dlia issledovaniia vrednykh nasekomykh v Zapadnoi Evrope i u nas v Rossii [Experimental stations for insect pest research in Western Europe and in Russia]*. Odessa.

Naylor, Simon. 2006. Nationalizing provincial weather: Meteorology in nineteenth-century Cornwall. *British Journal for the History of Science* 39:407–433.

Odessa Zemstvo Executive Board. 1879. *Stenograficheskii otchet o zasedanii komissii, sostoiashchei pri Odesskoi zemskoi uprave dlia razrabotki voprosa o vrednykh nasekomykh* [Stenographic report on the meetings of the commission affiliated with the Odessa Zemstvo Executive Board for discussing the question of insect pests]. Odessa.

Palladino, Paolo. 1996. *Entomology, ecology and agriculture: The making of scientific careers in North America, 1885–1985*. Amsterdam: Harwood Academic.

Pearson, Thomas S. 1989. *Russian officialdom in crisis: Autocracy and local self-government, 1861–1900*. Cambridge: Cambridge University Press.

Porchinskii, Iosif. 1879. *O vrednykh nasekomykh Iuzhnoi Rossii* [On pest insects in Southern Russia]. St. Petersburg: Tip. V.F. Demakova.

Pot'e. 1846. Sposob dlia istrebleniia saranchi v iuzhnykh chastiakh Rossii [Means to exterminate locusts in the southern parts of Russia]. *Zhurnal Ministerstva Gosudarstvennykh Imushchestv* 18:250–256; 2nd pagination.

Regional Congress of the Zemstvo Representatives of eight provinces of Southern Russia. 1882. *Trudy oblastnogo s'ezda predstavitelei zemstv 8 gubernii Iuzhnoi Rossii s 10 po 19 fevralia 1882 g. v Khar'kove po voprosu ob istreblenii khlebnogo zhuka i drugikh vrednykh v sel'skom khoziaistve nasekomykh* [Proceedings of the Regional Congress of the Zemstvo Representatives of eight provinces of Southern Russia from 10 to 19 February 1882 in Khar'kov on the question of exterminating the grain beetle and other insect pests harmful to agriculture]. Khar'kov.

Regional (3rd) Congress of Zemstvo Representatives from 8 Provinces of Southern Russia. 1883. *Trudy oblastnogo (III) s'ezda predstavitelei zemstv 8 gubernii Iuzhnoi Rossii po voprosu ob istreblenii khlebnogo zhuka i drugikh vrednykh nasekomykh s 10 po 21 fevralia 1883 v Khar'kove* [Proceedings of the Regional (3rd) Congress of Zemstvo representatives from 8 Provinces of Southern Russia on the question of the extermination of the grain beetle and other insect pests, 10–21 February 1883 in Khar'kov]. Khar'kov.

Regional Congress of Zemstvo Representatives from Southern Russian Provinces. 1885. *Trudy oblastnogo s'ezda predstavitelei zemstv iuzhnykh gubernii Rossii dlia izyskaniia mer bor'by s vrednymi nasekomymi i protiv epizootii s 11 po 20 fevralia 1885 g. v Khar'kove* [Proceedings of the Regional Congress of Zemstvo Representatives from Southern Russian Provinces on the question of insect pest control and against epizootic, 11-20 February 1885 in Khar'kov]. Khar'kov.

RES (Russian Entomological Society). 1872. *Protokoly obshchikh sobranii i otchet Russkogo entomologicheskogo obshchestva za 1871* [Minutes of the general meetings and the report of the Russian Entomological Society for 1871]. St. Petersburg.

RES (Russian Entomological Society). 1873. *Izvlechenie iz protokolov obshchikh sobranii Russkogo entomologicheskogo obshchestva za 1872 god* [Extracts from the Minutes of the general meetings of the Russian Entomological Society for 1872]. St. Petersburg.

RES (Russian Entomological Society). 1893. *Sostav Russkogo entomologicheskogo obshchestva k 1 dekabria 1892* [The membership of the Russian Entomological Society by December 1, 1892]. St. Petersburg.

Rul'e, Karl F., and August Farenkol'. 1847. *O rzhanom cherve, poedavshem ozim' v 1846 godu na prostranstve osmnadtsati gubernii. [About the grain worm that ate winter crops in eighteen provinces in 1846].* Moscow: Universitetskaia tipografiia.

Russia. 1881. No. 59608. 11 maia 1879. Vysochaishe utverzhdennoe polozhenie Komiteta Ministrov, ob'iavlennoe Senatu ministrom vnutrennikh del 22-go togo zhe maia [May 11, 1879, Statement of the Committee of Ministers approved by His Royal Highness and announced to the Senate by the Minister of the Interior on May 22]. *Polnoe sobranie zakonov Rossiiskoi Imperii, vtoroe sobranie [Complete Collection of Laws of the Russian Empire, 2nd Collection]* 54 (1): 362.

Russia. 1885. No. 125. 3 maia 1881. Vysochaishe utverzhdennoe polozhenie Komiteta Ministrov [May 3, 1881, Statement of the Committee of Ministers approved by His Royal Highness]. *Polnoe sobranie zakonov Rossiiskoi Imperii, tret'e sobranie, [Complete Collection of Laws of the Russian Empire, 3rd Collection]* 1: 56–57.

Russia. Ministry of State Domains. 1841. O dostavlenii v III departament ekzempliarov vrednykh nasekomykh [On conveying insect pest specimens to the 3rd Department]. *Zhurnal Ministerstva Gosudarstvennykh imushchestv [Journal of the Ministry of State Domains]* 1 (2): lix–lx.

Russia. Ministry of State Domains. 1847. Zemledel'cheskaia khronika [Agricultural Chronicle]. *Zhurnal Ministerstva Gosudarstvennykh Imushchestv* 24:165–168; 5th pagination.

Russia. Ministry of State Domains. 1850. Rasporiazhenie po Departamentu sel'skogo khoziaistva 15 avgusta 1850 [Instructions of the Department of Agriculture, August 15, 1850]. *Zhurnal Ministerstva Gosudarstvennykh Imushchestv* 57 (10–12): 43; 1st pagination.

Schwerdtfeger, F. 1972. Julius Theodor Christian Ratzeburg, 1801–1871. Vater der Forstentomologie, Wegbereiter der angewandten Entomologie. *Zeitschrift für Angewandte Entomologie* 70:260–266.

Secord, Anne. 1994. Corresponding interests: Artisans and gentlemen in nineteenth-century natural history. *British Journal for the History of Science* 27:383–408.

Semenov-Tian-Shanskii, Andrei P. 1910. Izdaniia obshchestva i nauchno-literaturnaia deiatel'nost' ego chlenov [Publications of the Society and scholarly and literary pursuits of its members]. *Trudy Russkogo entomologicheskogo obshchestva* 39:19–32.

Simashko, Iu. I. 1861. Ob otkrytii i uchrezhdenii Russkogo entomologicheskogo obshchestva [On the establishment of the Russian Entomological Society]. *Trudy Russkogo entomologicheskogo obshchestva* 1:49–50; 2nd pagination.

Skarzhinskii, Viktor. 1846. Skazanie o saranche [A legend about locusts]. *Zapiski Imperatorskogo Obshchestva sel'skogo khoziaistva iuzhnoi Rossii [Notes of the Imperial Agricultural Society of Southern Russia]* 3:131–146.

Smith, Ray F., Thomas E. Mittler, Carroll N. Smith, eds. 1973. *History of Entomology*. Palo Alto: Annual Reviews.

Sorensen, Conner. 1988. The rise of government sponsored applied entomology, 1840–1870. *Agricultural History* 62:98–115.

Sorensen, Willis Conner. 1995. *Brethren of the net: American entomology, 1840–1880*. Tuscaloosa: University of Alabama Press.

Special consultation of Zemstvo Representatives. 1881. *Khlebnyi zhuk. Protokoly i doklady osobogo soveshchaniia predstavitelei zemstv, sozvannogo dlia obsuzhdeniia voprosa o khebnom zhuke v Odesse 28 fevralia—6 marta 1881* [The grain beetle. Minutes and papers of the special consultations among Zemstvo Representatives convened for discussing the question of the grain beetle in Odessa on February 28–March 6, 1881]. Odessa.

Sukhova, Natalia G. 1993. Petr Ivanovich Koeppen kak geograf (k 200-letiiu so dnia rozhdeniia) [Petr Ivanovich Koeppen as a geographer (commemorating his bicentennial)]. *Izvestiia Russkogo geograficheskogo obshchestva* 5:1–11.

Sukhova, Natalia G., and Olga A. Krasnikova. 2002. K biografii P. I. Koeppena [Contributing to the biography of P. I. Koeppen]. In *Deiateli russkoi nauki XIX–XX vekov [Figures of Russian nineteenth and twentieth-century science]*, ed. M. Khartanovich, vol. 2, 31–61. St. Petersburg: Nestor-Historia Publishing House.

Terrall, Mary. 2010. Following insects around: Tools and techniques of eighteenth-century natural history. *British Journal for the History of Science* 43:573–588.

Vetter, Jeremy. 2011. Lay observers, telegraph lines, and Kansas weather: The field network as a mode of knowledge production. *Science in Context* 24:259–280.

Vucinich, Alexander. 1970. *Science in Russian culture, 1861–1917*. Stanford: Stanford University Press.

Zabarinskii, Petr A. 1887. *Entomologicheskie issledovaniia i ekskursii, proizvedennye v Bessarabskoi, Ekaterinoslavskoi, Poltavskoi, Tavricheskoi i Khersonskoi guberniiakh v 1887 g. [Entomological research and trips made in the Bessarabia, Ekaterinoslav, Poltava, Taurida and Kherson provinces in 1887]*. Odessa: Slavianskaia tipografiia N. Khrisogelos.

Marina V. Loskutova is an associate professor at the History Department at the National Research University-Higher School of Economics, St. Petersburg campus, Russia. She works on the history of exploration of natural resources and the interaction between academic communities, state administration, and the wider public in the nineteenth-century Russian Empire.

Anastasia A. Fedotova is a senior researcher at the St. Petersburg Branch of the Institute for the History of Science and Technology, Russian Academy of Sciences. She received her PhD in the history of plant ecology in Russia. Her research interests include the role of scientific knowledge in the rationalization of agriculture and forestry and history of applied biology from the mid-nineteenth to mid-twentieth century in the Russian Empire and the USSR.

Chapter 9
Nutrition Science and the Practice of Animal Feeding in Germany, 1850–1880

Brendan Matz

Introduction

During the second half of the nineteenth century, economic incentives and opportunities associated with rapid industrialization moved animal husbandry to the center of German agriculture. Livestock production became an important source of profit for farmers, and meat and milk products made their way to an increasing number of German tables. At the same time, animal husbandry became a significant area for scientific research and debate. In this chapter, I examine one of the fields of inquiry that was driven by and contributed to this shift in the relationship between plant and animal cultivation in Germany, the science of animal nutrition.[1] In addition to providing an overview of the field, including the core set of questions that animated scientists, I examine the efforts made by academics to make their work useful and relevant to practitioners. I argue that these translational efforts, which included the publication of manuals and articles for farmers and the dissemination of annotated tables of scientific data and feeding standards, did not go unheeded by German practitioners. The evidence considered here suggests that academic scientists provided practitioners with resources that proved valuable to some, especially elite practitioners. However, the sources also indicate that German farmers interested in improving or refining their feeding strategies had other resources available to them, namely publications and social forums where the "art" of animal feeding took precedence over its scientific aspects. During the period considered, German practitioners of animal feeding maintained a legitimate claim to expert knowledge about their animals and the specific places those animals inhabited.

[1] A sustained engagement with animal breeding and veterinary science, two other important fields of research related to animal production, lies outside the scope of this chapter.

B. Matz (✉)
Department of Technology, Culture and Society, New York University—Polytechnic School of Engineering, Brooklyn, NY 11201 USA
e-mail: brendan.matz@gmail.com

D. Phillips, S. Kingsland (eds.), *New Perspectives on the History of Life Sciences and Agriculture,* Archimedes 40, DOI 10.1007/978-3-319-12185-7_9

Historians have not explored the history of German agricultural animal feeding and its relationship to nutrition science in-depth, so this chapter inevitably involves some preliminary groundwork. The standard general histories of nutrition science emphasize human nutrition and the laboratory context,[2] and a comprehensive history of German animal husbandry written by a participant historian focuses on technical developments without considering the complex relationship between scientists and practitioners.[3] Two important exceptions are the work of Volker Klemm and Mark Finlay. Both have considered important aspects of animal feeding and nutrition research as part of broader projects.[4] In addition to their foundational work in this area, which has been invaluable for giving coherence to my narrative, recent scholarship in the history of nineteenth-century German agricultural science has begun to illuminate the rich historical terrain to be explored at the intersection of science and art, theory and practice. However, most of this work has focused on theories of heredity, genetics, and plant and animal breeding.[5] The story of animal feeding brings other dimensions of agricultural science that traditionally have been viewed as less practice dependent, namely animal chemistry and physiology, into the conversation. Although it does not address animal feeding in detail and focuses on a different historical period, Frank Uekötter's recent synthesis of German agriculture during the twentieth century is particularly helpful for understanding the changes considered here. Uekötter's analysis calls attention to the diversity of agricultural knowledge within specific historical contexts. I have found it helpful to think of animal feeding as part of a "knowledge system" that contains an array of "knowledge resources."[6] Academic scientists created some of these resources, but the reaction of practitioners to them varied and other useful resources emerged solely out of practice or drew from other domains of credible expertise. Animal feeding was a time-consuming, arduous, and messy enterprise that defied management by strictly "scientific" means.

The chapter is divided into three sections. Section 9.1 provides some background on the history of animal feeding in Germany to around 1850, which is designed to set a baseline for tracking the changes that follow as well as to identify continuities across time. Section 9.2 examines the rise of extensive scientific interest in animal feeding that began around mid-century. The year 1880 serves as the chronological end point of the chapter because it marks the beginning of a new era for nutrition research and its relationship to practice. This discussion relies on a sample of the

---

[2] See Carpenter (1995). McCollum (1957). An important edited volume on the history of nutrition science also does not engage with the animal feeding context of nutrition research. See Cunningham and Kamminga (1995).

[3] Comberg (1984).

[4] Finlay (1992). Klemm (1992). Klemm also coauthored a series of helpful bibliographic sketches of important figures in German agricultural science, including animal nutrition researchers. See Müller and Klemm (1988).

[5] For German plant breeding, see Wieland (2004). Harwood (2005), *Technology's dilemma.* My own work on German and American animal breeding considers the relationship between science and practice in that area of research. See Matz (2011). See also, Wood and Orel (2001).

[6] Uekötter (2010, pp. 24–36).

voluminous scientific and "advice" literature on animal feeding and nutrition published at the time, most of which has received little attention from historians of science and technology. The handbooks, tables, and feeding standards provide a link to Sect. 9.3, which considers the impact of expert scientific advice on practitioners and the persistence of alternatives.

Animal Feeding in Germany to 1850

Prior to the nineteenth century, the cultivation of livestock in the German lands was deemed subordinate to plant agriculture, and farmers did not tend to set aside land for growing cattle, sheep, and pig feed. Livestock were viewed as a "necessary evil": They provided manure and power for work, but they required more attention than they were worth.[7] By the end of the eighteenth century, many German farmers began to pay greater attention to farm animals and their manure. Countless agricultural treatises emphasized the benefits of making improvements in the standard three-field system by integrating livestock into it. Rather than viewing farm animals as an unavoidable burden, they began to recognize the potential animal husbandry had for improving grain-based agriculture. With more extensive and deliberate use of manures, grain production could be increased significantly. Yet this goal could only be achieved through improvements in the overall feeding and care regimen for livestock. To address the feeding issue, they began to plant feed crops—such as clover, legumes, beets, and potatoes—in the field that traditionally had remained fallow.[8]

The fodder crops presented opportunities for practitioners to make livestock better serve their farm enterprises, but the new feeding regimens also ushered in challenges. Questions about what foods to use and in what amounts became commonplace among farmers, spurring on experts to provide more comprehensive advice on feeding animals.[9] One of the most influential of these advisers was Albrecht Thaer. Trained as a medical doctor, Thaer developed an interest in gardening and farming and became a central figure in German agricultural research and education during the first half of the nineteenth century.[10] Dissatisfied with the lack of practical engagement with agriculture within the university system, he set up a private school on a large estate in Möglin outside Berlin in 1806. His observations and experiments conducted at Möglin lay the foundation for guidelines for feeding animals centered around two principles: the integration of plant and animal agriculture and the "hay value."

[7] Abel (1967, p. 239).

[8] Abel (1967, pp. 306–310).

[9] Gohren (1872, p. 8).

[10] For biographical information on Thaer and a discussion of his importance in the history of German agricultural science, see Klemm and Meyer (1968).

According to Thaer, plant and animal agriculture should not be approached as separate issues because farms only flourished when they properly integrated the two. In his *Grundsätze der rationellen Landwirthschaft* (*Principles of Rational Agriculture*), he wrote: "The increase in the number of animals raises the productivity of the land because of improved dung recovery, and the increase in plant production raises the demand for animals." Describing the reciprocal influence of plants and animals on the farm as a "great balance-wheel," he advised purchasing and managing livestock with the goal of optimizing both the production of useful animals and the generation of plant fertilizer.[11] Using hay as the standard of measurement (because it was the most common livestock feed at the time), he set out to establish the equivalent value of various feeds for achieving these goals. For example, according to his calculations, 100 lb of hay had the same practical value for feeding adult cattle as 44 lb of grain, 200 lb of potatoes, and 400 lb of green clover. This approach to feeding was supposed to simplify the day-to-day operation for practitioners, helping them choose from an expanding class of fodder crops, calculate basic feed requirements given specific farm conditions, and estimate manure output.[12]

Although Thaer grounded his approach to feeding in the everyday realities of practice, he did not neglect the natural sciences. Indeed, the model for higher agricultural education that he established at Möglin, which became the standard approach in Germany into the 1860s, stressed the importance of bringing together science and practice as "equal and complementary partners."[13] His feeding trials were guided by the prevailing chemical understanding of the day, which proposed that plants possess a vital force that enables them to derive nourishment from finely divided organic matter in the soil, or humus.[14] His students and other academics working within the academy system attempted to refine his feeding methods through additional observation and experimentation. For example, August von Weckherlin, who served as the director of the Agricultural Academy in Hohenheim from 1837 to 1845, continued to emphasize the close integration of plant and animal agriculture and to operate within Thaer's hay valuation framework.[15] However, he also sought to bring this approach to feeding in line with new developments in chemistry and veterinary science as well as his own practical experience and field experiments. In his widely read and acclaimed animal husbandry manual from 1846, he drew from an expanding scientific literature on the chemical makeup of different animal feeds to develop practical advice for farmers assumed not to have a background in these fields. An important element of this translational effort was the construction of tables of feed equivalents that assisted practitioners in making individual judgments on the farm that were grounded in the latest scientific understandings.[16] These sug-

[11] Thaer (1809, p. 252).

[12] Thaer (1809, pp. 252 and 258–285). See also, Klemm (1992, p. 140); Müller and Klemm (1988, pp. 49 and 112–113).

[13] Harwood (2005, pp. 79 and 84). Quotation at 84.

[14] Klemm (1992, p. 132) and Brock (1997, pp. 146–147).

[15] For Weckherlin, see Hermann (1980).

[16] Weckherlin (1846, pp. 98–193).

gestions for what and how much to feed various animals for specific purposes exhibited the commitment to science and practice that characterized the agricultural research and education system in Germany at mid-century.

Nutrition Science and Intensive Production Ascendant

While the integration of plant and animal agriculture was of primary importance in Thaer's system, he did see the potential of improved animal husbandry to augment the economic situation in the German states. He viewed the Spanish Merino breed of sheep, whose fine fleece was highly coveted internationally, as particularly promising and wrote a textbook for sheep breeders under the auspices of the Prussian government.[17] During the first half of the nineteenth century, his vision came to fruition, as Saxony and Prussia became important world centers for the breeding of Merinos and the export of fine wool.[18] But with the exception of those in regions where fine-wool sheep husbandry took root and flourished, few farmers found themselves in a position to invest in improved breeds and feeding strategies for competitive trade. By the time of the publication of Weckherlin's manual in 1846, however, market conditions had begun to shift in favor of more intensive animal husbandry.

Although the socioeconomic and political causes of this shift are complex, it can be attributed in brief to the increase in demand for animal products in the cities, as the interrelated processes of industrialization, urbanization, and population growth took effect. At the beginning of the nineteenth century, there was a legal separation between rural and urban economies. But with the introduction of liberal free trade statutes beginning in 1810, the boundary between country and city became more porous. Moreover, when people left the countryside to earn money wages, they became dependent on these new urban market systems. Steadily rising incomes allowed city workers to improve their diets and purchase more expensive food commodities, such as meat and milk. This increase in demand for animal products led to incentives and opportunities for farmers to produce more.[19]

Scientists and progressive practitioners interested in increasing meat and milk production saw improved feeding as one viable strategy for achieving this goal. The hay value approach to feeding, which had been refined through the work of figures like Weckherlin, emerged as a target of their criticism. The tables that appeared in feeding handbooks and in agricultural journals were rarely in agreement, which presented problems for farmers looking for definitive guidance in their day-to-day

[17] See Thaer (1811), *Handbuch*. At page three, Thaer sought to convince his readers that fine-wool sheep could be a highly profitable investment and encouraged them to visit the wool markets in Berlin and Breslau as proof.

[18] Körte (1862) and Wood and Orel (2001, pp. 152–170).

[19] Teuteberg (2007) and Achilles (1993, pp. 212–214 and 252–255).

operations.[20] Meadow hay, which served as the standard of measurement, was subject to great variation in its quality. Moreover, the hay value approach did not take into consideration the constituents of the various feeds.[21] Comparisons between hay and legumes, for example, did not factor in the important difference in protein content. Only systematic chemical analysis could address such issues.

Improvements in methods of organic analysis, many of which occurred in the 1830s through the labors of Justus von Liebig and his Giessen School, allowed chemists to determine precisely the chemical constituents of a given sample.[22] For animal feeding, this meant that components such as fats, carbohydrates, proteins, and inorganic materials could now be isolated and measured in a quantity of hay or other fodder crop. Chemical analysis also facilitated the study of the differential effects of feed components on animal growth and health. The French chemist Jean-Baptiste Boussingault, who set up an experimental estate in Alsace in 1836, made early advances in this area by determining the nitrogen content of a large number of foods and evaluating the effects of nitrogen intake on livestock. His experiments showed that nitrogenous constituents of food were extremely important in feeding, leading him to develop tables for farmers that rated foods (still in relation to a hay standard) in terms of these constituents.[23] Feeding standards, or norms, for the different types of farm animals followed, becoming an integral part of how knowledge about animal nutrition was disseminated in a useful form to farmers.

In the German context, the earliest feeding tables were based on research conducted at the first agricultural experiment stations. As Mark Finlay has shown, these institutions grew out of a perceived need on the part of groups of farmers and estate owners to pursue scientific research that answered practical questions.[24] Although there were some investigations being conducted at the more prominent agricultural academies,[25] the work done there was insufficient to meet the needs of a growing number of practitioners seeking to make use of valuable knowledge and, at the same time, protect themselves from the errors of too ambitious theorizing. From the very beginning, the analysis and chemical evaluation of feeds was an important part of the experiment station agenda. At the first German agricultural experiment station, which was established at Möckern in Saxony in 1851, the pursuit of answers to practical animal feeding questions was written into the bylaws. Research conducted in the "closest connection to practical efforts of various kinds" was to include the following: "…the constituents of plants and their effect on the animal organism,

---

[20] Weckherlin himself referred to the disagreements about the hay value approach and the ongoing efforts to make it more accurate and consistent. See Weckherlin (1846, p. 174).

[21] Müller and Klemm (1988, pp. 112–113).

[22] Brock (1997, pp. 37–71).

[23] On Boussingault, see McCosh (1984). The historical section of the report on the Rothamsted Experiments provides a useful examination of Boussingault's work with nitrogenous feeds. See Lawes and Gilbert (1895, pp. 252–260).

[24] Finlay (1992), "Science, practice, and politics"; Finlay (1988).

[25] See, e.g., Haubner (1837). Haubner performed his experimental work on digestion in ruminants while at the Eldena Agricultural Academy.

especially in terms of feeding, [and] the analysis and evaluation of feeds for the different goals of animal feeding."[26]

Reports of the research conducted during the early years of Möckern's operation (from 1851 to 1857) reveal a wide range of practice-oriented inquiries into animal feeding and nutrition. In some respects, these early experiments were similar to the investigations carried out during the first half of the nineteenth century. Working in the Thaer tradition, researchers compared the effects of different feeds on livestock with an eye to the integration of plant and animal agriculture, often paying as much attention to the relationship between feeding and dung output as they did on meat and milk production. However, key differences began to emerge as well. The experiments involved increasingly detailed chemical analysis of the feeds used and, in the case of milk, the product itself. As part of this chemical analysis, the amounts of nitrogenous versus nonnitrogenous constituents were carefully determined, and the use of this type of analysis to foster more intensive production began to be considered. An important component of this gradual shift in emphasis towards intensive feeding was the evaluation of so-called power feeds, such as coarse grain meal, seeds, and oil cakes. For example, in one series of experiments, station researchers observed the effect of adding rapeseed cakes to a hay-based diet on the production of milk in cows. They came to the conclusion that nitrogen-rich foods like rapeseed did in fact have a significant impact on milk yields and offered tentative suggestions for the ratio of nitrogenous to nonnitrogenous components for intensive feeding.[27]

As quantitative data of this kind accumulated, it became the foundation for the new feeding standards. Hubert Grouven, an agricultural chemist who served as the inaugural director of the agricultural experiment station at Salzmünde (also in Saxony) from 1859 to 1866, made the first attempt to systematically review the experimental data on animal feeding generated at Möckern and elsewhere. In a collection of lectures published in 1859, while he was still operating a private chemical laboratory on his family estate near Cologne, he proposed daily-feeding norms for the various farm animal types based on their weight and intended use. For instance, a milk cow weighing 1400 lb required the following: 30 lb of "dry matter," 4 lb of protein, 1.3 lb of fat, and 16.7 lb of carbohydrates. The ratio of nitrogenous to nonnitrogenous components for this diet was given as 1:5. For the intensive fattening of cattle weighing 1400 lb, he suggested an even more nitrogen-rich recipe with a ratio of 1:4.4.[28] Grouven first introduced his standards in lectures given before agricultural societies in and around Cologne, but he later expanded on them in a more thorough consideration of the experimental literature on animal feeding.[29] He continued to pursue research into nutrition at Salzmünde,[30] but Möckern's own first director, Emil Wolff, would soon become the most prominent expert advisor on animal feeding.

[26] Crusius (1857, p. xii).

[27] Crusius (1857, p. xii). On the use of power feeds, see Svoboda (1915).

[28] Grouven (1859, p. 591).

[29] Grouven (1863, pp. 356–366).

[30] See Grouven (1866, pp. 187–190).

Wolff, who left Möckern in 1854 to become professor of chemistry at the Agricultural Academy in Hohenheim, introduced an innovative approach to feeding that garnered widespread and prolonged attention. Trained as a chemist, he gained practical agricultural experience early in his career and became committed to finding ways to make scientific research intelligible and useful for practitioners.[31] His treatise from 1861, *Die landwirthschaftliche Fütterungslehre und die Theorie der menschlichen Ernährung* (*Agricultural Animal Feeding and the Theory of Human Nutrition*), reflected this perspective. Recognizing that "our science is not yet developed enough to offer short and definite tenets as guidelines for farmers," he nevertheless saw the value of using data collected from "carefully carried out feeding experiments" to produce a series of tables to be used by practitioners. These tables were meant to allow them to register "at a glance" all the important experimental results pertaining to the quantity and quality of feeds.[32] In this way, he took on the role of exacting arbiter of the voluminous data being generated within various investigative settings. Experiments pertaining to the digestion and nutritional value of feeds drew particular attention from Wolff. Indeed, he dedicated the entire first part of the treatise to a detailed examination of the chemistry, anatomy, and physiology of digestion.[33] The earlier experimental method, which focused primarily on chemical analysis, told investigators little about what was going on inside the animal. To answer questions about what components of specific feeds and feed mixtures were actually digested and put to productive use by the body, a meticulous accounting of the food taken in and the waste products going out was necessary. Wolff himself conducted a version of this type of experiment at the Agricultural Experiment Station at Hohenheim in an effort to identify the digestibility of a variety of common feeds for sheep.[34] The digestibility tables that he constructed from experimental averages listed typical feeds, such as alfalfa and sorghum, and provided specific numbers for the chemical makeup of each, including fiber content.[35]

In addition to these tables, Wolff translated his data into detailed feeding standards. He first published these in the 1864 volume of *Mentzel und v. Lengerke's Verbesserter Landwirthschaftlicher Hülfs-und Schreib-Kalender*, an important reference work and calendar for German farmers that contained a wealth of practical advice on all aspects of agriculture. The journal continued to publish the Wolff standards with little modification until 1896, and Wolff promoted them himself through his publications written expressly for practitioners.[36] Unlike Grouven, who

[31] For Wolff, see Müller and Klemm (1988, pp. 114–121).

[32] Wolff (1861, pp. v–vi). Wolff's effort to draw a connection between animal and human nutrition research is significant and deserves historical attention.

[33] Wolff (1861, pp. 363–388).

[34] Wolff (1870, pp. 57–111).

[35] See, e.g., Wolff (1861, pp. 460–464).

[36] Literature written specifically for farmers was not uncommon in the eighteenth century, but this literature took on a new quality in the nineteenth century. For example, the traditional farmers' calendars began to serve more specific educational goals. See Haushofer (1963, pp. 79–80). The Mentzel and v. Lengerke calendar illustrates this process.

calculated daily diets based on the total amount of gross material consumed, he based his standards on the amount of digestible materials in an average sample of a specific feed or feed mixture. For the fattening of oxen and cows at a standard weight of 1000 lb, he advised a daily feeding regimen consisting of 23.5 lb of total dried "organic matter." This total was to include 6 lb of fiber, 3.2 lb of digestible nitrogenous substances, and 14.3 lb of digestible nonnitrogenous material (1.2 lb of fat). To arrive at these figures, he used fiber as an estimate of the indigestible portion of feed and maintained a set ratio of 1:4 for nitrogenous to nonnitrogenous elements. Along with the breakdown of optimal feed components, he offered a series of specific daily diets based on digestibility data for each animal type and feeding goal. Twelve such formulas were given for cattle fed for meat production, including the following: 10 lb of meadow hay, 5 lb of wheat straw, 62 lb of turnips, 4 lb bean meal, 2 lb of flax seed, and 2.5 lb of rye bran.[37] Wolff's standards provided clear-cut formulas for putting the new theories of nutrition into practice and stood alongside similar practical tools in the calendar, including a method of fertilizer preparation and a chart for determining the sugar content of beets.[38]

Wolff's attention to physiological questions points to a broader shift towards the bridging of chemistry and physiology that took place in German agricultural science during the 1860s and 1870s. This transition can be explained in part by the new ways of thinking about animal nutrition introduced by Liebig. In his 1842 treatise *Die Their-Chemie, oder die organische Chemie in ihrer Anwendung auf Physiologie und Pathologie* (*Animal Chemistry, or Organic Chemistry in Its Application to Physiology and Pathology*), he theorized that foods containing nitrogen played such an important role in animal nutrition because they comprised the organized tissue of the body and provided all of the energy for muscular work. These plastic aliments were contrasted with the respiratory, or nonnitrogenous, aliments, which were thought to undergo oxidation during the process of respiration to produce animal heat.[39] While much of Liebig's nutritional theory remained speculative, his work raised an array of compelling questions that inspired other research at the intersection of chemistry and physiology. At the Munich Physiological Institute, the physiologist Theodor Bischoff and his assistant and Liebig student Carl Voit performed a series of groundbreaking nutritional experiments on dogs between 1857 and 1859. Guided in part by Liebig's theory, these experiments determined with remarkable precision the intake and outgo of nitrogen in carnivorous animals under specific conditions, including physical exertion. Later variations on the experimental design, which employed an enclosed respiration chamber designed by Max von Pettenkoffer, allowed them to measure respiratory gaseous exchanges as well.[40] By the late 1860s, Voit's research showed that animal metabolism was more complex than Liebig's theory had allowed, with some protein being oxidized during respira-

[37] Mentzel (1869, pp. 10–17).

[38] Mentzel (1869, pp. 18, 27).

[39] Liebig (1842, p. 120).

[40] Holmes (1988).

tion along with carbohydrates and fats. However, Voit continued to maintain the distinction between plastic and respiratory aliments.[41]

The physiological chemistry pursued by Liebig and the Munich Institute had only indirect bearing on agriculture, but Wilhelm Henneberg, another Liebig student, soon put the new theories and methods to use in agricultural research. Beginning in the 1860s and continuing through the 1870s, agricultural scientists at the experiment stations exerted greater influence on the direction of research and shifted priorities away from the achievement of immediate practical goals towards the examination of basic scientific questions.[42] Henneberg, who became the director of the Agricultural Experiment Station at Weende near Göttingen in 1857, played a major role in this shift by introducing a research program using livestock as experimental organisms that incorporated the sophisticated methods and instruments employed at Munich. He stressed precision and quantification, the use of instruments like the Pettenkoffer respiration apparatus, the control of environmental variables in a laboratory setting, and attention to fundamental questions about animal metabolism.[43] The findings of the experiments he conducted with his assistant Friedrich Stohmann were directed at "farmers and physiologists," but the authors made little effort to translate their findings into practical suggestions or guidelines.[44]

Figures like Henneberg challenged Thaer's earlier emphasis on the close integration of science and practice, as well as his insistence on the integration of plant and animal agriculture. Their research agenda was specialized and directed at fundamental questions, and the system of agricultural higher education many of them advocated called for Liebig's more science-intensive curriculum. The agricultural academies built on Thaer's model began to receive competition from agricultural institutes like the one established at the University of Halle in 1862, and by 1870 nearly all of the agricultural academies had either disappeared or been folded into universities.[45] During the period considered, agricultural science rose to great prominence in Germany, and the science of animal nutrition was one of its most successful branches.[46] Nevertheless, the rise to prominence of nutrition science in Germany did not signal the disappearance of more practice-oriented approaches. In the final section, we focus on the relationship between the scientific advice proffered by academics and the everyday practice of feeding.

---

[41] Brock (1997, pp. 69–74).

[42] Finlay (1992, pp. 126–133).

[43] For Henneberg, see Lehmann (1890).

[44] Henneberg and Stohmann (1860). The second volume of the treatise appeared in 1864.

[45] On the rise of the university institutes, see Harwood (2005, pp. 80–84).

[46] In the years following 1880, experimental research into animal feeding changed, with academic scientists beginning to examine the thermodynamic foundations of metabolic processes. The relationship between nutrition science and agricultural practice also entered a new era. See Treitel (2008). Rosen (1959).

The Practice of Animal Feeding

The literary productions of academics clearly found a readership among some practitioners. To take Emil Wolff as an example, his manual written for farmers, *Die rationelle Fütterung der landwirthschaftlichen Nutzthiere* (*The Rational Feeding of Livestock*), sold 11,000 copies between its date of first publication in 1874 and the issuance of a third edition in 1881. Moreover, Wolff's practical feeding norms, which were included in abbreviated form in the manual, found an even wider readership through their publication each year in a widely circulated calendar and almanac for practitioners.[47] Although some of Henneberg's writings were likely less accessible, given their attention to method and fundamental scientific questions, he wrote with an understanding that he was entering into a conversation with a large community of both academics and practitioners who had a keen interest in topics such as, "Feeding by hay value or by chemical principles."[48] A cursory survey of some of the agricultural journals of the day confirms that discussion of scientific, or rational, animal feeding was a ubiquitous presence on their pages. Not only in *Die landwirtschaftlichen Versuchs-stationen* (*The Agricultural Experiment Stations*), a journal where one would expect to find this type of reportage but also in more practice-oriented publications such as the weekly *Annalen der Landwirthschaft* (*Annals of Agriculture*), the yearly *Landwirthschaftliche Jahrbücher* (*Agricultural Yearbooks*), and the monthly *Landwirthschaftliches Centralblatt für Deutschland* (*Central Agricultural Newspaper for Germany*). Looking more closely at the last of these, primarily because it was the most "practical" of the three, one finds that in the 1865 volume ten substantive articles on animal feeding appeared. Most of these engaged with the chemical and physiological aspects of feeding, including a long piece on the digestibility of whole grains in pigs. In addition to these feeding-specific articles, there were reports of the chemical analysis of fodder crops.

The content of these German agricultural journals suggests that practitioners had a large amount of scientific information at their disposal, but we are still left with the question of what they thought about nutrition science and how it was put to use at the level of the estate or farm. A good starting point for answering these questions is the reports of the *Versammlung deutscher Land- und Forstwirthe* (*Assembly of German Foresters and Farmers*). Founded in Dresden in 1837, the organization drew inspiration from the work of Thaer and brought together representatives from both the world of academic science and practitioners with various backgrounds. To achieve its main goal of fostering personal interaction and the exchange of ideas between German farmers, the association held a series of "travelling conferences" in cities throughout the German-speaking lands. The meeting in Königsberg in 1863, at the high point of the association's popularity, drew 3307 participants.[49] The official report of the 1865 Dresden event provides a window into who the attendees

[47] See Wolff (1881, p. vii).

[48] Henneberg and Stohmann (1860, p. v).

[49] Haushofer (1936, pp. 76–77).

were and what they discussed. A survey of the membership list shows that although practitioners of various kinds belonged to the association, the majority of members were elite practitioners from large estates, state officials, and academics. In terms of geographic representation, most of the members came from Prussia, with the province of Saxony providing the largest number. However, the southern and western lands and Austria were also represented.[50] The topics encompassed all of the important branches of agriculture, and each topic was addressed in a separate session.

In the conference session on animal husbandry, a number of feeding-related questions were posed to the participants. For example, the group considered the degree to which various animal types digested plant fiber, the role fats played in relation to carbohydrates, and the effects of nitrogen-rich diets on animal form. They also addressed directly the issue of new feeding standards based on chemical and physiological principles. In response to the question of whether the new norms had "already found broad acceptance in practice," it was recognized by the participants that feeding based on the chemical components of feeds had been found to be useful by many and had even garnered some support among "small landholders." However, the sentiment was also expressed that a more straightforward method was desirable, one whereby the "nutritional value again could be expressed in a simple number." It was even suggested that hay values be retained and included along with the new standards.[51] At the gathering in Vienna in 1868, the participants returned to this theme of feeding standards. Again, there was a general acceptance of the importance of chemical and physiological research but with reservations about how to make it useful in practice. A number of practical questions were raised but left unanswered, such as whether farmers could point to experience in practice of specific diets leading to greater meat production.[52]

While these reports suggest that some German farmers—especially those from larger estates in Prussia—engaged seriously with the advice literature written by academics and even put it to use in feeding their animals, the sources also indicate that practitioners did not accept scientific advice unquestioningly. For one, academics did not always present a united front and often disagreed on key issues, including the best way to convey scientific knowledge to practitioners. For instance, Martin Wilckens, a professor of physiology and animal husbandry in Vienna, challenged the advisability of offering standard tables and feeding norms at all, arguing that the use of average numbers for the components of feeds and for their digestibility had "great statistical value but absolutely none for the individual cases to which they were applied."[53] Wilckens' viewpoint found support from one of the most prominent figures in German agriculture at the time, director of the Agricultural In-

[50] *Versammlung deutscher Land-und Forstwirthe* (1866, pp. 27–61).

[51] *Versammlung deutscher Land-und Forstwirthe* (1866, p. 196).

[52] *Versammlung deutscher Land-und Forstwirthe* (1869, pp. 201–204).

[53] Wilckens (1878, pp. 845–846).

stitute at Halle Julius Kühn, who also called for greater attention to specific experimental results and individual variation in livestock.[54] Differences in the viewpoints of academics can be accounted for in part by the different institutional settings in which they operated. Jonathan Harwood's work reveals the variety of institutional forms that agricultural scientists inhabited and the many constituencies they served. Although the agricultural disciplines, including animal feeding, became more science oriented during the period considered, many academics still felt pressure to maintain a close connection to the world of practice, especially in areas where small farms and less elite practitioners predominated.[55]

Yet even if we accept that many of the ideas and practices of nutritional science gained widespread acceptance among farmers, this does not mean that they had access to only one kind of knowledge resource. As Frank Uekötter's research illustrates, the history of the influence of science and technology on agriculture can be analyzed productively in terms of a "system of knowledge" that includes the contributions of academic scientists but is not circumscribed by them. The feeding of different species of animals for specific production goals was an extremely complex undertaking with numerous variables that were difficult to control. Academic advisors had little or nothing to say about many of these, including the important issue of the palatability of feeds. When academics did provide specific suggestions, they often admitted the need for skilled implementation by an experienced practitioner. For example, in his discussion of feeding norms in the fourth edition of his feeding manual, Wolff indicated that in practice one should not "too anxiously try to attain the exact numbers." Instead, the numbers were to serve as a "clue" to the practitioners who would then make adjustments and refinements based on their own experience or the experience of others.[56] Practitioners looking for advice from other farmers found resources at their disposal that addressed issues of immediate practical importance. For example, Heinrich Richter, a "practical farmer in Dahlen," wrote a manual on the various feed mixtures for milk cows. It too conveyed information about the "chemical components of feeds," but it did so from the level of the farm itself, where actual ingredients needed to be combined in specific ways to make a recipe work.[57] The more practice-oriented feeding manuals emphasized the equal standing of science and practice in the tradition of Thaer, rather than the application of basic scientific knowledge advocated by Liebig. As an example of one of these manuals informed its reader: "…the work of the practitioner essentially consists of independently applying and exploiting the achievements of science for the use and glory of practice, such that through his own research and effort he transforms the treasures of science into practical gain."[58] Another handbook put the relationship

[54] Kühn (1878, p. 127).

[55] Harwood (2005, pp. 111–174).

[56] Wolff (1885, pp. 143–144).

[57] Richter (1859).

[58] Gohren (1872, p. vii).

between science and practice in similar terms: "The farmer can only derive value from [scientific research] if he works and strives along with it."[59]

The work of the experienced animal husbandman at the level of the estate or farm was frequently talked about in terms of "art (Kunst)." Alan Marcus, in his discussion of the American context, has referred to art in the agricultural sense as a type of knowledge encompassing individual rather than general cases, lived experience rather than theoretical generalities.[60] Like all examples of what historians of science have referred to as "tacit knowledge," such knowledge is difficult to codify and grows out of the training of the senses as much as the intellect.[61] Within the culture of animal husbandry in Germany, practitioners were initiated into the art of feeding not only through the printed knowledge resources written by other farmers or practice-friendly academics. Those who could not afford to pay others to do the hard work on the farm also learned how to feed their animals by doing it from day to day over many years, looking closely at their own animals and those of their neighbors, and sharing experiences with others by word of mouth about what did and did not work. In other words, did a specific feeding recipe actually produce the results intended? Did the pig put on weight quickly? Did the cow produce more milk this year?

According to most expert advisors, even those with the closest relationship to the feeding stalls, art proved most powerful when balanced with science. To this point, the practical breeder and academic Hermann Settegast wrote in 1868: "Like agriculture on the whole, animal husbandry is both an art and a science."[62] However, the art of animal feeding, like its closely related pursuit the art of breeding, was fostered within agricultural circles that did not necessarily bear a strong impression from scientific expertise. Some of these forums for the cultivation and refinement of art drew participants from throughout the German lands. As previously discussed, the *Versammlung deutscher Land- und Forstwirthe* brought practitioners together at travelling conferences to share ideas and practices. Beyond the formal discussions that were transcribed, practitioners likely exchanged anecdotes, recipes, and subjective impressions that did not make it into the historical record. Beginning with the conference in Dresden in 1865, conference attendees could also participate in an agricultural exhibition where animals were judged based on their individual and breed characteristics. These public displays of animals provided animal feeders with visual models of animals transformed through careful breeding and feeding.[63] The tactile and visual acuity necessary to produce a prize-winning animal was of-

[59] Settegast (1872, p. 14), *Die landwirthschaftliche Fütterungslehre.*

[60] Marcus (1985, pp. 19–22).

[61] The literature on tacit knowledge in the history of science and technology is expansive. For a classic discussion of the concept of tacit knowing, see Polanyi (1966). On tacit knowledge in animal husbandry, see Grasseni (2007, pp. 47–66).

[62] Settegast (1868, p. vi).

[63] See Peterson (1883). Peterson indicates that one of the main goals of animal exhibitions was to encourage farmers to emulate the animal forms they encountered.

ten a topic of discussion in the practical literature. For example, one East Prussian estate owner praised the practical breeders of England who "through many years of individual, practical engagement with the breeding and feeding of animals and countless, splendid exhibitions that are visited with undivided attention...have an outstanding knowledge of animal bodies and an ability to determine their characteristics and performance capability."[64]

These "national" conferences and exhibitions played an important role in the circulation of art-related knowledge, especially after the establishment of the *Deutsche Landwirtschafts-Gesellschaft* (German Agricultural Society) in 1884.[65] But most forums for the cultivation of animal husbandry were regional or local and served the specific needs of groups of practitioners. Between 1815 and 1848, agricultural societies spread throughout the German states, forming what one historian has called an "almost entirely seamless network."[66] These societies, among other things, introduced farmers to new approaches to animal husbandry and often facilitated the introduction of livestock breeds from abroad.[67] In the 1870s and 1880s, more specialized arrangements emerged for the promotion of animal husbandry. These breeders' associations focused primarily on creating breed standards, selecting animals for breeding purposes, and maintaining registries of the approved animals that included detailed genealogical information.[68]

Although feeding was not the main focus of the associations, decisions about what type of animal to promote in a given area had a significant impact on the practice of feeding. Some farmers and estate owners embraced the high-performing imported breeds, which in some areas were used to dramatically transform the character of local animals. For instance, in the Miesbach district of Bavaria, practitioners introduced large and easily fattened Simmentaler bulls from Switzerland and bred them to the smaller regional animals to such an extent that the characteristics of the latter largely disappeared. An agricultural surveyor commissioned by the Bavarian government to report on the progress of animal breeding in the German states observed in 1864: "The old Miesbacher cattle disappears...and the products of crossing, which represent the form and color of the Simmentaler type, take its place."[69] Yet many of the early breeders' associations focused on the consolidation and improvement of regionally identified livestock strains, which exhibited a certain degree of uniformity by virtue of their geographic isolation over a long period. These animals were known for being well adapted to the climate and environmental conditions of specific regions and were therefore embraced as good all-purpose livestock for small and medium-sized farms.[70] While imported breeds

[64] Witt (1865, p. 296).

[65] On the *Deutsche Landwirthschafts-Gesellschaft,* or the DLG, see Hansen and Fischer (1936).

[66] Haushofer (1963, p. 79), *Die deutsche Landwirtschaft im technischen Zeitalter.*

[67] May (1856, pp. 86–89); and Schlögl (1954, p. 246).

[68] Martiny (1883).

[69] Göring (1864, p. 33).

[70] Comberg (1984, pp. 520–654) and Martiny (1883, pp. 154–183).

like the British Shorthorn and the Spanish Merino exhibited remarkable production characteristics and came to embody rational and scientific animal husbandry, they required special feeding regimens, did not satisfy the diverse needs of peasant agriculture, and did not adapt well to some climates.[71] The demanding feeding regimen for improved breeds often involved "power feeds" produced off the farm from the fruits and seeds of blossoming plants, which added additional expense to the upkeep of livestock.[72]

Conclusion

Between 1850 and 1880, German academic scientists took increasing interest in animal feeding and nutrition. They conducted research at the new agricultural experiment stations and also in the academy and university setting, and their work led to new understandings of the chemistry of feeds as well the physiological processes of livestock metabolism. Academic scientists sought to make this knowledge accessible to practitioners by lecturing, writing treatises and practical manuals, and organizing important data and experimental results into tables. They also published standards and norms based on the differential effects of feed components on meat and milk production and the degree of digestibility of feeds. Within the sphere of formal agricultural education, they placed greater emphasis on scientific proficiency and introduced new methods and tools for the measurement of the intake and outgo of the chemical components of feeds and respiratory gas exchange.

Some agricultural practitioners saw the principles of scientific animal feeding as useful and translatable to practice, such as the group of predominantly elite farmers associated with the Assembly of German Foresters and Farmers. The emphasis on the chemical components of specific feeds and feed mixtures and their degree of digestibility made sense to them as a useful strategy for increasing the productivity of their animals. The rising demand for meat and milk products in expanding industrial cities created incentives for this kind of innovation. Many of the farmers who relied on animals bred for increased productivity also embraced the tables, feeding standards, and scientific advice literature. However, practitioners who found nutrition science useful did not encounter a discrete body of knowledge ready-made for application. Academic scientists like Emil Wolff tried to work from averages and consensus data, but the scientific community still disagreed on many crucial points.

Beyond the knowledge resources offered by academic scientists, practitioners had access to other kinds of advice. They could delve into a more practice-oriented literature that engaged with questions often left unexplored by academics, and they could meet face to face at national conventions and local and regional association gatherings to exchange experiences and feed recipes. At animal exhibitions and

[71] See May (1875, pp. 49–74).

[72] These power feeds had become so common by the 1880s that microscopic methods were developed to help practitioners determine their purity and authenticity. See Benecke (1886).

through daily farming practice, they also had the opportunity to cultivate the art-related aspects of feeding, including skilled eyes and hands for identifying a properly fattened steer or a promising milk producer. Few practitioners would have argued that practice required no scientific rationalization, but farmers who viewed animal feeding as an applied science were equally rare. Adherents of Thaer's approach to the farming enterprise did not disappear, and strong arguments continued to be made for the equal partnership of science and practice. Indeed, the hay value standards themselves continued to be used by practitioners who found that they worked for them, despite the opinion of many academics that they were archaic and unmodern. The integration of plant and animal agriculture also did not disappear. The promise of success for many practitioners, especially the nonelite, lay in intimate local knowledge of their animals and the plants they ate as well as the local climate, geography, and market conditions, rather than in the embrace of scientific theories, animals, and feeds from off the farm.

References

Abel, Wilhelm. 1967. *Geschichte der deutschen Landwirtschaft vom frühen Mittelalter bis zum 19. Jahrhundert*. Stuttgart: Verlag Eugen Ulmer.

Achilles, Walter. 1993. *Deutsche Agrargeschichte im Zeitalter der Reformen und der Industrialisierung*. Stuttgart: Verlag Eugen Ulmer.

Benecke, Franz. 1886. *Anleitung zur mikroskopischen Untersuchung der Kraftfuttermittel auf Verfälschungen und Verunreinigungen*. Berlin: Verlag von Paul Parey.

Brock, William H. 1997. *Justus von Liebig: The chemical gatekeeper*. Cambridge: Cambridge University Press.

Carpenter, Kenneth J. 1995. *Protein and energy: A study of changing ideas in nutrition*. Cambridge: Cambridge University Press.

Comberg, Gustav. 1984. *Die deutsche Tierzucht im 19. Und 20. Jahrhundert*. Stuttgart: Verlag Eugen Ulmer.

Crusius, Wilhelm, ed. 1857. *Agriculturchemische Untersuchungen und deren Ergebnisse angestellt und gesammelt bei der landwirthschaftlichen Versuchstation zu Möckern in den Jahren 1851 bis 1857*. Leipzig: Verlag von Georg Wigand.

Cunningham, Andrew, and Harmke Kamminga, ed. 1995. *The science and culture of nutrition, 1840–1940*. Amsterdam: Editions Rodopi.

Finlay, Mark. 1988. The German agricultural experiment stations and the beginnings of American agricultural research. *Agricultural History* 62:41–50.

Finlay, Mark. 1992. Science, practice and politics: German agricultural experiment stations in the nineteenth century. PhD dissertation, Iowa State University.

Gohren, Theodor von. 1872. *Die Naturgesetze der Fütterung der landwirthschaftlichen Nutzthiere*. Leipzig: Verlag von C.L. Hirschfeld.

Göring, P. J. 1864. *Wanderbeiträge zur Thierzucht*. Erlangen: Verlag von A. Deichert.

Grasseni, Cristina. 2007. Good looking: Learning to be a cattle breeder. In *Skilled visions: Between apprenticeship and standards*, ed. Cristina Grasseni, 47–66. New York: Berghahn Books.

Grouven, Hubert. 1859. *Vorträge über Agricultur-Chemie mit besonderer Rücksicht auf Thier-und Pflanzen- Physiologie*. Köln: F.C. Eifeu's Konigl. Hof-Buch-und Kunsthandlung.

Grouven, Hubert. 1863. *Kritische Darstellung aller Fütterungsversuche und der deraus für den praktischen Gebrauch resultirenden chemischen Fütterungs-Normen*. Köln: Druck und Verlag von Wilh. Hassel.

Grouven, Hubert. 1866. *Salzmünde: Eine landwirthschaftliche Monographie*. Berlin: Wiegandt und Hempel.

Hansen, Johannes, and Gustav Fischer. 1936. *Geschichte der deutschen Landwirtschafts-Gesellschaft*. Berlin: Deutsche Verlagsgesellschaft.

Harwood, Jonathan. 2005. *Technology's dilemma: Agricultural colleges between science and practice in Germany, 1860–1934*. Bern: Peter Lang.

Haubner, Gottlieb Karl. 1837. *Über die Magenverdauung der Wiederkäuer nach Versuchen*. Anclam: Verlag von W. Dietze.

Haushofer, Heinz. 1963. *Die deutsche Landwirthschaft im technischen Zeitalter*. Stuttgart: Verlag Eugen Ulmer.

Henneberg, Wilhelm, and Friedrich Stohmann. 1860. *Beiträge zur Begründung einer rationellen Fütterung der Wiederkäuer: Praktisch-landwirthschaftliche und physiologische Untersuchungen für Landwirthe und Physiologen, Erstes Heft*. Braunschweig: C. A. Schwetschke und Sohn.

Hermann, Klaus. 1980. August von Weckherlin. In *Lebensbilder aus Schwaben und Franken*, ed. Robert Uhland, vol. 14, 192–218. Stuttgart: W. Kohlhammer Verlag.

Holmes, Frederic L. 1988. The formation of the Munich school of metabolism. In *The investigative enterprise: Experimental physiology in nineteenth-century medicine*, ed. William Coleman and Frederic L. Holmes, 179–210. Berkeley: University of California Press.

Klemm, Volker. 1992. *Agarwissenschaften in Deutschland: Geschichte-Tradition von den Anfängen bis 1945*. St. Katharinen: Scripta Mercaturae Verlag.

Klemm, Volker, and Günther Meyer. 1968. *Albrecht Daniel Thaer: Pioneer der Landwirtschaftswissenschaften in Deutschland*. Halle: Niemeyer.

Körte, A. 1862. *Das deutsche Merinoschaf*. Breslau: Verlag von Joh. Urban Kern.

Kühn, Julius. 1878. *Die zweckmässigste Ernährung des Rindviehs vom Wissenschaftlichen und praktischen Gesichtspunkte, Siebte Auflage*. Dresden: G. Schönfeld's Buchhandlung.

Lawes, John Bennet, and J. Henry Gilbert, 1895. *The Rothamsted experiments*. Edinburgh: William Blackwood and Sons.

Lehmann, Franz. 1890. Wilhelm Henneberg. *Journal für Landwirthschaft* 38:503–546.

Liebig, Justus von. 1842. *Die Thier-Chemie, oder die organische Chemie in ihrer Anwendung auf Physiologie und Pathologie*. Brauschweig: Verlag von Friedrich Vieweg und Sohn.

Marcus, Alan I. 1985. *Agricultural science and the quest for legitimacy: Farmers, agricultural colleges, and experiment stations, 1870–1890*. Ames: Iowa State.

Martiny, Benno. 1883. *Die Zucht-stammbücher aller Länder*. Bremen: Druck und Verlag von M. Heinsius.

Matz, Brendan. 2011. Crafting heredity: The art and science of livestock breeding in Germany and the United States, 1860–1914. Ph.D. dissertation, Yale University.

May, Georg. 1856. *Die Vieh-Stämme und Schläge, und der Zustand der Rindvieh-Zucht Bayerns*. Landshut: Krüll'sche Universitäts-Buchhandlung.

May, Georg. 1875. *Die Erfolge der englishen Shorthornzucht in Deutschland: Ein historisch-statistischer Beitrag zur Shorthornzucht*. Berlin: Verlag von Wiegandt, Hempel, und Parey.

McCollum, Elmer. 1957. *A history of nutrition: The sequence of ideas in nutrition investigations*. Boston: Houghton Mifflin.

McCosh, Frederick W. J. 1984. *Boussingault: Chemist and agriculturalist*. Dordrecht: D. Reidel.

Mentzel, Oskar, ed. 1869. *Mentzel und v. Lengerke's Verbesserter Landwirthschaftlicher Hülfs- und Schreib-Kalender auf das Jahr 1869, Zweiter Theil*. Berlin: Wiegandt & Hempel.

Müller, Hans-Heinrich, and Volker Klemm. 1988. *Im Dienste der Ceres: Streiflichter zu Leben und Werk bedeutender deutscher Landwirthe und Wissenschaftler*. Leipzig: Urania Verlag.

Peterson, C.1883. *Die landwirthschaftlichen Thier-Austellungen*. Bremen: Druck und Verlag von M. Heinsius.

Polanyi, Michael. 1966. *The tacit dimension*. Chicago: University of Chicago Press.

Richter, Heinrich. 1859. *Futtermischungen für Milchkühe mit Rucksicht aud die chemischen Bestandtheile der Futtermittel*. Dresden: Schönfeld'schen Buchhandlung.

Rosen, George. 1959. The conservation of energy and the study of metabolism. In *The historical development of physiological thought*, ed. Chandler Brooks and Paul Cranefield, 243–263. New York: Hafner Publishing Co.
Schlögl, Alois, 1954. *Bayerische Agrargeschichte*. München: Bayerischer Landwirtschaftsverlag.
Settegast, Hermann. 1868. *Die Thierzucht.* Breslau: Verlag von Wilh. Gottl. Korn.
Settegast, Hermann. 1872. *Die landwirthschaftliche Fütterungslehre.* Breslau: Verlag von Wilh. Gottl. Korn.
Svoboda, Hanno. 1915. *Die Erzeugung und Verwendung der Kraftfuttermittel.* Wien: A. Hartleben.
Teuteberg, Hans Jürgen. 2007. Urbanization and nutrition: Historical research reconsidered. In *Food and the city in Europe Since 1800*, ed. Peter Atkins, Peter Lummel, and Derek Oddy, 13–24. Aldershot: Ashgate.
Thaer, Albrecht. 1809. *Grundsätze der rationellen Landwirthschaft, Erster Band.* Berlin: In der Realschulbuchhandlung.
Thaer, Albrecht. 1811. *Handbuch für die feinwollige Schafzucht auf Befehl des Königl. Preuss. Ministerium des Innern.* Berlin: Ville und Krammer.
Treitel, Corinna. 2008. Max Rubner and the biopolitics of rational nutrition. *Central European History* 41:1–25.
Uekötter, Frank. 2010. *Die Wahrheit ist auf dem Feld: Eine Wissensgeschichte der deutschen Landwirtschaft.* Göttingen: Vandenhoeck & Ruprecht.
Versammlung deutscher Land-und Forstwirthe. 1866. *Amtlicher Bericht über die 25. Versammlung deutscher Land-und Forstwirthe zu Dresden, vom 25. Juni bis 1. July 1865.* Dresden: G. Schönfeld's Buchhandlung
Versammlung deutscher Land-und Forstwirthe. 1869. *Amtlicher Bericht über die 26. Versammlung deutscher Land-und Forstwirthe zu Wien, vom 31. August bis 5. September 1868.* Wien: Druck und Commissions Verlag von Carl Gerold's Sohn.
Weckherlin, August von. 1846. *Die landwirthschaftliche Thierproduktion, Erster Theil.* Stuttgart: J.G. Cotta'scher Verlag.
Wieland, Thomas. 2004. *Wir beherrschen den pfanzlichen Organismus besser…: wissenschaftliche Pflanzenzüchtung in Deutschland, 1889–1945.* München: Deutsches Museum.
Wilckens, Martin. 1878. *Form und Leben der landwirthschaftlichen Hausthiere.* Wien: Wilhelm Braumüller.
Witt, R.M. 1865. Beiträge zur Kenntnis und Beurtheilung des Aeusseren des Thierkörpers zumeist der englischen Kunstracen. *Jahrbuch der deutschen Viehzucht* 2:293–315.
Wolff, Emil. 1861. *Die landwirthschaftliche Fütterungslehre und die Theorie der menschlichen Ernährung.* Stuttgart: J.C. Cotta'scher Verlag.
Wolff, Emil. 1870. *Die landwirthschaftlich-chemische Versuchsstation Hohenheim: Deren Einrichtungen und Thätigkeit in den Jahren 1866 bis 1870.* Berlin: Wiegandt & Hempel.
Wolff, Emil. 1881. *Die rationelle Fütterung der landwirthschaftlichen Nutzthiere: Auf Grundlage der neueren thierphysiologischen Forschungen, Dritte Auflage.* Berlin: Verlag von Paul Parey.
Wolff, Emil. 1885. *Die rationelle Fütterung der landwirthschaftlichen Nutzthiere: Auf Grundlage der neueren thierphysiologischen Forschungen, Vierte Auflage.* Berlin: Verlag von Paul Parey.
Wood, Roger, and Vitzslav Orel. 2001. *Genetic prehistory in selective breeding: A prelude to Mendel.* Oxford: Oxford University Press.

Brendan Matz teaches in the Technology, Culture, and Society department at NYU-Poly in Brooklyn, USA. He is writing a book tentatively titled *Crafting Heredity: Animal Breeding in Germany and the United States, 1860–1929,* on the history of agricultural animal breeding and the scientific study of heredity in Germany and the USA. His interests include the history of heredity and genetics, history of nutrition science, and the sciences of the mind and brain. Research for the chapter in this volume was conducted while he was a Haas Postdoctoral Fellow at the Chemical Heritage Foundation 2011–2012.

Chapter 10
Artificial or Biological? Nature, Fertilizer, and the German Origins of Organic Agriculture

Corinna Treitel

Introduction

Among its many points of origin, organic agriculture has a particularly important root in the German life reform movement.[1] Founded in the last third of the nineteenth century, the movement included vegetarians, naturopaths, nudists, anti-vaccinationists, and others dedicated to ameliorating the negative effects of industrial modernity through practices designed to make modern lifestyles more natural. Eating a natural diet, one low or completely lacking in meat but rich in whole grains as well as fresh fruits and vegetables, was an early priority that quickly spawned the dream of remaking German agriculture along more natural lines. The core elements of this vision were nicely captured in a 1911 cover illustration from the life reform periodical *Vegetarian Lookout* (Fig. 10.1).

While sowing seeds, this nude farmer signaled his proximity to nature by his bare-chested exposure to the elements and his barefooted contact with the land. Except for the dog in the distance, moreover, the conspicuous absence of animals suggested that this manly labor occurred not on a common German "mixed" farm, where livestock were ubiquitous, but in a garden filled only with plants. The bulging muscles and full beard, finally, testified to the healthful effects on the body of raising and eating one's own plant foods. Images such as these pointed to the central role that life reformers accorded agriculture in their vision of nature around 1900. Like the more "natural" diet they propounded in other venues, a more "natural"

[1] For an international overview, see Lockeretz (2007). Country-specific studies include, for Britain, Conford (2001); Barton (2001); Matless (2001). For the United States, Beeman and Pritchard (2001); and for Japan, Moen (1995). The best overview of the German story is Vogt (2000a).

C. Treitel (✉)
Department of History, Washington University in St. Louis, Campus Box 1062, 1 Brookings Drive, 63130, St. Louis, MO, USA
e-mail: ctreitel@wustl.edu

D. Phillips, S. Kingsland (eds.), *New Perspectives on the History of Life Sciences and Agriculture,* Archimedes 40, DOI 10.1007/978-3-319-12185-7_10

Fig. 10.1 Cover of *Vegetarian Lookout* in 1911

style of agriculture—one that decentered or eliminated animals while prioritizing plants—became for life reformers the necessary foundation of both individual and social health.

Natural, however, had no fixed meaning in the life reform context and natural agriculture only slowly acquired the ecological focus that is central to organic methods today. When Eduard Baltzer, one of the movement's early pioneers, first began to criticize German agriculture as unnatural in the 1870s, his focus was human hunger. The animal-centered methods practiced by Germany's farmers left too many people malnourished and created a dangerous social problem whose best resolution lay with mass conversion to the natural lifestyle. Baltzer urged consumers to become vegetarians and, in a twist that will be surprising to modern readers, exhorted farmers to embrace the new "artificial" fertilizers being developed by German chemists. Four decades later, the next generation of life reformers had begun to articulate a quite different vision for rendering agriculture more natural, one that problematized these artificial fertilizers while calling for a more ecological approach to nutrition, both plant and human. As the life reformer Gustav Simons quipped in 1911, "[w]e humans are…only plants in the garden of nature," by which he meant that humans and plants belonged to a single environment in which the health and survival of both were intimately connected through fertilizer.[2] By equating humans with plants and encasing both in nature, Simons expressed here a new conceptual framework within which life reform versions of organic agriculture were first articulated.

There were both biological and political aspects to this process of giving natural agriculture an ecological orientation. From 1900 or so onwards, life reform farmers articulated an important variant of what Lynn Nyhart has called the "biological perspective." Nyhart's study focused on the popular realm of nineteenth-century German natural history, especially zoology, and its central claim that "the organism [was] a living being embedded in nature, whose survival depended on its ability to interact successfully with both its physical environment and the other organisms around it." The morally charged message accompanying the biological perspective, that even in an age of industrial modernity humans must find ways to live in harmony with nature or face catastrophe, Nyhart suggests, also became a pillar of twentieth-century German ecology.[3] Life reform agriculture after Simons, this essay argues, provided an important alternative venue for developing this biological perspective and its ecological call to arms. *Biologisch* (*biological*), a key term among the organic pioneers, signaled their conviction that natural agriculture must rest not on artificial fertilizer but nurturing living systems already in the soil. For nature as much as for civilization, they claimed, the choice between artificial and biological methods was nothing less than a choice between death and life.[4]

At the same time, life reform farmers gave their biological perspective political meanings largely absent from the popular zoologists examined by Nyhart. Before the ecological shift encapsulated in Simons' quip that "humans are…only plants in the garden of nature," life reformers had belonged largely to the liberal project. A dedicated republican in 1848, for example, Baltzer insisted until his death in 1887 that the natural lifestyle was a necessary step on the road to German democracy and freedom. Life reform farmers after him retained Baltzer's dream of a more natural agriculture but jettisoned his democratic frame. In response to agricultural crisis, war, and defeat, they moved politically rightwards and infused their new cultivation methods with hyper-nationalist and even anti-Semitic dreams of a new agrarian,

---

[2] Simons (1911, p. 11).

[3] Nyhart (2009, p. 2).

[4] Even today, indeed, "biologisch" remains the German word of choice for organics, yet carries its own distinctive meanings accumulated during the complex history sketched below.

post-monarchical, and post-democratic future. Largely ignored or downplayed by previous scholars, this shift to the right, I argue, was also central to German attempts before 1939 to farm according to nature.[5]

This rightward shift and its connection to the biological perspective is the major theme running through the following essay. The first section charts how and why life reformers embraced artificial fertilizers in the 1870s, only to problematize them by 1914. After a brief review of the war years, the second section considers the scientific and political factors that stimulated the emergence of an early form of German organic agriculture, "biological agriculture," in the 1920s–1930s. Throughout, this essay seeks to play on Wendell Berry's insight that "eating is an agricultural act" by asking how agricultural acts in late nineteenth- and early twentieth-century Germany took on political and ultimately biopolitical meanings that rooted national health, stability, and power in the control of human and plant nutrition.[6]

Hunger, Health, and Chemistry

For the first generation of life reformers, German agriculture failed the test of social efficiency. "Let's calculate socially!" demanded Eduard Baltzer in 1873. Not only did it cost ten times as much to feed a meat eater as a vegetarian, he reckoned, but also meat eating wasted "animal property [and] induces a colossal waste of other kinds of property as well as punishing the spendthrift with…a lamentably shorter lifespan."Here, "other kinds of property" meant primarily water and land, natural resources increasingly used to raise cash crops with little to no nutritional value for people: hops, grapes, and potatoes for making beer, wine, and schnapps, say, or sugar beets for sweetening and tobacco for smoking.[7] Animals raised for slaughter, channeled natural resources for human consumption, of course, but with breathtaking inefficiency. Whereas one acre of land devoted to growing cattle fodder would eventually feed only one meat-eating person, reallocating the same acre to growing crops for direct human consumption would feed at least ten vegetarians.[8] Contemporary agriculture, in other words, failed in its primary aim, which was to provide food for humans cheaply and plentifully. Germans paid the price in chronic hunger, poor health, shortened lifespan, poverty, and emigration, all of which tore the social

[5] Assessing the politics of early organic agriculture has been a fraught enterprise, particularly in the German context. Anna Bramwell, for instance, argued that German ecologism expressed a political orientation beyond left and right. See Bramwell (1985, 1989). For important critiques, see Stephens (2001) and Olsen (1990). Other historians simply elide the political history of organic agriculture before the Nazi era. See especially Vogt (2000a). The rightward shift documented here, it is worth noting, was well in line with international developments. For the closely linked British case, see the book by Conford noted above as well as Conford (2005, 2008).

[6] Berry (1990, p. 145).

[7] Baltzer (1873, pp. 44–45 and 87–88).

[8] Baltzer (1903, p. 21). This was a favorite trope of vegetarians, who credited the idea to Alexander von Humboldt.

fabric and drained national wealth and power. Agriculture, Baltzer concluded, had become "unnatural" and had to be remade to serve the common good.

In the 1870s, when Baltzer made his social calculations, no one yet knew exactly what a more natural agriculture might entail, but its association with the forces of agricultural modernization was strong. Quoting Henry Charles Carey, an American economist popular with social liberals across Europe, Baltzer framed the substitution of plant for animal foods as a sure sign of agricultural progress.[9] At the same time, Baltzer's suggestions for reform also encompassed a variety of measures that had little to do with animals. These included:

- Communalizing water and arable land
- Mechanizing farm work
- Replacing large estates with small farmer-occupied plots
- Acclimatizing nutritious foreign crops for domestic cultivation
- Using state controls to ensure that farmers concentrated on raising crops for direct human consumption
- Establishing a state-run system of agronomical schools and research stations[10]

These proposals echoed those being made by progressives well outside the incipient life reform movement in response to the crisis engulfing German agriculture in the 1870s. As cheap corn and wheat from the Americas and Russia flooded German markets, cereal prices dropped by at least 10%, despite protective tariffs, while large-scale migration from rural to urban areas led to a rise in rural wages.[11] The resulting agricultural depression put German farmers in a tight squeeze and provided a golden opportunity for modernizers of all kinds, including life reformers such as Baltzer, to think German agriculture anew.

Progressives both in and out of life reform circles quickly honed in on the problem of animals. The question of slaughter aside, animals provided crucial labor as well as manure to Germany's mixed farms. Animals, however, did not necessarily use the farm's resources efficiently. Some portion of land had to be devoted to raising fodder, while human labor was required to feed animals as well as to collect their manure and then apply it back to the fields.[12] Baltzer had no objection to the use of farm animals for labor, as long as they were kindly handled, but objected to animal manure as a poor use of resources.[13] Scientists and farmers with no stake in life reform agreed. Although not a life reformer, for instance, the agronomist William Löbe captured Baltzer's attention with a wheat cultivation technique that reduced the need for animal manure.[14]

---

[9] Carey (1860, pp. 315–316). Carey was a major influence on Baltzer. See especially Baltzer, *Reform*, 7–35.

[10] Baltzer (1873, p. 37, 41–45, and 98–101).

[11] For an overview, see Zanden (1991).

[12] Zanden (1991, p. 232).

[13] Baltzer (1873, p. 44).

[14] Löbe (1856). This was an annotated translation of Samuel Smith's *Lois Weedon Husbandry* (1856), which was in turn a modern adaptation of Jethro Tull's *New Horse-Houghing Husbandry* (1731). The cultivation technique promoted by all three writers involved deep ploughing and no animal fertilizer.

In this quest to decenter animals, *Kunstdüngung* (*artificial fertilizer*) eventually took pride of place. Artificial fertilizer had been pioneered in the 1840s by the chemist Justus von Liebig, who showed that animal manure alone could not maintain soil fertility on a modern farm and advised adding a variety of supplementary fertilizers to restore soil nutrients artificially. For Liebig, these supplementary fertilizers divided into two categories: organic in the sense that they originated from plants or animals (e.g., bone meal, night soil, and urine) and inorganic in the sense that they came from ores and minerals (e.g., superphosphates and potassium silicate). When farmers began to experiment with Liebig's methods over the following decades, they changed his nomenclature by dividing fertilizers into "natural" (farmyard dung) and "artificial" (everything else, whether organic or inorganic, naturally occurring or chemically synthesized). By the late nineteenth century, these artificial fertilizers had come to include nitrates extracted from Chilean mines (Chile saltpeter or the sodium nitrate $NaNO_3$), guano (Peruvian bird excrement), bone meal (made from the crushed bones of slaughtered animals), ammonia compounds, potash, and superphosphates.[15]

Artificial fertilizers beckoned Baltzer as the perfect replacement for animal manure and the ideal tool for rendering German agriculture more natural. His conversion to the cause came via a Saxon farmer named Roeder, who reported in an 1872 issue of *Der Chemische Ackermann,* Germany's leading journal of applied agricultural chemistry, on 20 years of field trials with the new substances. Several decades before, Roeder had purchased land used in just the ways that Baltzer excoriated: for growing crops destined for the on-site distillery and the feeding troughs of the farm's cows. Depleted by years of over farming, the soil required massive inputs of manure, some of which had to be purchased. Realizing that he could no longer break even with mixed farming, Roeder sold almost all the livestock, shut the distillery, and devoted himself to plant husbandry. Experimenting with a variety of artificial fertilizers—among them Chile nitrates, latrine runoff, processed animal remains (mainly bone, hair, and hide), potassium salts, and superphosphates—Roeder succeeded in restoring his soil chemistry, boosting crop yield, and dramatically raising monetary profits. Artificial fertilizers identified with the help of modern chemistry, Roeder promised, could now free farmers from their dependence on animals and "boost production while lowering consumption."[16] As the epitome of scientifically informed farming in the 1870s, Roeder's report proved to Baltzer that farms dedicated to raising plant foods with little to no animal manure were both possible and profitable.[17]

[15] For a list of commonly used artificial fertilizers, see (1908) *Dünger und Düngung*. For Liebig's agricultural chemistry, see Brock (1997, Chap. 6). For Liebig's discussion, see Liebig (1843, pp. 226–279).

[16] Roeder (1872). Roeder's article appeared in *Der Chemische Ackersmann*, edited by Adolf Stöckhardt. An agricultural chemist with a global reputation, Stöckhardt used this journal to bring Liebig's agricultural chemistry to practicing farmers. Brock (1997, p. 169).

[17] Baltzer, *Reform,* 27. Either Baltzer had not read the fine print or did not object to the fact that Roeder, following Liebig's own recommendations, used factory-generated animal remains in his fertilizer mix.

This openness to scientific modernity continued among life reformers in the 1880s, when two more solutions to the animal problem captured the imagination of agricultural progressives: green manures (*Gründüngung*) and Chinese methods for recycling waste. The vegetarian leader Maximilian Klein typified the era. Green manures, crop rotation, and the fallow system supplemented with artificial fertilizers and human waste treated according to Chinese methods, he claimed in lectures around the country, could provide all the soil fertility that German farmers needed.[18] Green manures were cover crops allowed to grow for a short time before being ploughed back into the ground to restore soil nutrients and they were being promoted vigorously in the 1880s by Albert Schultz-Lüpitz, a prominent leader of the newly formed *Deutsche Landwirtschaftsgesellschaft* (German Agricultural Society, f. 1885), an association explicitly dedicated to bringing the insights of modern science and technology to bear on German agriculture.[19] Klein's reference to Chinese methods, meanwhile, had roots in both German chemistry and European orientalist literature. Liebig had considered the usefulness of night soil as early as the 1840s. In his popular *Letters on Modern Agriculture* (1859), moreover, he explicitly praised Chinese farmers for making it a habit to return from selling their produce at urban markets with an equal volume of city waste: human and animal excrement, ashes, the remains of slaughtered animals, and so on.[20] Klein's comments probably also drew on the writings of the French agronomist Eugene Simon, whose reports on Chinese agriculture in the 1880s excited Western modernizers everywhere. Painting a picture of small peasant farmers endlessly recycling human waste as fertilizer while feeding millions plentifully, his 1885 work *La cité chinoise* became a standard text among European radicals by century's turn.[21]

Although this broad consensus between life reformers and scientific modernizers on the saving power of artificial fertilizer might seem surprising in retrospect, it was in fact the product of a common starting point: European hunger. The "hungry forties," an era of widespread crop failures that helped precipitate sociopolitical crises from the Irish potato famine of 1845 to the political revolutions of 1848, had been defining experiences for this generation, propelling Liebig into agricultural chemistry and Baltzer into life reform. For them, national survival hinged on feeding the soil to feed the people. With none other than Liebig warning that Germans' mixed farms overworked the soil in a grand *Raubsystem* (*system of exploitation*) that must necessarily end in hunger and civilizational collapse, it was not so outlandish for life reformers to pillory farmers for engaging in *Raubbau* (*overworking the soil*) and envision a future of Chinese-like growers who would farm intensively, eat vegetarian, and recycle efficiently.[22]

[18] Klein (1889, p. 32). The first edition came out in 1885. Green manures, crop rotation, and the fallow system are all widely used by organic farmers today. Roeder also used green manures. Roeder (1872, p. 76).

[19] Uekötter (2011, p. 163).

[20] Liebig (1859, pp. 244–245).

[21] Carpenter and Kropotkin, for instance, drew on the book in their own utopian writings.

[22] Brock (1997); Klein (1889, pp. 32–33).

If life reformers concerned about hunger thus enthusiastically adopted new ideas about farming from modernizers outside the life reform movement, it was their growing dedication to national health and wealth that eventually led them to confront the costs of artificial fertilizers and embark on a long quest to develop ostensibly more "natural" ones instead. By the 1880s, field trials had convinced agricultural chemists that higher crop yields hinged on adding potassium-, phosphorus-, and especially nitrogen-containing compounds to soil through mineral fertilizers.[23] Fixated on the relation of human hunger to mixed farming, early life reformers had displayed no qualms about artificial fertilizers.[24] This changed in the late 1880s, when a new generation of life reformers began to identify the negative physiological and economic effects associated with some of these fertilizers, particularly nitrogenous ones. Here, Julius Hensel and his case for *ground stone fertilizer* (*Steinmehldüngung*) led the way. Hensel studied at the University of Berlin with Eilard Mitscherlich and Heinrich Röse, both chemists with deep interests in mineralogy and geology, before turning to popular medical and scientific writing to support himself.[25] Reflecting the mineralogical interests of his mentors, his widely read and translated work *New Macrobiotics* (1881) rejected germ theory in favor of the humoral view that disease resulted from poor chemical composition of the blood. He drew particular attention to the dangers of an oversupply of ammonia, a nitrogen-containing compound produced by the breakdown of protein, and an undersupply of trace minerals, especially iron, sulfur, and calcium.[26] An essay on *Bread from Stones* published a few years later in the bourgeois journal *Ueber Land und Meer* (*Over Land and Sea*) extended these mineralogical ideas about human health to agriculture and led to a book, *Life: Its Foundation and How to Sustain It* (1885), which proved so successful that Hensel published an expanded second edition in 1890.

Hensel's target now became agricultural chemists for their proselytizing on behalf of two types of artificial fertilizer rich in nitrogen, guano from Peru and nitrates from Chile. Not only did German farmers now waste vast sums of money buying these fertilizers from abroad but also their tendency to overfertilize with nitrogen had as its consequence a tendency to underfertilize with other minerals. This produced weak and malnourished plants susceptible to pests, frost, and heavy rains; animals and humans who consumed these crops became themselves malnourished and prone to disease. The declining quality of cow's milk encapsulated the entire problematic chain: plants under-fertilized with lime, a calcium-rich compound, were

---

[23] For an excellent overview of mineral fertilizers in this period, see Smil (2001, Chaps. 1–3).

[24] First-generation life reformers did, however, occasionally worry that crops fertilized with farmyard dung might be of lower quality and thus damaging to human health. See Hahn (1869, p. 19); Klein (1889, pp. 31–32); Baltzer (1873, p. 44).

[25] Hensel (1890, p. iv).

[26] Hensel called poor blood composition *Blutentmischung,* a humoral term later popularized by Heinrich Lahmann, and *Dyskrasia,* a traditional Hippocratic term. As an antitode to poor blood, Hensel offered a recipe for "Hensel's tonic," a mixture with no nitrogen but plenty of iron, sulphur, and calcium. Hensel (1881, p. 86, 120–122, 129, and 180–181). Hensel's relationship to the life reform movement was unclear. His use of such terms as "macrobiotics" and "poor blood" indicate that he may have belonged to naturopathic circles, but he also criticized vegetarians for being impractical. Hensel (1890, pp. 500–502).

fed to cows, who produced low-calcium milk drunk by children, who then proved overly susceptible to pox, scarlet fever, scrofula, diphtheria, and other illnesses. Hensel's solution to this problem was to advocate green manures for delivering low doses of nitrogen and a mixture of primitive rocks easily available in Germany—gypsum (a sulfate), volcanic rock (rich in silica), and dolomite (rich in magnesium and calcium)—for adding the requisite minerals.[27] "How rich, how strong and how healthy will we Germans be when we make our mountains tributary to yield new soil from which new wholesome cereals may be formed," he wrote a few years later in his bestselling book *Bread from Stones*. "We need then no more send our savings to Russia, to Hungary, to America, but will make our way through life by our strong elbows and with German courage, and shall keep off our adversaries."[28]

Hensel's heterodox views about fertilizer scandalized agricultural chemists, but what elevated him beyond the category of mere cranks was that his vagrancy attracted immediate public attention, much of it positive. When a firm in Bubenheim began selling a "universal fertilizer" made according to Hensel's recipe, farmers embraced it with enthusiasm, attracted both by its low cost and the positive reviews it quickly garnered from fellow farmers. This pushed scientists and popularizers into print. Otto Zacharias, one of Germany's foremost scientific popularizers, called Hensel's scientific errors "hair raising" in the *Leipziger Zeitung* in 1890, followed 2 years later by Paul Wagner, a leading agricultural chemist, who delivered a scathing critique of Hensel's theories in *Deutsche landwirtschaftliche Presse*.[29] In 1893, Hensel's fertilizer even engaged the attention, mostly negative, of the Prussian parliament.[30] How are we to explain these reactions? For one thing, Hensel's market success and the sharp rebukes from Wagner, Zaccharias, and others suggested the relative weakness of scientific experts in the field of agricultural reform.[31] Hensel's claims on behalf of ground stone fertilizer, after all, rested on the same chemical model of soil fertility deployed by the experts and thus challenged them on their own turf. At the same time, however, his pooh-poohing about nitrogen raised uncomfortable questions that chemists were reluctant to face. Nitrogenous fertilizers certainly boosted crop yield, Hensel acknowledged, but at what cost to human and plant health? And given that Germany now led the world in importing Chile nitrates, what economic and political vulnerabilities did this introduce to national life?[32] However imperfectly, Hensel's work raised doubts about artificial fertilizers, as well as the scientific disciplines and industries advocating them, that lingered for decades.

---

[27] Hensel (1890, p. iii, 479, 483, 491, 504, and 509–510).

[28] Hensel (1894, pp. 94–95). From the life reform perspective, German chemists had an unhealthy obsession with nitrogen. While nutritional physiologists urged Germans to eat more protein (the only macronutrient containing nitrogen), especially from meat, and agricultural chemists urged Germans to fertilize more with nitrogenous substances, all in the pursuit of national health and fitness, life reformers stood out for vigorously advocating low-protein low-meat diets and low-nitrogen fertilizers, also in pursuit of national health and fitness.

[29] Zacharias (1890, p. 298); Wagner (1892, pp. 979–980). For Zacharias as popularizer, see Daum (2002, pp. 401–402). Wagner also directed the Agricultural Research Station in Darmstadt.

[30] [1894] *Mineraldünger.*

[31] Uekötter (2011, pp. 146–159).

[32] Smil (2001, p. 48).

Among life reform farmers and gardeners, Hensel's fertilizer provoked immediate interest and mixed evaluations. Karl Utermöhlen, an early settler at the life reform colony *Heimgarten,* praised Liebig and Hensel as the father and son of nutritional chemistry and blasted artificial fertilizers for producing low-quality plants whose consumption by animals and humans led to disease. Drawing on field trials at *Heimgarten,* the life reform colony *Eden,* and various private gardens, Utermöhlen reported that combining Hensel's fertilizer with animal manure and even small amounts of the infamous artificial fertilizer Chile nitrate produced high-quality crops in abundance.[33] A few years later, in contrast, his fellow *Heimgarten* settler Julius Sponheimer called field trials with Hensel's fertilizer a "fiasco." Still interested in finding a replacement for farmyard dung, however, Sponheimer recommended that life reformers continue experimenting with Chinese and Japanese practices for recycling night soil and urban waste.[34] What all of this suggests is that life reformers at century's turn took a flexible approach to the fertilizer question and, indeed, proved willing to use a variety of fertilizers, including artificial ones, in their quest to make animal manure obsolete.

As much as Hensel's tirades against artificial fertilizers incensed agricultural chemists and received mixed reviews among life reformers, his claims on behalf of the physiological and economic benefits of ground stone fertilizers resonated widely with other audiences. Within medicine, his theories about the role played by trace minerals in health inspired the physician Heinrich Lahmann to create a very successful dietary therapy that involved low-protein/low-nitrogen meals full of fresh mineral-rich fruits and vegetables.[35] Unlike his colleagues in regular medicine, moreover, Lahmann followed Hensel in attending closely to the agricultural aspects of medical practice. "Agriculture," Lahmann wrote in 1892, "suffers from wrong nutritional theories." An insistence on putting meat at the center of German diet, he meant, encouraged raising, growing, or importing beef and grain, while what Germans needed most, fresh fruits and vegetables, was least produced by German farmers. So frustrated did he become with local supplies, indeed, that in 1894 he bought an estate to produce food for his sanatorium. Demonstrating the close links that continued in this period between academic scientists and life reformers, Lahmann also hired Friedrich Falke, later one of Germany's most prominent scientific agronomists, to remake the farmland to his specifications, something that presumably also included his oft-repeated support for fertilizing with Hensel's mix and recycled urban waste.[36] Hensel's ground stone fertilizers also proved popular

[33] Utermöhlen(1895, p. 3, 13–14, 23, and 33).

[34] Sponheimer (1905, pp. 51–57). The *Deutsche Landwirtschaftsgesellschaft* also had a special committee devoted to studying the prospects for recycling urban waste. Uekötter (2011, p. 163).

[35] Lahmann's model of pathogenesis also followed Hensel's in its stress on how diet caused good or poor blood composition.

[36] Lahmann's estate was Gut Friedrichsthal. Lienert (2002, p. 45). Lahmann had little hope that agricultural reform would help poor Germans right away, so his books also included detailed recommendations for self-provisioning and civic action. He urged city dwellers to go out on weekends and holidays to gather wild greens in meadows and open fields. Chicory, dandelion, sorrel, borage, mustard, and even the young leaves of stinging nettles, he noted, were high in minerals and free

among a wide political spectrum of post-liberal reformers. August Bebel, Germany's leading socialist politician, included a discussion of Hensel's fertilizer, urban waste recycling, and the moral as well as material costs of meat eating in his utopian book *Women and Socialism* (33rd edition, 1900). Like Baltzer, Bebel saw issues of social and economic efficiency at stake in such topics.[37]

Within the life reform movement itself, in contrast, the project of naturalizing agriculture moved decisively under the control of *völkisch* (hyper-nationalist and anti-Semitic) farmers. Heinrich Bauernfeind, now hailed as one of the prophets of organic agriculture, complained bitterly that national wealth was being frittered away on Chile nitrates, when German mountains could provide all the "nature fertilizer" (*Naturdünger*) that farmers would ever need.[38] Having rebranded Hensel's fertilizer as natural, Bauernfeind also reframed it politically in the language of xenophobia and anti-Semitism. His "Ground Stone Song" rhapsodized:

> Ground stone fertilizer does not come from overseas
> From Jews, from abroad it does not come here
> It comes to me and you from mountains
> Truly: it comes to us from side to side
> We warn, we warn the farmer
> From terrible enemies who lie in wait.

True to type, Bauernfeind even included a recipe for ground stone fertilizer that listed the geographical source of each stone in his mix: Most came from Germany, a few from Bohemia, Norway, or Switzerland, and none from further abroad.[39]Among life reformers in the early years of the twentieth century, Bauernfeind's *völkisch* embrace of Hensel's mix as "natural" was unexceptional. Gustav Simons, who experimented many years with organic composts at the life reform colony *Eden,* distinguished sharply between artificial and natural fertilizers in his 1904 article on "Race and Diet." "Artificial fertilizer," he warned "causes degeneration of the soil, plants, cattle, [and] men," while "natural fertilizers" —which included Hensel's mix, peat, animal manure, night soil, and recycled urban waste—promoted high-quality plants and thus high-quality humans.[40] Appearing in the *völkisch* journal *Kraft und Schönheit* (*Power and Beauty*) and followed by an article on Jews as foreigners to German culture, readers could not have failed to glean the racial significance with which these natural fertilizers had now been imbued.

---

to all. Since fresh fruits were harder to acquire, he called for social pressure to encourage civic authorities to plant public orchards. Lahmann (1892, pp. 154–155 and 158–159).

[37] Bebel (1900, pp. 355–356, 390–392).

[38] Vogt names both Bauernfeind and Gustav Simons as pioneers of ecological agriculture, but does not explore their political valence. Vogt (2000a, pp. 64–65).

[39] Bauernfeind (1912, p. 8, 27, and 39). Bauernfeind claimed not to use anything but ground stone fertilizer on his plants. Bauernfeind (1908, p. 8). Simons, in contrast, did admit to using animal manure.

[40] Simons (1904, p. 20 and 159). Elsewhere, he declined to foreswear the artificial fertilizer Chile nitrate wholly, but worried that overuse harmed crop quality. Simons (1911, p. 25).

In the four decades between Baltzer's charge that German agriculture was unnatural and the paeans of fin-de-siécle life reformers to natural fertilizer, the scientific and political meanings of "natural" had changed dramatically. For Baltzer, a social liberal, "natural" touched only incidentally on the somatic qualities of plants and animals and not at all on the geographical provenance of fertilizer. Rather, the unnatural state of German agriculture stemmed from social structures and habits that rendered the nation unable to feed itself plentifully. Naturalizing German agriculture thus entailed removing the inefficiencies associated with animals by introducing artificial fertilizers being identified and developed with the help of agricultural chemistry.

The enthusiasm with which life reformers initially regarded artificial fertilizers, however, waned in the late 1880s as new economic and physiological concerns came to the fore. For Hensel and those who came after him, "artificial" became a term of opprobrium referring primarily to imported nitrogenous fertilizers, especially Peruvian guano and Chile nitrate. These undercut German independence, critics charged, while poisoning German soil and blood. From here, it was but a short step to the xenophobic and anti-Semitic themes introduced by Bauernfeind and Simons. "Natural" came in their hands to refer not to social structures and habits but to the geographical sources and physiological effects of ground stones on German bodies. In their quest to reform fertilizer habits, in short, life reformers also began to articulate a proto-ecological perspective that embedded Germans in nature and pinned their survival on Germans feeding themselves from and within their own natural environment. Given its origins in the mountains rather than manure, of course, ground stone fertilizer was in the nineteenth-century sense of the word merely a special type of artificial fertilizer. Nonetheless, its special claims to Germanness, healthfulness, and naturalness would prove highly appealing for decades to come. Even more importantly, because ground stone fertilizer worked much better in *völkisch* ideology than on German farms, life reformers continued their quest to find an efficacious replacement that still met ideological needs. By the 1920s, as the next section shows, this effort precipitated a full-scale conversion to the biological perspective and the emergence of early organic agriculture.

Hunger, Health, and Life

Hunger came back to the fore in debates over German agriculture around 1900, when a small but influential group of figures sounded the alarm about the globalization of the nation's nutritional economy. Germany's reliance on international markets for food and fertilizer, they warned, put national security at risk. By 1913, imported grains accounted for one sixth of total domestic supplies, with wheat imports alone comprising one third of domestic wheat stock.[41] To make matters worse, German farmers had become avid users of artificial fertilizers, many of them purchased

41 Hunt (1974, p. 316).

abroad. By 1900, Germany was the world's leading importer of Chile nitrates, to which much of the nation's recent gain in crop yields was credited.[42] "Can Germany alone," as Gustav Simons put it in 1907, "feed today's Germans?"[43] His emphatic no was born out during the First World War, when the Allied blockade turned the nation's dependence on imports into a fatal liability. The threat of wartime hunger followed by the real hunger that Germans suffered during the blockade, indeed, profoundly reshaped the debate over German agriculture. Ensuring Germany's food security by domestic means now became a national priority.

In this struggle to eliminate the nation's nutritional vulnerabilities, life reformers found themselves in both consonance and dissonance with the new regime of Weimar Germany. Among experts, the consensus view after 1919 was that autarky necessitated full-scale conversion to intensive farming, which meant, among other things, embracing a new generation of synthetic nitrogenous fertilizers produced right at home by giants of the German chemical industry such as BASF. Official monomania about the saving power of German chemistry, however, soon hit a wall. Synthetic fertilizer use rose precipitously in the 1920s, it was true, but harvest yields dropped while soil-related problems proliferated. By the early 1930s, even experts had begun to wonder if German chemistry held the key to German food security.[44]

Into this breach stepped life reformers, who began in the mid-1920s to move beyond Hensel's fertilizer. In an effort to achieve the consensus goal, autarky, but sidestep the consensus method, intensive farming, they invented a new style of agriculture whose careful attention to soil ecology marked it as an important expression of the biological perspective. These inventors, moreover, embedded their practices in hyper-nationalist programs of racial renewal that linked soil ecology to the imperial designs of Germany's radical right. This section considers these scientific and political shifts by spotlighting one main variant of interwar agriculture: Ewald Könemann's *biological soil cultivation* (*biologische Bodenkultur*). Although there were other pioneers of interwar organic agriculture, Könemann presents a revealing and understudied case. A pioneer of life reform agriculture during the interwar years, he expressed a world view typical of these pioneers, one that jumbled new biological concepts with radical political fantasies. Könemann, moreover, remained a dominant force in the movement until his death in 1976, when the West German environmental movement finally absorbed the ecological methods that he had invented 50 years before into a new political context.[45]

---

[42] Chile nitrate delivered nitrogen at 30 times the concentration of animal manure. Smil (2001, p. 46 and 48).

[43] Simons, *Die deutsche Volksernährung*, 96.

[44] Uekötter (2011, pp. 183–214). Vogt (2000, pp. 31–36). Smil (2001, Chaps. 4–5). In English, see also Uekötter (2006); and Vogt (2007).

[45] Scholarly attention has focused on the biodynamic methods associated with Rudolf Steiner and a small group of anthroposophical farmers. The best overview is Vogt (2000a, Chap. 4). For opposing views on the relation of this group to the Nazi regime, see Vogt (2000b); Treitel (2009); and Staudenmaier (2010, pp. 226–252). Perhaps the more interesting question concerns the politics of these groups before 1933. Less studied but no less revealing were other pioneers of organic agriculture, including Wilhelm Büsselberg, Leberecht Migge, Walter Rudolph, and Friedrich Schöll.

Originally trained as a farm manager, Ewald Könemann suffered wounds during the First World War that turned him towards naturopathy and natural diets. Contact with Richard Bloeck, a member of the life reform colony *Eden* already experimenting with alternative cultivation methods, convinced him that he could combine his old training in agriculture with his new passion for life reform.[46] Farming, teaching, and editing the early journal of the natural farming movement, *Bebauet die Erde* (*Cultivate the Earth*), enabled him by the mid-1930s to present a mature vision for German agricultural reform, one that combined practical details on how to farm according to nature with a more theoretical discussion of why Germans should do so.

In his 1939, book *Biological Soil Cultivation and the Fertilizer Economy,* Könemann opened with an apocalyptic yet ecological answer to the why. A century ago, with three-quarters of the German population still living on the land, manure and waste largely remained on the land and kept the "fertilizer economy" in equilibrium. "Into this pastoral," mourned Könemann, "now intrude world commerce, industry, and capital." Farms had ceased to be self-contained holdings supporting a family and instead become factories transforming their own soil into products for urban consumption. Just as cities were upending the rural social order by draining it of farmers, urban eaters were damaging the nation's very soil by robbing it of nutrients. Depleted, unhealthy soils produced poor, unhealthy plants, which in turn produced weak, unhealthy animals and people. Declaring nature's cycles broken, Könemann defined the task ahead as making "our fields rich and healthy again–eternally fertile." Dismissing intensive agriculture as a panacea whose large harvests hid long-term problems, Könemann urged readers to embrace a new kind of agriculture, biological agriculture, one that cultivated the soil according to nature's laws of life and growth. "Renewal," he proclaimed, "depends on a biological cultivation of the soil and [a biological] fertilizer economy."[47] The unsustainability of intensive farming supplied the why, in short, while biological agriculture supplied the how.

Könemann's vision for naturalizing agriculture built on the work of two biologists: Felix Löhnis and Raoul Francé. Both men had drawn attention before the war to the role of soil microorganisms in plant growth, warning that ignorance of soil biology caused the nation huge financial and nutritional losses.[48] Painting microorganisms as integral to agriculture, however, flew in the face of common wisdom. Popular audiences had only just begun to associate microorganisms, negatively, with dirt and disease, while agronomists still under the sway of Liebig's chemical model regarded plant nutrition as a simple matter of plants taking up fertilizer directly. Unable to find secure academic work, both men took their heterodoxy

---

With the exception of Rudolph, whose politics are unclear, all of these figures were political radicals. Migge was an anarchist, while Büsselberg and Schöll were Könemann's fellow travelers in *völkisch* circles.

[46] [1959] *Ewald Könemann 60 Jahre.*

[47] Könemann (1939, pp. 1–2).

[48] Löhnis (1906). See also Löhnis (1913, p. 377); Francé (1913, pp. 92–93). Könemann refers to Löhnis and Francé in several places. See, e.g., Könemann (1939, p. 35 and 42). For agricultural bacteriology, see Uekötter (2011, pp. 214–225); and Vogt (2000a, pp. 39–44).

elsewhere. Löhnis went to the USA, while Francé became one of Germany's most ubiquitous scientific popularizers.[49] And it was here in the popular realm that agricultural bacteriology, this new and poorly institutionalized branch of biology, finally blossomed.

Avid consumers as well as critics of German science, life reformers were electrified by Francé's message that civilization had entered a new phase in which biology rather than chemistry would lead.[50] In his book, *The Edaphon: Investigations into the Ecology of Soil Micro-organisms* (1913), for instance, Francé wrote lyrically about the *Edaphon,* "the life-community (*Lebensgemeinschaft*) of animal and plant organisms living in the soil," urging readers to apply its lessons to human society.[51] In *The Life of the Soil* (1922), he extended the Edaphon concept to agriculture and illustrated his assertions with hand-drawn images of microscope slides that contrasted living soil with dead dirt. Widely reprinted in life reform literature, one slide portrayed fertile soil as an Edaphon of wriggling microorganisms that fixed nitrogen, composed organic matter, and dissolved minerals, all activities that delivered essential nutrients to plant roots. The companion slide, in contrast, showed a mixture of inert minerals that roots took up imperfectly, if at all. Fertile soil was alive with the biological power of microorganisms, these illustrations implied, while infertile soil collapsed towards death under the chemical weight of minerals.[52] Francé's pantheistic vision of nature as infused with life captured the imagination of millions of people who had abandoned traditional religious beliefs in favor of new worldviews fashioned from an eclectic range of building blocks, from East Asian Buddhism to homegrown German biology. Life reformers had been at the forefront of this development since the 1860s at least and Francé had moved freely between life reform and scientific circles since the early twentieth century.[53]

Given these overlapping cultural spheres and life reformers' on-going discomfort with artificial fertilizer, it was unsurprising that when Francé consigned the reigning paradigm of agricultural chemistry to history and declared a new age of soil biology in 1922, Könemann and like-minded farmers saw their opportunity.[54] Agriculture, as Könemann put it, must rest on a basic Francéan principle: "Everything lives–even the soil. Countless micro-organisms sustain the soil and keep it fertile, ready to be cultivated, and well structured. All maintenance must take this into account." Adapting Francé's concept of the Edaphon, Könemann developed new techniques for taking care of the humus, the fertile top soil in which the Edaphon supposedly thrived. He recommended shallow turning and frequent mulching

[49] Uekötter (2011, p. 216). In 1903, Francé helped found the *Kosmos Gesellschaft,* a major popular scientific association, and several of the volumes that he wrote for its press, the *Franckh'sche Verlagshandlung* in Stuttgart, became best sellers. Daum (2002, pp. 184–188); Roth (2000, p. 62).

[50] For Francé, as an early apostle of biotechnology, see Bud (1993, pp. 62–63). For Francé's impact on forest ecology, see Wilson (2012, pp. 190–194).

[51] Francé (1913, p. 5).

[52] Francé (1922, p. 27 and 39). Könemann reprinted some of these "slides" in his 1939 book. See, e.g., Könemann (1939), plate between pages 32 and 33.

[53] Roth (2000, pp. 59–60).

[54] Francé (1922, p. 27, 39, 67, and 70).

to enable “the little helpers” to do their work. An elaborate illustration showed that these little helpers lived in linked strata, from algae and fungi at the top to bacteria, protozoa, threadworms, and rhizopods in the middle to rotifer, small bacteria, and earthworms at the bottom.[55] Deep plowing and synthetic fertilizers damaged this delicate living world, while biological methods sustained it through shallow turning, frequent mulching, and, most importantly, natural fertilizers. These included old-style fertilizers such as farmyard dung and green manure as well as new-style composts made from rural, urban, and industrial waste. To this end, Könemann designed a self-composting toilet that harnessed the power of microorganisms to convert human waste into living food for the Edaphon. He also extolled *Edaphon-Edelerde* (*Edaphon vital earth*), a soil inoculant patented by Francé. Containing the same microbiological mix as a healthy Edaphon, this inoculant could be injected into piles of urban waste to transform them quickly, hygienically, and odorlessly into living soil.[56] And, although Könemann did not foreswear commercial fertilizers completely, he expressed deep reservations about the new nitrogenous fertilizers of synthetic chemistry. These boosted harvest yield, he admitted, but only at high cost to plant and soil quality.[57]

Könemann's comments about quality pointed to what made this new style of agriculture biological, natural, and, in his mind, superior. It was, in the first place, biological because of the way it fed crops. Rather than nourishing plants directly, as the mineral fertilizers of synthetic chemistry did, biological fertilizers fed the Edaphon and the Edaphon fed the plants. Nourishment came not from the mechanical application of inert minerals, but living processes unfolding in the soil itself. What made biological agriculture natural, in turn, was its respect for natural law. “We remain totally in the bosom of nature,” Francé had warned in 1920, “and can never break her laws.”[58] With its mineral fertilizers, heavy machinery, and obsession with volume, intensive agriculture left eroded, acidified, and depleted soil in its wake, a sure sign that nature's laws had been broken. Biological agriculture, in contrast, kept the soil fertile by using nature's laws to properly manage the living community within it. The superiority of biological agriculture, finally, was manifest in its product. In contrast to the tasteless, weak, and nutrient-poor plants produced by intensive agriculture, biological products allegedly tasted better, lasted longer, and were more nutritious. In support of the last claim, Könemann invoked studies by Ragnar Berg, a prominent diet reformer who had started his career with Heinrich Lahmann. A chemist by training, Berg had found that green cabbage fertilized with ammonium sulfate, the most common synthetic fertilizer at the time, contained

[55] Könemann (1939, p. 98).

[56] Könemann (1939, pp. 235–236 and plate 14).

[57] Könemann (1939, pp. 186–188, 235–236, 264–265, and 285–286). For his extended discussion of fertilizers, see 171–251. For his model self-composting toilet, see plate 14 between pp. 216 and 217. Note that Könemann's ideal was farming without animals and animal manure, but in practice biological agriculture did use animal manure and prescribed treating farm animals with “comraderly attention” (p. 165).

[58] Francé (1920, p. 276).

more nitrogen but less protein and trace minerals than biologically cultivated cabbage. Chemically fertilized heads, Könemann warned, might be large, but inside they were "degenerate and sick" and their consumption endangered human and animal health alike. In an analogy that betrayed his life reform origins, he likened the current obsession with chemical fertilizers to an unnatural diet centered on red meat and white bread with a few canned vegetables on the side. Life-giving elements—vitamins for people, trace nutrients for plants—were missing.[59]

Könemann's biological agriculture also carried a radical politics, one that he built on Francé's vision of nature. In *The Life in the Fields* (1922), the book that inspired Könemann, Francé had drawn vast political consequences from Europe's ignorance of soil biology. All the disasters of modernity—war, imperialism, emigration, proletarianization, industrialization, and revolution—stemmed from Europe's inability to produce enough food for itself, he claimed.[60] This was a failure of European civilization, to be sure, but primarily a *biological* rather than cultural one. The human brain was the foundation of civilization, after all, but was itself nothing more than "an adaptation to its environment" and, like all evolutionary adaptations, could become a liability when environments changed. Current biological catastrophes, from a falling birth rate to increasing soil exhaustion, signaled that this turning point was now at hand. "[Man] can never overcome nor bring nature to completion," Francé had warned elsewhere. "He can never grasp her totally.... When he considers what he knows about the earth's history, he is scared, anxious, and humbled: The evolution of the world proceeds, and I–I was not her goal and am not her endpoint, I am only a piece of a totality that I cannot comprehend, that is reflected in me, and that calls me 'Nature'." Francé's ecological vision of civilization also extended to the European himself as nothing more than "a native of his own land, the special product of this region of the earth, just like the Alpine mountains or the beech trees that grace their foothills."[61]

Könemann adopted this Francéan habit of ecologizing Europeans and embedded it in a hyper-nationalist frame. *Bebauet die Erde,* the pioneering journal of German organic agriculture that he edited from the mid-1920s until the 1950s, gave away its *völkisch* commitments on its cover, which always sported old Germanic terms to indicate the month (e.g., *Julmond* for July). In an early and influential essay entitled *Agriculture Without Animals: Natural Soil Cultivation* (1925), similarly, Könemann adopted unconventional orthographic habits popular in *völkisch* circles.[62] Perhaps most tellingly, finally, Könemann voluntarily adapted his dream of German farmers growing German food on German soil nourished with German waste to the political designs of the Nazi leadership after 1933. As he remarked in an essay, welcoming

[59] Könemann (1939, p. 59 and 87).

[60] Francé (1922, pp. 47–48).

[61] Francé (1920, p. 269, 276–280).

[62] For example, he substituted the supposedly more Germanic "f" for the "v" of conventional German usage. *Viehloser Ackerbau* (*agriculture without animals*) thus became *fiehloser Ackerbau*. Könemann (1925). The Nazi regime also considered adopting this orthography a few years later. See Birken-Bertsch and Markner (2000).

the new minister of agriculture Richard Darré that year, the task ahead must be "the native (*bodenständig*) reshaping of our entire existence through colonization of German space. The way to do this is through settling [and biologically] cultivating German soil, through organically mastering German living space (*Lebensraum*)."[63] Terms such as *bodenständig* (*rooted in the soil or landscape*) and *Lebensraum* (*living space*) belonged at this point to Nazi plans for racializing Germany and Germanizing eastern Europe.[64] The little helpers in the soil, his word choice implied, could serve the project of national reconstruction and expansion. In Könemann's hands, in short, biological agriculture became compatible with the racial and imperial designs of the Third Reich.

Conclusion

From Baltzer to Könemann, the dream of making German agriculture more natural endured but its changing scientific and political sources meant that "natural" stayed in flux. Baltzer had extolled artificial fertilizer developed by German chemistry as the path towards nature and democracy, only to be challenged by Hensel, Simons, and Bauernfeind, who excoriated imported artificial fertilizer for causing disease and crippling national independence. It was more natural, more healthful, and economically more rational to feed Germans with plants fed on German stones, they claimed, only to be stymied by crops that refused to grow well with stones alone. Inspired by Francé's lyrical vision of the Edaphon, Könemann finally brought ideology and practical success into line by breaking with the chemical model. Adapting the heterodox science of agricultural bacteriology, Könemann invented an early form of organic agriculture, biological agriculture, that claimed to be more natural because it followed the soil's own laws of life while enabling Germans to maintain their nutritional independence in an age of looming war and racial conflict. As all of this suggests, by embracing the view that humans were only plants in nature's garden, life reformers after Simons turned their movement into an important site for the continuation as well as politicization of the biological perspective in the twentieth century.

---

63 Könemann (1933, p. 9 and 32).

64 *Bodenständig* was popularized by Alwin Seifert, a landscape architect and cultural conservative closely tied to the Nazi regime. For Seifert and his use of the term, see Rollins (1995, p. 503). As for *Lebensraum,* this concept has a long and checkered history. In the Nazi context, it connoted creating living space for Germans through the conquest and racial reconstruction of Eastern Europe. It is important to note, however, that the term carried other meanings before and after the Third Reich. For its coinage by the geographer Friedrich Ratzel, see Smith (1980). For Darré's use of the term, see Gerhard (2005). For the repurposing of this term by West German nature conservationists in the 1950s, see Chaney (2008, p. 117).

References

1894. Mineraldünger. *Anlage zu den Stenographischen Berichten über die Verhandlungen des Hauses der Abgeordneten*, vol. I, 67–68. Berlin: W. Moser.

1908. Dünger und Düngung. *Meyers Grosses Konversations-Lexikon*, vol. 5, 6th ed., 277–280. Leipzig: Bibliographisches Institut.

1959. Ewald Könemann 60 Jahre. *Organischer Landbau* 5: 97.

Baltzer, Eduard. 1873. *Ideen zur socialen Reform*. Nordhausen: Ferd. Förstemann.

Baltzer, Eduard. 1903. *Die Reform der Volkswirtschaft vom Standpunkte der natürlichen Lebensweise*. Leipzig: H. Hartung & Sohn.

Barton, Gregory. 2001. Sir Albert Howard and the forestry roots of the organic farming movement. *Agricultural History* 75:168–187.

Bauernfeind, E. Heinrich. 1908. *Natur- und Kunstdüngung*. Lorch: Karl Rohm.

Bauernfeind, E. Heinrich. 1912. *Die Vorteile der Aschen-, Mineral-, Steinmehl- oder Naturdüngung und die Nachteile der jetzt üblichen Mist- und Kunstdüngung*. Weidenbach-Triesdorf: Selbstverlag.

Bebel, August. 1900. *Die Frau und der Sozialismus*. Stuttgart: Dietz.

Beeman, Randall S., and James A. Pritchard. 2001. *A green and permanent land: Ecology and agriculture in the twentieth century*. Lawrence: University of Kansas Press.

Berry, Wendell. 1990. *What are people for?* New York: North Point Press.

Birken-Bertsch, Hanno, and Reinhard Markner. 2000. *Rechtschreibreform und Nationalsozialismus: Ein Kapitel aus der politischen Geschichte der deutschen Sprache*. Göttingen: Wallstein Verlag.

Bramwell, Anna. 1985. *Blood and soil: Richard Walther Darré and Hitler's 'Green Party'*. Abbotsbrook: The Kensal House.

Bramwell, Anna. 1989. *Ecology in the twentieth century: A history*. New Haven: Yale University Press.

Brock, William H. 1997. *Justus von Liebig: The chemical gatekeeper*. Cambridge: Cambridge University Press.

Bud, Robert. 1993. *The uses of life: A history of biotechnology*. Cambridge: Cambridge University Press.

Carey, H.C. 1860. *Principles of social science*, vol. III. Philadelphia: J. B. Lippincott.

Chaney, Sandra. 2008. *Nature of the miracle years: Conservation in West Germany 1945–1975*. New York: Berghahn Books.

Conford, Philip. 2001. *The origins of the organic movement*. Edinburgh: Floris Books.

Conford, Philip. 2005. Organic society: Agriculture and radical politics in the career of Gerard Wallop, Ninth Earl of Portsmouth. *Agricultural History Review* 53:78–96.

Conford, Philip. 2008. Saturated with biological metaphors: Professor John Macmurray and the politics of the organic movement. *Contemporary British History* 22:317–334.

Daum, Andreas. 2002. *Wissenschaftspopularisierung im 19. Jahrhundert: Bürgerliche Kultur, naturwissenschaftliche Bildung und die deutsche Öffentlichkeit 1848–1914*. Munich: R. Oldenbourg.

Francé, R. H. 1913. *Das Edaphon: Untersuchungen zur Oekologie der bodenbewohnenden Mikroorganismen*. Munich: Deutsche mikrologische Gesellschaft.

Francé, R. H. 1920. *Die Gewalten der Erde: Eine Geschichte der Entfaltung des Lebens*. Berlin: Ullstein.

Francé, R. H. 1922. *Das Leben im Ackerboden*. Stuttgart: Kosmos, Gesellschaft der Naturfreunde.

Gerhard, Gesine. 2005. Breeding pigs and people for the Third Reich: Richard Walther Darré's agrarian ideology. In *How green were the Nazis? Nature, environment, and nation in the Third Reich*, ed. Franz-Josef Brüggemeier, Mark Cioc, and Thomas Zeller, 129–146. Athens: Ohio University Press.

Hahn, Theodor, ed. 1869. *Der Vegetarianismus, seine wissenschaftliche Begründung und seine Bedeutung für das leibliche, geistige und sittliche Wohl des Einzelnen, wie der gesammten Menschheit. Ein Beitrag zur Lösung der socialen Frage*. Berlin: Theobald Grieben.

Hensel, Julius. 1881. *Neue Makrobiotik, oder die Kunst Seuchen zu verhüten und zu heilen*. Stuttgart: Selbstverlag.

Hensel, Julius. 1890. *Das Leben: Seine Grundlagen und die Mittel zu seiner Erhaltung physikalisch erläutert zum praktischen Nutzen für Ackerbau, Forstwirthschaft, Heilkunde und allgemeine Wohlfahrt*. Philadelphia: Boericke & Tafel.

Hensel, Julius. 1894. *Bread from stones: A new rational system of land fertilization and physical regeneration*. Philadelphia: A J Tafel.

Hunt, James C. 1974. Peasants, grain tariffs, and meat quotas: Imperial German protectionism reexamined. *Central European History* 7 (4): 311–331.

Klein, M. L. G. 1889. *Die harmonische Lebensweise (Vegetarismus). Die Grundbedingung zur Erlangung von Gesundheit, Wohlstand und moralisch-ästhetischer Gemütsanlage*. Berlin: Verlag von Max Breitkreuz.

Könemann, Ewald. 1925. Fiehloser Ackerbau-natürliche Bodenbearbeitung. *Tao: Monatsblätter für Verinnerlichung und Selbstgestaltung* 11:1–20.

Könemann, Ewald. 1933. *Aufgaben der Kolonisation in Deutschland*. Oberellen bei Eisenach: Verlag Bebauet die Erde.

Könemann, Ewald. 1939. *Biologische Bodenkultur und Düngewirtschaft*. Tutzing: E Siebeneicher.

Lahmann, Heinrich. 1892. *Die diätetische Blutentmischung (Dysämie) als Grundursache aller Krankheiten. Ein Beitrag zur Lehre von der Krankheitsdisposition und Krankheitsverhütung. Ein Buch sowohl für jene, welche gesund machen, als auch für solches die gesund bleiben oder werden wollen*. Leipzig: Otto Spamer.

Liebig, Justus von. 1843. *Die Chemie in ihrer Anwendung auf Agricultur und Physiologie*. Braunschweig: Friedrich Vieweg.

Liebig, Justus von. 1859. *Naturwissenschaftliche Briefe über moderne Landwirtschaft*. Leipzig: C F Winter.

Lienert, Marina. 2002. *Naturheilkundiges Dresden*. Dresden: Elbhang-Kurier Verlag.

Löbe, William. 1856. *Der höchste Ernte-Ertrag. Beschreibung einer neuen und bewährten Culturmethode des Weizens*. Leipzig: Otto Spamer.

Lockeretz, William, ed. 2007. *Organic farming: An international history*. Oxfordshire: CAB International.

Löhnis, Felix. 1906. *Einführung in die Bakteriologie für Landwirte*. Leipzig: Hugo Voigt.

Löhnis, Felix. 1913. *Vorlesungen über landwirtschaftliche Bakteriologie*. Berlin: Gebrüder Borntraeger.

Matless, David. 2001. Bodies made of grass made of earth made of bodies: Organicism, diet and national health in mid-twentieth-century England. *Journal of Historical Geography* 27:355–376.

Moen, Darrell. 1995. The emergent culture of the Japanese organic farming movement and its implications for political economy. Ph.D. dissertation, University of Wisconsin.

Nyhart, Lynn. 2009. *Modern nature: The rise of the biological perspective in German*. Chicago: University of Chicago Press.

Olsen, Jonathan. 1990. *Nature and nationalism: Right-wing ecology and the politics of identity in contemporary Germany*. New York: St. Martin's Press.

Roeder. 1872. Zehn Jahre Mist- und zwanzig, resp. funfzehn Jahre künstliche Düngung. *Der chemische Ackersmann. Naturkündliches Zeitblatt für deutsche Landwirthe* 18:74–76, 137–147.

Rollins, William. 1995. Whose landscape? Technology, Fascism, and enviromentalism on the National Socialist autobahn. *Annals of the Association of American Geographers* 85:494–520.

Roth, René Romain. 2000. *Raoul H Francé and the doctrine of life*. Bloomington: First Books Library.

Simons, Gustav. 1904. Rasse und Ernährung. *Kraft und Schönheit* 4:156–159.

Simons, Gustav. 1911. *Bodendüngung, Pflanzenwachstum-Menschengesundheit. Ein Ratgeber für denkende Gartenfreunde*. Berlin: Verlag Lebenskunst-Heilkunst.

Smil, Vaclav. 2001. *Enriching the earth: Fritz Haber, Carl Bosch, and the transformation of world food production*. Cambridge: MIT.
Smith, W. D. 1980. Friedrich Ratzel and the origins of Lebensraum. *German Studies Review* 3:51–68.
Sponheimer, J. 1905. *Der Vegetarismus eine wirtschaftliche Notwendigkeit. Versuch einer wissenschaftlichen Begründung des Vegetarismus*. Berlin: Verlag Lebensreform.
Staudenmaier, Peter. 2010. Between Occultism and Fascism: Anthroposophy and the politics of race in Germany and Italy, 1900–1945. Ph.D. dissertation, Cornell University.
Stephens, Piers H. G. 2001. Blood not soil: Anna Bramwell and the myth of Hitler's Green Party. *Organization and Environment* 14:173–187.
Treitel, Corinna. 2009. Nature and the Nazi diet. *Food and Foodways* 17:139–158.
Uekötter, Frank. 2006. Know your soil: Transitions in farmers' and scientists' knowledge in Germany. In *Soils and societies: Perspectives from environmental history*, ed. J. R. McNeill and Verena Winiwater, 322–340. Strond: White Horse.
Uekötter, Frank. 2011. *Die Wahrheit ist auf dem Feld: Eine Wissensgeschichte der deutschen Landwirtschaft*. Göttingen: Vandenhoeck & Ruprecht.
Utermöhlen, Karl ed. 1895. *Über die Quellen der Fruchtbarkeit. Eine Anleitung zur dauernden Verbesserung unseres Acker- und Garten-Bodens nach dem grundgelegenden Forschungen Justus von Liebigs und Julius Hensels*, 2nd ed. Liebau in Schlesien: Verlag von Förster & Wedel (J. Heisig).
Vogt, Gunther. 2000a. *Entstehung und Entwicklung des ökologischen Landbaus im deutschsprachigen Raum*. Bad Dürkheim: Stiftung Ökologie und Landbau.
Vogt, Gunther. 2000b. Ökologischer Landbau im Dritten Reich. *Zeitschrift für Agrargeschichte* 48:161–180.
Vogt, Gunther. 2007. The origins of organic farming. In *Organic farming: An international history*, ed. W. Lockeretz, 9–29. Oxfordshire: CAB International.
Wagner, Paul. 1892. Eine Mahnung an die Landwirte. *Deutsche Landwirtschaftliche Presse* XIX, 979–980. Nov. 26.
Wilson, Jeffrey. 2012. *The German forest: Nature, identity, and the contestation of a national symbol, 1871–1914*. Toronto: University of Toronto Press.
Zacharias, Otto. 1890. Chemische Schöpfungsgeschichte. *Leipziger Zeitung (Wissenschaftliche Beilage)* 75:297–298.
Zanden, J. L. van. 1991. The first green revolution: The growth of production and productivity in European agriculture 1870–1914. *Economic History Review* 44 (2): 215–239.

Corinna Treitel is associate professor in the Department of History, Washington University in St. Louis, USA. She is a German historian interested in the ways that science, culture, and politics have intersected since the late eighteenth century. Her first book, *A Science for the Soul: Occultism and the Genesis of the German Modern* (2004), examined the German occult movement as an aspect of European modernism. Her current research concerns the history of food and nutrition in Central Europe. She is working on a new book exploring the history of natural foods and farming in German-speaking Europe since the early nineteenth century.

Chapter 11
Science, Promotion, and Scandal: Soil Bacteriology, Legume Inoculation, and the American Campaign for Soil Improvement in the Progressive Era

Mark R. Finlay

Introduction

In his 1904 address before the American Microscopical Society, the American plant pathologist Thomas J. Burrill delivered a message that his audience must have been glad to hear: The microscope had been more important than "any other piece of mechanism whatever in promoting and prolonging life."[1] Burrill backed this claim by highlighting the recent discovery of soil microbes, and the evidence that "all nutrition as applied to man seems to be ultimately conditioned upon the activity of certain micro-organisms in the soil." Arguing that "soil fertility and man's virility are closely related," Burrill had found confirmation among the farmers in his native Illinois. Those who lived on poor soils, he observed, "take unconsciously a slower step, require more time in which to transact business, and have less relish for physical or mental activity," while those who live on good soils were part of a "progressive, strong, hopeful, and happy populace."[2] Good soils, he continued, included bacteria that only the microscopists could see, "wonder-working little creatures" that were the "fertility producers [and] advance agents in the making of a farm."[3]

[1] Burrill (1904, p. 426).

[2] Burrill (1904, p. 434).

[3] Burrill (1904, pp. 422–429).

Mark R. Finlay is deceased. The email address below is that of his wife, Kelly Applegate.

M. R. Finlay (✉)
Department of History, Armstrong Atlantic State University, Savannah, GA, USA
e-mail: kappleg8@comcast.net

D. Phillips, S. Kingsland (eds.), *New Perspectives on the History of Life Sciences and Agriculture,* Archimedes 40, DOI 10.1007/978-3-319-12185-7_11

A key question, then, was how to manage and multiply the beneficial kinds of soil bacteria. Burrill had good news. Thanks to the recent discovery of legumes' ability to fix atmospheric nitrogen through the bacteria found on their root nodules, scientists had unveiled an apparent panacea: packages of pure bacterial cultures known as "legume inoculants." Appearing amid great fanfare in 1904, and distributed widely through the United States Department of Agriculture (USDA), these packages contained millions of nitrifying bacteria that seemed to offer a simple means to improve the quality and quantity of soil bacteria, to enhance soil fertility, and to raise farmers' incomes. Speaking at the peak of the Progressive Era, Burrill's address demonstrated the confidence of the American life scientists who believed the applications of their research could improve society as a whole.[4]

Burrill's address also sheds light on the emerging disciplines of soil bacteriology and applied botany. Earlier notions that tied fertility to the chemical and physical nature of soils had often proven unsatisfactory on scientific and economic grounds. Thus, advances in soil bacteriology, and scientific understanding of soils as living, organic, and interdependent environments became a powerful force in the Progressive Era. Thanks to their late nineteenth-century contributions to medicine, public health, and hygiene, bacteriologists had already claimed that they offered solutions to many urban problems. They now moved into the countryside, confident that bacteriology and applied botany could solve rural problems as well. Life scientists could claim that only they had the expertise to explain the importance of the microbial world in modern agriculture, and only they knew how to maximize and utilize soil bacteria's beneficial attributes. Bacteriologists were especially well situated—conceptually and institutionally—to see soil organisms as an untapped agricultural and economic resource. As historian Eric Kupferberg has put it, many experts came to see effective management of soil bacteria as the "apotheosis of scientific farming."[5]

Although one prominent historian has suggested that the "United States government took only a minor interest in this new field," much of the early work in soil bacteriology took place at the USDA.[6] This history fits well within a larger historiography of that institution, and with the historiography of Progressive Era science in general. Historians of science, especially the late Philip Pauly, have highlighted the aggressive efforts of USDA scientists at the turn of the century to harness the life sciences for the social good. As Pauly explained, government scientists believed that they could "rationalize and accelerate" progress.[7] Particularly under the leadership of Secretary of Agriculture James "Tama Jim" Wilson—the longest-serving Cabinet secretary in American history—USDA bureaucrats aimed to make the department a center of research in the life sciences. Determined to end the department's reputation as simply a source of free seed samples, patronage jobs, and answers to simple farming questions, Wilson wanted his scientists to wrestle with larger problems. Aided by the disproportionate power of rural legislators in Congress and their hopes

[4] Burrill (1904, p. 434).

[5] Kupferberg (2001, p. 163). See also Gossel (1988, pp. 1875–1900).

[6] Rossiter (1979, p. 235).

[7] See Pauly (2000, pp. 80–84); and Pauly (1994).

to address the populists' demands of the 1890s, Wilson managed to win lump-sum appropriations that funded specific scientific research projects—including several in the realm of soil bacteriology. Wilson also worked to turn the department into an institution that resembled a university in terms of its hiring practices, expectations for research, and control of the avenues of publication.[8] As one scholar has put it, the USDA under Wilson portrayed the "ideal bureaucrat" as one who was both a "researcher and a scholar."[9] He and his colleagues also expected USDA employees to be committed to political neutrality, scientific integrity, and delivering the lessons, practices, and vocabulary of modern science to the nation's farmers.[10]

The USDA aggressively expanded its agenda at the turn of the last century. A prominent example was the new team of "plant explorers," officially known as the Section for Seed and Plant Introduction (SPI). Formed in 1897, the plant explorers embarked on a mission to improve American agriculture, and society in general, through the systematic introduction of new agricultural crops. While it may be possible to criticize their efforts as a manifestation of botanical imperialism, or as a stepping stone in the rise of industrialized forms of agriculture, the plant explorers saw their work as a sincere effort to improve the standard of living for all.[11] According to one, Walter Swingle, the SPI crew hoped "to accomplish much good to the human race," either through the "pursuit of truth for its own sake or to benefit mankind."[12] Better known for their quest for tropical and other exotic plants that might thrive in the USA and its new territories, the plant explorers also found that both ordinary legumes and unseen bacteria offered additional weapons for their arsenal. The plant explorers soon moved into the new Bureau of Plant Industry (BPI), an agency formed in 1900 that linked several scientists united in the belief that coordinated work in bacteriology, botany, and other life sciences could improve crop production and address social concerns.[13]

Two men at the center of these efforts, Walter T. Swingle and George T. Moore, both born in 1871, were in many ways typical of the USDA's life scientists of this era. Swingle had entered Kansas State Agricultural College at the age of 16, where he became a protégé of the botanist William A. Kellerman and was a fellow student of his future USDA colleague, David Fairchild. Fairchild first joined the USDA in 1889, Swingle did so in 1891, and both became founding employees of the SPI in 1897.[14] Moore had earned a doctorate from Harvard, had ties to the scientific social circles at the Marine Biological Laboratory in Woods Hole, Massachusetts,

[8] Carpenter (2001, pp. 212–216). See also Hoing (1964); and Coppin (1990).

[9] Carpenter (2001, p. 216).

[10] For more scholarship on USDA scientists' efforts to address broader concerns, see Smith-Howard (2003), pp. 13–32); Hersey (2011); and Kimmelman (1983).

[11] For more on the plant explorers, see Pauly (2007, pp. 125–29); and Jones (2004).

[12] University of Miami, Department of Archives and Special Collections, Walter Tennyson Swingle (hereafter WTS) Papers, Box 30; WTS to Father, 17 December 1897.

[13] Pauly (2009, p. 84); and Stevenson (1954).

[14] Seifriz (1953); "Biographical Note" in finding aid to the Walter Tennyson Swingle Collection, University of Miami.

and before he reached the age of 30, he had become head of the botany department at Dartmouth College. Once they reached the nation's capital, both Swingle and Moore (as well as Fairchild and other USDA colleagues) joined Washington's influential Cosmos Club, becoming part of a community of scientists active beyond the halls of the USDA building.[15] As their stature and confidence in the Progressive Era ethos of improvement increased, they were more than ready to offer commentary on social issues beyond their areas of immediate expertise. Moore then gained national prominence when, with much fanfare, he announced that he had selflessly donated his patented methods for distributing bacterial cultures to the American people. Swingle, and especially Moore, seemed to represent prototypical USDA scientists, those whose work in the field and in the laboratory was destined to benefit the entire nation.

This history took a surprising turn in 1905 when Moore became the focal point of an apparent scandal. As described in more detail below, Moore's vigorous promotion of the only partially proven legume inoculants reflected poorly on the department's new scientific agenda and standards of professional ethics. When it appeared that Moore was trying to profit from his discovery, he was pressured to resign and quickly faded from the national scene. But the episode had other significance as well, for it pressured USDA scientists to recast their work in ways that addressed threats to the department's reputation for integrity and the expectation that it promoted scientific advances that had been validated as effective and worthwhile. In the end, the sudden rise and collapse of the USDA's soil bacteriology projects proved to be a reflection of both the prospects and the limitations of hopes for widespread social improvement through Progressive Era science.

The Context of the "Nitrogen Question"

Since the origins of human manipulation of the environment through agriculture, maintaining soil fertility has remained among farmers' foremost challenges. In his 1798 *Essay on Population,* Thomas Malthus argued that it would become impossible for agricultural workers to produce enough food to keep up with the growing population. Malthus saw hunger and disease as inevitable, but less even pessimistic scholars also understood that the problem of maintaining soil fertility was a limiting factor in social prosperity. By the mid-nineteenth century, scientists learned more of the nitrogen cycle, and artificial nitrogenous fertilizers seemed to help bend that constraint. Fertilizers quickly became almost essential for commercial agriculture, and much of the global economy depended upon farmers' ability to find and purchase new sources of nitrogen. The issue became especially pertinent in the late nineteenth century as depletion of the nitrate mines of Chile and the guano beds of the South Pacific seemed imminent, as monopoly interests emerged to dominate the commercial fertilizer industry, and as the perceived closing of the American frontier

[15] On the Cosmos Club, see Pauly (2009, pp. 53–54); Flack (1975).

signaled that virgin lands would not offer a "safety valve" for farmers who depleted their soils.[16]

Breakthrough discoveries came in 1886, when Hermann Hellriegel and his colleague Hermann Wilfarth solved the ancient riddle of legumes' ability to utilize atmospheric nitrogen. The German scientists proved that legumes simply would not grow in sterile soils lacking bacteria, regardless of the type or quantity of fertilizers applied. However, when they applied a soil extract taken from a field previously cultivated with legumes, they found that peas produced impressive amounts of nitrogenous matter, even in sterile and unfertilized soils. The German scientists concluded that the nutrients could have come through the symbiotic work of bacterial microbes that lived on the root nodules of healthy legumes and transferred fixed atmospheric nitrogen into nitrates that plants could absorb.[17]

This was not simply another nuance in the scientific understanding of crops and soils. In the words of one witness to Hellriegel's public announcement, his paper was a "sensation" and "bravos" greeted the scientist when he left the podium.[18] The news opened the possibility of boundless natural soil fertility, perhaps eliminating the need for frequent soil analyses and costly chemical fertilizers. Scientists soon suggested that "diseased" soils might be improved by the application of small amounts of the desired microscopic organisms, a form of inoculation similar to how humans can be protected from dangerous disease.[19] Friedrich Nobbe and his colleague Lorenz Hiltner led the search to capitalize legume inoculation in the early 1890s.[20] By 1896, Nobbe and Hiltner had entered into an agreement with an emerging giant in the German chemical industry, later known as Hoechst, to produce their bacterial cultures on a commercial scale under the trade name "Nitragin."[21] The following spring, in 1897, the rival chemical firm Bayer introduced its own inoculant, "Alinit," which reportedly could assimilate atmospheric nitrogen for both legumes and nonlegumes. If successful, this product would be even more revolutionary, for it could conceivably make all agricultural lands permanently fertile.[22] Dozens of German experiment station scientists and practicing farmers lined up to test these potential panaceas, while an American scientist predicted that "germ" and "vest

[16] Gorman (2013, pp. 68–69); Smil (2001, pp. 8–12).

[17] Summaries of Hellriegel and Wilfarth's work are in Hellriegel (1887) and Clark (1895). Hellriegel and Wilfarth (1888).

[18] Archives of the Rothamsted Experimental Station, Harpenden, England, John Henry Gilbert to John Bennett Lawes, 28 September 1886. Further praise of Hellriegel's research is in Springer (1892).

[19] In 1884, the British writer Maxwell Masters predicted that future farmers would be able to grow as much with a "pinch" of the appropriate "ferment-producing germs" as with a ton of other fertilizers. See Masters (1884), p. 17).

[20] Hartmann et al. (2008).

[21] Farbewerke vorm. Meister, Lucius,and Brüning (1898). Townsend (1897). See also Aikman (1896).

[22] Geheimes Staatsarchiv Preussischer Kulturbesitz (hereafter GSBK), Berlin, I. HA Rep. 87B, Ministerium für Landwirthschaft, Domänen u. Forsten, Nr. 13236, Henry W. Böttinger to Ministry, 2 June 1897.

pocket" fertilizers would become the wave of the future.[23] Although early results were disappointing, many popular press writers hailed the potential of this discovery; as one wrote, it "lowered the boundary between gods and men."[24]

Crookes, Nitragin, and the American Reaction

Hellriegel's discoveries and the German patents for legume inoculants soon attained even greater significance. One hundred years after Malthus had made similar predictions, the British chemist William Crookes drove the issue home in his September 1898 presidential address before the British Association for the Advancement of Science. There Crookes warned that the population of the world's "wheat-eating peoples" was growing faster than farmers' abilities to meet the demand. Crookes predicted "grave peril" for the "civilized nations" if scientists failed to find new sources of nitrogen to replace the amounts lost perennially through grain cultivation. While Crookes seemed less concerned about the fate of the "rice-eating peoples," he cautioned that "the great Caucasian race could cease to be foremost in the world and will be squeezed out of existence by races to whom wheaten bread is not the staff of life." Presenting a thorough and reasonable analysis of the global production of wheat and the predicted demand for nitrogenous fertilizers, Crookes concluded that the world's supply of free nitrogen could be exhausted by the year 1931.[25] While Crookes was actually an optimist, and assumed that scientists would find solutions to the "nitrogen question" in time, the immediate effect was to provoke a great deal of public anxiety.[26] As historians such as Corinna Treitel have shown, Crookes's warnings fit closely within the era's debates over "biopolitics," or the notion that a nation's biological and nutritional health was a reflection of its geopolitical power.[27]

Crookes was virtually silent on how legumes and legume inoculants might have an impact on this issue, but those in the USDA were not. Walter Swingle, the

[23] For German examples, see GSBK, I. HA Rep. 87B, Ministerium für Landwirthschaft, Domänen u. Forsten, Nr. 13236, "Anbau Versuche mit Leguminosenimpfung unter Anwendung der Knöllchen Bakterien,"; Proceedings of the Curatorium der Königliche Pflanzenphysiologischen Versuchsstation, 10 August 1895; and Proceedings of the Curatorium der Königliche Pflanzenphysiologischen Versuchsstation, 21 April 1894. See also Sächsisches Hauptstaatsarchiv, Dresden; Ministerium des Innern, Nr. 15678, Vol. III, Stutzer to Ministry, 31 January 1898, and Storck to Ministry, 25 October 1898. For an early American example, see Duggar (1897).

[24] Townsend, "Nitragin," p. 202. See also (Anonymous. 1905h) "Inoculation of the earth"; Johnson (1900). On disappointing results, see GSBK, I. HA Rep. 87B, Ministerium für Landwirthschaft, Domänen u. Forsten, Nr. 13236, Stutzer to Ministry, 31 January 1898, and Storck to Ministry, 25 October 1898.

[25] Crookes (1898). See also Brock (2008, pp. 375–87).

[26] On the response to Crookes's speech, see (1898a) "Nitrogen and wheat" and (1898b) "Answering an alarmist"; (Anonymous. 1898) "The world's supply of wheat"; Noyes (1898); Davis (1899): "Crookes vs. Atkinson, Dodge, et. al."

[27] See Treitel (2008); and Dickinson (2004). Although these scholars focus on Germany, parallels with the USA are made clear.

USDA plant explorer then stationed in France, quickly dismissed Crookes's warnings because nitrogen-fixing legumes could provide an answer. Spurred directly by Crookes's speech, Swingle soon scoured European seed catalogs, experimental farms, and laboratories in search of promising varieties of sweet clover, peas, and other legumes. By early 1899, his travels had taken him to Italy, Greece, Turkey, Algeria, and Tunis as well. The hunt turned up scores of new legume varieties, including several that "seem to be a quite different character from any nitrogen collectors we are now growing in the South."[28] Swingle also called for American agricultural experiment stations to ramp up research on nitrogen-fixing plants, giving them "a real problem to work with."[29]

Swingle was even more excited about the possibility of legume inoculation. He had already visited the laboratory of Nitragin's developer, Friedrich Nobbe, and in 1898 he enthusiastically reported the news from France that "Alinit is very hot stuff," especially because many claimed it might be proven to work on nonleguminous crops as well. Swingle recognized that, if true, the inoculant would "revolutionize the culture of cereals in the dry rich lands of the [American] West."[30] Swingle also aimed to track down samples of the soils that had been planted in promising legumes, for these were sure to contain "the right kind of bacteria" for further study.[31] Upon his return to the USA, Swingle and his colleagues continued their study of soil bacteriology and a possible biological solution to the nitrogen question. "No doubt about it," fellow plant explorer David Fairchild wrote to Swingle in 1900, "the earth bacteria are going to be recognized as the most important factors in agriculture."[32] Research budgets at the USDA for soil, botanic, and bacteriological investigations exploded and in 1901 Swingle took charge of the department's soil bacteria work as head of the BPI's new Division of Plant Physiology.[33]

In his new role, Swingle immediately tried to lure the Dartmouth bacteriologist George T. Moore to lead the project. But Moore—who had already gained some prominence for his research on the relationship between disease and algae in urban water supplies—found himself comfortable in academia, and he especially enjoyed his connections with the emerging summertime retreat of American biologists at Woods Hole, Massachusetts. He was reluctant to alter his budding career.[34] But

[28] Swingle Papers, Box 15, WTS to DF, 18 Oct 1898. For his complete 1898–99 itinerary, see Swingle Papers, Box 33, 8 December 1903. In 1901, the USDA again dispatched plant explorers to North Africa, men who returned with another 105 leguminous species. Library of the Missouri Botanic Garden, St. Louis, George Thomas Moore Papers, (hereafter Moore Papers), Box 1, WTS to George T. Moore, (hereafter GTM), 21 May 1901.

[29] Swingle Papers, Box 15; WTS to David Fairchild (hereafter DF), 11 September 1898.

[30] Swingle Papers, Letterbook 2; WTS to O. F. Cook, 30 October 1898, and WTS to O. F. Cook, 22 November 1898.

[31] Swingle Papers, Box 15; WTS to O. F. Cook, 21 October 1898.

[32] DF to WTS, 20 November 1900, Box 33, Swingle Papers. Fairchild (1938, pp. 196–197).

[33] For instance, American agricultural experiment stations employed zero bacteriologists in 1900, but 18 just 5 years later. See True (1937, p. 137).

[34] Kleinman (2010) and (Anonymous. 1965) "George Thomas Moore." For more on the Woods Hole laboratories as the summer home of a vibrant community of the nation's leading biologists,

speaking the Progressive Era language of improvement, Swingle insisted that "the nitrogen problem is a *big one* and promises *great results* both from a scientific and from a humanitarian standpoint."[35] As additional enticements, Swingle promised that Moore could maintain his summer ties with Woods Hole, he boasted of the doubling of the USDA's budget for botanical work in just 1 year, and he implied that Moore could hire research assistants as needed. In short, Swingle argued, there is "no such fund or organization anywhere in *the entire world*" that resembled the USDA.[36] Moore finally accepted the job after deciding that he could become "enthusiastic" about research on nitrifying bacteria.[37]

Swingle laid down Moore's first task—to prepare an improved legume inoculant—even before Moore had signed his USDA contract. Swingle sent clover samples to Moore while still on the Dartmouth campus, and he also asked Fairchild to seize the seeds of some promising North African legumes before they came onto the market—before the USDA could be accused of stealing a commercial product. Meanwhile, Moore gathered microscopes, soil sterilizers, and other research apparatus at Woods Hole, and Swingle, Fairchild, and Moore rendezvoused in Massachusetts to work out laboratory methodologies.[38] As his own research intensified, Moore demanded that "No effort must be spared" to get his hands on viable specimens of the German legume inoculants like Nitragin.[39]

Moore's quest also became a priority for the BPI and USDA as a whole. Even as the research was in its early stages, Moore's supervisor, BPI head Beverly Galloway, announced that soil bacteriology, with its potential to increase the nation's store of fertile soils, was "particularly" important, and he urged his scientists to quickly develop methods to multiply the proper nitrifying organisms and to distribute beneficial bacteria to farmers cheaply and effectively.[40] Moore soon confidently announced that an answer to the nitrogen question lay at hand. As Moore explained, he and his colleagues had "perfected" methods of artificial inoculation that were far superior to the Germans'.[41] After a visit to Nobbe's and other European scientists' laboratories in late 1903, Moore was even more convinced of the superiority of his methodology. He applied for a patent in May 1903, and in March 1904, those rights officially came under the USDA's control.

---

see Pauly (1988).

[35] Moore Papers, Box 1, WTS to GTM, 6 March 1901. Emphasis in original.

[36] Moore Papers, Box 1, WTS to GTM, 6 March 1901. Emphasis in original. For more on the USDA building research facilities and staffing that resembled a university, see Carpenter (2001, pp. 221–226).

[37] Swingle Papers, Box 23, GTM to WTS, 11 March 1901.

[38] Moore Papers, Box 1, WTS to GTM, 29 March 1901; National Archives, RG54, E26W, Swingle Letterbook, WTS to GTM 4 May 1901; and WTS to DF, 16 July 1901; and Swingle papers, Box 24, GTM to WTS, 10 July 1901; and A.F. Woods to WTS, 14 August 1901.

[39] Swingle Papers, Box 24, GTM to WTS, 27 August 1901. Emphasis in original.

[40] Galloway (1902, p. 56).

[41] Moore (1903). Another report in the volume asserts that the government cultures were "at least five times as great as the nitrogen-gathering power of the ordinary forms found in nature." See Wilson (1903, p. 21).

At the same time, soil improvement was becoming a national project that involved much of rural America. Significantly, Moore framed his research interests in botany and soil bacteriology in ways that matched the USDA's mission to become a national center of research in the life sciences. In short, he asked ordinary farmers to do much of the field trial work and generate positive publicity for his discovery. The government's policy was to send envelopes of desiccated bacteria stored on cotton balls, with instructions to first mix a packet of chemicals (provided in the package) with rainwater, add the bacteria, stir and wait 24 h, add another provided chemical, moisten legume seeds in the solution, allow the seeds to dry, sow them as normal onto fields, and keep records of their results. Remarkably, over 3500 farmers and other amateur researchers (out of some 12,500 who received samples) reported their results to the USDA. In Moore's words, this was "one of the largest experiments of this nature ever undertaken by any country."[42]

These field experiments could not have been successful without widespread public interest in soil improvement. Newspaper headlines told of "Bacteria from Uncle Sam," delivered at no cost, and even the *New York Times* placed news of Moore's bacteria distributions on its front page.[43] In addition, a wide range of popular writers, many of whom were no experts in the life sciences, generated the enthusiasm necessary for soil bacteriology and legume inoculation to seem essential issues. In 1903, for instance, the American novelist Theodore Dreiser, then earning his living as a freelance journalist, penned a magazine article that argued the soil question was one of the most important of his era. Poor soils, he believed, led inevitably to malnourished, poorly clothed, and ill-housed farmers, people who were unable to adjust to the social demands of modern life.[44] The well-known journalist Ray Stannard Baker visited Nobbe's laboratory in rural Saxony and returned with a most glowing report of the cutting-edge research facilities, including large trees that had been growing in "water cultures" of fertilized solutions that contained no soil, for over 20 years. Baker enthusiastically described Nobbe's legume inoculation ideas as an answer to Crookes's warnings of a nitrogen famine, and he praised the USDA's emerging plans to bring bacterial cultures into areas "deficient in nodule-forming germs."[45] Again alluding to Crookes's warnings, other evangelists of soil bacteriology celebrated underground microbes as ones that could help cure disease, conquer world hunger, and save the European races from collapse. The most optimistic writers believed that scientists might soon find a way to have nitrogen-fixing bacteria grow on the roots of nonleguminous crops.[46] Another writer framed the issue in terms of economic efficiency; soil bacteria could reduce labor costs by eliminating

[42] Moore and Robinson (1905a, p. 31).

[43] (1903) "Bacteria from Uncle Sam"; ad in New York Times, 8 April 1904.

[44] Dreiser (1903).

[45] Baker (1903). Baker also observed a "suggestion of intelligence" in soil bacteria, for they knew enough to behave differently in relation to the amount of nitrogen in the soil.

[46] Johnson (1900); Wood (1903); Clarke-Nuttall (1902/1903).

the demand for imported fertilizers and perhaps obviate the necessity for tedious crop rotations.[47]

Meanwhile, Moore and his colleagues intensified their well-orchestrated campaign to promote legume inoculants. Far from fearing the infamous muckraking journalists of the Progressive Era, USDA officials knew how to plant stories and how to manipulate the press in order to publicize its agenda, bringing attention to the department's timely response to the nitrogen problem. Through David Fairchild, Moore's BPI colleague and soon one of Alexander Graham Bell's sons-in-law, Moore delivered a talk and magic lantern show before prominent guests in the Bell home. It proved so popular that the inventor asked for an encore lecture the following week. There, Gilbert Grosvenor, another of Bell's sons-in-law and head of the National Geographic Society, offered to arrange publicity for Moore's new methods through both *National Geographic* and *Century* magazines.[48] Moore agreed to the idea, provided his USDA photographs, and maintained editorial approval for such articles. These reports praised Moore in no uncertain terms. The *National Geographic* piece declared that rumors of the nitrogen famine were "greatly exaggerated" and reminded readers that Moore had "generously deeded [his patent] to the American people." The *Century* article was similarly bold, featured a large portrait of Moore, and concluded with the promise "there is no section of country which will not profit from Dr. Moore's discovery."[49] Other popular magazines and newspapers also touted the USDA scientist, consistently praising him as an altruist who willingly passed up the chance for commercial wealth in order to make it possible for farmers to finally tap into the atmospheric nitrogen that bathed their fields.[50]

The ambitious bacteriologists and botanists at the USDA's BPI fully believed in the importance of their work. In the words of BPI chief Beverly T. Galloway, "to be a scientist is to be a man of affairs," and he pushed this colleagues to pursue an aggressive form of science that had recognizable and utilitarian public benefits.[51] Moore, Swingle, Fairchild, and others followed this mantra, and clearly were willing to become evangelists for their discipline and on various social issues issue of the day.[52] USDA scientists also spoke of the how their work could improve society. Galloway, for instance, argued that "mankind as a whole is bettered, and the struggle

[47] Schneider (1903).

[48] Moore Papers, Alexander Graham Bell to GTM, 18 March 1904; and Gilbert H. Grosvenor to GTM, 25 May 1904.

[49] Gilbert Grosvenor (1904a), "Inoculating the ground"; Grosvenor (1904b), "Inoculating the ground," p. 839.

[50] (Anonymous. 1905h) "The inoculation of the earth."

[51] Galloway (1904, p. 13).

[52] David Fairchild had proposed that a new international language could arise to help coordinate research efforts in the sciences. Swingle Papers, Box 33, DF to WTS, 10 August 1901. Swingle became a proponent of an artificial language called pasigraphy, which was based upon a common and overlaying set of symbols, like Chinese, which he hoped could eventually become a "nearly perfect language" and facilitate international communication. Swingle (1905). He later wrote in favor of another technique to improve international scientific communication—the metric system. See Swingle (1909).

for life is made less a burden" through applications of the work of modern botanists, and he was particularly convinced that no other branch of science would bring more benefit to mankind than bacteriology.[53] George Moore was even more blunt when he urged that more "able-bodied men" become research botanists in order to combat the perception that botany was "somewhat of an effeminate calling."[54] In a commencement talk that focused on the USDA's accomplishments, he urged his audience to be "as proud of having the finest and most efficient Department of Agriculture in the world as he is of nation's possessing the biggest gunboat."[55] Moore situated his own legume inoculation work in the context of larger national concerns, because in effect legume inoculation expanded the nation's cultivated acreage. He also envisioned an expanding role for USDA research, for it could also lead to improvements in the industries that depended upon microorganisms, such as brewing, wine making, and cheese making. Interestingly, Moore gave these optimistic talks even as the foundation that he had built at the USDA was beginning to crumble.

George Moore and the National Nitro-Culture Company

The excitement over legume inoculants and a patent protecting the Moore method soon attracted entrepreneurs. In July 1904, two investors in the National Nitro-Culture Company (NN-CC) of West Chester, Pennsylvania, visited Moore at his Woods Hole summer retreat. They soon offered him a job at more than double his USDA salary, promising he would need to work only a few hours each day. Moore initially declined the offer, but unsurprisingly, he did hope to parlay it into a higher salary.[56] His supervisors promised to help, but it is telling that BPI chief Galloway framed the issue in ways that reflected the USDA's commitment to a broader agenda. In fact, Galloway had already delivered a major speech in which he stressed the importance of scientific integrity and the need for government scientists to ensure that "trust imposed on us has been fully and honestly respected."[57] There is a "certain prestige" that comes with government work, Galloway believed, and he warned Moore that joining a commercial enterprise would cause him to lose "at once [his] caste as a scientific man."[58] For a while, Moore resisted the temptation.

---

[53] Galloway (1902, pp. 49–59).

[54] Moore (1905).

[55] Moore Papers, Box 1, GTM, "The creation and development of plant industries by the government," [undated commencement speech, c. 1904 or 1905].

[56] National Archives, RG 54, A. F. Woods Letterbooks, Book 2, A. F. Woods to GTM, 22 July 1904 and A. F. Woods to GTM, 29 July 1904.

[57] Galloway (1904, p. 12).

[58] National Archives, RG 54, E1, Box 38, [Beverly T. Galloway] to GTM, 2 August 1904. In any event, Moore's salary increased to US$ 3000 in early 1905, a 67 %increase over his salary when his employment began in late 1901. See National Archives, RG 54, A. F. Woods Letterbooks, Book 2, A. W. Woods to B. T. Galloway, 20 March 1905.

As Moore was deciding whether to stay at the USDA, he continued to push "nitro-cultures," as legume inoculants were becoming known, through his Washington connections. Two USDA publications that appeared in January 1905 were examples. Although these contained sensible caveats and presented a sober analysis of the facts, the louder message was summarized in one of the pamphlet's subtitles: "The Successful Use of Artificial Cultures by Practical Farmers."[59] An unpublished draft of one document included especially hyperbolic language: If the product could be developed for nonleguminous crops, it announced, "We will have something valuable almost beyond comprehension."[60] In any case, the published report asserted that Moore and his new inoculant had succeeded beyond all expectations.[61] The longer bulletin included about 150 favorable testimonials from farmers, not one that was unfavorable, and tellingly, no reports from professional botanists or experiment station scientists. The government's warnings, presented in a section entitled "when to expect failure with inoculation" were buried on page 29 of one pamphlet, and it brushed aside data indicating a failure rate of over 25 %.[62]

The publicity campaign of 1904 and Moore's publications of January 1905 exacerbated the storm of demand for the government's "nitro-cultures." Farmers flooded BPI offices with some 40,000 requests for free samples, up to 1000 per day.[63] Harvard's prominent botanist and Moore's former professor, William Farlow, begged Moore for a "'coke' of your leguminous 'tonic'" as a special favor.[64] The USDA struggled to find a way to deal with the thousands of correspondents—including diplomats and agricultural leaders from Canada, France, Germany, Russia, Australia, and Martinique—who were disappointed to learn that demand exceeded supply; soon only those who funneled their requests through influential Congressmen and Cabinet officers enjoyed much success. But this too was a problem, for it came into conflict with the department's commitment to focus on research and get out of the business of supplying samples through political patronage. BPI chief Galloway and Assistant Chief Albert Woods both feared that widespread demand for free bacterial samples cost excessive time and money, and also detracted from the bureau's broader mission. As Galloway put it, "I believe that the general and promiscuous distribution should be stopped" and replaced with a more "systematic" approach.[65]

[59] Moore and Robinson (1905b); and Moore and Robinson (1905a).

[60] National Archives, RG 54, A. F. Woods Letterbooks, Book 1, Draft of Report "Nitrogen-Fixing Bacteria,"Book 1, [undated, but likely late 1904].

[61] Preface to Moore & Robinson (1900b), 5.

[62] Moore & Robinson (1900b), 45. The data in this report were based on a summary of 2502 replies received by 15 November 1904. The data in the second report were based on a summary of 3540 replies received by 31 December 1904. The failure rate had improved slightly, now only 21 %. As before, these testimonials conspicuously did not include a single comment from the hundreds of people who reported that the inoculant provided no benefit at all.

[63] National Archives, RG 54, A. F. Woods Letterbooks, Book 1, GTM to Assistant Secretary, 18 April 1905.

[64] Moore Papers, Box 1, W. G. Farlow to GTM, 29 January 1905.

[65] National Archives, RG 54, A. F. Woods Letterbooks, Book 1, A. F. Woods to GTM, 18 November 1904, and B. T. Galloway to A. F. Woods, 21 January 1905.

The government eventually ceased its free distributions, but that only increased the demand at the private firms. Meanwhile, the NN-CC had secured endorsements from experiment station scientists, such as George Washington Carver, by offering them the privilege to work as wholesalers in selling the commercial product to farmers and dealers.[66] Business at the NN-CC was booming, enough to make West Chester's postmaster complain of his suddenly increased workload.

Meanwhile, Moore was becoming something of a national hero. In September 1904, the *Washington Post* asserted that if there were more men like him in the government, there would be less graft.[67] His work on purifying drinking water also continued to keep him in the limelight, a useful example of how the agricultural sciences could also be helpful in solving urban problems.[68] In January 1905, an organization in Ventura, California, sent to Moore an unsolicited "resolution" that praised him for his "generosity…and patriotism in presenting the [inoculation] discovery to the American people."[69] In March 1905, *Scientific American* published back-to-back articles in the same issue on the two distinct areas of Moore's research.[70] That spring, Moore's alma mater, Wabash College, congratulated him on the "services rendered to the American people" and had him deliver a commencement address, just 11 years after he had walked the same campus as an undergraduate.[71]

Nevertheless, a few skeptics began to question the efficacy of legume inoculants in general and the Moore patent in particular. In many ways, the rural press proved to be more cautious and sensible about the legume inoculation hoopla than the mass market journals and USDA publications. *The Country Gentlemen,* for instance, published in 1904 a sober and thoroughly footnoted review of the scientific literature, concluding that artificial inoculation would be effective only under specific soil conditions and was "in no wise a 'cure-all'."[72] *Wallace's Farmer* also commented on the unwelcome trend of popular magazine writers trying to become the distributors of agricultural knowledge, and strongly urged Midwestern farmers to focus simply on planting legumes on a regular basis.[73] For his part, the eminent Cornell professor Liberty Hyde Bailey suggested that legume inoculation might have value, but he also warned of sensationalized reports in the press.[74]

---

[66] George Washington Carver Papers Microfilms, Tuskegee Institute, Tuskegee, Alabama, Reel 2, Edward H. Jacob to George Washington Carver, 18 December 1904. Jacob promised to "amply compensate" Carver for his promotional work.

[67] Washington Post, 28 September 1904.

[68] Washington Post, 6 January 1905.

[69] "Resolution No. 17," to GTM, Box 1, Moore Papers.

[70] (Anonymous. 1905c) "Bacterial soil inoculation for vegetables"; and (Anonymous. 1905a) "An important discovery in the purification of contaminated water."

[71] H. Z. McLaw to GTM, 10 May 1905, Box 1, Moore Papers; and Indianapolis News, 14 June 1905.

[72] (Anonymous. 1904) "Soil inoculation." Also (Anonymous. 1905g) "Soil inoculation: What it can and cannot accomplish."

[73] (Anonymous. 1905f) "Soil inoculation"; (Anonymous. 1905d) "Government bacteria."

[74] Bailey (1905).

The journal *National Farmer and Stockman* and its lead correspondent, Alva Agee, led the charge. In his initial analysis, published in January 1905, Agee admitted that legume inoculation seemed promising, although he cautioned that for countless farmers, the soil naturally contained enough bacteria to make artificial cultures redundant and a complete waste of money. At first, Agee complained only that reckless publicity over the potential of legume inoculation was causing "harm…to the department [of agriculture] and to the public."[75] The journalist also warned that hype about the simplicity of "vest-pocket fertilizers" would "warp" farmers' independence and tempt them to ignore the traditional skills necessary to successfully manage their own soils and crops.[76] Agee also questioned why the government seemed determined to enter a market that could better be left to the private sector. In addition, the journal published letters from subscribers who complained that Moore's USDA publication declined to print the negative reports about inoculants that they had sent in.

The matter reached a new level in April 1905, following the appearance of an article by freelance journalist Raymond Porter in *Pearson's Magazine*. Although Moore had personally vetted and approved the author's draft manuscript, Porter's essay was perhaps the most hyperbolic yet. Subtitled "The Wonderful New Discovery Enabling Farmers to Do Away with Nitrogen Fertilizers," the article once again presented Moore's process as a solution to Crookes's nitrogen question and as an improvement over Nobbe's method. The issue was "beyond dispute," Porter added, "for the United States Agricultural Department itself says so."[77] Porter asserted that bacterial fertilizers promised tenfold productivity increases, and also (although illogically) both increased profits for farmers and lowered prices for consumers. He excitedly asserted that inoculants could bring both farmers and the nation benefits that were "almost beyond comprehension."[78] "In the opinion of agricultural scientists," Porter concluded, "not in the history of the Department of Agriculture has there been a more promising development."[79]

This article impelled Agee to challenge what he called the USDA's "campaign of advertising that was unique in the history of scientific agriculture."[80] He first made two investigative trips to the NN-CC's offices in Pennsylvania, where he found suspicious packages sent from the USDA by 1-day mail, and that the company had neither a bacteriology laboratory nor a bacteriologist. There he learned that the firm's business model was a rather simple one: Following directions on the labels of the USDA's free packages, it multiplied the bacteria and then repackaged them for commercial sale. Agee's reporting also took him to Washington, where he called upon Secretary Wilson, the Assistant Secretary of Agriculture, Willet Hays, and Moore himself. Each had a reasonable explanation for the department's position, so

[75] Agee (1905b) "Farm facts and fancies."

[76] Agee (1905a) "Bacteria talk."

[77] Porter (1905, p. 398).

[78] Porter (1905).

[79] Porter (1905, p. 403).

[80] Agee (1905f). "The booming of nitro-culture."

Agee left these interviews with renewed faith in the honesty of the government's research. In April 1905, both in private correspondence and in his publications, Agee explicitly cleared Moore of any impropriety.[81]

Yet Agee soon returned to the story, ever more curious about how the USDA scientists could "indulge in such a queer departure from the standards of scientific men." The state experiment stations, he noted, were now in the uncomfortable position of cautiously questioning the unconfirmed claims that emanated from USDA headquarters in Washington. The whole matter, Agee believed, threatened to bring agricultural science "as a whole into disrepute."[82] Agee's editorials urged the USDA to "stop worshipping at the shrine of publicity" and return to the kind of work that earns real respectability.[83] Then on 15 July 1905, Agee and his publisher delivered their evidence to President Roosevelt.[84] Agee soon presented copies of a November 1904 document that clearly showed how the company promised the scientist 141 shares, and his wife 43 shares, of the company's total of 250 shares, in exchange for Moore's commitment to deliver the methods of his "secret process" as well as any subsequent improvements exclusively to the NN-CC. Valued at $ 100 per share, the deal would have made the Moores quite wealthy indeed.[85]

Moore soon confessed to the whole affair. He admitted that, indeed, he had been offered a lucrative post with the NN-CC, and that, while still awaiting a pay raise from the USDA, he had supplied the firm with the bacterial cultures it needed to begin operations, taught its employees how to multiply the cultures, and supplied the firm with suggestions on how the products might be marketed.[86] He also admitted that his vigorous promotion of inoculants—in speeches and as the lead author of the two USDA reports—led to the rapid exhaustion of government supplies and drove up prices on the commercial market. It also became clear that the NN-CC had transferred stock shares to the name of Moore's wife as a way to secure the scientist's connection with the firm in case the USDA salary increase did not materialize. Moore also confessed that he had covered this up when confronted in April, then testifying that "neither directly nor indirectly do I hold stock in any of these companies." Moore dutifully submitted a letter of resignation on the next day.[87] Secretary Wilson promptly accepted the resignation, something that evidently came as a surprise to Moore.

[81] Agee (1905f, p. 125). In a telegram sent to Assistant Secretary Hays (at the Cosmos Club), Agee assured Hays that he was "entirely confident of your doctor's integrity." National Archives, RG 16, E8, Box 29, Alva Agee to Willet Hays, 22 April 1905.

[82] Agee (1905c). "Farm facts and fancies."

[83] Agee (1905d) "Farm facts and fancies."

[84] National Archives, RG 54, E1, Box 38, T. D. Harman to Theodore Roosevelt, 15 July 1905.

[85] National Archives, RG 16, Microfilm 440, James Wilson to the President, 10 October 1905.

[86] National Archives, RG 54, A. F. Woods Letterbooks, Book 2, A. F. Woods to B. T. Galloway, 20 March 1905, pp. 156–159.

[87] National Archives, RG 16, Letterbooks of the Secretary of Agriculture, Book 105, James Wilson to B. F. Barnes, 18 and 28 July 1905. Many of the details are repeated in Agee (1905e) "Nitroculture discredited"; and (1905) "New department scandal."

The episode brought ignominy to a department determined to stake a claim as a leader in American science. "Another idol shattered!" cried one newspaper editorial; "we had thought that the real scientist was a man above the sordid things in life!"[88] Other newspaper headlines screamed "UGLY Charge Made against Government Employee," and "Nitro-Culture Graft!"[89] A few other newspaper writers came to Moore's defense, including the comment in the *Washington Post* that "probably no young scientist of the age has gained such a wide and enviable reputation as Dr. Moore." His "manly qualities and integrity" were well known, and the *Post* reported that Moore had been embarrassed by all the attention he had received.[90] Another paper pointed out that he could have made millions of dollars from his discoveries, and with a personal story that was "absolutely quixotic," it would be impossible for any jury to convict him.[91]

But the Moore episode was especially noticeable because it coincided with several other alleged scandals that rocked the USDA that summer. Secretary Wilson also faced accusations of leaking crop reports to favored investors, of lavish expenditures through the Weather Bureau, of offering his son privileged access to gold lands in Alaska, and of profiting through the printing of meat inspection labels.[92] Added together, these episodes challenged the USDA leaders' long push to establish a reputation for integrity and for socially useful scientific research. Three days after Moore's resignation, President Roosevelt called Secretary Wilson to his vacation home in Oyster Bay, New York. The press reported that Wilson's resignation was imminent, although the Secretary explained to a concerned farmer that "my fighting blood is up" and "I certainly shall clean house."[93] In the end, Justice Department officials concluded Moore's actions did not amount to a criminal offense; as Attorney General William Moody put it, federal laws "do not cover all classes of wrongs."[94] Several of Moore's allies, including Harvard University President Charles Eliot, lobbied to have him reinstated.[95] But Secretary Wilson had no interest; he told President Roosevelt that Moore had repeatedly deceived him and had "violated…the basic principles of ethics which should prevail in a scientific corps as in the Department."[96]

---

88 Savannah Morning News, 29 July 1905.

89 Sandusky [OH] Star-Journal, 28 July 1905; Galveston Daily News, 28 July 1905. In its editorials, The National Stockman and Farmer showed some sympathy for Moore's temptation to capitalize on his work. It found greater fault with the overall "rottenness" of the USDA, especially its "policy" of seeking notoriety at the expense of scientific rigor. See (Anonymous. 1905e) "Secretary Wilson's responsibility."

90 Washington Post, 6 August 1905.

91 [Chicago] Inter Ocean, 31 July 1905.

92 Washington Post, 6 August 1905.

93 Iowa State University Special Collections, James Wilson Papers, Box 4, James Wilson to C. D. Boardman, 22 August 1905. See also (Anonymous. 1905b) "Another department scandal"; Hoing (1964, pp. 176–182).

94 Alexandria Gazette and Virginia Advertiser, 2 August 1905; New York Times, 20 August 1905.

95 National Archives, RG 54, E 1, Box 38, Charles W. Eliot to President Theodore Roosevelt, 7 October 1905.

96 National Archives, RG 16, Microfilm 440, James Wilson to the President, 10 October 1905. There had been an earlier offer to reinstate Moore to the payroll, but that was soon rescinded. See

Government Policy after Moore

But the damage had been done, and USDA leaders mobilized to restore the department's reputation for scientific integrity.[97] Evidently concerned that he could be implicated for profiting from his research, Swingle instructed his father to process legal papers to be sure that his name had been removed from the ownership of a commercial date farm in southern California.[98] Meanwhile, assistant secretary of Agriculture Hays demanded sales information from the NN-CC, hoping to defuse allegations that the company had made "millions" of dollars though its association with Moore.[99] As that was underway, Hays urged farmers to not depend "too much" on commercial inoculants until more test results were in. BPI chief Galloway also worked to distance the USDA from the NN-CC on the grounds that the firm lacked a "sufficient scientific basis."[100] He soon added that other firms were no better, as virtually all of the inoculating material on the market was "practically worthless."[101] The Department also quickly abandoned Moore's method of shipping bacteria stored on cotton, and launched a new method that used hermetically sealed test tubes.[102] Moore's massive public experiment project also came to an end. USDA officials now made the case that they had completed all of the necessary trials, and that it now was time to allow the private sector take over the legume inoculation industry.[103]

The episode also gave the state agricultural experiment stations the chance to reassert their legitimacy. Scientists working on the land grant university campuses had felt overlooked in an era dominated by Secretary Wilson and the other imperious Washington bureaucrats who attempted to centralize research and control spending in the agricultural sciences. Tensions became sharper in 1905 and 1906, as battles over the Adams Act—which promised greater research budgets for state-lev-

National Archives, RG 54, E 1, Box 38, [Beverly T. Galloway] to GTM, 1 September 1905; [Beverly T. Galloway] to GTM, 8 September 1905.

[97] Moore evidently took a job with the NN-CC through 1906, but he managed to reestablish his academic credentials mainly through his connections at Woods Hole. See Marine Biological Laboratory (1907) and Marine Biological Laboratory (1909, p. 19). Moore joined the faculty at Washington University in 1909, and he served as director of the Missouri Botanical Garden from 1912 to 1953. See Kleinman (2010).

[98] Swingle Papers, Box 30, WTS to father, 4 August 1905. There was good reason for concern. Swingle had invested—and urged his friends and family to do so as well—in a date farm that was virtually across the street from a USDA date research facility, one that worked primarily on improving the specimens that Swingle himself had imported from the Mediterranean and where Swingle had been a research advisor.

[99] National Archives, RG 54, E28, Letterbook 55, W. M. Hays to E. Jacobs, 19 August 1905.

[100] National Archives, RG 54, E5, Letterbook 70, B. T. Galloway to A. F. Woods, 21 August 1905.

[101] National Archives, RG 54, E5, Letterbook 79, B. T. Galloway to Charles F. Curtiss, 7 February 1906. Similar examples include National Archives, RG 54, E5, Letterbook 85, B T Galloway to R. M. Winans, 30 March 1906.

[102] Kellerman and Robinson (1905).

[103] National Archives, RG 16, Microfilm No. 440, Reel 53, James Wilson to William F. Atkinson, 28 August 1905.

el experiment stations—were fought in Washington. When the bill passed, research dollars shifted to the state level, which to some extent diminished the spotlight that had shone upon USDA scientists.[104]

The inoculation problem also gave state experiment station scientists the chance to monitor the commercial market. Tellingly, many of these reports were written in a new key, in a dry style that lacked the color and enthusiasm of earlier articles in the popular magazines. C. G. Hopkins of the Illinois station, for instance, complained that the entire industry had been built on "erroneous and misleading" use of the work of experiment stations and the USDA, and he warned many of the claims regarding inoculation are "greatly exaggerated and overestimated."[105] The Pennsylvania Agricultural Experiment Station reported results that showed farmers gained only the slightest possible advantage by using artificial cultures.[106] A New York Agricultural Experiment Station study revealed that government nitro-cultures had a brief shelf life, were "exceedingly unreliable," and were in any case unnecessary for a large majority of farms. Several stations soon confirmed the New York findings: nitro-cultures prepared according to the Moore method were difficult for farmers to prepare and ineffective in the field. This led to the "inevitable" conclusion that artificial inoculation was generally "unremunerative and unwise."[107] A scientist at the Texas Agricultural and Mechanical College reported that "farmers are being victimized," both by "worthless" inoculants and by exaggerated claims that had them applying inoculants to fields where they were unnecessary.[108] German studies similarly revealed what they called "several cases of swindle" in the US market.[109]

Back in Washington, the USDA's position became noticeably less visible and less vocal. Oddly enough, the bacteriologist who replaced Moore, Karl F. Kellerman, was the son of Swingle's and Fairchild's mentor at Kansas State. But his approach to the legume inoculation was cautious, and his laboratory continually focused its efforts on basic soil bacteriology research and routine market regulation.[110] Tellingly, Kellerman's publications prominently explained "when inoculations fail" on the first page, and he bluntly discussed the futility of attempting to use inoculants on nonleguminous crops.[111] Also symptomatic of the new approach, the USDA now refused to release photographs of its bacterial research to the popular press.[112] The

[104] Rosenberg (1964).

[105] Hopkins quoted in Kupferberg (2001, p. 216).

[106] Butz (1905).

[107] Harding and Prucha (1905); Harding and Prucha (1906); Voorhees and Lipman (1907, p. 105); and Starnes (1905, p. 101).

[108] El Paso Herald, 21 February 1906.

[109] GSPK, I. HA Rep. 87B, Ministerium für Landwirthschaft, Domänen, und Forsten, Nr. 13238, Schneidewind to Ministry, 5 January 1907.

[110] National Archives, RG 54, E2, T. R. Robinson to Karl F. Kellerman, 16 September 1908.

[111] Kellerman and Robinson (1908); and Kellerman (1910).

[112] Moore Papers, Box 1, Waldemar Kaempffert to GTM, 8 July 1907. The journalist Kaempffert, hoping to write an article for *Cosmopolitan* or *Century* magazine, approached Moore for photographs of legume inoculants after he had been denied access to USDA photographs.

BPI gradually ended its distribution of bacteria samples, and by 1915, its official policy was simply to investigate commercial firms and publish the names of any manufacturer whose inoculants fell below an acceptable standard.[113]

The Wider Meanings of Soil Bacteriology

For many people, the rise and fall of Moore and the NN-CC did not diminish hopes that soil bacteria could lead to social improvement. Some authors read implications into the phenomenon that went far beyond the science of symbiosis in the legumes and bacteria, confident that this could also reveal fundamental interconnections in the entire organic world. Such ideas struck a chord in the early decades of the twentieth century, as scientific discoveries and social reform movements seemed interconnected. As recent scholarship on the emerging Country Life Movement suggests, reformers believed that farmers had particular abilities and responsibilities to maintain soil fertility.[114] Soil preservation became a moral issue, and reformers wanted farmers to gain access to valid information on how manipulations of the unseen soil microbes affected the larger world. That too had social implications, for improved respect for farmers' status and intelligence could stanch the flow of valued rural citizens from farms to cities. W. S. Harwood's 1906 text *The New Earth* boldly articulated similar views and defended the country's investments in agricultural science as "conspicuous evidence that there is something else in America besides greed and graft."[115] Explaining that the "Old Earth" was in a state of decay, with desolate, untidy homes, indebted farmers, "deadening isolation," and "deepening hate," Harwood promised that agricultural science could lead to a "New Earth" of prosperous, progressive, modern, well-kept homesteads, where families enjoyed books, music, and culture.[116] Harwood's book explicitly connected rural reforms with the soil inoculation industry: Advances in the farmers' ability to gain wealth from atmospheric nitrogen, he said, was the surest way to keep "advanced tillers" on the farm.[117]

Meanwhile, the new commercial legume inoculation firms—offering products with creative names like Nitrogerm, Soilvita, Nod-o-gen, Stimugerm, Farmogerm, and UneedR—often embraced a similar rhetoric. For instance, an Indiana company co-opted the anticorporate language of agrarian radicals and promised that

[113] Powell (1927, p. 10). See also National Archives, RG 54, E2, Box 81, W. J. Spillman, J. M. Westgate, and Karl F. Kellerman, "Report of the Committee upon Methods of Legume Inoculation," and several letters of Karl Kellerman, August 1914. As the quality of legume inoculants became more reliable, experiment stations in Wyoming, Michigan, Missouri, Oregon, and Wisconsin sold their own bottled bacteria and used the proceeds to fund further research. See Kupferberg (2001, p. 216); and Leonard (1932).

[114] Peters and Morgan (2004).

[115] Harwood (1906, p. 334).

[116] Harwood (1906, pp. 1–5).

[117] Harwood, (1906, pp. 26–27).

bacteria—which it labeled "little helpers"—enabled farmers to earn great wealth "without the interference of any trust, syndicate, or other combine."[118] A California firm explained that farmers had a responsibility to prepare "country homes" for the "busy little people," or the "little famil[ies]" of legume bacteria that do work necessary for the very "survival" of humankind.[119] The American firm that claimed to follow the Nobbe–Hiltner process of manufacturing Nitragin made an allusion to Moore—an "optimistic bacteriologist, who probably meant well"—but whose methods of bacterial distribution via cotton balls had failed. This company's advertisements accented another social dimension, promising "more pleasures, more comforts, more owners, and fewer tenants" through bottled bacteria.[120] Inoculation firms also touched upon some early examples of ecological thinking, promoting soil bacteria as "natural" alternatives to chemical fertilizers. Campaigns that championed farmers' role in preserving and maintaining the environment also included explicit and implicit attacks on agricultural chemical and fertilizer monopolies. The American Nitragin Company's promotional literature asked "Why should we not use Nature's way when it is more economical and easier than the ways provided by man?"[121]

Conclusion

By 1910 or so, the notion that legume inoculations were panaceas for soil and social problems had diminished. While field trials had demonstrated that many farms could benefit from artificial inoculation, most indicated that the expense and effort were unnecessary in fields that had been recently planted in legumes. The situation reached a symbolic turning point in 1911, when Moore's successor, Karl Kellerman, admitted that soil bacteriology research had stagnated. Indeed, as historian Margaret Rossiter has indicated, American soil bacteriologists had become frustrated with the low levels of institutional support for their work and turned to their European colleagues for help.[122] Confessing that recent research "sometimes gives us baffling results," Kellerman expressed little of the confidence that had been evident earlier in the century. The only hope for soil bacteriologists during this "critical period," he explained, was to return to the laboratories of European scientists.[123] Even George Moore reentered the debate, now 7 years removed from the episode that had forced his resignation. In a speech he delivered in 1912, Moore said he regretted the days when "no theory [was] too bizarre, no miracle too improbable, so long as we [fell]

[118] Smith (1913, pp. 14–15).

[119] (Anonymous. 1915) Country homes for busy little people.

[120] (Anonymous. 1912) "Old farms made new."

[121] Advertisement for "Nitragin" Company, Milwaukee, circa 1912, from the University of California at Santa Barbara, Special Collections.

[122] Rossiter (1979, pp. 235–36).

[123] National Archives, RG 54, E2, Box 81, K. F. Kellerman to William A. Taylor, 9 January 1911.

back upon the soil bacteria to account for it." Perhaps symptomatic of an end to the Progressive Era faith that experts had uncovered the many mysteries of the soil, Moore now said that scientists' arrogance of the past had to end. Another lesson he drew from that era was that there was no reason for "any particular glorification of the biologist"—perhaps including himself—for the unknowns of science would always outnumber the knowns.[124]

Looking back at the turn-of-the-twentieth-century discussions of the "nitrogen famine" and legume inoculations sheds light on a time when soil bacteria fit within the era's ethos of improvement. Reformers found reason to hope that soil bacteria could provide an almost miraculous and inexpensive source of valuable nitrogen, and thus an attractive alternative to farming systems based upon chemical inputs. Promoters also touted soil bacteria because of their social implications, for they seemed to be a tool that could help keep desirable rural citizens on the farm. These ideals also affected the aggressive life scientists at the USDA. Walter Swingle mobilized quickly when he had the chance to respond to the "nitrogen question," and George Moore's more intensive research work reflected the department's call for a confident and valuable form of public science. Moore's improprieties eventually embarrassed the department, although his superiors evidently tolerated them before the summer of 1905 because of the publicity that the legume inoculation program generated. Such attention was also valuable for the scientists themselves, who were eager to claim that applications of botany, bacteriology, and plant introductions could improve society. These multiple ambitions ran into hurdles, however. In the end, the invisible, unreliable, perishable, and often unnecessary microbes of the legume inoculants brought hopes of both soil improvement and social improvement, but they also very nearly shook the USDA's reputation for scientific integrity.

Debates over Crookes's "nitrogen question" came to a sudden halt when the Haber–Bosch process, developed between 1909 and 1914, made artificial fixation of atmospheric nitrogen possible. The Haber–Bosch process soon became reified and ubiquitous, fundamental to the ever-expanding agricultural production of almost all industrialized nations. The many legume inoculation firms that had emerged in the first decade of the twentieth century represented in some ways a last gasp of the small-scale nitrogen producers, and most soon disappeared. While some advocates still pushed for alternatives to chemical inputs—such as "green manures," or the deliberate expansion of legumes' place in crop rotations—such campaigns typically did not challenge a paradigm that for decades has been centered on just a few grain crops and annual inputs of millions of tons of artificial ammonia fertilizers.[125] Soil inoculants are still available, but the worldwide demand has remained steady for some time.[126] The USDA plant explorers who scoured the earth for plant

---

[124] Moore (1912).

[125] Cafer du Plessis (2009).

[126] Krimsky and Wrubel (1996). In recent years, only about 15 % of the soybean crop has been treated with traditional commercial inoculants. Meanwhile, efforts to apply genetic engineering to nitrogen fixation have also reached limitations: Fixating microbes consume energy (via carbohydrates) that decrease the amount available to produce crop yields.

introductions scored a success of another kind when soybeans emerged triumphant among the hundreds of legume varieties they had tested. By 1973, when US production of soybeans surpassed that of wheat, Crookes's warnings about famine among the wheat-eating peoples seemed ironic and antiquated.[127] But this history is not finished. Agricultural experts have improved legume inoculation technologies over the past several decades, driven largely by hopes to deliver leguminous proteins and natural nitrogenous fertilizers to the developing world.[128] Especially in the 1970s, researchers sought what one scholar calls the "holy grail" of the biological nitrogen fixation: the possibility of using genetic engineering techniques to developing nitrogen-fixing cereals.[129] Until that breakthrough occurs, however, most of the agricultural establishment will remain tied to manufactured nitrogen fertilizers. Indeed, in recent times the greater issue has not been shortages of nitrogen, but an overabundance of the nutrient in the form of agricultural runoff and other pollutants.[130]

References

1898a. Nitrogen and wheat. *New York Times*, September 8.
1898b. Answering an alarmist. *New York Times*, September 25.
1903. Bacteria from Uncle Sam. Trenton *Times*, August 8.
1905. New department scandal. New *York Times*, July 28.
Agee, Alva. 1905a. Bacteria talk. *National Stockman and Farmer* 28:1551.
Agee, Alva. 1905b. Farm facts and fancies. *National Stockman and Farmer* 28:1394.
Agee, Alva. 1905c. Farm facts and fancies. *National Stockman and Farmer* 29:196–197.
Agee, Alva. 1905d. Farm facts and fancies. *National Stockman and Farmer* 29:500.
Agee, Alva. 1905e. Nitro-culture discredited. *National Stockman and Farmer* 29:562.
Agee, Alva. 1905f. The booming of nitro-culture. *National Stockman and Farmer* 29:122–125.
Aikman, C. M. 1896. Nitragin: An important advance in the science of agriculture. *Contemporary Review* 70:210–214.
Anonymous. 1898. The world's supply of wheat. *Scientific American* 79:210–211.
Anonymous. 1904. Soil inoculation. *Country Gentleman* 69:555.
Anonymous. 1905a. An important discovery in the purification of contaminated water. *Scientific American* 92:219.
Anonymous. 1905b. Another department scandal. *Country Gentleman* 70:706.
Anonymous. 1905c. Bacterial soil inoculation for vegetables. *Scientific American* 92:218–19.
Anonymous. 1905d. Government bacteria. *Wallace's Farmer* 30:627.
Anonymous. 1905e. Secretary Wilson's responsibility. *National Stockman and Farmer* 29:563.
Anonymous. 1905f. Soil inoculation. *Wallace's Farmer* 30:683.
Anonymous. 1905g. Soil inoculation: What it can and cannot accomplish. *Country Gentleman* 70:324
Anonymous. 1905h. The inoculation of the earth. *Current Literature* 38:358.
Anonymous. 1912. Old farms made new. Brochure for the German-American Nitragin Company, Milwaukee.

[127] Pauly (2007, p. 128).
[128] Brockwell (1981).
[129] Simmonds (2007, p. 6).
[130] Gorman (2013, pp. 1–4); and Smil (2001, pp. 177–97).

Anonymous. 1915. *Country homes for busy little people*. San Francisco: Western Soil and Bacteria Co.

Anonymous. 1965. George Thomas Moore. *National cyclopædia of American biography*, Vol. 47., 680. New York: James T. White.

Bailey, Liberty Hyde. 1905. Nitrogen and agriculture. *Country Calendar* 1:27–28.

Baker, Ray Stannard. 1903. The scientist and the food problem. *Harper's Magazine* 107:932–937.

Brock, William. 2008. *William Crookes (1832–1919) and the commercialization of science*. Aldershot: Ashgate, 2008.

Brockwell, J. 1981. A strategy for legume nodulation research in developing regions of the Old World. *Plant and Soil* 58 (1–3): 367–382.

Burrill, T. J. 1904. Micro-organisms of soil and human welfare. *Science* 20:426–434.

Butz, George E. 1905. Nitro-cultures. *National Stockman and Farmer* 29:698.

Cafer du Plessis, Elizabeth. 2009. Meatless days and sleepless nights: Food, agriculture, and environment in World War I America. PhD Diss., Indiana University, USA.

Carpenter, Daniel P. 2001. *The forging of bureaucratic autonomy: Reputations, networks, and policy innovation in executive agencies, 1862–1928*. Princeton: Princeton Univ Press.

Clark, Ernest. 1895. Hermann Hellriegel. *Journal of the Royal Agricultural Society* 56:764–67.

Clarke-Nuttall, G. 1902/1903. A botanical discovery and its possibilities. *Longman's Magazine* 41:119–125.

Coppin, Clayton. 1990. James Wilson and Harvey Wiley: The dilemma of bureaucratic entrepreneurship. *Agricultural History* 64:167–181.

Crookes, William. 1898. Address of the President before the British Association for the Advancement of Science, Bristol, 1898. *Science* 8:561–575, 601–612.

Davis, C. Wood. 1899. Wheat: Crookes vs. Atkinson, Dodge, et. al. *The Forum* 27:101–13.

Dickinson, Edward Ross. 2004. Biopolitics, fascism, democracy: Some reflections on our discourse about 'modernity'. *Central European History* 37:1–48.

Dreiser, Theodore. 1903. The problem of the soil. *The Era* 12:239–249.

Duggar, J. F. 1897. *Soil inoculation for leguminous plants*. Alabama Agricultural Experiment Station Bulletin #87. Montgomery: Brown.

Fairchild, David. 1938. *The world was my garden: Travels of a plant explorer*. New York: Scribners.

Farbewerke vorm. Meister, Lucius & Brüning. 1898. *Die Boden-Impfung für Leguminosen mit Reincultivirten Bakterien.* Hoechst am Main: A. A. Wagner.

Flack, James Kilpatrick. 1975. *Desideratum in Washington: The intellectual community in the capital city, 1870–1900*. Cambridge: Schenkman.

Galloway, Beverly T. 1902. Applied botany, retrospective and prospective. *Science* 16:49–59.

Galloway, Beverly T. 1904. The twentieth century botany. *Science* 19:11–19.

Gorman, Hugh S. 2013. *The story of N: A social history of the nitrogen cycle and the challenge of sustainability*. New Brunswick: Rutgers Univ Press.

Gossel, Patricia Peck. 1988. The emergence of American bacteriology, 1875–1900. PhD Diss., Johns Hopkins University.

Grosvenor, Gilbert H. 1904a. Inoculating the ground. *National Geographic* 15:225–228.

Grosvenor, Gilbert H. 1904b. Inoculating the ground: A remarkable discovery in scientific agriculture. *Century Magazine* 68:831–839.

Harding, H. A., and M. J. Prucha. 1905. The quality of commercial cultures for legumes. New York Agricultural Experiment Station Bulletin #270.

Harding, H. A., and M. J. Prucha. 1906. Value of commercial cultures for legumes. *Scientific American* 94:227.

Hartmann, Anton, Michael Rothballer, and Michael Schmid. 2008. Lorenz Hiltner, a pioneer in rhizosphere microbial ecology and soil bacteriology research. *Plant and Soil* 312:7–14.

Harwood, W. S. 1906. *The new earth: A recital of the triumphs of modern agriculture in America*. New York: Macmillan.

Hellriegel, H. 1887. Welche Stickstoffquellen stehen der Pflanze zu gebote? *Landwirtschaftlicher Versuchsstationen* 33:464–465.

Hellriegel, H., and Wilfarth, H. 1888. *Untersuchungen über Stickstoffernährung der Gramineen und Leguminosen*. Berlin: Kayssler.
Hersey, Mark. 2011. What we need is a crop ecologist': Ecology and agricultural science in Progressive-Era America. *Agricultural History* 85:297–321.
Hoing, Willard L. 1964. James Wilson as Secretary of Agriculture, 1897–1913. PhD Diss., University of Wisconsin.
Johnson, Maurice L. 1900. Microbes: Are they inherently pathogenic? *Westminster Review* 154:324–333.
Jones, Jeffrey J. 2004. The world was our garden: U. S. plant introduction, empire, and industrial agri(culture), 1898–1948. PhD Diss., Purdue University.
Kellerman, Karl F. 1910. *Methods of legume inoculation*, USDA BPI Circular #63. Washington: Government Printing Office.
Kellerman, Karl F., and T. R. Robinson. 1905. *Inoculation of legumes*, Farmers' Bulletin #240. Washington: Government Printing Office.
Kellerman, Karl F., and T. R. Robinson. 1908. *Progress in legume inoculation*. Farmers' Bulletin #315. Washington: Government Printing Office.
Kimmelman, Barbara A. 1983. The American Breeders' Association: Genetics and eugenics in an agricultural context, 1903–13. *Social Studies of Science* 13:163–204.
Kleinman, K. 2010. George Thomas Moore: Botanist, (Phycologist), Administrator. *Archives of Natural History*. 37 (1): 173–174.
Krimsky, Sheldon, and Roger Wrubel. 1996. *Agricultural biotechnology and the environment: Science, policy, and social issues*. Urbana: University of Illinois Press.
Kupferberg, Eric D. 2001. The expertise of germs: Practice, language, and authority in American bacteriology, 1899–1924. Ph.D. dissertation, Massachusetts Institute of Technology, USA.
Leonard, Lewis T. 1932. The commercial inoculant business in the United States. *Proceedings of the Second International Conference of Soil Science, Commission*. Vol. 3, 74–82. Moscow: State Publishing House of Agricultural, Cooperative, and Collective Farm Literature.
Marine Biological Laboratory. 1907. *Tenth report, for the years 1903–1906*. Woods Hole [sic]: n.p.
Marine Biological Laboratory. 1909. *Eleventh report, for the years 1907–1908*. Woods Hole: n.p.
Masters, Maxwell T. 1884. *Plant life on the farm*. New York: Orange Judd.
Moore, G. T. 1903. Bacteria and the nitrogen problem. *Yearbook of the U.S. Department of Agriculture, 1902*. 333–342. Washington: Government Printing Office.
Moore, G. T. 1905. Applied botany and its dependence upon scientific research. *Science* 21:321–33.
Moore, G. T. 1912. Microorganisms of the soil. *Science* 36:609–616.
Moore, G. T., and T. R. Robinson. 1905a. *Beneficial bacteria for leguminous crops*, USDA Farmers' Bulletin #214. Washington: Government Printing Office.
Moore, G. T., and T. R. Robinson. 1905b. *Soil inoculation for legumes: Reports on the successful use of artificial cultures by practical farmers*, USDA BPI Bulletin #71. Washington: Government Printing Office.
Noyes, A. D. 1898. The predicted wheat famine. *Nation* 67:237–238.
Pauly, Philip J. 1988. Summer resort and scientific discipline: Woods Hole and the structure of American biology. In *The American development of biology*, ed. R. Rainger, K. R. Benson, and J. Maienschein, 121–150. Philadelphia: University of Pennsylvania Press.
Pauly, Philip J. 1994. Modernist practice in American biology. In *Modernist impulses in the human sciences, 1870–1930*, ed. Dorothy Ross, 272–289. Baltimore: Johns Hopkins University Press.
Pauly, Philip J. 2000. *Biologists and the promise of American life: From Meriwether Lewis to Alfred Kinsey*. Princeton: Princeton University Press.
Pauly, Philip J. 2007. *Fruits and plains: The horticultural transformation of America*. Cambridge: Harvard University Press.
Peters, Scott J., and P. A. Morgan. 2004. The Country Life Commission: Reconsidering a milestone in American agricultural history. *Agricultural History* 78:289–316.
Porter, Raymond. 1905. Vaccinating the ground. *Pearson's Magazine* 17:398–403.

Powell, Fred W. 1927. *The Bureau of Plant Industry: Its history, activities and organization.* Baltimore: Johns Hopkins University Press.
Rosenberg, Charles E. 1964. The Adams Act: Politics and the cause of scientific research. *Agricultural History* 38:3–12.
Rossiter, Margaret W. 1979. The organization of the agricultural sciences. In *The organization of knowledge in modern America, 1860–1920*, ed. Alexandra Oleson and John Voss, 211–248. Baltimore: Johns Hopkins University Press.
Schneider, Albert. 1903. Bacteria in modern economic agriculture. *Popular Science* 62:333–343.
Seifriz, William. 1953. Walter T. Swingle: 1871–1952. *Science* 118:288–89.
Simmonds, Jeanette. 2007. Community matters: A history of biological nitrogen fixation and nodulation research, 1965 to 1995. Ph.D. dissertation, Rensselaer Polytechnic Institute.
Smil, Vaclav. 2001. *Enriching the earth: Fritz Haber, Carl Bosch, and the transformation of world food production.* Cambridge: MIT Press.
Smith, Isaac A. 1913. Soy beans and secrets of legume inoculation. n.c., n.p.
Smith-Howard, Kendra. 2013. *Pure and modern milk: An environmental history since 1900.* New York: Oxford University Press.
Springer, Alfred. 1892. Micro-organisms of the soil. *Nature* 46 (1198): 576–579.
Starnes, Hugh N. 1905. *Some field notes on soil inoculation.* Georgia Agricultural Experiment Station Bulletin #71. Georgia Experiment Station.
Stevenson, John A. 1954. Plants, problems, and personalities: The genesis of the Bureau of Plant Industry. *Agricultural History* 28:155–162.
Swingle, W. T. 1905. Suggestions in favor of pasigraphy. *The Monist* 15:148–150.
Swingle, W. T. 1909. A simple to metric conversion table: Changing metric units into English equivalents. *Scientific American Supplement* 68 (1733): 190.
Townsend, C. F. 1897. Nitragin. *Knowledge* 20:201–202.
Treitel, Corinna. 2008. Max Rubner and the biopolitics of rational nutrition. *Central European History* 41:1–25.
True, Alfred C. 1937. *History of agricultural experimentation and research in the United States, 1607–1925, including a history of the United States Department of Agriculture.* USDA Miscellaneous Publication #251. Washington: Government Printing Office.
Voorhees, Edward B., and Jacob G. Lipman. 1907. *A review of investigations in soil bacteriology.* USDA Office of Experiment Stations Bulletin #94. Washington: Government Printing Office.
Wilson, James. 1903. Report of the Secretary. *Yearbook of the U.S. Department of Agriculture, 1902*. 9–124. Washington: Government Printing Office.
Wood, Eugene. 1903. The good bacteria. *Everybody's Magazine* 9:613–616.

Mark R. Finlay was professor of history at Armstrong Atlantic State University, USA, and assistant dean of the college. He was co-winner of the Hiebig-Wohler Friendship Prize in 1995, on the basis of his dissertation on German agricultural experiment stations. His book, *Growing American Rubber: Strategic Plants and the Politics of National Security* (2009), received the Theodore Saloutos Memorial Prize for best book published in the field of agricultural history.

Chapter 12
Mold Cultures: Traditional Industry and Microbial Studies in Early Twentieth-Century Japan

Victoria Lee

Introduction

Fermentation phenomena, both as life processes and as technologies, hold special significance in Japanese scientific culture. They take pride of place as an area of expertise where the country leads in contemporary biotechnology, and are prominent in daily life in producing commonplace foods such as miso (fermented soybean paste) and *nattō* (fermented soybeans) in people's homes, and had been a field of industrial specialization since medieval times in sake and other brewing houses. This chapter illuminates an early period in the creation of this scientific culture by looking at how Japanese scientists in universities, technical colleges, and government research institutes, and expert workers in the brewing industry studied microbes at the turn of the twentieth century, in the context of widespread and state-supported campaigns to modernize the indigenous brewing industries.

Japanese workers did not have the concept of "microbes" before the late nineteenth century. However, they had ways of understanding and handling microorganisms—visible en masse as mold formations—as part of essential steps in brewing not only sake and other liquors but also soy sauce and miso. The making of the rice mold *kōji* was the first step in brewing these products, and *kōji* making had been a lucrative monopoly industry since the thirteenth century. By the end of the nineteenth century, the preparation of dried mold spores to seed rice for making *kōji* had become a distinct sector from *kōji* making. These preparations were known as *moyashi* or *tanekōji* and would be sold either to *kōji* makers, or directly to those sake or soy sauce houses that made their own *kōji*. Around the same time in Europe, while bacteriologists developed new techniques for isolating and culturing microorganisms and rarefying their products to make vaccines, experts in brewing

V. Lee (✉)
Max Planck Institute for the History of Science, Boltzmannstraße 22, 14195 Berlin, Germany
e-mail: vlee@mpiwg-berlin.mpg.de

D. Phillips, S. Kingsland (eds.), *New Perspectives on the History of Life Sciences and Agriculture,* Archimedes 40, DOI 10.1007/978-3-319-12185-7_12

developed similar methods that would allow brewers to increase their control.[1] The novel techniques of pure culture were equally important to scientists, allowing them to preserve, collect, and classify individual microbial strains. This chapter focuses on the introduction of the technique of pure culture to the Japanese brewing industry at the turn of the century, and explores its implications for how both brewers—particularly *tanekōji* makers—and scientists worked with microbes.

How experts implemented pure culture in the Japanese fermentation industries opens a window onto the relationship between the modernization of traditional industry and the institutionalization of Western science in the Meiji period (1868–1912). The emergence of a new set of institutional structures for scientific research was driven by the combined efforts of local industrial leaders, prefectural government officials, and the Meiji state to improve Japanese industries, in order to increase their competitiveness both domestically and for the purpose of export.[2] Those industries that had existed in Japan before the Meiji period and continued to exist since—such as textiles, dyes, and pottery, to name a few—are some of the most important and most overlooked areas in modern Japanese science.[3] Brewing contributed by far the highest values of production among the entire manufacturing sector at the end of the nineteenth century, easily surpassing weaving and raw silk, as well as providing the government with its largest source of tax revenue and consuming a sixth of the rice harvested annually.[4] This chapter traces one aspect of how, in the late Meiji period, Japanese experts imported and adapted Western science in order to take a more systematic approach to processes of manufacturing. As the raison d'être that gave science its rationale in Meiji Japan, the practical application of foreign ideas to Japanese industries underlay the expansion and dynamism of institutions of agricultural and engineering science in the country.

The underexplored narrative of the significance of traditional industry exposes a different side to the formation of modern Japanese science than that seen in the dominant historiographical approach, which portrays the institutionalization of science in Japan as primarily a story of rapid transfer from the West, under the policies of a strong state and constituting an abrupt break from the past.[5] In this

[1] Japanese names in this chapter are written in conventional order with the surname first, but the reference list uses English-language word order. On the development of pure yeast culture and the introduction of the technique to the German beer industry, see Ceccatti (2001).

[2] After the 1880s, the national government shifted the focus of their policies from transplanting Western industries to supporting the "traditional" sectors that constituted the bulk of the economy. See Morris-Suzuki (1994, pp. 98–103).

[3] Recent work on the significance of traditional industry in twentieth-century Japan includes Nakaoka (2006); Clancey (2006); Onaga (2012). On their importance to the modern Japanese economy, see Tanimoto (2006a).

[4] Fujiwara (1999, p. 185); Tanimoto (2006b, p. 301).

[5] In the history of technology and science, as well as in the economic literature, Japan has often been accorded "exceptional" status in East Asia for breaking rapidly with a traditional past or possessing traditions that were closer to European than Asian countries. On rapid science and technology transfer under state promotion by the Meiji government and then under Ministry of International Trade and Industry (MITI) in the postwar period, the authoritative works remain Bartholomew (1989), and Johnson (1982). On Western-style traditions in early modern Japan, there is

chapter, I suggest that local industry helped to shape a relatively autonomous tradition of seeing microbes as living workers as much as pathogens in Japan: a view that, through large and lasting institutions, remained powerful far into the twentieth century.

Brewing in the Improvement of Traditional Industry

In January 1901, roughly 2 years after returning from a spell abroad at the *Versuchs- und Lehranstalt für Brauerei* (Experimental and Training Institute for Brewing) in Berlin and his appointment as professor and chair of agricultural products in the Department of Agricultural Chemistry of the College of Agriculture at Tokyo Imperial University, Kozai Yoshinao delivered a lecture titled "On the Improvement of Sake Brewing" at the Tokyo Chemical Society (*Tōkyō Kagakkai*), the prime meeting place for chemical industrialists and scientists to exchange information.[6] Kozai began by describing to the audience the basic nature of brewing processes. What was first needed in the making of sake was *kōji,* the most essential material in the brewing of sake. It was made by steaming rice, adding *tanekōji* (dried mold spores to seed the rice), dampening it, and then placing it inside the warm, humid "*kōji* room" to allow the growth of a kind of mold. When the *kōji* was ready, one made *moto,* the material needed to ferment sake. It was made by mixing *kōji* with steamed rice and water, and then making it ferment through various kinds of manual art. A process of great time and labor that took about 18 days, it was the most difficult part of the *tōji,* or head brewer's job.[7] Kozai went on to explain these two materials' functions in chemical terms: *kōji* changed starch into sugar by the power of mold, whereas *moto* changed sugar into alcohol and therefore contained abundant amounts of yeast (here called *iisuto,* signifying a foreign term). After applying these materials to steamed water and rice, the resulting liquid would be squeezed out through a fine press into a large cask and, having been clarified, placed into storage. Because the various mold and bacterial germs mixed in with the liquid would continue to provoke constant transformation, storage was when there was particular danger of spoilage, so one needed to be careful and from time to time perform *hiire* ("putting in fire," a heating process).

However, in the first place the *tanekōji* used to seed the *kōji* was already not pure. Then during *kōji* making, various bacteria and other kinds of mold in the air would multiply, so the resulting *kōji* would also not be pure. The kind of germs differed with different *kōji,* but they were organisms that were harmful or ineffectual for brewing. Then one would use this impure *kōji* to make *moto,* when more organisms

a large literature on *rangaku* ("Dutch studies," or studies of Western science and medicine based on translations of Dutch texts); for one example, see Jannetta (2007).

[6] The lecture is published as Kozai (1901a).

[7] The *tōji* was part of a seasonal labor force that would come to the brewing house in the winter months.

would enter through the air and water and breed. The yeast would gradually win the struggle for existence, but this was not to say that the other, plentifully mixed-in microbes would necessarily die out. Some would die when the alcohol was produced, but some would simply be latent, and when their enemy yeast rested in work, which was when the yeast completed the task of fermenting, they would strengthen and cause all kinds of damage. In other words, not only might the sake fall ill from infectious diseases when brewing but also the buds of disease were already in the raw materials. In such cases, there was not simply the danger of spoiling; it was *natural* for the sake to spoil. To improve sake brewing, it was necessary to remove every kind of harmful microbe. Kozai encouraged brewers to stop making the complicated *moto* entirely and instead to use pure cultures of well-chosen yeast.

Kozai would have encountered the pure-cultured yeast method at close quarters during his time in Berlin.[8] By the time Kozai gave this lecture, there had been two decades of movements in Japan to improve the sake industry by the application of scientific principles (*gakuri ōyō*).[9] The earliest movements emerged in the traditional brewing centers of the Kansai region in western Japan, with individual wealthy brewing improvers who began to publish manuals and trade magazines in order to spread enlightenment and educate other brewers about science, as well as to disseminate practical methods for achieving high-quality, standardized, mass-produced goods similar to those of the famed brewing districts. The brewing improvers were well read in the chemical and bacteriological research on brewing that had been published in the 1870s by *oyatoi gaikokujin*—"hired foreigners" that the Meiji government brought over from Europe as consultants, who in this case trained Japanese scientists for government ministries and public projects at the new colleges, while helping government factories to transplant beer and wine technology in Japan.[10] These early manuals, which took knowledge that had previously been secret and experiential and published it in the language of chemistry, were regional brewers' defensive responses in a difficult time as Japanese-produced liquor came under a new tax system and the industry was threatened by Western imports. They were also the reaction of local brewers against the early Meiji state-led industrial campaigns that focused solely on heavy, chemical and military-related industries,

[8] On the German context, see Ceccatti (2001).

[9] For a detailed account, see Fujiwara (1999, pp. 148–255).

[10] Among the most prominent in the field of brewing was British chemist Robert Atkinson, who taught applied chemistry at the Kaisei Gakkō (the predecessor of Tokyo Imperial University) and who described the technological processes of sake brewing at length, publishing his descriptions in a well-known pamphlet *The Chemistry of Sake Brewing* in 1881. Others who researched on sake included German bacteriologist Oskar Korschelt who worked with the Kaitakushi Brewery to help produce Sapporo Lager, and German botanist Hermann Ahlburg who like Korschelt taught at the Tokyo Medical School. For details, see Hasegawa (2001, pp. 71–100); Fujiwara (1999, pp. 49–85). Furukawa Yasu, *Dentō sangyō kara kindai sangyō e. Meiji no kagaku to kagaku kōgyō* (*From traditional industry to modern industry. Meiji-period chemistry and chemical industry*; Unpublished lecture, Hokkaido University, July 27, 2000) notes that the first generation of Japanese chemists continued to invest heavily in problems related to traditional industry. Of all the articles published in the first 10 years (1880–1890) of the *Journal of the Tokyo Chemical Society,* roughly half were on traditional products; Furukawa counts 42 of 88.

as well as on certain traditional products such as raw silk, tea, vegetable oils, and pottery that had quickly become important exports shaped to foreign tastes.[11]

By the 1890s, government officials in Tokyo too saw the economic imperative to nurture and develop the larger traditional industries across the country to ease balance of payments, chiefly through export to other Asian countries.[12] Following the reports of Maeda Masana, an official at the Ministry of Agriculture and Commerce, the Japanese state shifted its focus from transplanted to traditional industries and began a concentrated effort to build on the activities of local voluntary movements. In 1884, the central government issued a set of uniform rules on the formation of trade associations across the nation, at the same time leaving local industrialists and prefectural authorities to take the initiative and to enforce the new regulations. From that point on, the number of trade associations multiplied. Brewing improvers worked together with prominent Japanese scientists at the Imperial College of Engineering and technical advisers in the Ministry of Agriculture and Commerce as well as local brewing notables to set up industrial associations, along with trade magazines and experiment stations.

The large, established breweries in Kansai sought improvement using new Western methods to scale up production and make their goods competitive for export. In this sense, they shared the goals of the central government, although the latter also retained their focus on sake brewing as the largest source of tax revenue. Small sake breweries around the country, such as in Tōhoku in eastern Japan, subsequently picked up the movements begun in the famed Kansai districts, for different reasons.[13] They relied on new forms of technical communication such as brewing manuals to help them copy celebrated techniques (such as those of the Nada sake brewing district east of the port of Kobe), and to standardize the quality of their own goods. Their concern was to survive the competition against the expansion of larger brewers into local markets and to push the quality of their product above home-brewed sake. Success was uneven: trade magazines oriented to small-scale brewers complained about the "stubborn" *tōji*.[14] In Imazu in Nada, brewers hired a technical advisor from the Ministry of Agriculture and Commerce to run a series of trials, but the sake spoiled, which caused enormous loss. Many Nada brewers became suspicious of the value of Western science for Japanese brewing, keeping "scholars' methods" at arm's length.[15]

[11] The list of industries is from Furukawa, "Dentō sangyō kara kindai sangyō e."

[12] Sugihara (1994); Morris-Suzuki (1994, pp. 71–104).

[13] Brewing was a complex and layered industry: a number of wealthy brewers (among the nouveau riche of the early modern era) and large factories had established themselves over the course of the Tokugawa period (1603–1868), especially in the Nada region west of Kyoto and Osaka for sake and Noda and Chōshi east of Tokyo for soy sauce, who served urban markets. However, many breweries were small, recently founded and sold to local markets. Beyond commercial goods, a substantial amount of sake consumed was unrefined, home-brewed *nigorizake* drunk to ease a day's heavy labor at the farm. See Tanimoto (2006b, pp. 301–305).

[14] Outside of the specialist brewing districts, the *tōji* employed in the multitude of small, new brewing houses were of mixed quality in terms of skills. Fujiwara (1999, p. 165).

[15] Fujiwara (1999, pp. 194–199, 312–313.)

Such disappointments left a continuing legacy. Kozai himself was aware of these problems when he gave another lecture in April 1901 before a nationwide association of brewers and brewing experts, upon receiving the Medal of Honor at the Tenth Sake Tasting Meet held outside Tokyo in Saitama.[16] If the industry had been unable to artificially prevent damage by measuring natural influences on brewing processes, it was because knowledge was insufficient and the industry was immature, he said. Even he did not have all of the answers. But now Kozai spoke of the brewers who were paying large sums of money to cart in water from the wells of Nishinomiya, one of the renowned brewing regions of the Kansai area that was especially famed for its water, and he wondered if there were not sources of suitable water available in a more convenient location. Rather than investing in water or rice, Kozai pointed to microbial materials as the most important within the brewing process, remarking that it was impossible to brew good sake (*junshu*) with impure *tanekōji* or *moto*. Kozai's pure yeast method was lifted directly from the model of beer and Emil Hansen's extensive research on yeast varieties. He explained that different types of yeast (*kōbo*) had different physiologies, and only by selecting a superior kind of yeast could one guarantee a good brew. Kozai, along with other sake improvement advocates, saw a number of changes as part of the tide of modernity. He extolled the advantages of labor- and cost-saving technologies and the ability to produce uniform sake from uniform yeast. In addition, he thought that professionalism of the industry should rely on academic knowledge as well as practical experience, as was the case in comparable professions such as medicine and law.

Since the end of the Sino-Japanese War, when the Japanese government raised the liquor tax to fund further military preparation, peripheral eastern brewers in Kantō and Tōhoku had become the most vocal in calling for the government to establish a national Brewing Experiment Station in Tokyo, hoping that the research and training it would provide to brewers would alleviate the challenges they faced vis-à-vis the large breweries in Kansai.[17] Kozai declared that those opposed to a Brewing Experiment Station were selfish people who wanted to prevent competitors from making good sake. In this respect, scientists saw themselves as playing a key role in raising production levels by encouraging a high level of knowledge across the entire country. Moreover, Kozai pointed to how science-based technical training was superior to relying on the uncommunicated power of skilled craftsmanship that was held by the *tōji*. Brewery owners in Tōhoku were hiring highly skilled *tōji* from Tanba in Kansai as part of the regional improvement movements.[18] Kozai said that even if the materials were excellent, if the architect was not properly trained he would not build a good house; similarly, a badly trained *tōji* would not make good *moto,* and even if it were good, it might be accidental.

Around this time, Tokyo Imperial University's agriculture and engineering colleges trained students for work in government, and more significantly for industry, new technical schools trained many technicians for sectors such as pottery or tex-

[16] Kozai (1901b).

[17] Fujiwara (1999, p. 350).

[18] Fujiwara (1999, pp. 202–205).

tiles. The largest of these were the higher technical schools, for which the government created a set of national standards in 1903 to encourage local elites in major centers of manufacturing to establish such schools, though some cities had done so already.[19] The government issued a similar ordinance defining national standards for Industrial Experiment Stations, which conducted surveys of local rural industries and hosted training courses. Brewing was similar to other "traditional," rural industries in that most companies were too small scale to support scientific experimentation, so that experts saw government institutions as necessary in order to shoulder the development of better materials, techniques, and tools.[20] Moreover, science advocates hoped to improve the sake industry by attempting to "First Improve (*kairyō*) the Heads of *Tōji*," as one especially well-received opinion in one of the many trade magazines *Jōkai* (*Brewing Ocean*) put it.[21] Brewery owners were often preoccupied with account books and abacuses, entrusting technological matters entirely to *tōji*, so what was the point of lecturing to the owners? Promoters hoped too that small businesses could survive by hiring a scientifically educated *tōji*.

In February 1902, the recently appointed director of the Tokyo Industrial Experiment Station, Takayama Jintarō, published an opinion in the first issue of *Jōkai* on "The Necessity of a Brewing Experiment Station."[22] It was clear that Japan needed a government-run institution for improvement, as it should not be left entirely to the tradesmen to shoulder the initiative for an industry so important to the national income. From the perspective of chemical industry, Takayama explained, there were two different kinds of industries. On the one hand, there were industries that had their origins in the distant past and were unique. On the other, there were "imitation" industries taken from the West after the Meiji Restoration. The former included porcelain, lacquerware, sake, and soy sauce; the latter included matches, beer, cement, glass, soap, Western-style paper, and alkali products. The vitality had been tamed in the former and naturally these industries tended to adhere to old customs and hesitate to seek improvement. But among them, Takayama reminded his brewer readers, there were industries such as porcelain and lacquerware that had won worldwide renown for their manufacturing methods, and the shape and design of their products. It was well known that these products had won great acclaim at the international exhibitions in Paris in 1867 and Vienna in 1873.

Takayama did not remind brewers how badly sake—a product possessing "not a single merit"—had been received at the same exhibitions. In the 1870s, the majority of Japanese writers had thought sake to be unhealthy, in contrast to the benefits of wine and beer. The imported liquors that "civilized" people drank flooded into the cities of Yokohama and Kobe under the unequal treaties and could supposedly make farmers and soldiers work more efficiently.[23] Takayama reflected now, three decades later, that while the Meiji state campaigns under the slogan of "develop

[19] Morris-Suzuki (1994, pp. 100–103).

[20] Kimoto (1902).

[21] Sawamura (1903).

[22] Takayama (1902).

[23] Fujiwara (1999, pp. 20–48).

industry and promote enterprise" (*shokusan kōgyō*) had successfully transplanted Western industries—not only stemming imports but even exporting products overseas—Japan should not lose unique products like sake. Takayama urged the application of modern scientific principles to the sake industry, even though sake brewing was an enterprise that was mostly household scale.

Microbes made frequent appearances in *Jōkai* in the early 1900s. *Jōkai,* a Tokyo-based trade magazine, was one of the publications that aimed to disseminate enlightenment to medium or small-scale breweries. In the Q&A columns, brewers asked questions such as, what kind of soy sauce fermentation microbes existed? Moreover, what were the great enemies of sake brewing "*bakuteriya*"? A technical expert replied that there were a multitude of different types of bacteria existing in the raw materials, and not all of them were bad or strong, nor was it easy to distinguish between them whether by outward appearance or under a microscope. He described and illustrated the appearance under a microscope of common bacteria that caused the main types of spoilage: incomplete fermentation (too sweet), nonvolatile acid fermentation (lactic acid fermentation), and volatile acid fermentation (butyric acid fermentation).[24] One lecture published by the magazine explained the importance of hygiene in brewing: some still believed that sake was the work of gods, the speaker began, for science had not completely opened the country. But disinfection (by new chemical means) was essential, and those hands who worked in the brewing houses must also keep clean clothes and clean bodies, as if they were in a sacred place: it was no coincidence that good sake was produced in places that adhered to such customs.[25] Another lecture explained methods for inspecting the number of bacteria in materials such as *kōji, moto* or machines without a microscope, for microscopes were extremely expensive items, though they were certainly handy if one used them all of the time. Bacteria were invisible to the naked eye, but could be seen if propagated as a colony. The lecture explained the sampling and culturing methods, which required other specialized equipment such as flasks, filter paper, test tubes, petri dishes, and pipettes.[26]

The Ministry of Agriculture and Commerce and the Ministry of Finance finalized the decision to establish a national Brewing Experiment Station in Tokyo in July 1902, and soon it was under sole jurisdiction of the latter, sake brewing being the largest source of tax revenue in Japan. Among the initial Investigation Committee were Tejima Seiichi (principal of the Tokyo Higher Technical School), Takayama Jintarō (director of the Industrial Experiment Station), Kozai Yoshinao (professor at the College of Agriculture of Tokyo Imperial University), and Yabe Kikuji (official appraiser at the Ministry of Finance, and who had isolated the first sake yeasts in 1893).[27] Medical bacteriology bore the most glamorous and highest visibility achievements in Japanese science in the early 1900s, and pathogenic thinking also dominated the earliest microbial ideas in brewing science. At the same time, the

---

[24] Ono (1902).

[25] Yamagata (1902).

[26] Kimoto (1903).

[27] Ōtsuka (1979, p. 1).

large-scale efforts of government officials, scientists, and industrialists to improve traditional industry—with their origins in movements from below at the initiative of local governments and rural industries—constituted one of the most dynamic, powerful, and wide-reaching scientific trends of the period.

Transformation in the *Tanekōji* Industry

Konno Seiji, the original founder of today's *moyashi* (the traditional term for *tanekōji*) companies Akita Konno Shōten and Kobe-based Konno Shōten in the Meiji period, was a pivotal figure in bringing pure culture of *tanekōji* into the brewing industry. Born in 1882, Seiji was one of the sons of a brewing family in Kariwano in Akita Prefecture, the oldest after his elder brother died when Seiji was five. In the snowy town on the Sea of Japan side of Honshū, Seiji's father was the kimono-clad *tōji* of the family's soy sauce factory. However, a fire entirely destroyed the factory when Konno Seiji was young.[28] After Seiji completed his studies at Akita Middle School, he left the cold northern prefecture to study the scientific principles of brewing. At the time, the only college in Japan that had a Brewing Department was Osaka Higher Technical School, located in the heart of the metropolitan merchant capital.[29]

Surrounded by the traditional brewing districts of Kansai, Osaka Higher Technical School's Brewing Department trained technicians from breweries all over the country. The Brewing Department had been established in 1897 in response to calls from brewers to create an independent subject for brewing, unlike at Tokyo Higher Technical School where training in the use of microscopes for example was under the Applied Chemistry Department. It was rumored that the manager of Osaka Beer (the predecessor of Asahi Beer) prodded the prefectural government's decision by buttonholing a high-ranking official after a nationwide meeting of the Association of Sake Brewing.[30] At the school, Konno Seiji studied under Brewing Department head Tsuboi Sentarō. Tsuboi, a chemist and microbiologist who had graduated from the Imperial College of Engineering (later the Faculty of Engineering of Tokyo Imperial University), saw the department's research as bringing scientific ideals and actual practice (*jicchi*) close together.[31] At the time, Tsuboi's laboratory was working on the pure culture of *tanekōji* as well as yeast for application in industry. Tsuboi's advertisement in the back pages of *Jōkai* in 1902 asking brewers to buy pure-cultured *moyashi* made by his college laboratory joined those of established, commercial *moyashi* makers licensed by the Ministry of Agriculture and Com-

[28] Konno Eiichi (nephew of Konno Seiji), Konno Hiroshi (president of Akita Konno Shōten and grandnephew of Konno Seiji) and Konno Kenji (former president of Akita Konno Shōten and nephew of Konno Seiji), interview by author, Kariwano, Daisen-shi, Akita-ken, February 20, 2012.

[29] Akita Konno shōten kabushiki kaisha (2011, pp. 26–27).

[30] Hyakushūnen kinen jigyōkai (1996, pp. 10–13).

[31] Tsuboi (1903).

merce, who variously claimed that their particular pure-cultured *tanekōji,* the fruit of laborious research efforts and enthusiastically tested technology, drew high praise in the "twentieth-century brewing world."[32] In the picture that both Tsuboi and commercial *moyashi* makers painted in their advertisements, the application of science placed their product at the cutting edge of the industry.

Konno Seiji graduated in the spring of 1905 and entered Kawamata Shōyu, one of the largest soy sauce companies in western Japan whose factory was part of the chimneyed cityscape of Sakai, south of Osaka. As chief technician of Kawamata, Konno was busily occupied with the mechanization of the factory.[33] He was a man so obsessed with the precision of watches that he would make charts of how late each one ran to record its reliability, checking its performance in horizontal and vertical directions.[34] Apprentices remember that he kept the factory very clean.[35] While directing the newly opened Kawamata Shōyu Brewing Experiment Station, that autumn Konno Seiji isolated an excellent *kōji* microbe "*Kawamata kin*" which the company began to use for their soy sauce. In 1909, Seiji isolated a microbe suitable for sake, and the following year while keeping his position at Kawamata, he and two brothers, Shigezō and Kenkichi, opened a shop Konno Shōten in Kyoto and began selling "*sake moyashi Konno kin*" and other microbes as pure-cultured *tanekōji* to *kōji* makers and to sake, miso, and soy sauce companies. The shop soon moved back to Osaka and opened another department for selling tools and machinery, many of which Seiji played a leading role in developing and patenting at Kawamata. Konno Shōten also published a trade journal *Jōzōkai* (*Brewing World*), and later opened a further soy sauce *moyashi* branch in Sakai and a sake *moyashi* branch in Nada.[36]

These developments were underway well before the national Brewing Experiment Station in Tokyo developed a method for the pure culture of *kōji* microbes, on which the related Brewing Society published their first report in 1911.[37] By then, other companies were already rapidly adopting the use of pure culture. The largest soy sauce companies in Kantō, such as Noda Shōyu (later Kikkōman), Yamasa Shōyu, and Higeta Shōyu, also began to make *tanekōji* in-house by the 1910s.[38] In fact, the head technician at Higeta Shōyu had interned under Konno Seiji at Kawamata before he first began isolating and pure-culturing *kōji* microbes for soy sauce

[32] [1902] Advertisements, *Jōkai* 10.

[33] Akita Konno shōten (2011, pp. 27–28).

[34] Akita Konno shōten (2011, pp. 28–29).

[35] Kawamata kabushiki kaisha (2000, p. 90).

[36] Akita Konno shōten (2011, p. 26); Kawamata kabushiki kaisha (2000, pp. 88–89).

[37] Murakami (1986, p. 35); Chikudō (1911a, 1911b).

[38] Murai (1989, p. 40); Fukuoka-ken shōyu kumiai (1979, p. 158). Kikkōman also claims that they it pioneered pure-cultured *tanekōji* in 1904 and that the practice spread out from there. Nakadai (1995, p. 4).

tanekōji at Higeta in 1912.[39] Subsequently, new specialist *tanekōji* companies appeared in the late 1910s and 1920s to supply smaller soy sauce makers.[40]

Unlike the yeasts that the Brewing Society worked to maintain, the distribution of *kōji* microbes was already under the private monopoly of *tanekōji* companies, which specialized in preparing what were dried microbial spores that would seed *kōji* making elsewhere. The *tanekōji* sector had distant roots in the medieval *kōjiza* ("*kōji* groups") who held lucrative monopolies over *kōji* making and thereby controlled the source of the entire medieval brewing economy. In the medieval and early modern period, the shogunate and domains usually banned *kōji* making by unlicensed houses, partly to regulate tax collection but also to minimize brewing activity in order to suppress wastage of valuable rice. Where the monopoly system weakened, as had happened in the past, specialist *kōji* makers continued to supply brewing houses which preferred not to make *kōji* in-house.[41]

In the Meiji period, as the sake improvement movement grew, more and more brewing companies requested *tanekōji* from specialist makers rather than making the starter in-house, and by the end of the period almost no sake brewers in Nada were making *tanekōji* themselves, though in-house manufacture was still prevalent in regions on the periphery of the sake economy.[42] Now, it was common practice to shake off the spores of the *kōji* from a good brew, dry them, and use them as seeds in the next round of *kōji* making; the spores in this case were called *tomokōji*. However, the original spore starter—at this point sold by *tanekōji* houses distinct from specialist *kōji* makers—was tricky to generate.[43] In theory, if a sake or soy sauce company built a *kōji* room, it was possible to make original starter spontaneously, by putting steamed soybean and ground wheat (to grow suitable microbes for soy sauce) or steamed rice (for sake) in an open *kōji* box, then leaving it on the shelf of the *kōji* room to wait for mold to enter from the air. In a longstanding *kōji* room, plenty of good *kōji* microbes should have settled there and be floating in the air. But how could one get to that point, and where did *kōji* microbes come from? Moreover, how could one maintain good *kōji* after finding it? With successive culturing any good *kōji* would become old, produce fewer spores and become contaminated by other molds like *kekabi* or *kumonosukabi*. The color of the spores would darken and turn black, the mold would have a strange smell, and the taste of the sake or soy sauce would worsen.[44] Secrecy helped to preserve the *tanekōji* sector.

It is likely that *tanekōji* companies who cultured *kōji* microbes pure were in the minority in the early twentieth century. For example, Kōjiya Sanzaemon in Kyoto,

---

[39] Yamazaki (1974, p. 2).

[40] Murai (1989, p. 40).

[41] One such incident occurred in Kyoto in the mid-fifteenth century, when the shogunate attempted to revoke the monopoly law in response to wider discontent, and the struggle with Kitano Tenmangu Shrine, which dominated *kōji* making in and around the capital, resulted in most of the shrine burning down. Koizumi (1984, p. 105).

[42] Murakami (1986, p. 35).

[43] Murakami (1986, p. 34).

[44] Kawamata kabushiki kaisha (2000, pp. 86–88).

who sold *moyashi* under the label Biokku and claimed lineage from a *kōjiza* licensed by the Ashikaga shogunate (1336–1573), adopted pure culture technology much later in 1951.[45] In the post-World War II period, however, the number of *moyashi* companies shrank. Compared with the hundred or so that existed at the beginning of the twentieth century, by the last decades of the twentieth century there were fifteen *tanekōji* companies of which six distributed nationally. Though the companies were all small scale with fewer than 50 employees each, the concentration of the industry implies the level of technology needed to stay competitive.[46]

Konno Seiji's nephew Konno Kenji became head of Akita Konno Shōten much later, by which time the main branch of the company had moved back to Kariwano due to rice shortages in Kansai during Second World War. Kenji has a childhood memory of watching an apprentice of Seiji's, Ueno Shiejirō, "making *genkin,*" or generating the original starter microbes.[47] Ueno had not attended a technical school and had learned these methods under Konno Seiji during the war. Starting with a mass of spores floating in water, Ueno used a syringe to deposit a droplet of the spore mixture into a container of pure water, repeating until he had a very dilute mixture. Then Ueno would draw a mark on the cover glass and look at the sterile colored liquid through the microscope, to see whether there was only a single spore on the dish. If there was, he sterilized a piece of filter paper by splashing alcohol on it and used it to suck up the spore. He expanded the single spore into a colony by culturing it on rice grains, in other words making *kōji* within a flask, using wide-bottomed flasks that Konno Seiji had specially designed to increase the area for culturing. Then he subjected the pure colony to testing, investigating properties such as the formation of proteases, amylases, acid-resistant amylases, and so on. Repeating the procedure with hundreds of single spores taken from the same original sample, Ueno selected the strongest resulting colony, and then taking the spores from that colony repeated the entire process. In this manner, by thoroughly investigating weak and strong microbes using the single-spore method, only the microbes with the very best qualities would be propagated and made into *genkin.* Droplet by droplet, taking spores wrapped in single droplets, one could cultivate them and bring up their descendants.

The single-spore method was also crucial to preserving the selected microbe. Otherwise, when successively cultured, the strain would quickly degrade.[48] What protected the products of *tanekōji* makers who employed pure culture methods was partly their reputation, and also the fact that other makers did not have the technology to maintain the strains even if they physically possessed them. The expense of maintaining high-quality strains also meant that brewing factories increasingly preferred to purchase *tanekōji* from specialist makers.

---

45 Murai (1989, p. 42).

46 Figures are for the year 1985. Murakami (1986, p. 36).

47 Konno Eiichi, Konno Hiroshi and Konno Kenji, interview by author.

48 Konno Eiichi, Konno Hiroshi and Konno Kenji, interview by author.

The reason for the swift adoption of pure culture technologies in the *tanekōji* industry was that they were upgraded versions of technologies that *tanekōji* makers had already been using. Since the products that they sold were dried microbial spores, they had long-held practices for identifying and isolating "good" cultures, mainly relying on sensory means. By inspecting the color of the spores, one could tell what kind of mold the *kōji* consisted of as well as how old the *kōji* was, as the yellow or yellow-green spores tended to darken to brown with successive culturing and with time. From the smell, one could tell how dry the spores were and the method of production.[49] This varied widely between makers, for example, in the geographical source of the ash used, how they stacked the *kōji* boxes during culturing, and the way and degree to which they dried the spores.[50] If the *moyashi* maker put the spores in his mouth, he could make similar distinctions through their taste and hardness. He could also check them for bacterial colonies. Finally, the maker could actually make *kōji* and see how smoothly it went, and then ask breweries to try out the *tanekōji* and see how the sake, soy sauce, or miso tasted. By these means, the *moyashi* maker could select the best spores. Makers had also, since the late thirteenth century, attempted to store, maintain, and propagate the mold cultures as purely as possible by adding special ash.[51] In a report in 1903, Tsuboi Sentarō noted if one went to the places where *kōji* was made, sometimes the workers would first sprinkle camellia ash onto the rice, and then after mixing in the *kōji* on which were stuck all kinds of microbes and bacteria and bringing the whole thing into the *kōji* room, somehow only the mold microbes would multiply.[52] New scientific methods allowed *tanekōji* makers to fulfill the same aims with a much higher degree of control.

Most importantly, *tanekōji* makers and academic scientists did not only share common tools and techniques but also shared similar intellectual concerns. University laboratories, government-run experiment stations, and the thousands of brewing houses who specialized in *tanekōji, kōji,* sake, or soy sauce shared concerns for isolating, identifying, and preserving individual microbial strains, and investigating their properties. Academic scientists depended upon *tanekōji* makers and other brewers to provide them with their objects of study, the microbes, which they then studied and preserved in a similar fashion. It was the shared intellectual concerns between academia and industry that helped to drive the adoption of new technologies in a dynamic private sector, a sector which in turn shaped the way academic researchers thought about problems and the research objects they used.

[49] Murakami (1986, pp. 42–43).

[50] Murakami (1986, p. 35).

[51] Murakami (1986, pp. 32–33).

[52] Tsuboi (1903, p. 47).

Notions of *Kin*

As academic scientists adapted the techniques of Western microbiology to the microorganisms they found in Japan, the concerns of indigenous brewing shaped the way they organized their work profoundly.[53] Saitō Kendō and Takahashi Teizō later became key figures in building microbial collections in Osaka and Tokyo and establishing a tradition of microbial studies. The material context that local industry provided for their work gave the way they thought about and referred to microbes a highly distinctive slant. They came to attach meaning to microbes as useful organisms as well as living beings with complex physiologies. Their approach to microbes had an important influence on microbiology in the interwar period.

Back in 1902, the Investigation Committee for the Establishment of a Brewing Experiment Station asked botanist Saitō Kendō to investigate the microbes around a possible site for a modern brewing laboratory in Takinogawa Village in Tokyo.[54] Saitō was a fresh graduate from the Faculty of Science at Tokyo Imperial University and a mold specialist. Since at the time nobody in Japan was an expert on "fermentation microbes" (*hakkōkin*), Saitō copied methods from German, Danish, and Japanese books on brewing science. His research was primarily taxonomic in aim, with the goal of elucidating and classifying new microorganisms. Much of it was pioneering work because "at the time the kinds of wild fungi produced in Japan were completely unknown."[55]

The Tokyo Tax Office and Inspectorate published a collection of scientific reports from Saitō Kendō in 1905. Saitō specifically looked for organisms with properties that mapped onto stages in the brewing process.[56] Scientists knew that in soy sauce brewing, during *kōji* making and when the *moromi* (fermentation mash) matured, microbes with starch- and protein-decomposing ability caused dramatic changes in the *kōji* and raw materials. As the *moromi* matured, it contained large amounts of acid, in particular lactic acid which was present in commercial soy sauce, and so Saitō sought microbes that produced acid. Small amounts of alcohol in the mature *moromi* were likely to be responsible for soy sauce's distinctive aroma, so he looked for microbes that produced alcohol. Even if he did not find microbes with these three properties, he reasoned, microbes that could survive in *moromi* with such a high concentration of table salt must have interestingly complex functions. Rather than transferring droplets of *moromi* directly onto the colloid-based culture medium for investigation, Saitō created an intermediate step with a medium that more closely mimicked the conditions of the soy sauce itself, *kōji* water with 17 % table

[53] Gradually, it also became clear that one of the materials they studied, *kōji,* was specific to Japanese breweries and did not appear to be similar to anything in the wild. The classification of *kōji* microbes has at times been a site of intense debate among microbiologists. The origins of *kōji* microbes are not clearly known.

[54] Saitō (1949, p. 224). Saitō recounts the date as 1901, slightly earlier than the date given by accounts from the Brewing Experiment Station.

[55] Saitō (1949, p. 224).

[56] Saitō (1905, pp. 1–24

salt (the same concentration as commercial Yamasa soy sauce from the Chōshi region in Chiba Prefecture east of Tokyo), where he first let a drop of *moromi* blossom for a few days. By seeking microbes as functional steps within the soy sauce brewing process, Saitō's study yielded two new species of lactic acid-producing bacteria and one new species of yeast. He argued that other microbes played symbiotic roles with the main *kōji* microbe *Aspergillus oryzae* and the soy sauce yeast *Saccharomyces soja,* and if brewers mismanaged this symbiosis the brewing would proceed sluggishly.[57] On the other hand, he also found a host of new species of molds and bacteria that were "useless" in soy-sauce brewing.

Saitō found more new species as he sampled microbes in the air and water in a sake factory in Kumagaya in neighboring Saitama Prefecture in December, the month when sake brewing typically began.[58] In the hot, humid *kōji* room that incubated *kōji,* and in the cool fermentation room where *moto,* rice, and water were left to ferment as *moromi,* organisms and spores in the air settled onto petri dishes that Saitō placed around the rooms twice a day. *Kōji* microbes were plentiful in the *kōji* room, and various molds in the fermentation room. But some of the yeasts there were not good yeasts, for when cultured in *kōji* water they easily suppressed sake yeast (*Saccharomyces sake*). Saitō found species of lactobacilli that seemed to multiply symbiotically with cultures of sake yeast in *kōji* water. Afterward he spent several years extending his studies on "microbes floating in air," knowing that their number and variety changed vastly with the seasons and day-by-day with weather, and that different crowds and spaces drew different microbes.[59]

Over time, Saitō came to view microbes as sensitive and localized organisms that were broader than simply pathogens (*byōgenkin*), germs (*baikin*), or bacteria (*saikin*). Likewise, the Japanese term for microbe, *kin,* came to refer as much to fungi (*kinrui,* or more technically *shijōkin*). For Saitō, Japanese brewing microbes were a part of the diversity of "useful fermentation microbes" (*yūyō hakkōkin*) pro-

---

[57] Saitō Kendō discovered and named *Saccharomyces soja*. Yabe Kikuji named *Saccharomyces sake*. As for *Aspergillus oryzae,* Hermann Ahlburg working at the Tokyo Medical School isolated a microbe from rice *kōji* which he judged was different from the known *Aspergillus flavus* Link in 1876 and named it *Eurotium oryzae*; its classification was soon changed to *A. oryzae* (Ahlburg) Cohn. A decade later Oskar Kellner, working at the Komaba Agricultural College (later the College of Agriculture, Tokyo Imperial University), sent a sample of Japanese *kōji* to microbiologist Carl Wehmer working at the University of Hannover, who isolated a strain from it and gave the strain the same name as Ahlburg's. Later, Wehmer sent his strain to be preserved in the Netherlands as CBS No. 102.07, and in the USA as Thom No.113. But Ahlburg's original strain went instead to the *tanekōji* specialists Nihon Jōzō Kōgyō in Bunkyō-ku, where it was sold under their *moyashi* brand Marufuku Yukijirushi. Brewers in Japan tended to use a different system, dividing the strains into what were commonly called "yellow" *kōji* microbes for sake, soy sauce, and vinegar and "black" *kōji* microbes for shōchū, alcohol, and awamori, and then subdividing them into species including *Aspergillus sojae* that were not necessarily internationally or nationally recognized by scientists. The common Japanese name for the *kōji* microbe became *kōji -kin* after about 1895 when Kozai Yoshinao and his student Yabe Kikuji first used the term. See Murakami (1977); Murakami (1986, pp. 47–48, 57–58).

[58] Saitō (1905, pp. 25–39).

[59] Saitō (1909).

duced in Asia, which Europeans had first introduced to the world through their expeditions in the 1880s.[60] Whether Eastern or Western, whether old or new—where human culture flourished, one could find the technological operations and skills of fermentation. Yet in Asia, fungi were used to saccharify (unlike in Europe, where malt was used) and yeast to ferment, creating the special properties of the liquor of each brewing region. Saitō arranged his 1906 descriptive survey by product, because a particular kind of microbe was responsible for each product. Beginning with *Aspergillus,* he described *A. oryzae* used for Japanese sake, *A. wentii* for the soy sauce of Java, *A. luchuensis* for awamori of the Ryukyu Islands, and *A. batatae* for the sweet potato liquor of Hachijō Island. He moved on to *Monascus purpureus* for Taiwanese "red *kōji* liquor," followed by the "Chinese yeast" that consisted of various *Mucor* and *Rhizopus* fungi (commonly called *kekabi* or "hairy mold" and *kumonosukabi* or "spiderweb mold," respectively, as they were often found in spoiled vegetables) used in China, Cochinchina, India, Java, and Taiwan, and which operated as mold starters like Japanese *tanekōji*. He had studied some of these himself using samples sent to Tokyo from the Chinese quarter in Kobe.[61]

In 1911, Saitō Kendō left the Brewing Experiment Station to take a position at the South Manchuria Railway Central Laboratory, where he became director of the entire laboratory in 1922. There he continued to study and gather microbial strains used in the local fermentation industries for the laboratory's collection, until he moved back to the home islands in 1927 as lecturer in the Brewing Department at Osaka Higher Technical School (the school later became the Faculty of Engineering, Osaka Imperial University).[62] He sent many new specimens for preservation to the *Centraalbureau voor Schimmelcultures* (CBS) type culture collection in the Netherlands.[63] He came to see microbial collections as being like a garden, and along with the gathering and classification of tropical plants, he thought that there should be botanical gardens that corresponded to the microbial world, that would gather specimens from the tropics to the north and south poles and allow research on their theory and applications. He regretted that only Dutch and Japanese researchers were interested in useful rather than pathogenic tropical bacteria.[64] In the twentieth century, Saitō Kendō was one of the most important contributors to Japan's microbial culture collections.[65]

---

[60] Saitō Kendō (1906).

[61] Saitō (1905, pp. 41–47).

[62] Hasegawa (1996, p. 5).

[63] For example, see Saitō's contributions in Lodder and Kreger-Van Rij (1952).

[64] Saitō (1949, pp. 76–80).

[65] In 1956, the Institute for Fermentation, Osaka (IFO) was the largest culture collection in Japan, and held 1303 molds, 754 yeasts, 239 bacteria, 81 actinomycetes, and 14 protozoa. The figures are from Foster (1961, p. 444). The IFO collection included substantial contributions from the Manchuria collection, as well as from the Government Research Institute of Formosa collection supervised by agricultural chemist Nakazawa Ryōji, formerly Saitō's colleague at the Brewing Experiment Station in the 1910s. See Hasegawa (1996, pp. 5–6); Higher Education & Science Bureau, Ministry of Education, Japan (1953).

Another of the earliest large-scale microbial culture collections was at the Brewing Experiment Station, which opened in Takinogawa in 1904. It held strains contributed by Saitō Kendō, Nakazawa Ryōji, and agricultural chemist Takahashi Teizō.[66] Takahashi Teizō was a non-regular staff member at the Brewing Experiment Station alongside his position as assistant professor in the Department of Agricultural Chemistry in the College of Agriculture at Tokyo Imperial University (Tōdai). A student of Kozai Yoshinao, as faculty he took over Kozai's seminar in agricultural products, until his own new seminar in brewing science and microbial physiology (later renamed fermentation science) split from it in 1924.[67] Takahashi copied strains from the Brewing Experiment Station to develop a smaller, parallel collection in the Department of Agricultural Chemistry for his research and teaching.[68] Because of institutional developments that separated the Department of Agricultural Science from the Department of Agricultural Chemistry, agricultural chemistry (*nōgei kagaku,* "chemistry for the agricultural arts") had come to bear the role of transmitting basic chemistry to Japanese agriculture, and had a much more theoretical orientation than one might expect.[69] Moreover, the department had also begun to include substantial microbiological teaching focused on brewing microbes.[70]

Takahashi continued Kozai Yoshinao's project to encourage the use of pure-cultured yeast in sake brewing. He studied methods of yeast preservation developed in Europe, and confirmed which methods worked best for sake yeast. His research questions on yeast were simultaneously problems of practical relevance to the brewing industry and ways to distinguish between different varieties for classification. In extensive studies, he and other members of the Brewing Experiment Station isolated 62 strains from *moto* (technically called "*shubo,*" a term probably derived from the neologism for yeast, "*kōbo*") collected from various brewing districts, and characterized each strain's morphology, whether or not it formed spores or a film, what kind of colony it formed, and—more relevantly to brewing—its ability to fer-

[66] See Hasegawa (1996, p. 5).

[67] Hasegawa (1996, pp. 6–7). When the seminar split up in 1924, microbial chemist Yabuta Teijirō took over the seminar in agricultural products. Sakaguchi (1998).

[68] Hasegawa (1996, p. 7).

[69] The two departments had been set up separately in the 1870s, and when the two departments separated again after briefly uniting in 1886, basic soil science with geology as its foundation was taught in agricultural science, while soil science similar to that envisioned by Justus von Liebig was taught in agricultural chemistry. Plant nutrition was subsumed under the agricultural chemistry department's seminars in fertilizer science and plant physiology. As a result of the splitting of the two departments, topics directly related to agricultural production, such as soils, fertilizers, plant nutrition, and animal feed, came to take a marginal place within agricultural chemistry. Agricultural chemistry became a broad, heterogeneous discipline embracing chemistry, biology, and microbiology. Kumazawa (2003).

[70] Initially, agricultural chemistry at the College of Agriculture as taught by *oyatoi gaikokujin* had been based on Justus von Liebig's original vision centered on fertilizers and soil science, expanding into plant and animal physiology. One *oyatoi gaikokujin* Oscar Leow introduced the teaching of bacteriology and fermentation science, and subsequently Kozai Yoshinao was intrigued by the development of bacteriology under Pasteur in France while he studied in Berlin; through their teaching, agricultural chemistry came to include microbiology as well. Kumazawa (2003).

ment sugar, what kind of acids it produced, how much it assimilated amino acids, and its ability to liquefy gelatin.[71] Takahashi also carried out large-scale investigations on sake disease-causing *hiochi* microbes (*hiochi kin*), the name originating from the term for spoilage when sake "dropped" the fire put in during *hiire*.[72]

Kōji microbes only slowly came to take more prominence in Takahashi's collections, for studies on them were greatly outnumbered by yeast studies for which the investigative techniques of isolating, culturing, testing, and preserving had already been developed in Europe. Takahashi isolated strains of *A. oryzae* from different *tanekōji* samples for sake, and *kōji* for soy sauce and tamari (a kind of rich soy sauce). A variety of strains existed even within one kind of sake, but the physiological differences were most striking between industries: The fungi used in the sake industry formed more sugars, and the fungi used in the soy sauce and tamari industries formed more amino acids, chemicals associated with flavor.[73] It was clear that the configurations of amylolytic and proteolytic enzymes that these different *kōji* microbes made were related to how the microbes had been selected and propagated for creating specific good products in the breweries.

In 1906, a new sake trade association the Brewing Society (*Jōzo Kyōkai*) opened at the same site as the Brewing Experiment Station, and began to distribute pure-cultured yeast strains to sake factories.[74] The "*Kyōkai* yeasts" were neatly organized as numbered strains for the nation's brewers to select and order, and they had been isolated and chosen from samples of *shubo* or *moromi* sent from breweries or local tax inspection offices across the country.[75] In fact, the yeasts were not very successful, though Takahashi visited numerous factories to oversee attempts to use them.[76] In 1909, scientists at the Brewing Experiment Station developed the "fast *moto*" method whereby pure-cultured yeast would simply be added to *moto* to speed up the process, but the *moto* foamed too quickly and produced sake that tasted poorly, a fact brewing scientists later explained by the complex and at the time hazily understood role of lactobacilli in the ecology of the *moto*.[77] Japanese sake brewing traditionally had not preserved yeast-containing samples from the last brew to seed the next, as German beer brewers had or as they themselves did for *kōji,* instead wild yeast entered spontaneously and multiplied during the long *moto*-making process. Industrial surveys show that until the late 1960s, the majority of brewers in Nada were using the natural *moto* method.[78]

---

[71] Takahashi (1902); Takahashi et al. (1914).

[72] Takahashi and Sakaguchi (1958, p. 8, item 30).

[73] The different research materials reflected the fact that sake companies purchased *tanekōji* from specialist houses, whereas the largest soy sauce companies maintained their own *tanekōji*. Takahashi (1909); Takahashi and Yamamoto (1913).

[74] Nihon Jōzō Kyōkai (1975, p. 3, 107–111).

[75] Akiyama (1977, p. 397); Takahashi et al. (1914, p. 4).

[76] Akiyama (1977, p. 396); Takahashi (1904). Fujiwara (1999) gives an account of the program but overlooks its failure in this period.

[77] Akiyama (1977, pp. 396–397).

[78] Akiyama (1977, p. 403).

In 1925, Takahashi Teizō's research laboratory at Tōdai took a sharp shift from practical studies of brewing to fundamental studies of organic acid fermentations, especially of fungi. There was material continuity with earlier studies as he concentrated on molds such as *Rhizopus* that were already employed in industry in Japan or other Asian countries, and were present in his culture collections. There was also intellectual continuity in the vision of microbes as complex objects for the study of life and effectors of chemical change. Mostly, they were biochemical studies to determine what acids were synthesized from or converted into other acids by microbes.[79] Takahashi's students too worked on related problems of the microbial fermentation of organic acids, which were relatively unusual topics worldwide at the time.[80] Their work established a pattern of fermentation research at Tōdai that lasted into the post-World War II period.[81] By that time, agricultural chemistry with its integration of microbiological and chemical studies had become a mammoth discipline, an umbrella for work not only on brewing but also on pharmaceuticals, food, fine chemicals, and theoretical work in microbial physiology and genetics.[82] The discipline provided an institutional frame in which fermentation would continue to shape the development of both pure and applied research.

Conclusion

The late Meiji-period adaptation of Western science to the indigenous brewing industries left an impact on microbiological research far beyond the Meiji period. Technologies to generate and propagate *genkin,* the original starter microbes with which *kōji* mold and the brewing of all sake, soy sauce, or miso began, were refined long before microbiology emerged in the late nineteenth century. For *tanekōji* makers, their ability to control the forms of life that constituted their product was critical to their reputation and survival. The standardization of new, precise procedures of pure culture for the microbes of the Japanese brewing industries relied on the tinkering of both Western-trained Japanese scientists in technical colleges and universities and expert workers in the brewing industry, and exchange between them. National government-supported institutions placed an emphasis on "Western" techniques even in the traditional brewing industry, while local industrialists and the technical colleges that served them focused on upgrading the technological practices that they knew best. Because of this, the *tanekōji* sector provided a way of modernizing the brewing industry in the area of microbiology, and the small scale of these companies belied the level of specialization and technology they possessed. The dynamism of smaller players in industry has been neglected historiographically in favor of efforts initiated by government officials and university-trained scientists such as yeast pure culture. In this respect, the pure culture of *kōji* was one of the

[79] See Takahashi and Sakaguchi (1958).

[80] Sakaguchi (1998, p. 198).

[81] Beppu (1987).

[82] Sakaguchi (1998, p. 195).

most important and pervasive technological changes in the brewing industry, and echoes the continuity between tradition and modernity encapsulated in the common motto of the *tanekōji* industry, *onko chishin* (*find new wisdom through cherishing the old*).

Techniques of handling and studying microorganisms were standardized in this exchange. However, looking beyond standard techniques such as pure culture, local material traditions deeply shaped the practice of microbiological research in Japan. Scientists saw their categories of investigation through the organization of processes by which brewers in local industry operated. Indigenous industry helped to create a relatively autonomous microbiological tradition in Japan that treated microbes as useful resources to be manipulated in spite of their variation and sensitivity as living objects, and saw microbes as effectors of chemical change and complex objects for the study of life. The significance of these processes in other countries in Asia and in the Japanese empire doubled their meaning to Japanese researchers, especially as they were working in part to develop traditional products for export to Asia. What was left after this era of the scientific improvement of traditional industry was a particular way of seeing microbes as living workers and not only pathogens, which was distinctive and powerful, because of the way scientists came to know microbes.

References

1902. Advertisements, *Jōkai* 10.

Akita Konno shōten kabushiki kaisha, ed. 2011. *Konno Moyashi 101nen no ayumi* [History of 101 years of Konno Moyashi]. Daisen, Akita: Akita Konno shōten kabushiki kaisha.

Akiyama, Yūichi. 1977. Sake to kōbo [Sake and yeast]. In *Nihon no sake no rekishi: Sakazukuri no ayumi to kenkyū* [History of sake in Japan: Sake brewing history and research], ed. Benzaburō Katō, 393–440. Tokyo: Kyōwa hakkō kōgyō.

Bartholomew, James. 1989. *The formation of science in Japan: Building a research tradition*. New Haven: Yale University Press.

Beppu, Teruhiko. 1987. Biseibutsugaku no kiso to ōyō [Basic and applied in microbiology]. In *Nōgei kagaku no 100nen* [100 years of agricultural chemistry], ed. Nihon nōgei kagakkai, 136–142. Tokyo: Nihon nōgei kagakkai.

Ceccatti, John. 2001. Science in the brewery: Pure yeast culture and the transformation of brewing practices in Germany at the end of the nineteenth century. Ph.D. dissertation, University of Chicago.

Chikudō, Shō. 1911a. Tanekōji ni tsuite (1) [On tanekōji (1)]. *Jōzō kyōkaishi* 6 (7): 47–52.

Chikudō, Shō. 1911b. Tanekōji ni tsuite (2) [On tanekōji (2)]. *Jōzō kyōkaishi* 6 (8): 32–41.

Clancey, Gregory. 2006. *Earthquake nation: The cultural politics of Japanese seismicity, 1868–1930*. Berkeley: University of California Press.

Foster, J.W. 1961. A view of microbiological science in Japan. *Applied Microbiology* 9:434–451.

Fujiwara, Takao. 1999. *Kindai nihonshuzōgyōshi* [History of the modern sake industry]. Kyoto: Mineruva shobō.

Fukuoka-ken shōyu kumiai, ed. 1979. *Fukuoka-ken shōyu kumiai nanajūnenshi* [Seventy-year history of the Fukuoka Prefecture Soy Sauce Association]. Fukuoka: Fukuoka-ken shōyu kōgyō kyōdō kumiai, Fukuoka-ken shōyu jōzō kyōdō kumiai.

Hasegawa, Takeji. 1996. Nihon no biseibutsu kabu hozon jigyō 1. Yōran jidai [Microbial strain preservation projects in Japan 1. Early period]. *Microbiology and Culture Collections* 12 (1): 1–10.

Hasegawa, Takeji. 2001. Zatsuroku biseibutsugakushi [Miscellaneous records on the history of microbiology]. Ōsaka: Hakkō kenkyūjo.
Higher Education and Science Bureau, Ministry of Education, Japan. 1953. *A general catalogue of the cultures of micro-organisms maintained in the Japanese collections*. Tokyo: Monbushō daigaku gakujutsu kyoku.
Hyakushūnen kinen jigyōkai, ed. 1996. *Hyakunenshi: Ōsaka daigaku kōgakubu jōzō, hakkō, biseibutsukōgakuka* [100-Year Magazine: Department of Brewing Science/Fermentation Science/Microbial Engineering, Faculty of Engineering, Osaka University]. Suita: Ōsaka daigaku kōgakubu jōzō, hakkō, biseibutsukōgakuka.
Jannetta, Ann. 2007. *The vaccinators: Smallpox, medical knowledge, and the 'opening' of Japan*. Stanford: Stanford University Press.
Johnson, Chalmers. 1982. *MITI and the Japanese miracle: The growth of industrial policy, 1925–1975*. Stanford: Stanford University Press.
Kawamata Kabushiki Kaisha, ed. 2000. *Murasaki: Sakai no shōyuya Kawamata, Daishō 200nen no ayumi* [Murasaki: 200-Year history of Kawamata/Daishō, soy sauce brewer of Sakai]. Sakai: Kawamori Mikio.
Kimoto, Chōtarō. 1902. Nihonshu jōzō kairyō ni tsuite no kibō [Hopes regarding sake brewing improvement]. *Jōkai* 3:6–8.
Kimoto, Chōtarō. 1903. Shuzō genryōchū 'bakuteriya' no kensahō [Inspection methods for bacteria in the raw materials of sake manufacture]. *Jōkai* 11:17–23.
Koizumi, Takao. 1984. *Kōji kabi to kōji no hanashi* [Kōji mold and the story of kōji]. Tokyo: Kōrin.
Kozai, Yoshinao. 1901a. Nihonshu jōzō no kairyō ni tsuite [On the improvement of sake brewing]. *Jōzō zasshi* 305:23–27.
Kozai, Yoshinao. 1901b. Nihonshu jōzōhō no kairyō to kenkyū kikan no seibi ni tsuki [On the improvement of sake brewing and the construction of a research facility]. *Jōzō zasshi* 308:4–11.
Kumazawa, Kikuo. 2003. Riibihi to Nihon no nōgaku—Riibihi tanjō 200nen ni saishite [Liebig and agricultural science in Japan—on the occasion of the 200th anniversary of Liebig's birth]. *Hiryō kagaku* 25:1–60.
Lodder, J., and N. J. W. Kreger-Van Rij. 1952. *The yeasts: A taxonomic study*. Amsterdam: North-Holland Pub.
Morris-Suzuki, Tessa. 1994. *The technological transformation of Japan: From the seventeenth to the twenty-first century*. Cambridge: Cambridge University Press.
Murai, Toyozō. 1989. Tanekōji konjaku monogatari [Story of tanekōji past and present]. *Shushi kenkyū* 7:39–44.
Murakami, Hideya. 1977. Sake to kōji [Sake and kōji]. In *Nihon no sake no rekishi: Sakazukuri no ayumi to kenkyū* [History of sake in Japan: Sake brewing history and research], ed. Benzaburō Katō, 319-350. Tokyo: Kyōwa hakkō kōgyō.
Murakami, Hideya. 1986. *Kōjigaku* [Kōji science]. Tokyo: Nihon jōzō kyōkai.
Nakadai, Tadanobu. 1995. Kikkōman ni okeru shōyu jōzō gijutsu kaihatsu no rekishi [History of the development of soy-sauce brewing technology at Kikkōman]. *Chiba-ken kōgyō rekishi shiryō chōsa hōkokusho* 4:1–11.
Nakaoka, Tetsurō. 2006. *Nihon kindai gijutsu no keisei: 'Dentō' to 'kindai' no dainamikusu* [The formation of modern Japanese technology: The dynamics of 'traditional' and 'modern']. Tokyo: Asahi shinbunsha.
Nihon jōzō kyōkai, ed. 1975. *Nihon jyōzō kyōkai nanajūnenshi* [Seventy-year history of the Japan Brewing Society]. Tokyo: Nihon jōzō kyōkai.
Onaga, Lisa. 2012. Silkworm, science and nation: A sericultural history of genetics in modern Japan. Ph.D. dissertation, Cornell University.
Ono, Ryōzō. 1902. Shuzō jō no daiteki 'bakuteriya' ni tsuki kotau [Answer regarding 'bacteria,' the great enemy of sake brewing]. *Jōkai* 10:34–36.
Ōtsuka, Kenichi. 1979. Kenkyū no rekishi to tenbō [Overview and history of research]. In *Jōzō shikenjo sōritsu 75 shūnen kinen kōen* [Memorial lectures for the 75th anniversary of the establishment of the Brewing Experiment Station]. *Jōzō shikenjo hōkoku* 151:1–6.

Saitō, Kendō. 1905. *Tōkyō zeimusho kantokkyoku Saitō Kendō chōsa* [Surveys by Saitō Kendō for the Tokyo Tax Office and Inspectorate]. Tokyo: Tōkyō zeimusho kantokkyoku.

Saitō, Kendō. 1906. Higashi Ajia no yūyō hakkōkin [Useful fermentation microbes of East Asia]. *Tōyō gakugei zasshi* 23(303):507–520.

Saitō, Kendō. 1909. Taikichū no fuyū kinshi [Microbes and spores floating in the air]. *Tōyō gakugei zasshi* 26(331):154–163.

Saitō, Kendō. 1949. *Hakkō biseibukki* [An account of fermentation microorganisms]. Ōsaka: Fuminsha.

Sakaguchi, Kin'ichirō. 1998. *Hakkō to shugaku* [Brewing and liquor science]. Tokyo: Iwanami shoten.

Sawamura, Makoto. 1903. Mazu tōji no zunō o kairyō su beshi [We should first improve the heads of tōji]. *Jōkai* 12:12–14.

Sugihara, Kaoru. 1994. The development of an informational infrastructure in Meiji Japan. In *Information acumen: The understanding and use of knowledge in modern business*, ed. Lisa Bud-Frierman, 75–97. London: Routledge.

Takahashi, Teizō. 1902. Kōbo no chozōhō [Method of preservation of yeast]. *Jōkai* 10:16–26.

Takahashi, Teizō. 1904. Yo ga iwayuru yodan shikomi shijō kakuchi shijō no kekka gaihyō [Short criticism on the results of test brewing of four-stage mash preparation in several districts]. *Jōkai* 30:49–61.

Takahashi, Teizo. 1909. A preliminary note on the varieties of *Aspergillus Oryzae*. *Journal of the College of Agriculture, Imperial University of Tokyo* 1:137–140.

Takahashi, Teizo, and Kin-ichiro Sakaguchi. 1958. Summaries of papers. Tokyo: Committee of Commemorative Meeting of 35 Year's Anniversary of Professor Kin-ichiro Sakaguchi.

Takahashi, Teizo, and Takeharu Yamamoto. 1913. On the physiological differences of the varieties of *Aspergillus Oryzae* employed in the three main industries in Japan, namely saké-, shôyu-, and tamari manufacture. *Journal of the College of Agriculture, Imperial University of Tokyo* 5:151–161.

Takahashi, Teizō, Kamajirō Eda, Junshirō Okumura, Takeharu Yamamoto, Ryōji Nakazawa, Kōzō Shibukawa, and Matao Yukawa. 1914. Seishu kōbo no henshu ni tsuite [On varieties of sake yeast]. *Jōzō shikenjo hōkoku* 54:1–66.

Takayama, Jintarō. 1902. Jōzō shikenjo no hitsuyō [The necessity of a brewing experiment station]. *Jōkai* 1:4–7.

Tanimoto, Masayuki, ed. 2006a. *The role of tradition in Japan's industrialisation: Another path to industrialisation*. Oxford: Oxford University Press.

Tanimoto, Masayuki. 2006b. Capital accumulation and the local economy: Brewers and local notables. In *The role of tradition in Japan's industrialisation: Another path to industrialisation*, ed. Masayuki Tanimoto, 301–322. Oxford: Oxford University Press.

Tsuboi, Sentarō. 1903. Jōzōkai no kakumei jigyō [Revolutionary projects in the brewing world]. *Jōkai* 23:45–48.

Yamagata, Unokichi. 1902. Jōzōgyō to seiketsu [The brewing industry and cleanliness]. *Jokai* 11:16–21.

Yamazaki, Yoshikazu. 1974. Higeta Shōyu gijutsu enkaku no kaisō [Reflections on the history of Higeta Shōyu technology], *Chōmi kagaku* 21 (5): 2–6.

Victoria Lee is a postdoctoral fellow at the Max Planck Institute for the History of Science in Berlin, Germany. She completed her Ph.D. in history of science at Princeton University with a dissertation titled "The Arts of the Microbial World: Biosynthetic Technologies in Twentieth-Century Japan." Her work focuses on the history of fermentation and the applied life sciences in modern Japan, with related interests in the histories of microbes, craft, food, resources, and biotechnology.

Chapter 13
The Coproduction of Station Morphology and Agricultural Management in the Tropics: Transformations in Botany at the Botanical Garden at Buitenzorg, Java 1880–1904

Robert-Jan Wille

Introduction

In 1892, the Botanical Garden of Buitenzorg in the Netherlands Indies (Bogor in modern-day Indonesia) celebrated its 75th anniversary. Ten professors from the German-speaking world who had in the decade before visited the Dutch colonial Garden sent the director Melchior Treub (1851–1910) a certificate that honored the institute's growing role in global science, an example for other institutes in the tropics:

> Aus kleinen Anfängen sich mühsam emporarbeitend hat er [the Garden RW] in diesem Zeitraum eine mehr und mehr wachsende Bedeutung für die Wissenschaft erlangt und steht jetzt als ein leuchtendes Vorbild da für ähnliche wissenschaftliche Anstalten der Tropenwelt.[1]
> Starting from modest beginnings, [the Botanical Garden] has raised itself up in this period to reach a position of ever-growing importance for science and now stands as a shining example for similar scientific institutions of the tropical world.

Among these ten were influential biologists like Karl von Goebel (1855–1932), Ernst Stahl (1848–1919), and Gottlieb Haberlandt (1854–1945).

Buitenzorg as an institution was growing too. A few years later, the eminent naturalist Ernst Haeckel (1834–1919) visited the Garden and was amazed by its expanding bureaucracy, which was the result of plantation owners spending more money and paying Bogor civil servants to work for Buitenzorg:

> Welchen Umfang in Folge dessen der erhöhte Geschäftsverkehr im Bureau des Bogor-Institutes angenommen hat, geht aus folgenden Thatsachen hervor: Im Jahre 1893 wurden 1927 amtliche Briefe versendet, im Jahre 1895 schon 2350 und 1897 endlich 4302. Die

[1] Lotsy (1912, p. 19).

R.-J. Wille (✉)
Scholar in residence, Deutsches Museum, Museuminsel 1, 80538 Munich, Germany
e-mail: r.wille@deutsches-museum.de

D. Phillips, S. Kingsland (eds.), *New Perspectives on the History of Life Sciences and Agriculture,* Archimedes 40, DOI 10.1007/978-3-319-12185-7_13

> Zahl der unentgeltlich an Pflanzer abgegebenen Samenpflanzen, Ableger, Samen u. s. w. stieg in denselben drei Jahren von 1159 auf 1663 und 2294.[2]
> From the following facts it becomes clear to what extent the Bureau of the Bogor Institute has increased its business activities: in 1893, 1927 official letters were sent; already in 1895 the number was 2350, and, finally, in 1897, 4302. The number of free seedlings, cuttings, seeds, etc., sent to plantation owners rose in the same three years from 1159 to 1163 and 2294.

By 1900, Buitenzorg was in many ways a larger research entity than the average Dutch science and medical faculties combined: It had more staff (including hundreds of Javanese laborers in the garden and administrators) and relatively more money to spend on specific research areas.[3]

It was not a normal hortus botanicus. First of all, it was a garden in the tropics. European visitors who expected a garden full of flowers and Victorian hot houses were in for a surprise. When in 1904 the Colorado ecologist Francis Ramaley (1870–1942) visited the institute as an official guest (the fifth American to do so),[4] he wrote:

> In a properly appointed botanical garden most people expect to see also some hot houses for orchids and a tank with warmed water for tropical water lilies and lotus. Should an ordinary mortal, or even a botanist, be dropped from a balloon into the middle of the garden

[2] Haeckel (1901, p. 70).

[3] One can conclude this by comparing in the supplements to the Dutch parliamentary debates, the *Handelingen* (*van de Eerste en Tweede Kamer der Staten-Generaal*) and especially the yearly budgets of the Netherlands (Ministry of the Interior; division of Education, Arts and Sciences, universities) and the Dutch Indies (Ministry of Education, Religion and Trade, after 1904 Ministry of Agriculture) that are attached to the debates. In 1900, the five faculties of the University of Leiden, the most important university of the Netherlands received 839,788 guilders (including three state museums), the other two universities of Utrecht and Groningen got half of the money; Buitenzorg by itself only received 20 % of the amount of the latter two. But its Indonesian staff (administration and labor in the garden) was cheaper and was not put on the budget completely and transparently, the institute had a research task that can be compared with that of a "half faculty" (focusing on the life sciences and chemistry, but without teaching obligations), only salaries formed part of the budget (housing and materials were budgeted somewhere else), and, more important, it received sums from the private industry as well. In 1904, however, when Buitenzorg became its own Department of Agriculture, it received circa 7,000,000 guilders. But now more was included in the budget, including all the money for coffee, cinchona, and other cultures. For a comparison between the United States Department of Agriculture (USDA) and Buitenzorg, see also: Zangger (2011, p. 384).

[4] The first was David Fairchild (1869–1945) in 1896, who worked at the USDA, but whose visit was sponsored by a wealthy millionaire. In 1923, he was to found a tropical garden named "the Kampong," named after a typical Javanese village, in Coconut Grove, Florida. The second was the Harvard student Edwin Mead Wilcox (1876–1931) in 1899, also paid for by private funding, as a preparation for a function at the new Soledad facility, which in the end he turned down. He became director of the Oklahoma Agricultural Experiment Station and later on moved to Alabama, Nebraska, and the Dominican Republic. The third was Charles Lester Marlatt (1863–1954), an entomologist from the USDA and the fourth was Charles Sprague Sargent (1841–1927), director of Harvard's Arnold Arboretum. After Ramaley's visit, eight more Americans would visit Buitenzorg before the First World War. On Fairchild and Wilcox, see Raby (2012, vol. 38, pp. 85–86). For a list of all visitors to the Buitenzorg laboratory in the 50 years after its foundation, see, Dammerman, "A history of the visitors' laboratory."

> at Buitenzorg, he would, for a time, hardly appreciate that he was in a botanical garden.… The plants are mostly trees, no warm tanks are necessary, and there are cool houses instead of hot houses.[5]

But it was also an unusual tropical garden. What separated Buitenzorg Garden from the many other tropical gardens, especially the many satellites of Kew Gardens in the British Empire, were not its trees, but its laboratories. These laboratories formed the main attraction for foreign visitors and they inspired many other national and colonial governments to acquire tropical laboratories in the tropics as well. Ramaley told his American audience:

> It is the wish of the director of the gardens that botanists from all countries should make use of the garden for study. At his suggestion the government erected some years ago a commodious laboratory for the exclusive use of visiting men of science.… With the increased importance of the tropics which has come in recent years, there should be a greater interest developed in the study of tropical life. It is much to be desired that our own botanists make use of this and other tropical gardens in order that we may not remain behind other nations in this important branch of natural science.[6]

Five years after the visit, Ramaley would found a field station in the Rocky Mountains.[7] He must have been inspired by his visit to the Buitenzorg botanical gardens: He visited not only the central garden but also the mountain garden of Tjibodas (present-day Cibodas), a division of Buitenzorg at the slope of the volcanic mountain of Gedéh (Gunung Gede), where another visitor's station had been created in the 1890s.

The gardens have received some attention from historians. But those who have read the studies that deal with this institute may get the impression that there have been two different gardens with the name of Buitenzorg. The same Garden has been presented either as a botanical station used by a network of foreign and predominantly German ecological botanists, as Eugene Cittadino did in his 1991 monograph, or as a state institute that tried to monopolize the scientific management of agriculture in the Dutch Empire, as Harro Maat, Suzanne Moon, Andrew Goss, and Wim van der Schoor did in studies that have appeared since 2001.[8] With the exception of van der Schoor, Buitenzorg's scientific work and its politics have only been studied in fragments. Either the colonial tropics of Java formed the general contours for the detailed study of new scientific practices (Cittadino) or science was treated as a black box in the close analysis of colonial agricultural policy in the Dutch Indies (Maat, Moon, and Goss).[9]

I will show that biological practice and agricultural management in the last decades of the nineteenth century formed two sides of the same Javanese coin. Both

---

[5] Ramaley (1905, p. 579).

[6] Ramaley (1905, p. 589).

[7] On Ramaley, see Vetter (2011); Vetter (2012).

[8] Cittadino (1991); Maat (2001); Moon (2007); Goss (2011); van der Schoor (2012).

[9] Roersch van der Hoogte and Pieters put it even more firmly: "With regard to the history of the Botanical Garden of Buitenzorg, there is no adequate scientific-historical survey," Roersch van der Hoogte and Pieters (2013, p. 93).

were coproduced, at the microlevel of scientific institutionalization and at the mesolevel of imperial policy.[10] Buitenzorg was able to grow so fast because its staff members used the Garden's function as an academic biological station with academic visitors from all over the world (like Naples, Wimereux, and Woods Hole) to strengthen its other function as a state division producing scientific expertise on agriculture, forestry, and other forms of economic biology. And vice versa: The staff used the Garden's function as an official state department to lure more academic visitors.[11]

By focusing on Buitenzorg as both a botanical station and an imperial department of agriculture, I have ignored the third part: the "garden" proper. This partly reflects the ideas put forward by the historical actors in this chapter. In 1899, the name "botanical-agronomic institute" (hyphenated to denote its dual character) was considered by its director Melchior Treub to be a better name for the Buitenzorg complex than "'s Lands Plantentuin" (State Botanical Garden).[12] However, for the sake of tradition he kept to its old name. Moreover, the institute still carried out the traditional tasks of a garden: sending, receiving, and storing seeds and plants.[13] In fact, academic "stationism" (the global movement of scientists building scientific stations) and agricultural improvement were ideologies and practices that were placed on top of these traditional horticultural routines.

Whereas Buitenzorg around 1885 was a garden with a laboratory, more than a decade later it had turned into a laboratory complex with a large "field" surrounding it. This field consisted of multiple gardens, experimental fields, and townscapes. Who was responsible for this change from a classical garden to an academic-agricultural laboratory? This was a new type of colonial scientist: the university morphologist who saw agriculture as a means to a different goal—the strengthening of a unified academic biology. Lynn Nyhart has analyzed several generations of German zoologists in the nineteenth century and has shown the continuing dominance of the approach (*Richtung*) of morphology in Germany, a biology that focused on the development (*Entwicklung*) of form, which became evolutionary in the 1860s and which in the zoology departments (but not the anatomy departments) centered around microscopes, cells, and soft tissues. She has also shown how a generation of evolutionary morphologists born around the 1850s expanded this morphological orientation in all kinds of directions: Many left the alma mater to found research

[10] That the sociological label "co-production" matches the practice institutes like Buitenzorg is shown in Storey (2004).

[11] For more on Naples and Wimereux, see de Bont (2009).

[12] "State," meaning both "belonging to the political dependency of the Dutch Indies" and "directed by the colonial government." "Land" can be translated as "Land" too, and with the "'s" being an abbreviation of "des," an archaic genitive form of "de" (the), the whole could also literally be translated as "Plant Garden of the Land." By "land" the Dutch Indies is meant, not the Dutch Empire as a whole. The Dutch did not use epithets like "imperial" and if it had been a pan-imperial institute, the institute would probably be either "koninklijk" (royal) or "nationaal," with a seat in the Netherlands proper.

[13] Treub (1899, pp. 2–3).

institutes somewhere else. I want to extend that "expanding evolutionary morphology" model to the Dutch Empire and to (tropical) botany.[14]

Transplanted outside their original ecology of the university laboratory, these evolutionary morphologists were anxious to prove their new status as hired experts. They actively made use of the colonial environment and its natural and social disasters like plant diseases, pests, food-related problems, and even volcanic explosions. In the process, they adapted themselves. These morphologists ended up as both producers and servants of an expanding colonial bureaucracy.

A Laboratory Botanist Is Sent Out to the Indies

As Buitenzorg's director between 1880 and 1909, Melchior Treub was responsible for the transformation of European morphology into a project of imperial unified science. In 1873, Treub finished his dissertation at Leiden University, in which he had tried to prove the hypothesis of Anton de Bary (1831–1888) that lichens were composites of algae and parasitic fungi. It was the subject of a faculty prize contest and he was awarded a gold medal for it. After that, he was promoted to the position of lab assistant of the botanical laboratory, a new function in the Dutch education system.[15]

Treub was a silent laboratory reformer. As a student, he socialized with the zoologists who under the German zoologist Emil Selenka (1842–1902) had become Leiden's Darwinists. Their morphological research program was expanding fast, both intellectually and spatially, in the same way that it had done in Germany.[16] With a brand-new zootomic laboratory on the horizon, morphologists were leading a national movement for a marine laboratory on the Dutch coast, copying the zoological station in Naples (1872) and closely cooperating with other universities like Utrecht.[17]

In botany, however, the hortus and the herbarium were still the *loci principales:* Professor Willem Suringar (1832–1898) was an evolutionary skeptic, and although an expert in botanical microscopy, he was not very active in spreading the gospel of the morphological laboratory, preferring systematics instead. Because of this, another student of Suringar, Hugo de Vries (1848–1935), had left Leiden to pursue his luck in Germany and later in Amsterdam.[18] As the new lab assistant, however, Treub was to make evolutionary morphology an acceptable discipline in Leiden in the

[14] Nyhart (1995, p. 23).

[15] Treub (1873). The other lab assistant in Leiden was Hoek. Utrecht would soon follow with the appointment of lab assistants.

[16] Nyhart (1995). For an analysis of the growth of evolutionary morphology in the anatomical discipline in the medical faculties, see Rooy (2011).

[17] I am preparing a dissertation on the growth of "stationism" in zoology and botany in the Netherlands and the Dutch Indies between 1872 and 1904. See also Robert-Jan Wille (2009).

[18] Zevenhuizen (2008, p. 68).

1870s. He convinced Suringar that microscopical anatomy, developmental biology, and physiology were important tools for the study of the "lower plants" and that a careful study of genealogical relationships was necessary for studying the pros and cons of Darwinism.[19] In the meantime, the study of cryptogam development offered Treub what the study of invertebrates offered to his zoological friends: a window on the evolutionary history of life.

However, in the late 1870s it became clear that there was no future for Treub in the Netherlands: No chairs were about to change soon and a proposed lectorship for Treub in botanical physiology did not make it to the national budget.[20] Treub felt he had to accept the position of director of the Buitenzorg Gardens.[21] Luckily, a fellow student, William Burck (1848–1910) joined him, as vice-director, a new position. This extra position allowed them to divide the workload: Treub's friend would reorganize the herbarium and work on plant systematics: Treub could continue his work on the experimental morphology of cryptogams (plants that produce spores, not seeds).

The first 3 years were an isolated period. Treub felt himself to be in the "middle of nowhere." He should not be taken literally: Buitenzorg was a crowded town and the seat of the governor. Treub meant that he did not have the company of academics and natural scientists.[22] At first, he wanted to go back to Europe as soon as possible to take up a professorship. Later, that changed.

The Primacy of Morphology

It is interesting to see that Treub (and to a lesser extent Burck) in the first years in the tropics did not focus on the traditional tasks of (tropical) botanical gardens: economic botany, systematics, and acclimatization. Treub was focusing on issues that he and many European laboratory morphologists deemed important: the development of mosses, ferns, and orchids in order to study the evolutionary relationship between different groups of seed plants and spore plants. For Treub these were not "economic" plants, even though, as Denise Phillips shows in this volume (Chap. 2), Germans had been studying cryptogams in the context of soil rebuilding programs.

---

[19] Suringar wrote an introduction to one of Treub's works, *Si donc d'une part l'étude des fleurs, poursuivi d'une manière scrupuleuse et méthodique pourra mener au perfectionnement de l'arrangement des familles…il est hors de doute d'autre part que les études histiogéniques de tous les organes sont de nature à porter des lumières, là où les relations morphologiques sont obscures sans elles, et qu'elles méritent d'occuper une place très-honorable parmi les guides sérieux et confiables….* See Treub (1876, pp. vii–viii).

[20] National Archives, The Hague, 2.04.13, Ministry of the Interior, Department of Education, Higher Education, 1875–1918, 139, request in letter of Leiden faculty of sciences to Leiden university curators, 30 September 1879. No such item was ever put on the budget.

[21] Went (1916, p. 13).

[22] Artis Library Amsterdam, Hoek Papers, four letters in the early Buitenzorg years from Melchior Treub to Paulus Hoek, especially one sent on 3 January 1881.

Treub's studies were consciously aimed at a European academic public and did not directly serve the Indies government that hired him: Most of the time he worked (and in many instances continued his Leiden work) on the study of the development of lycopods, cycads, and orchids. He did this in the light of the evolutionary relationship between cryptogams and (flowering) seed plants.[23] Hugo de Vries saw Treub as one of the first in the tropics to merge Darwinian ideas with the research that Wilhelm Hofmeister (1824–1877) decades before had done on the relationship between lower and higher plants. According to De Vries, Treub was bridging the *kloof* (gap) between these groups.[24]

A specific subject of Treub—a subject he had adopted from De Bary—concerned the embryology and life cycle of lycopods (clubmosses). In 1885, he proudly sent a paper to be read at the Royal Dutch Academy of Sciences: He declared to his fellow members that he had finally succeeded in isolating the prothallium, the plant's sexual form that alternated with the "main" plant, a nonsexual form. Contrary to mosses, vascular plants like lycopods, ferns, and seed plants have dominant nonsexual generations, but contrary to seed plants where the sexual generation is internalized, the sexual generation of lycopods and ferns lives an isolated life and in the case of lycopods this form is hardly visible to the naked eye. Treub had found the prothallia under his microscope after carefully trying to cultivating them.[25]

It is remarkable that Treub was sponsored by the colonial state to investigate this kind of matter, since in British Ceylon at that time, German-oriented laboratory botanists like Daniel Morris (1844–1933) and most notably Henry Marshall Ward (1854–1906) did focus on economic plants (or on the lower organisms that threatened them). Morris and Ward worked on coffee rust, a plant disease caused by the fungus *Hemileia vastatrix,* which threatened the British imperial coffee culture.[26] The difference was of course that Morris and Ward were paid to work exclusively on the coffee rust and that Treub was free to choose whatever subject he wanted to pursue, so long he was able to do his managerial tasks.

The comparison between Treub and Ward is not arbitrary:In the late 1870s, the disease had spread to Dutch Sumatra and Java and was causing unrest. But the Institute of Buitenzorg did not pick it up immediately: Only in 1887 did Burck start to work on experiments with coffee rust. The gardens did research a few other plant diseases before 1887, but the two Dutch morphologists started only actively coordinating a unified phyto-pathological research program after they had finally settled in Buitenzorg.

---

[23] See the many publications that between 1880 and 1887 appeared in the *Annales du Jardin Botanique de Buitenzorg* and the *Verslagen en Mededeelingen van de Koninklijke Akademie der Wetenschappen,* hereafter *VMKAW,* for example: Treub 1884. For a more detailed list, see Lotsy (1912, pp. 27–31).

[24] Vries (1904).

[25] A summary of the paper that was read is found here: *VMKAW* 3–1, 1885, 189–193.

[26] The botanists studied the life cycles of fungi to promote "a new role for [Kew Gardens] as an imperial center for research into economic botany and agriculture." McCook (2011, p. 95). In 1881, a final report was published.

Several things happened in the period 1880–1887 that would make Treub change his mind about returning to Europe. In this period, the European colonies in Southeast Asia suddenly became promising sites of microscopical research. The tropical environment played a role in this change: For the Dutch Indies, it was a time of disaster and disease. Social and political effects of these changes in the working landscape of Java made Treub reconsider his initial plans and to remain at Buitenzorg. As will be discussed below, Treub fundamentally transformed the scientific infrastructure and mission of the Botanical Garden at Buitenzorg, at the same time changing the "sociotechnical landscape" of the Dutch Indies forever and putting a new layer of political sediment on top of the longue *durée* of the botanical history of the Dutch tropics.[27] But this new layer could not have been built without the continuing input from European "stationism."

From "Isolated" Garden to Station

A wake-up call for Treub was the visit of German botanist Hermann zu Solms-Laubach (1842–1915) in 1883. Solms-Laubach was a paleobotanist and a plant morphologist whom Treub had met previously in Leiden and in Naples and who worked in Göttingen. During their last meeting in Italy, as Treub was passing through on his journey to Buitenzorg, Solms-Laubach was invited to visit Java.[28] His visit motivated Treub to plan Buitenzorg as an international botanical station like the zoological station in Naples.[29]

Due to the warfare state that the Netherlands Indies was at the time, a lot was invested in military infrastructure.[30] Treub was able to occupy an abandoned military hospital, in which was built an improvised visitors' laboratory or *vreemdelingen-laboratorium* (foreigners' laboratory). He used the visit of Solms-Laubach to press the Batavia and The Hague governments to fund an exchange program. Thanks to Solms-Laubach, other Germans had been encouraged to go to Buitenzorg too, visits that were immediately used as showcases for Dutch backwardness: Why did Dutch scientists not visit the tropics? Treub and Solms-Laubach stressed the importance of

[27] See Chap. 1 and especially the diagram of Rip (2002, p. 9). On early modern Dutch science and its dependency on the colonial tropics, Cook (2007).

[28] Solms had sent in an article for publication in Treub's before-mentioned scientific journal *Annales,* a journal Treub had inherited from his predecessor and of which only one volume had appeared. Together with the articles written by staff members Treub and Burck, Solms's articles would transform the journal into a more European-style journal that captured all the disciplines of biology, not just taxonomy.

[29] Solms-Laubach (1911); Zeijlstra (1959, p. 54).

[30] "Warfare state" is a reference to Edgerton (2006). I use the term because in 1887 a quarter of the public expenditure of the Dutch Indies went to the War and Navy departments. The Netherlands was in a long-lasting colonial war of attrition with the Sultanate of Aceh at that time. See the data sampled by Jan Luiten van Zanden and Joost Mellegers and the latter's accompanying text and bibliography on the public finances of the Dutch Indies: Mellegers (n.d.).

tropical investigations in situ for the physiological and morphological study of life, to prevent a Europe-centered view on the laws of nature.[31]

In 1887, Treub went on leave to Europe. He and Solms-Laubach were invited to come to the British Association for the Advancement of Science that year in Manchester. There he met Ward and many German botanists, including his scientific hero De Bary. At this conference, Treub probably realized that research stays in the tropics were not always a hindrance to an academic career: On the contrary, they could be launching pads for one.[32]

During his 1887 sabbatical, Treub not only visited the BAAS but also the Netherlands, where he attended the meetings of the Royal Academy of Science (he was a member since 1879). Treub convinced the Academy and the Dutch state to fund Dutch students to come to Buitenzorg. They felt that more "Dutchmen" should go to Buitenzorg and that botanists had the same right to an exchange program as the zoologists, having had an official state-funded exchange program with Naples since 1874.[33] In the years that followed, Treub was able to convince the German, Austrian, Belgian, Swiss, and Russian governments as well to fund researchers and students to come to Buitenzorg.[34]

The British and Dutch metropolitan conferences may have stimulated Treub to think about the new prestige of microscopical research in the colonies: A phase of "academic drift" (i.e., keeping to the standards and subjects of European morphology in an area that was not perfectly adapted to an academic moral economy) was followed by a phase of epistemic expansion (i.e., modifying an academic research program through extending them into areas of political economy, the "unacademic").[35] Agriculture was one of the areas that Treub now finally started to embrace, and he finally took on the diseases that the tropical environment graciously offered to him.

The Natural Environment Helps a Little

In addition to plant diseases, there was growing interest in tropical diseases that afflicted humans. During his Dutch year, Treub was able to join the first edition of the *Nederlandsch Natuur- en Geneeskundig Congres* (NNGC), a new national organization for physicians and natural scientists partly modeled on the British Association for the Advancement of Science: In 1879, still in Leiden, Treub himself had

[31] Solms-Laubach (1884). See also: *Verslag* [*omtrent den staat*] *van 's Lands Plantentuin* [*en de daarbij behoorende inrichtingen, betreffende het jaar*], hereafter *VLP*, 1884, 12.

[32] Williamson (1896, p. 189).

[33] Archive of North Holland at Haarlem, Royal Dutch Academy of Sciences archives 64, 129, map on Buitenzorg fund, letter Treub to Academy, 19 January 1885 and letter Treub and the three botany professors of Leiden, Utrecht, and Groningen in name of the Academy to the Minister of the Interior, 20 August 1887. On the Dutch seat in Naples since 1872, see Wille (2009).

[34] Went (1915, p. 12. For the case of Switzerland, see Zangger (2011, pp. 385–386).

[35] Elzinga (1997); Harwood (2010); Kaiserfeld (2013).

called for such an organization.[36] At the NNGC meeting in Amsterdam, Treub saw that physicians who had returned from the Indies and who had done microscopical studies to find a cure for the beri-beri disease had received heroes' welcomes. Beri-beri was a mystery disease that had killed many Dutch soldiers who were fighting an imperial war against the Sultanate of Aceh, a war that had been going on for more than a decade. If there was ever a time for steering more state funding into colonial laboratories, it was now: The Indies government was considering building a bacteriological laboratory in the vicinity of Batavia.[37]

As if plant and human diseases and wars were not enough, "Nature" had given another opportunity to Treub to convince his fellow Dutchmen of the importance of detailed scientific investigation of the archipelago. In 1883, the island of Krakatau exploded in a series of volcanic eruptions that killed tens of thousands of people. The Dutch government sent Rogier Verbeek to investigate; Verbeek published a report on the eruption in the following years. In 1886, he led an expedition to the remains of the islands around Krakatau, taking Treub with him. In a meeting of the Royal Dutch Academy of Sciences in January 1888, Treub presented a report on this expedition. He presented Krakatau as an opportunity to study the recolonization of an empty island by vegetation: first algae, then lower plants, and then the higher plants and animals. It was nature experimenting with evolution and botanical colonization.[38]

Science in the Indies was more than just collecting and exploring, thought Treub. In the year that Krakatau exploded, the Colonial World Exhibition had been organized in Amsterdam. In the aftermath, the Indies government had reserved 10,000 guilders for "natural exploration" by Dutch individuals.[39] Treub had already been thinking about what such work should involve. He had convinced the Batavia government that the Indies offered more than just "exploration" and needed to be the subject of detailed study "as a whole," although the colonial government was not willing to pay for it yet. Detailed study meant comparative study of "lower organisms" (algae; invertebrates and plant cells) under the microscope, not the macroscopic collection of higher organisms. Treub had also supported plans of another biologist in the archipelago to found a marine zoological station.

But the Indies government could not decide between supporting the building of local institutions or encouraging expeditions by European academics.[40] So, during

[36] Treub (1879).

[37] See the lectures (including a small presentation of Treub on tropical plant budding) in Stokvis et al. (1888). Treub reviewed this conference and applauded its scientific nationalism: Treub (1887a).

[38] See *VMKAW* 3–5, 1889, 4–5. It led to this article: Treub (1888). For more on the role of "nature's experiments," see Kohler (2002, pp. 212–251).

[39] See the national budget attached to the parliamentary debates: *Handelingen,* 1882–1883, appendix B, 23, Chap. II, department V, item 58.

[40] National Archives The Hague, 2.02.01, Ministry of Colonies, *verbalen* 1850–1900, 4032, *verbaal* 12 February 1887, containing reports and correspondence between The Hague and Batavia/Buitenzorg on the support of science in the Dutch Indies, including advice by Treub. On the nature of the Dutch colonial archives and the historical system of *verbalen* in relationship to knowledge and power formation within the imperial state, see Stoler (2010, pp. 8–15).

his stay in the Netherlands, Treub founded an organization that had as a goal a coherent program of scientific exploration of the Dutch Indies, where laboratory scientists tried to take over the initiative of geographers, ethnographers, and taxonomists of the State Museum for Natural History.[41]

Both the new Buitenzorg fund of the Academy and this new exploration committee confirmed Treub's position as gatekeeper to a detailed laboratory study of the Dutch Indies. From 1888 onwards, he used this function to expand his botanical gardens, by first convincing the Dutch government in 1888 to send an extra pharmacologist to Buitenzorg, then by convincing the Indies government to pay for another botanist and a chemist to study plant diseases from 1890 onwards, for which new laboratories were built. The simultaneous crises in agriculture in the Netherlands and the Dutch Indies made politicians and entrepreneurs turn their heads in the direction of Germany, where experiment stations had been built.[42] When Dutch sugar planters in Java asked a German expert for advice, he referred them to Treub, who had an international reputation by then, possibly thanks to the fact that his visitors' station had been advertised in Germany. The planters pressed the Dutch Parliament to invest in the extension of Buitenzorg.[43]

The new biologist who was to accompany the chemist to Buitenzorg was Jacob Marinus Janse (1860–1938) and he was one of the first Dutch *botanists* to have visited the zoological station of Naples with a stipend. Before going to Java, he would visit the station for a second time. That was important, because next to doing agricultural (phyto-pathological) research the new botanist had an important second task: Buitenzorg's visitor laboratory came under his supervision.

The Place of Buitenzorg

In 1892, when the Garden celebrated its 75th anniversary and a book with maps was produced in Dutch and German, Treub had reorganized his "station" into six departments, five of them led by academics with PhDs, including Treub, Burck, and Janse. In the meantime, the 34th guest arrived at the Buitenzorg visitors' station: Since 1883, 15 had come from Germany, including his own evolutionary zoology professor Selenka (who had moved back from Leiden to his homeland); 11 Dutchmen, both from the Dutch Indies and from the metropolis; 4 Russians; 2 Brits; 1 from Austria-Hungary and 1 Swede.[44]

What kind of place did they find? When Treub first took over the directorship of Buitenzorg, it already consisted of three different gardens. In 1880, there was a central garden, a traditional hortus with an herbarium, situated in the center of Buitenzorg village, next to the palace of the governor of the Dutch Indies. There

[41] For a formal history of this society, see Pulle (1940).

[42] Maat (2001, pp. 58–71).

[43] *Handelingen,* 1889–1890, 95.

[44] Treub (1893); Treub (1892). On the foundation of the Garden, see Weber (2012).

was a garden in Tjikeumeuh, now part of Bogor, organized around an agricultural extension college and further up the mountains, there was Tjibodas, the cooler acclimatization zone for European plants.

But a lot had changed in 1892. The central Garden was now seen as the main area for "pure research," with a series of botanical and pharmacological laboratories, a museum, a herbarium, the office, a photographical bureau, the library, and the visitors' laboratory, which had all been built or appropriated in the 4 years before.[45]

The second garden of Tjikeumeuh was presented as the official *cultuurtuin* (Dutch), *Versuchsgarten* (German), or experimental garden: the domain of the chemist. Around 1900, it accommodated agricultural experiments and had as its main building an agricultural chemical laboratory. The map of the garden shows its experimental nature: Instead of the meandering paths of the central garden, the *cultuurtuin* shows a grid pattern.

The third garden of Tjibodas—the one that Ramaley visited also—was in Tjibodas, on the slopes of Mount Gedeh (now Gunung Gede), a volcano. This "mountain garden," the *bergtuin* or *Gebirgsgarten,* was more than a day's ride and a long climb to the east. The garden was now used for the collecting and research of plants of the cooler mountainous zone in situ. Under Treub, a large area of "primeval forest" was acquired and left untouched. In 1891, a field station had been built there, to allow scientists to study its ecology.[46] According to the biologist F. A. F. C. Went (1863–1935), it was the Dutch Indies' first "monument of nature."[47] It was also used for retreat, the sanatorium of Sindanglaya being in the neighborhood of Tjibodas. Foreign visitors were especially welcomed.

In 1898, 24 Europeans worked at the laboratories in the three gardens, with 15 members having academic science degrees.[48] More than 150 visitors who had stayed almost 700 months at Buitenzorg in the period before the Great War would bring their research findings to institutes in Germany (52 visitors), The Netherlands (21 visitors), Russia (19 visitors), Austria (16 visitors), and the USA (13 visitors).[49] Not all of them were sent by universities: Many were civil servants, with the United States Department of Agriculture (USDA) being an important employer among them. The relationship with the American department proved to be especially helpful to Treub: In 1902, he visited the USA himself, taking the USDA as a model for

[45] Solms-Laubach (1884).

[46] Dammerman (1945).

[47] Went (1915, p. 13). Because of the cooler mountain climate, this natural monument had a certain "Europeanness"; it reminded Treub of the outdoors of the Dutch interior (the province of Gelderland). Treub (1881).

[48] Went (1898).

[49] Only five Frenchmen and four British visitors had come; they had their own imperial institutes. Noteworthy are the visitors from the smaller countries of Switzerland (five), Sweden (four), and Belgium (four). Some visitors went straight to other colonies: There were visitors from German Cameroon, British Fiji, and Belgian Congo. University cities that sent more than five scientists: Berlin, Jena, Munich, and St. Petersburg (half of them from the university, half of them via the Imperial Academy of Sciences). Dammerman, "A history of the visitors' laboratory."

a new technical department of agriculture in the Dutch Indies organized around the Garden, which would come into existence in 1904.[50]

Some of the international visitors to the laboratories (including those who were hired on contract basis by the Dutch government) moved on to build or expand botanical and agricultural laboratories in their home countries or in their colonial empires, from the Rocky Mountains to islands and countries of the Caribbean and from the Congo River to the Usambaras of Tanganyika.[51]

Codevelopment as a Morphological Research Theme

Morphologists like Treub had in the meantime stopped focusing on cryptogams and phylogeny and had developed new research themes, coming to terms with all the problems in tropical agriculture. In the working landscape of Java, morphologists constructed subjects like plant growth, soil science, pest control, entomology, and fisheries research as parts of a larger biological framework, with chemistry, physiology, and veterinary science as handmaidens to the study of botanical and zoological growth.[52]

The work at Buitenzorg was strengthened by the foreigners visiting the laboratories who brought new knowledge to the site. Rostock's Karl von Goebel (1855–1932) studied the developmental biology of liverworts living on trees; the USDA's David Fairchild (1869–1954) studied the food relationship between termites and the fungi in their nests; and Zürich's Alfred Ernst (1875–1968) studied the volcanic ecology of Krakatau.

Sometimes a whole new subfield was constructed, where taxonomists, morphologists, ecologists and experimenters, both botanists and entomologists found each other. In particular, the study of ants (myrmecology) was an important research subject: Treub himself studied the relationship between ants and plants, and many visitors were, like Burck, experts in "myrmecophilous" plants, such as the

---

[50] [Anon 1902]: "Professor M. Treub, Director of the Botanical Garden at Buitenzorg, Java, was a visitor at the garden [i.e. in New York] during a few days in mid-November and again toward the end of the month. In addition to the inspection of some of the other botanical institutions of America, Professor Treub made a study of the Bureau of Plant Industry of the U.S. Department of Agriculture. The entire botanical and agricultural needs of the island of Java, with its twenty-four millions of inhabitants, are cared for in the Buitenzorg Garden, which is thus in effect a department of agriculture of the Dutch government for the island. Very important arrangements for future exchanges of seeds, specimens, and books were made with him." For more on the modeling of the department on the USDA, see Goss (2011, p. 89).

[51] With the Rocky Mountains a reference is made to Ramaley; with the Caribbean to Fairchild and Wilcox; with the Usambaras to Albrecht Zimmermann (more about him further in the text). The Belgian Léon Pynaert (1876–1968) was the first director of the Congolese botanical garden of Equatorville (now Eala-Mbandeka) and stayed at Buitenzorg to prepare his directorship in the tropics.

[52] For Kohler's definition of working landscapes as "environments in which humans are a dominant presence," see Kohler (2011, p. 222).

Austrian-Hungarian Gottlieb Haberlandt (1854–1945) and the German George Karsten (1863–1937). Kiev's Vladimir Karavaiev (1864–1934) studied the anatomy of tropical ants. Without Buitenzorg, William Morton Wheeler (1865–1937) would never have been able to write his 1910 work, *Ants: Their Structure, Development and Behavior*. It is filled with references to work done by Buitenzorg visitors.[53]

Morphologists and some plant physiologists all over the world were interested in the adaptation of one species to another, but especially so in Buitenzorg. It would not take long before this interest would develop into a new research program: ecology. The first international definition of "ecology" in 1893 was the study of adaptation of one species to another, both in the field and in the laboratory, that is to say, the study of the processes of codevelopment and coevolution.[54]

Research into codevelopment played an important role in Buitenzorg's annual reports and articles. Codevelopment was both studied in the field and in the laboratories, with the emphasis on the latter. In 1892, a publication list was produced for the period of Treub's directorship. Out of almost 100 pages, a third was devoted to morphology and *ontwikkelingsbiologie* (developmental biology) and another third was either systematics or chemistry and pharmacology in the service of laboratory botany. The rest was devoted to plant sociology and above all to "biology," a term that mainly referred to the study of symbiotic and parasitic relationships between plants and between plants and animals. Not only the relationship between crops and pests was studied but also the relationship between ant colonies and ant-attracting plants was to be a durable research theme.[55] The Garden of Buitenzorg formed a good environment for this kind of research, as Ramaley wrote in 1904:

> A moist climate, such as that of Buitenzorg, favors the growth of epiphytic or perched plants—also of parasites. Seeds or spores, carried by the wind or birds, find lodgment in the forks of trees. With plenty of moisture in the air and a constant warm temperature they grow luxuriantly. Thus it happens that trees are covered with moss. Even the very leaves are often marked with delicate patterns of moss and lichen. Orchids and ferns in great number are perched upon the horizontal branches and the smooth trunks also serve for the lodgment of many plants as well. Since Darwin's time everyone has known something about orchids: plants with curious flowers adapted to insect visits—flowers of handsome colors and strange shapes.... There plants inhabited by ants are sure to strike the attention of visitors. There are many of these so-called "myrmecophilous" plants in the garden at Buitenzorg.[56]

The study of mutual adaptation was not just a subject of great interest to the evolutionary morphologists but also—albeit for more economic reasons—was sold as an important subject to the state and the colonial planters. Treub succeeded in managing and maintaining state support of general morphology under the avatar of "applied science."[57]

[53] This Harvard entomologist never visited Buitenzorg, but was later asked to determine ants caught on the islands of Krakatau. Wheeler 1910; Wheeler 1924.

[54] Kohler (2002, p. 75).

[55] Treub (1982).

[56] Ramaley (1905, p. 579).

[57] See also van der Schoor (2012).

Agriculture as Applied and Transgressed Morphology

One form of codevelopment was of big interest to the planters and the colonial state: parasitism. After having written mainly about the developmental biology of cryptogams, ant-attracting plants, and other noneconomic plants, Treub little by little started to write about plant diseases. His first publication in economic biology had been on a case of parasitism in sugar canes (the "*sereh* disease"). It was published in 1885 in a new periodical that dealt exclusively with economic biology, named *Mededeelingen uit 's Lands Plantentuin*. The *Mededeelingen* series were constructed as the "practical" counterpart to the more "pure science"[58]-oriented journal *Annales*.[59] Treub also published a summary of his *sereh* research for his fellow academics in the *Annales* in 1887.[60]

The 1885 article was probably not an easy read for planters: Treub's report in *Mededeelingen* delved deeply into the academic discourse of embryology. It was an analysis of the interaction between the sugarcane *Saccharum officinarum,* the animal parasite and nematode worm *Heterodera schachtii,* and the fungus *Pythium* (that possibly aggravated the disease). He concluded that the diseased sugarcane was indeed infected and had not been degenerated into its wild ancestor, a theory that some others had put forward. It was a form of total biology: Treub dealt with development and evolution, botany and zoology.

Four years later, having returned from the Netherlands and having transformed himself from a single-minded morphologist into a managing director trying to tap into the resources of the plantation economy, he addressed the planters more directly and in a less esoteric fashion. He lectured to the Dutch Indies' *Maatschappij voor Nijverheid en Landbouw* (Society for Trade and Agriculture) where he reconstructed the study of plant diseases as the subject of a larger academic problem: the problem of the *Erscheinung der Symbiose* or appearance of symbiosis.

In 1878, De Bary had defined the problem as such: How can we explain the existence of "sustainable contracts" in nature between different life forms? These relations were "mutualist" (both parties benefited each other), "parasitic" (one party

[58] I will not deal with the constructive nature of "pure science" here. Both Goss and van der Schoor deal with this: Goss (2009); Goss (2011); van der Schoor (2012). See also the debate on Lewis Pyenson's use of the somewhat similar term "exact science" in imperial (Dutch) context: Pyenson (1993); Palladino and Worboys (1993); Pyenson (1989).

[59] Around 1900 Buitenzorg published many journals. Other journals were: *Teysmannia* (from 1890 onwards, a semipopular scientific garden journal); *Icones Bogorienses* (from 1897 onwards, containing images and descriptions of endemic plants); *Bulletin de l'Institut botanique de Buitenzorg* (from 1898 onwards; a journal containing selected translated articles from the *Mededeelingen*). From 1897 a series of taxonomic monographs was published for short-term visitors who did not have the time to do taxonomy themselves: *Flore de Buitenzorg*. Next to these journals, the Garden published an Annual Report every year (from 1868 onwards). The first volume of *Annales* was published in 1876, the first volume of *Mededeelingen* in 1885, a few months before Treub's contribution on sugarcane, which was the second installment in the series.

[60] Treub (1885); Treub (1887b).

receives benefit by damaging the other), or "commensalist" (one party benefits the other without receiving either benefit or damage). According to Treub, parasites did not suddenly appear as thieves in the night, but had had a long evolutionary history of adaptation based on *oefening* (exercise, practice, training). Aiming at the Dutch entrepreneurs in the audience, Treub used as a prime example the "enterprise of lichens," a composition of algae and fungi as he had demonstrated in his own dissertation.[61]

His message was twofold. First, parasitism could only be researched in the context of the larger research program of morphology and the study of symbiosis. Funding research on plant diseases automatically meant funding morphology, the former being presented as the applied form of the latter. Second, planters, state, and scientist should form a "mutualist" relationship themselves, in order to win the war on plant diseases. By sending out these kinds of messages between 1890 and 1900, Treub was able to expand his Buitenzorg station with more staff members and laboratories, paid for by the state and by the planters. Among them were chemists and entomologists, but most of them were laboratory botanists with degrees from Dutch (and German) universities who learned while working in the new environment of the Java plantations.

It is in the light of academic morphology having transformed into agriculturally applicable science that we must also understand the second installment in 1901 of *De dierlijke vijanden der koffiecultuur op Java,* the 44th volume of the *Mededeelingen* and a study on the threat of animal pests to Java's coffee culture. One of its authors was the Dutch zoologist Jacob Koningsberger (1867–1951), at that time chief of the agricultural zoological museum of the Garden and future minister of Colonial Affairs in The Hague.[62] The other author was Albrecht Zimmermann (1860–1930), botanist at the new Coffee Experiment Station of the Garden; he would later lead a state-owned experiment station in Amani, German East Africa, one of the more important science institutes in Africa during colonial times.[63] So, this volume was written by—as it turned out—ambitious men.[64]

Those who analyze the volume in isolation may be tempted to conclude that *Dierlijke vijanden* is above all an exercise in encyclopedic description: The volumes seem only to entail the dissemination and application of basic natural history. Chapters are organized around single groups of animals, and very few pages deal

[61] Treub (1889).

[62] On Koningsberger, see Goss (2009).

[63] Amani continued this status under British rule: Tilley (2011), Chaps. 3 and 4. For more on the German history of the institute, see: Bald and Bald (1972); Zimmerman (2006). Amani's main rival for the position of most prestigious colonial science project was Buitenzorg itself: Zimmerman (2006, p. 437).

[64] For a more extensive study of the experiment stations in the Dutch Indies in general, see van der Schoor (2012). For more on coffee research in laboratories and the field on the global level, see McCook (2011).

with general biology.[65] The first two or three paragraphs of every chapter introduce some common traits of the animals; the rest of the chapters describe the damage they deliver to the crop and possible remedies. Much is based on comparative literature study.

Historians who take some extra time to study the career of its authors may ask themselves whether this exercise is symbolic of a "demise" around 1900 of European laboratory morphology. In the 1880s and 1890s, Koningsberger and Zimmermann were students of Ambrosius Hubrecht (1853–1915) in Utrecht and Simon Schwendener (1829–1919) in Berlin respectively, two eminent professors in evolutionary animal and plant morphology. Hubrecht collected and compared embryos, looking for missing links and sorting them into trees and tables; Schwendener experimented with plant growth and adaptation.[66] In *Dierlijke vijanden,* their two students[67] seem to have forsaken their advanced laboratory training in developmental biology and instead were writing natural history compendia aimed at enlightening colonial entrepreneurs and farmers.

But Koningsberger and Zimmermann did not abandon a sinking ship of morphology.[68] On the contrary, they helped transform and expand Treub's original academic program of morphology in the Dutch Indies, integrating it in a larger project of unified biology. *Dierlijke vijanden* was a way of trying to maintain the position of Buitenzorg as the central center of expertise. In his introduction, Koningsberger writes about the necessity of providing others with information about animal pests and other threats to agriculture, but it is not just about providing facts and taxonomies in isolation. He shows that these kinds of taxonomies have another purpose as well: to show the planters that these "new" pests were part of a complex ecology that had changed quite dramatically with the deforestation due to the buildup of coffee plantations in the last decades. Koningsberger and Zimmermann pleaded for researching these kinds of new dynamics in the changing ecology of Java; taxonomy was just the beginning of such an effort, a foray into more detailed *laboratory* studies that should throw more light on the *field.*

[65] This formed part of a series of two: Koningsberger (1897); Koningsberger and Zimmermann (1901).

[66] Hubrecht and Schwendener were leading biologists in the Netherlands and Germany, respectively, hugely influencing the scientific agenda of their national landscapes. For more on Hubrecht, see Bowler (1996, pp. 181–183, 295–296); Nyhart (1995, p. 212); Hopwood (2005); Wille (2009). For more on Schwendener, see Cittadino (1991, p. 27 ff).

[67] Koningsberger wrote his dissertation with the botanical professor in Utrecht, Nicolaas Rauwenhoff (1826–1909), but Hubrecht acted as his zoological patron, both during his academic studies and his entomological career afterwards.

[68] The classical (and much contested) thesis of Allen (1975). For more on the history of institutional morphology and political machinations in the Netherlands proper between 1850 and 1900, see: Visser (1986); Theunissen and Donath (1986); Theunissen (2000), Chaps. 2, 3, 6, and 8; Jonge (2005); Rooy (2011).

Stations in the Tropics and Morphological Expansion

The story of Buitenzorg is not just the story of subjugation of agriculture to a black box of "pure science." It is a story of adaptation: Morphological practice was transformed by the ambition of the leading scientists to tap into the financial resources of a colonial plantation economy and to make use of the natural environment of the Indonesian archipelago.

The archipelago formed a political and natural habitat that was radically different from the European academic environment: the traditional hosts of morphology. Natural and social disasters in the archipelago (the import of the plant disease of coffee rust; explosion of Krakatau; the food-related human disease of beri-beri killing imperial soldiers) prompted state investment in surveys and research, giving opportunities for Dutch morphologists to claim expertise and finances.

The story of the adaption of the program of morphology to the tropics is not just a story of contingency and pragmatism: It is also a story of scientific Realpolitik and lobby politics. Morphologists were "advocates" of the first order. The European academic environment had already turned morphologists into experienced and conscious trespassers of neighboring territories: Since the 1860s, evolutionary morphologists in zoology and botany competed with two distinct groups of life scientists, claiming an independent and unified biology, not only in key cities like Berlin, Jena, Heidelberg, and London but also in the Dutch universities of Utrecht, Leiden, Groningen, and Amsterdam. With physiologists of medical laboratories, the professors in zoology and botany quarreled over the academic curriculum, allocation of university money, and laboratory space. With taxonomists and comparative anatomists of imperial state museums, Dutch university biologists fought over access to collections, state funding for expeditions, and access to the colonies, especially the Dutch Indies.[69]

What emerged was a wish for an independent space for academic morphological research, outside the reach of physiological laboratories and state museums. In the first stage, the focus had already been "relocated" from vertebrates and flowering plants to sea invertebrates and cryptogams. In the second stage, a new type of space emerged with the foundation of marine laboratories "in nature," paid by states and private parties and catering to academics: for example, in 1872, in Naples and Roscoff, France; in 1874, in Wimereux, France; in 1875, in Trieste in Austria-Hungary; in 1876, a "flying station" at different locations on the Dutch coast (later permanently settling in Den Helder); in 1881, in Sydney; in 1888, a British station was built in Plymouth and an American one in Woods Hole and in 1892 followed a German station on Helgoland. Some of these were catering to universities, others had a role in the management of state fisheries, but almost all locations were staffed

[69] Discussed in more detail in my dissertation, Wille, "The stationists, laboratory biology, imperialism, and the lobby for national science politics."

by people with academic degrees in the new biology, connecting their research to university fashions. They shared an "academic" ambition.[70]

With the foundation of Buitenzorg's visitors' station, modeled after Naples, and its institutional expansion thereafter, the idea of an academic laboratory in nature—"nature" meaning "outside the university city"—moved from marine zoology to tropical botany. Buitenzorg itself formed a model again for a new round of (botanical) stations: Amani in German East Africa (1902) and a series of stations founded in the "American tropics": the Harvard botanical station in Soledad at Cuba (1900) and Barro Colorado Island in the Panama Canal Zone (1923).[71]

And others were indirectly inspired by Buitenzorg. With the Monroe Doctrine in mind, in the 1890s, the New York Botanical Garden had wanted to become the "American Kew" by founding tropical gardens in the Caribbean, but when its director Nathaniel Lord Britton (1859–1934) hired the botanist Daniel Trembly MacDougal (1865–1958) as its director of laboratories, the Dutch garden became a focal point too.[72] In her dissertation on American station biology in the Caribbean tropics, Megan Raby writes about MacDougal that he was of the opinion "that a laboratory in the Caribbean could be for American botanists what the Dutch station at Buitenzorg, Java, was for the Europeans…. MacDougal encountered many researchers who had worked there, and their experiences apparently made a strong impression on him." Britton and MacDougal ended up by leasing a garden from the British in Jamaica: Under the name of Cinchona Station, it was later compared by contemporaries to Treub's institute.[73]

Buitenzorg was the link between a first generation of European zoology stations and a new generation of agricultural botanical stations in the tropics. The first stations at European coasts started as vehicles for a third way of doing biology, situated between metropolitan laboratories and natural history expeditions. Some of these European stations became launching pads for criticism of traditional evolutionary morphology: Experimentalist hardliners in Naples and the "new natural historians" and ecologists in the field stations of the American West thought the discipline either should focus more on experimental techniques of the laboratory or should pay more attention to processes in the field.[74]

However, their criticism had been based on the science of the European university that focused on the construction of evolutionary trees: In the agricultural

---

[70] Raf de Bont is preparing a monograph on the biology of the station movement in continental Europe. I would like to thank him for being able to read its manuscript.

[71] Zimmerman (2006, p. 437); Raby (2012, p. 7, 56).

[72] In 1896, Britton presented a list of 13 important foreign botanical gardens at the 1896 conference of the American Association for the Advancement of Science in Buffalo. He started this list with Buitenzorg, with the London's famous Kew Gardens in London as a runner-up. The other 11 foreign gardens were Berlin, Paris, Vienna, Geneva, Edinburgh, Dublin, Brussels, Port-of-Spain, Kingston, and Montreal. Went (1898); Britton (1896).

[73] Raby (2012, p. 35). Her first chapter deals with the movement in general and Cinchona in particular.

[74] Allen (1975); Kohler (2002).

stations in the tropics, morphologists had moved beyond this conception of evolutionary morphology. In the colonies, the impetus for cooperation was larger than competition: Instead of disciplinary "civil warfare," plant and animal morphology in the Dutch Indies became part of the larger project of a joint "war on plant diseases." Unified biology became more than just the study of the relationship between ontogeny and phylogeny by doing microscopic studies of cells. It encompassed the emergent relationships between plants and animals spread over the globe. In Buitenzorg, the morphologists did not "revolt" but adapted their program to a colonial environment. They put morphology—originally the most theoretical and least useful of the biologies—at the top of a pyramid, as a kind of "general" biology trickling down into "applied agricultural science." In the Dutch Indies laboratory, botany had transcended its academic origins, embraced experimental techniques and field practice, and had created a "symbiotic" morphology that had then become modular enough to be transferred between the different tropical empires of the West.[75] It was a general science of structure, development, and society.

Acknowledgments Apart from the editors, I would like to thank Cristine Webb for her extensive comments on this paper.

References

Allen, Garland E. 1975. *Life science in the twentieth century*. New York: Wiley.

Anon. 1902. Notes, news and comment. *Journal of the New York Botanical Garden* 3:224.

Bald, Detlef, and Gerhild Bald. 1972. *Das Forschungsinstitut Amani. Wirtschaft und Wissenschaft in der deutschen Kolonialpolitik, Ostafrika 1900–1918*. München: Weltforum.

Bowler, Peter J. 1996. *Life's splendid drama: Evolutionary biology and the reconstruction of life's ancestry, 1860–1940*. Chicago: University of Chicago Press.

Britton, Nathaniel Lord. 1896. Botanical gardens, origin and development. *Science* 88:284–293.

Cittadino, Eugene. 1991. *Nature as the laboratory: Darwinian plant ecology in the German Empire, 1880–1900*. Cambridge: Cambridge University Press.

Cook, Harold J. 2007. *Matters of exchange: Commerce, medicine, and science in the Dutch Golden Age*. New Haven: Yale University Press.

Cooper, Frederick, and Ann Laura Stoler. 1997. Between metropole and colony: Rethinking a research agenda. In *Tensions of empire: Colonial cultures in a bourgeois world*, ed. Frederick Cooper and Ann Laura Stoler, 1–56. Berkeley: University of California Press.

Dammerman, K. W. 1945. A history of the visitors' laboratory ('Treub Laboratorium') of the Botanic Gardens, Buitenzorg, 1884–1934. In *Science and scientists in the Netherlands Indies*, ed. Pieter Honig and Frans Verdoorn, 59–75. New York: Board for the Netherlands Indies, Surinam, and Curaçao.

De Bont, Raf. 2009. Between the laboratory and the deep blue sea: Space issues in the marine stations of Naples and Wimereux. *Social Studies of Science* 39 (2): 199–227.

Edgerton, David. 2006. *Warfare state: Britain, 1920–1970*. Cambridge: Cambridge University Press.

Elzinga, Aant. 1997. The science-society contract in historical transformation, with special reference to 'epistemic drift'. *Social Science Information* 36 (3): 411–445.

[75] Raby (2012, p. 37); Cooper and Stoler (1997, p. 13).

Goss, Andrew. 2009. Decent colonialism? Pure science and colonial ideology in the Netherlands East Indies, 1910–1929. *Journal of Southeast Asian Studies* 40 (1): 187–214.
Goss, Andrew. 2011. *The floracrats: State-sponsored science and the failure of the Enlightenment in Indonesia*. Madison: University of Wisconsin Press.
Harwood, Jonathan. 2010. Understanding academic drift: On the institutional dynamics of higher technical and professional education. *Minerva* 48 (4): 413–427.
Haeckel, Ernst. 1901. *Aus Insulinde: malayische Reisebriefe*. Bonn: Emil Strauss.
Hopwood, Nick. 2005. Visual standards and disciplinary change: Normal plates, tables and stages in embryology. *History of Science* 43:239–304.
Jonge, Hans de. 2005. Macht, machinaties en musea: Jan van der Hoeven, Hermann Schlegel en hun strijd om het Rijksmuseum van natuurlijke historie te Leiden. *Bijdragen en Mededelingen betreffende de Geschiedenis der Nederlanden* 120:177–206.
Kaiserfeld, Thomas. 2013. Why new hybrid organizations are formed: Historical perspectives on epistemic and academic drift. *Minerva* 51 (2): 171–194.
Kohler, Robert E. 2002. *Landscapes and labscapes: Exploring the lab-field border in biology*. Chicago: University of Chicago Press.
Kohler, Robert E. 2011. History of field science: Trends and prospects. In *Knowing global environments: New historical perspectives on the field sciences*, ed. J. Vetter, 212–240. New Brunswick: Rutgers University Press.
Koningsberger, Jacob Christiaan. 1897. *De dierlijke vijanden der koffiecultuur op Java I*. Batavia: Kolff.
Koningsberger, Jacob Christiaan, and Albrecht Zimmermann. 1901. *De dierlijke vijanden der koffiecultuur op Java II*. Batavia: Kolff.
Lotsy. J. P. 1912. *Levensbericht van Melchior Treub. Jaarboek van de Maatschappij der Nederlandse Letterkunde*. Leiden: Brill.
Maat, Harro. 2001. *Science cultivating practice: A history of agricultural science in the Netherlands and its colonies, 1863–1986*. Dordrecht: Springer.
McCook, Stuart. 2011. Managing monocultures: Coffee, the coffee rust, and the science of working landscapes. In *Knowing global environments: New historical perspectives on the field sciences*, ed. J. Vetter, 87–107. New Brunswick: Rutgers University Press.
Mellegers, Joost. n.d. Public finance of Indonesia 1817–1940. International Institute of Social History website. http://www.iisg.nl/indonesianeconomy/publicfinance.php. Accessed 10 July 2014.
Moon, Suzanne. 2007. *Technology and ethical idealism: A history of development in the Netherlands East Indies*. Leiden: CNWS.
Nyhart, Lynn K. 1995. *Biology takes form: Animal morphology and the German universities, 1800–1900*. Chicago: University of Chicago Press.
Palladino, Paolo, and Michael Worboys. 1993. Science and imperialism. *Isis* 84:91–102.
Pulle, A. A. 1940. *Overzicht van de lotgevallen en de werkzaamheden van de maatschappij in de eerste halve eeuw van haar bestaan*. Leiden: Brill.
Pyenson, Lewis. 1989. *Empire of reason: Exact sciences in Indonesia, 1840–1940*. Leiden: Brill.
Pyenson, Lewis. 1993. Cultural imperialism and exact sciences revisited. *Isis* 84:103–108.
Raby, Megan. 2012. Making biology tropical: American science in the Caribbean, 1898–1963. PhD Diss., University of Wisconsin.
Ramaley, Francis. 1905. The botanical garden at Buitenzorg, Java. *Popular Science Monthly* 67:579–589.
Rip, Arie. 2002. *Co-evolution of science, technology and society. Expert Review for the Bundesministerium Bilding und Forschung's Förderinitiative Politik, Wissenschaft und Gesellschaft* (Science Policy Studies). Enschede: University of Twente.
Roersch Van der Hoogte, Arjo, and Toine Pieters. 2013. From Javanese coca to Java Coca: An exemplary product of Dutch colonial agro-industrialism, 1880–1920. *Technology and Culture* 54 (1): 90–116.
Rooy, Laurens de. 2011. *Snijburcht: Lodewijk Bolk en de bloei van de Nederlandse anatomie*. Amsterdam: Amsterdam University Press.

Solms-Laubach, Hermann Graf zu. 1884. Der botanische Garten zu Buitenzorg. *Botanische Zeitung* 42:753–761; 769–780; 785–791.
Solms-Laubach, Hermann Graf zu. 1911. Melchior Treub. *Zeitschrift für Botanik* 3 (1): iii–vii.
Stokvis, B. J., J. Campert, and C. Kerbert. 1888. *Handelingen van het Eerste Nederlandsch Natuur- en Geneeskundig Congres*. Haarlem: De Erven F. Bohn.
Stoler, Ann L. 2010. *Along the archival grain: Epistemic anxieties and colonial common sense*. Princeton: Princeton University Press.
Storey, William K. 2004. Plants, power and development: Founding the Imperial Department of Agriculture for the West Indies, 1880–1914. In *States of knowledge: The co-production of science and social order*, ed. Sheila Jasanoff, 109–130. London: Routledge.
Theunissen, Bert. 2000. *Nut en nog eens nut: wetenschapsbeelden van Nederlandse natuuronderzoekers, 1800–1900*. Hilversum: Uitgeverij Verloren.
Theunissen, Bert, and M. J. Donath. 1986. De plaats van de morfologie in de Nederlandse zoölogie, 1880–1940: een terreinverkenning. *Gewina* 9:47–67.
Tilley, Helen. 2011. *Africa as a living laboratory: Empire, development, and the problem of scientific knowledge, 1870–1950*. Chicago: University of Chicago Press.
Treub, Melchior. 1873. *Onderzoekingen over de natuur der Lichenen*. Leiden: Van der Hoek.
Treub, Melchior. 1876. *Le méristème primitive de la racine dans les monocotyledons*. Leiden: Brill.
Treub, Melchior. 1879. Eene feest-vergadering. *De Gids* 43:128–156.
Treub, Melchior. 1881. Een tocht naar de bergtuinen van Tjibodas. *De Gids* 45:21.
Treub, Melchior. 1884. Études sur les Lycopodiacées. *Annales du Jardin Botanique de Buitenzorg* 4:107–138.
Treub, Melchior, 1885. *Onderzoeking over sereh-ziek suikerriet gedaan in 's Lands Plantentuin te Buitenzorg*. Batavia: Landsdrukkerij.
Treub, Melchior. 1887a. De openingsrede voor het eerste Nederlandsch Natuur- en Geneeskundig Congres. *De Gids* 51:336–348.
Treub, Melchior. 1887b. Quelques mots sur les effets du parasitisme de l'Héterodera javanica dans les racines de la canne Ã sucre. *Annales du Jardin Botanique de Buitenzorg* 6:93–98.
Treub, Melchior. 1888. Notices sur la nouvelle flore de Krakatau. *Annales du Jardin Botanique de Buitenzorg* 7:213–224.
Treub, Melchior. 1889. *Parasitisme en infectie in het plantenrijk. Voordracht den 5en Juni gehouden in de Vergadering der Maatschappij van Nijverheid en Landbouw in Nederlandsch Indië*. Batavia: Ogilvie & Co.
Treub, Melchior, ed. 1892. *Feestpublicatie ter gelegenheid van het 75-jarig bestaan van 's Lands Plantentuin, 18 mei 1817–18 mei 1892*. Batavia: Landsdrukkerij.
Treub, Melchior, ed. 1893. *Der botanische Garten s Lands Plantentuin' zu Buitenzorg auf Java: Festschrift zur Feier seines 75jährigen bestehens 1817–1892*. Leipzig: Engelmann.
Treub, Melchior. 1899. *Over de taak en den werkkring van 's Lands Plantentuin te Buitenzorg*. Buitenzorg: 's Lands Plantentuin.
Van der Schoor, Wim. 2012. Zuivere en toegepaste wetenschap in de tropen. Biologisch onderzoek aan particuliere proefstations in Nederlands-Indië, 1870–1940. PhD Diss., University of Utrecht.
Vetter, Jeremy. 2011. Rocky Mountain high science: Teaching, research and nature at field stations. In *Knowing global environments: New historical perspectives on the field sciences*, ed. J. Vetter, 108–134. New Brunswick: Rutgers University Press.
Vetter, Jeremy. 2012. Labs in the field? Rocky Mountain biological stations in the early twentieth century. *Journal for the History of Biology* 45:587–611.
Visser, Robert P. W. 1986. De ontwikkeling van de universitair biologische laboratoria. *Gewina* 9:255–265.
Vries, Hugo de. 1904. De oorsprong der bloemen. *De Gids* 68:527–528.
Weber, Andreas. 2012. *Hybrid ambition: Science, governance, and empire in the career of Caspar G. Reinwardt (1773–1854)*. Leiden: Leiden University Press.

Went, F. A. F. C. 1898. Dr. M. Treub als directeur van 's-Lands Plantentuin te Buitenzorg. *De Aarde en haar Volken* 34:93.
Went, F. A. F. C. 1915. *'s Lands Plantentuin te Buitenzorg. Onze koloniën 5*. Baarn: Hollandia-Drukkerij.
Went, F. A. F. C. 1916. *De plicht der vaderlandsche wetenschap jegens tropisch Nederland. Redevoering uitgesproken bij de opening der 143e algemeene vergadering van het Provinciaal Utrechtsch Genootschap van Kunsten en Wetenschappen*. Utrecht: Kemink & Zoon.
Wheeler, William Morton. 1910. *Ants: Their structure, development and behavior*. New York: Columbia University Press.
Wheeler, William Morton. 1924. Ants of Krakatau and other islands in the Sunda Strait. *Treubia* 5:239–258.
Wille, Robert-Jan. 2009. De nationale waarde van ontwikkeling. De patricische embryoloog Ambrosius Hubrecht en de noodzaak van de liberale staat om te investeren in onderwijs en wetenschap. *De Negentiende Eeuw* 33 (1): 34–52.
Williamson, William Crawford. 1896. *Reminiscences of a Yorkshire naturalist*. London: George Redway.
Zangger, Andreas. 2011. *Koloniale Schweiz: Ein Stück Globalgeschichte zwischen Europa und Südostasien (1860–1930)*. Bielefeld: Transcript.
Zeijlstra, H. H. 1959. *Melchior Treub: Pioneer of a new era in the history of the Malay Archipelago*. Amsterdam: Koninklijk Instituut voor de Tropen.
Zevenhuizen, Erik. 2008. *Vast in het spoor van Darwin: Biografie van Hugo de Vries*. Amsterdam: Atlas.
Zimmerman, Andrew. 2006. 'What do you really want in German East Africa, Herr Professor?' Counterinsurgency and the science effect in colonial Tanzania. *Comparative Studies in Society and History* 48:419–461.

Robert-Jan Wille Wille's essay is based on his doctoral dissertation entitled "The Stationists, Laboratory Biology, Imperialism, and the Lobby for National Science Politics," Radboud University Nijmegen, The Netherlands. His work explores the interaction between biology, nationalism, and the (colonial) state between 1870 and 1930, and adds to the growing literature on laboratory studies and the circulation of scientific knowledge and practice, as well as the growing literature on the impact of empire in the national culture and politics of the metropolis itself.

[illegible] 1998. Dr. M. [illegible] [illegible]
[illegible]
[illegible] Black.
[illegible]
[illegible]
[illegible]
[illegible]
[illegible]
[illegible]
[illegible]
[illegible]

[illegible]

Chapter 14
Regionalizing Knowledge: The Ecological Approach of the USDA Office of Dryland Agriculture on the Great Plains

Jeremy Vetter

Introduction

The environmental region, in the past and present, has served as a container for scaling up conclusions beyond the local but without the indefensible pretense of claiming that such knowledge could be considered universal. The region has been used to extend knowledge horizontally, by defining a larger area over which knowledge might be applicable.[1] Poised between the local and particular on the one hand, and the global and universal on the other, the region has been constructed as a middle ground: place based, yet spatially extended. Researching the environment regionally has thus conferred distinct advantages to those who have attempted it, even as such work has necessarily pulled them in both directions.

Field scientists themselves have felt this tension, in part, as a countervailing impulse to understand local places in ways similar to and overlapping with other local inhabitants, often stressing complex interrelationships, microcosmic holism, and irreducible particularity, alongside an equally strong, or possibly more powerful, impulse—through career and disciplinary pressures—to produce knowledge that will be credit worthy and valid within cosmopolitan science, which situates local places in larger systems of knowledge, and aims to draw conclusions that will have broader applicability. Perhaps no scientific discipline has faced this tension more persistently and acutely than ecology.[2] Many valuable studies in the history of

[1] Historians of science have not generally placed much emphasis on regions, but three suggestive examples focusing on regions in different countries include Smith (1987), Naylor (2010), and Phillips (2003).

[2] Worster (1985). In Worster's framing of the problem, though focused on history of ideas more than practices, a related tension between "imperial" (what I would call cosmopolitan scientific)

J. Vetter (✉)
Department of History, University of Arizona, Social Sciences Room 215, 1145 E. South Campus Dr., 85721 Tucson, AZ, USA
e-mail: jvetter@email.arizona.edu

D. Phillips, S. Kingsland (eds.), *New Perspectives on the History of Life Sciences and Agriculture,* Archimedes 40, DOI 10.1007/978-3-319-12185-7_14

ecology have already demonstrated the importance of local place to the production of knowledge, though they differ in the degree of emphasis they place on tensions with lab science or with universalizing scientific ambitions.[3] Moreover, as Sharon Kingsland has pointed out, "the formation of ecology as a discipline" in the USA has also involved "Americans' concurrent discussions about their own land."[4] Understanding local and regional places, for American field scientists, typically also meant debating their practical uses, and this chapter suggests a causal link between the emergence of ecological thinking and such practical questions.

Consider, as a leading family of instances, the vast domain of agriculture. The environmental context of farming has varied considerably from one place to the next, but patterned similarities across geographical space have led scientists and nonscientists alike to consider the region as a proper level for making knowledge claims. In the historical study of life sciences, agriculture, and the environment, it seems pertinent, then, to investigate not only the local production of knowledge but also the "regionalization" of knowledge. The main task of this chapter is to examine a case study of how knowledge was regionalized: the US Department of Agriculture's (USDA) Office of Dryland Agriculture, which was founded in 1905 and directed by agronomist E. C. (Ellery Channing) Chilcott (1859–1930), to generate agricultural knowledge for the Great Plains region of the USA.[5]

Compared to other regions, the deep historical roots of local environmental experience were simply not to be found to the same degree among the newly arriving farmers to the Great Plains, and the opportunity for agricultural scientists to solicit patronage for scientific research was thus all the more palpable. In its efforts to regionalize knowledge about the Great Plains, during the opening decades of the twentieth century, the USDA was responding not only to a tremendous public pressure but also to the desire to prove the value of scientific inquiry for producing reliable knowledge to guide the region's settlement and development. Moreover, the Office was able to draw upon what we might regard as an "ecological" approach to the study of the Great Plains environment. As I shall argue in this chapter, it was this distinctively ecological approach of the USDA's Office of Dryland Agriculture that both undergirded the robust scaling up of knowledge from the local to the regional

views of nature and "arcadian" views of nature drives the entire narrative of the history of ecology. For a recent volume of papers that collectively argue for the continuing relevance of place-based ecological approaches, along with a few historical contributions, see Billick and Price (2010).

[3] Examples include Cittadino (1993); Kingsland (1993); Klingle (1998); Schneider (2000); Way (2006). For the general framework of the lab-field borderlands, see Kohler (2002).

[4] Kingsland (2005, p. 2). She goes on to point out that "the same need to rationalize resource use that supported the conservation movement also supported research in ecology" (p. 4), a connection which is only heightened and amplified by the comparison with agricultural science research that I develop in this chapter.

[5] A basic, laudatory overview, of both the USDA Office and the state experiment stations who collaborated, is provided in Quisenberry (1977). Historiography of the dry-farming movement, more generally, is still dominated by the thorough and comprehensive work of Mary W. M. Hargreaves (1957). For later developments, see Hargreaves (1993). Brief treatments are provided in Hargreaves (1948); Hargreaves (1977).

and, at the same time, generated structural tensions with local collaborators, who were often associated with state agricultural experiment stations.

Farming is an intensely local activity, based on the particularities of place—even if sometimes it has also been more than that—and this has meant that even the most rigorously controlled and systematized scientific research on farming practices confronted its framers with the problem of how to make knowledge that was locally produced be deemed applicable elsewhere. However, for agricultural experiment stations in general, when conducting outdoor work sensitive to local environmental conditions, this typically meant rhetorical maneuvering to rationalize the use of the results to "stand for" a much larger region than the specific local site itself. After all, this was one of the major rationales behind the geographical dispersion of field stations across the country, which was often embedded in arguments for US federal government funding of state-level agricultural research stations (most prominently through the Hatch Act of 1887 and the Adams Act of 1906), as well as by states to justify the establishment of dispersed branch stations throughout their respective territories. But this did not by any means settle the problem; rather, it simply deferred it. How, then, could agricultural scientists affiliated with the USDA Office of Dryland Agriculture design their Great Plains research to produce knowledge that would be applicable to the region as a whole?

Setting Up the Office in Regional Context

The USDA Office of Dryland Agriculture emerged during a time of intense interest in the development of the Great Plains. Within the USDA itself, numerous bureaus and agencies had an interest in producing and circulating knowledge of this semi-arid region, which was bursting with new arrivals and whose settlement was being heavily promoted by state governments, railroads, and other land interests. Much of the initial Great Plains-oriented work was focused around finding and disseminating suitable crop and seed varieties for this environmentally challenging region, much like the USDA's work as a whole. Not only is the Great Plains—which, not incidentally, had recently come to be defined more widely as an environmental region around the very time that the office was founded—a vast land area by any measure, but debate over how the Great Plains region should be settled and developed had also reached a fever pitch by the early twentieth century.

Human action, too, had entered into a frenzied state, as thousands swarmed onto the Plains, more than at any other time in the region's history, to set up dryland farms, during one of the most spectacular land booms in world history. Many of those arriving farmers, either through previous farming experience elsewhere or through their limited experience on the Plains, did come to possess some knowledge of nature that was relevant to Great Plains agriculture. Even Chilcott himself—who, as we shall see, held up an especially high ideal for rigorous agricultural science—recognized this. Before his appointment to the USDA scientific research staff, and while still a professor of geology and agronomy at South Dakota Agricultural College, he called for better field research under semiarid conditions. But he also recognized the value of experiential knowledge. "The very best men in the agricultural colleges and experiment stations," he noted, "frankly admit that many practical farmers who know little or nothing of the methods adopted by scientists, nevertheless, have, by observation and practical experience, gained such a knowledge of the soil and its requirements for crop production that they can, from a simple examination of the soils with which they have become acquainted, make a reliable estimate of their crop-producing capacity."[6]

Within the USDA's Bureau of Plant Industry, the Office of Dry Land Agriculture was spun off as a separate subunit in 1905 from a preexisting agency under M. A. Carleton, which was responsible for "the work of introducing drought-resistant cereals into the semiarid districts of the West," when "it became apparent that investigations in cultivation methods and crop rotations should be carried on in connection with this work."[7] Chilcott himself, even before taking up his position at the USDA, lamented publicly that "nothing is being done in a practical, systematic, scientific way to test the value of crop rotation and the application of manures under ordinary field conditions." In his view, agricultural researchers thus deserved some of the blame for the common "contempt of book farming" often displayed by "practical farmers."[8]

Before exploring further how the Office of Dryland Agriculture defined the Great Plains region and then how it regionalized knowledge using an essentially ecological approach, as well as the structural tensions this generated with their state-level collaborators and others, we must acknowledge the elephant in the middle of the room—or, rather, the Hardy Webster Campbell in the middle of the Great Plains. If you have read anything about the Great Plains dry-farming boom of the early twentieth century, it is likely that the person whose name comes to mind is Campbell, the indefatigable promoter of what was called the "Campbell method" of "scientific agriculture" or "scientific soil culture" on the Great Plains.[9] By acknowledging Campbell's presence, not only can we tell a more complete story but his counterexample can also be used to bring out some of the distinctive features of the office's work. For, if Campbell's invocation of the language of science in his own books, magazines, and lectures was often intensely frustrating to Chilcott and his fellow agricultural scientists, it also provided a convenient foil. And given Campbell's notoriety in the public press, Chilcott and others believed it was essential to engage, sometimes critically, with Campbell's ideas.

[6] Chilcott (1903, p. 446).

[7] Galloway (1906, p. 236).

[8] Chilcott (1903, p. 448).

[9] Publications by or associated with Campbell included books, articles, and periodicals, among which perhaps the most well known was his *Soil culture manual,* which was published in a few different versions in the early twentieth century, e.g., Campbell (1907b). Among the many overviews of Campbell's life, one brief but solid synopsis is Hargreaves (1958). Laudatory popular accounts were also published during the early twentieth-century Great Plains dry-farming boom, including Cowan (1906).

The paradox in comparing Campbell with the USDA Office of Dryland Agriculture under Chilcott is that it confounds the conventional expectation of overlap between the categories of local, particular, and lay, on the one hand, and global, universal, and scientific, on the other. Indeed, it was Campbell, not Chilcott, who propounded a highly generalized, almost universal, set of methods that were supposed to apply to all farming everywhere in semiarid regions (and perhaps even in more humid regions!). Campbell presented—and Chilcott criticized—the "scientific soil culture" as essentially placeless. In Campbell's soil culture manuals, general principles of farming were deduced from model farms located at various points in the region, such as Hill City, Kansas, and Chase County and Holdrege, Nebraska, but without specifying any geographical limits or variations to those conclusions.[10] At times, Campbell even blurred the distinction between "scientific agriculture" on the Great Plains and the best farming practices for more humid regions, contending in 1907, when he appeared as the star figure at the first Dry Farming Congress in Denver, Colorado, that it should not even be called "dry farming" at all, since his system of scientific agriculture "indicates wet farming."[11]

This promotionalism and easy generalization of local results seemed like bad science to Chilcott. "Instead of getting at the facts and steadily extending the boundaries of knowledge," he complained in the 1911 USDA *Yearbook,* "there is a constant tendency to generalize broadly from any available information," a fault that could perhaps be ascribed not just to Campbell. But Chilcott left little doubt as to the primary target of his criticism, when he gave, as his example of "unwarranted statements which have received wide publicity...the one that a new and peculiar system of farming has been discovered or developed, which is of general application to all semiarid localities." In Chilcott's view, the very idea of a uniform "scientific soil culture system" for the Great Plains was wrong. "Agriculture has never been reduced to an exact science," he declared, "even under the most uniform and stable climatic conditions, and even though water is supplied artificially by irrigation. How utterly absurd it is, then, to suppose that dry farming can be reduced to a definite system."[12]

By contrast, Chilcott's own USDA Office argued fiercely for the importance of environmental variation, even within such a large and unusually uniform-appearing region as the Great Plains. The careful attention to variation in soils, temperatures, and even precipitation patterns that lay behind the USDA's work—a markedly ecological sensibility—was the key to the highly complex research system that the office designed and implemented under Chilcott. He and his collaborators insisted on the difficulty of generalizing too widely or quickly, and, in fact, they used this as a rhetorical cudgel for challenging the "Campbell system," whether implicitly or explicitly. Indeed, it was only the impressive array of well-funded research stations, superimposed on an ecologically defined environmental matrix that could, from the perspective of proponents of increased rigor in agricultural science such as Chilcott,

[10] Campbell (1907b, p. 67, 70, 72).

[11] Campbell (1907a, p. 111).

[12] Chilcott (1911, pp. 247–48).

produce reliable knowledge of the Great Plains region. This paradox could be found elsewhere in the history of the field and environmental sciences, but it was especially visible in a region and period in which most of the farmers were newcomers, and most of them unfamiliar with the distinctive environmental conditions of the semiarid Great Plains.

In selecting Chilcott to head up this new research agenda, even before it was made a separate office, the USDA was deliberately choosing someone with regional experience, since Chilcott had been previously serving as a professor at South Dakota Agricultural College. It is true that the college and its associated main agricultural experiment station was (and is) located in Brookings, near the eastern boundary of South Dakota, and thus at some distance from the Great Plains region, as defined by the Office of Dryland Agriculture.[13] Nevertheless, Chilcott could claim, in his own estimation, not only "seven years" in scientific experimentation at the South Dakota station, but also "a residence of twenty years in the semiarid regions," which he regarded as especially important in identifying significant problems—even as he also believed that more rigorous scientific methods would be required to produce reliable conclusions about them.[14] Likewise, the USDA's annual report of 1907 referred to "Mr. Chilcott's long experience, gained from a residence of about twenty-five years in this area, together with his frequent visits to all parts of it," once he began his USDA post.[15]

As these efforts at "regionalizing" Chilcott's experiential knowledge suggest, the USDA meant to go well beyond the borders of a single state. Indeed, the geographical purview of the new office was defined as "investigations in dry land agriculture and the correlation of all cooperative work of the Bureau of Plant Industry in the Great Plains area of the West."[16] This agenda matched up well with the one Chilcott had outlined just a few years earlier, in which he applied his aforementioned desire—for more systematic and scientific production of agronomic knowledge related to crop rotation, soil moisture conservation, and other practices—to the semiarid region of the continent, noting that "the nature of the problems involved" differed substantially "between the humid and the semiarid regions." More specifically, while "in the humid regions the most important object sought is the conservation of the soil fertility or plant food, ... in the arid and semiarid regions it is the conservation of soil moisture."[17] Among the semiarid regions of the western USA,

[13] Interestingly, this also meant that Chilcott's primary "Great Plains" residency was nearly a 100 miles to the east of Aberdeen, South Dakota, which was the place where Campbell gained the "experience…[from which he] formulated the basic procedures of his farming system," though Aberdeen was itself just barely west of the 98th meridian! See Hargreaves (1958, p. 63). As Hargreaves points out, Campbell had arrived in South Dakota in 1879, just in time for the Great Dakota Boom that peaked in the early 1880s (Hargreaves (1958, p. 62), thus suggesting that the dry-farming promoter himself had at least 20 years of practical experience in the region.

[14] Chilcott (1903, p. 450, 452).

[15] Galloway (1907, p. 319).

[16] Galloway (1906, p. 236).

[17] Chilcott (1903, p. 449).

the Office of Dryland Agriculture would focus on one specific and rather large region: the Great Plains.

It is worth noting, at this point, the sheer audacity of what was being proposed. In essence, the office defined a vast region of 330,000 square miles, only sparsely inhabited but, at that moment, being flooded with an unprecedented wave of farming immigrants, and they were proposing to produce and circulate knowledge of this immense domain at newly established local field stations widely dispersed throughout the area. These stations obviously could not be set up at mile-long intervals (or even 100-mile intervals), because the considerable expense of setting up and staffing each local station would limit the number of sites that could be supported and sustained. However, since they were intended to test an array of farming practices over many years using carefully controlled, standardized, and systematically recorded results and a uniform plan over the region, neither a survey approach (with its fleeting occupancy by traveling scientific experts) nor a lay-network approach (with its problematic dependence on untrained local collaborators) was even considered.[18] Ultimately, Chilcott and his collaborators faced the challenge of locating the stations, and subsequently directing their practices on the ground, in such a way that a widely dispersed set of a couple-dozen field stations would produce reliable knowledge that could be extended to the Great Plains region as a whole.

It seems unlikely that a similarly ambitious region-wide, highly systematized knowledge production enterprise had yet been established anywhere in the world at that time. One partial analogue was the US Weather Bureau (itself transferred to the USDA in the 1890s, from its initial supervision by the Army Signal Corps), though its ambitions were circumscribed mostly by the local collection of daily instrumental data, often by lay observers, including temperature, precipitation, and (in some cases during potential flood season) gauged river levels.[19] And certainly, the US national network of state experiment stations collectively constituted a large array of research locations, which often undertook similar experiments, but there was nothing like the scale of carefully coordinated, uniform investigative practices applied to every station across the country. Each state station operated under its own management, and the same research was not undertaken everywhere in a systematically unified fashion. There were some USDA-coordinated efforts—typically, the distribution of seed varieties for testing at a variety of locations thought to be potentially suitable—but no closely coordinated, trans-state, regional system of knowledge production. Within the Great Plains region, however, it was deemed possible for a single agency based on Washington, D.C., to es-

[18] For a discussion of this historical taxonomy of field practice, including lay networks and surveys (discussed here), stations (discussed elsewhere throughout this chapter) and quarries (unimportant in agricultural research but common in extractive disciplines such as paleontology and archaeology), see Vetter (2012).

[19] Some key works on the early history of US long-distance weather networks and the organization of the US Weather Bureau through the early twentieth century include Fleming (1990); Fleming (2000); and Monmonier (1988). For a case study including the Great Plains and focusing on the interactions between lay observers and government forecasters, see Vetter (2011).

tablish and maintain a network covering parts of ten states. The trope of the "natural laboratory" has often been invoked for all manner of field-based research, but, in the case of the Office of Dryland Agriculture, the idea seems to have achieved an impressive degree of regionalized reality: A gently sloping, relatively uniform, rectangle of 330,000 square miles, with one variable (temperature) increasing from north to south, and another variable (rainfall) increasing from west to east. It was like taking a two-dimensional graph plot and finding (or imagining) it existing in environmental reality.

Chilcott and his collaborators were, of course, correct to identify the temperature and rainfall gradients as tremendously important in governing the region's agricultural possibilities. Yet the Great Plains region, however defined, is not uniform—or even uniformly varying. To their credit, they also possessed a keen awareness of other complicating variables, such as soil and topographical differences, and they maintained an active interest in appreciating and accounting for them. A pair of early reports, for example, one from the office and the other from a related USDA division, focused on plant breeding, both derived from work at the Belle Fourche, South Dakota, branch field station, was located in an area of Pierre shale-derived soil "known as 'gumbo'…a heavy, black or gray, clay loam," a soil that posed particular challenges to farming, and thus was mentioned and discussed as such.[20] This peculiar soil was frankly described as "different from those existing in the greater part of the Great Plains region," because, despite its "high capacity for absorbing water," the gumbo soil "takes up water very slowly, so that during very heavy or long-continued rains there is considerable run-off." It thus cracked during prolonged dry periods, desiccating the subsoil and leading to damage due to freezing and tearing the roots of the crop plants.[21] Clearly, such local environmental variations would have to be acknowledged and accounted for. More broadly, Chilcott not only admitted the "great diversity" of Great Plains soils but also tried to deliberately locate the field stations in the region in order to represent different soil types, at least in "that portion of the area which is adapted to dry farming," and with an eye towards combining the results from all the diverse soil types to generate conclusions that "should apply to the Great Plains area as a whole, and those from individual stations, or from groups of stations, to extensive subdivisions of the area."[22]

An Ecological Style of Research

However, with only one or two dozen widely dispersed field stations (eventually), and with such a tremendous pressure to produce knowledge for the thousands of farmers then streaming into the Great Plains region, the most salient variables for specifying the extensibility and limits of the conclusions adduced—particularly

[20] Jensen (1910, p. 8).

[21] Dillman (1910, p. 11).

[22] Chilcott et al. (1915, p. 5).

when they seemed not to be applicable equally over the entire area—were often north–south (temperature) and west–east (rainfall). Ultimately, what drove the geographical design of the office's work was a desire to determine the best agricultural practices not only region-wide but also as they were affected by the region's environmental variability. It is perhaps in this respect that the research program of Chilcott and his collaborators differed most markedly from the contemporaneous approaches and claims of Campbell and other promoters of their own versions of "scientific agriculture." And it is this focus on how environmental variability could be mapped and integrated with local field stations dispersed throughout the Great Plains, as well as how the research at all of the stations was unified and systematized in order to relate that environmental variation to agricultural outcomes, which we might justifiably regard as an "ecological" approach, for its fundamental similarities to scientific practices in the newly emerging field of ecology of this period.

To be sure, it was very rare in the publications of the Office of Dryland Agriculture for Chilcott or his collaborators to relate their work explicitly to ecology or describe themselves as following an ecological approach. Yet such a research program could more easily be envisioned in an era when ecological thinking was becoming increasingly pervasive and even being born as a distinct (sub)discipline. Indeed, this type of thinking, though rarely labeled as "ecological," was probably much more widespread during the opening decades of the twentieth century than we have appreciated, particularly in the practical and applied sciences related to agriculture. It may be that "ecology" as a defined discipline was only the tip of a very big iceberg. Ecological thinking was prevalent in this period not only because of the felt need for field scientists to respond to the challenge posed by the rise of the indoor laboratory, though that was likely a factor, but more specifically to account more fully for how environmental variables were interrelated over space, and how they related to localized practices such as agriculture.[23]

For the Office of Dryland Agriculture, one key indicator of how ecological thinking permeated its work was the involvement of another unit of the USDA, which was at first called the "Physical Laboratory" during the early years of the office but which by 1909 had been renamed the department of "Physical Investigations." Led by L. J. Briggs, this unit's work during the early decades of the twentieth century was often closely tied to the USDA's work in dryland agriculture, starting with the 1907 report (i.e., the second fiscal year after the office was founded). By measuring "the physical factors influencing plant growth in the Great Plains area," Briggs and his collaborators had "for their object the determination of the influence of environmental factors at the different stations upon the growth and yield of the principal crop plants. Such observations when continued for a suitable term of years will show the normal conditions which prevail at representative stations throughout the

[23] On "[e]cologists' embrace of laboratory ideals and practices," see Kohler (2002, p. 86). For the more explicitly articulated face of the early symbiosis of ecology and agricultural science, see Hersey (2011).

area."[24] Notably, the USDA physical investigators had the ultimate authority over the details of this work, to decide the specific research plans that would be followed, not the local scientific staff. As the introductory statement to a 1910 report from the USDA agriculturist stationed in Belle Fourche, South Dakota—one of the first official USDA bulletins to provide data from the Office's work—noted, Briggs's agency had "general supervision over the physical measurements carried out at the dry-land stations" throughout the region.[25]

The recording of environmental variables by the Physical Investigations Department at these sites was designed to be considerably more elaborate than what prevailed at a typical US Weather Bureau instrumental observing site—and also closely integrated with the work of the Office of Dryland Agriculture, which included crop rotation experiments and other studies of agricultural practices. The initial report described the Physical Investigations work as follows: "Continuous automatic records of the temperature of the soil and air are kept at each of these stations. The humidity of the air and the evaporation from a water surface are also determined daily, and weekly measurements of the moisture content of the soil to a depth of 3 feet are made on representative plots of the different rotations."[26] These procedures are remarkably comparable to the emerging ecological research program of Frederic Clements and others during this same period, including precise measurement of physical processes that related different environmental components (soil, air, water, land) to plant growth and devoted increasing attention to local variations in soil moisture and changes over time (dynamics).[27]

The USDA agricultural scientists and the academic ecologists also shared more specific common interest in new ecological variables to study, including soil moisture, evaporation, and progressive plant growth.[28] Briggs, in his role as head of USDA "Physical Investigations," had by 1908 instructed the local representatives of the Office of Dryland Agriculture in the "[s]ystematic measurements" to be "made throughout the growing season," which were intended to "embrace as comprehensive a study of the conditions at the stations as the available time will permit, and include records of the temperature of the soil and air, the humidity of the air, the precipitation, and the evaporation from a water surface." Within this holistic, all-encompassing approach, special attention was focused on two especially elaborate and time-consuming measurements: The 3 foot soil moisture weekly records on the rotational plots, as noted above, and "evaporation from a

---

[24] Galloway (1907, p. 320).

[25] Jensen (1910, p. 3).

[26] Galloway (1907, p. 320).

[27] However, a significant difference was that the Chilcott, Briggs, and their collaborators viewed the key agent of change as human activity—whether in the form of plowing the soil itself or the multi-year crop rotations that were at the heart of the Office's research practice—whereas Clementsian ecologists would ultimately come to focus on processes of ecological succession toward climax in the absence of human disruption. See Clements (1916).

[28] For example, they were very similar to the variables studied by academic ecologists at sites such as the Desert Lab in Tucson, Arizona. On the history of the Desert Lab, see Kingsland (1993); McGinnies (1981); and Bowers (1990).

tank of water 8 feet in diameter, buried so that the surface of the water is level with the ground." For the latter, two years of attempts at evaporation tank measuring had already made it clear that the southern Great Plains experienced double the evaporation of the northern Great Plains "during the six summer months." In addition to the study of ecological variables at each site, USDA researchers also studied the agricultural vegetation itself, by measuring the "growth and composition of wheat and oat plants from week to week until harvested," which was believed to "show effectually the influence of environmental conditions upon the development of the crop, and furnish a basis for determining the controlling factors far more accurately than can be done from an analysis of the total yields at the end of the season."[29]

All of this measurement work, like academic ecology research, involved copious amounts of time, labor, expertise, and instrumental technology. The involvement of the USDA was crucial, not just in providing the political legitimacy to conduct research across such a large region spanning at least ten states but also for its capacity to devote considerably greater resources to the effort than the state governments of the sparsely populated and relatively non-wealthy Great Plains, even with their Hatch and Adams Act appropriations, could ever hope to support. By the end of its first decade, the Office of Dryland Agriculture had a "scientific staff…numbering about 30 men," most of whom it could deploy through the Great Plains region to particular field stations, with a few others to travel around or supervise from afar. And, as Chilcott never seemed to tire of pointing out in his later publications, the aggregate amount of work produced by his USDA was impressive. By 1915, for example, it involved 80 years of observations, using 1900 research plots and a grand total of 91,000 moisture determinations.[30] This equal mention of not only the agricultural plots themselves but also the environmental variables measured suggests that the ambitious "Physical Investigations" research agenda laid out by Briggs shortly after the founding of the Office had indeed become an integral part of its research work. This work had been touted, at the outset, likely by Briggs himself in a report incorporated into the USDA's larger annual report many years earlier, as having "excited the interest and approval of workers generally in the field of dry-farming investigations," when the USDA officers had determined the measuring activities to be "of sufficient importance to justify the entire time of a man at each station where dry-farming investigations are being carried on."[31] Above all, then, the valuable resources of the USDA seem to have been deployed at the field stations scattered throughout the Great Plains as much to measure ecological variables in a coordinated and systematized fashion, as to undertake the agricultural plot trials themselves.

[29] Galloway (1908).

[30] Chilcott et al. (1915, pp. 2–3).

[31] Galloway (1908, p. 318).

Cooperation and Conflict Between State and Federal Scientists

With such a strong desire to coordinate research at the various field stations, it was perhaps inevitable that structural conflict would arise between Chilcott as the central director at USDA headquarters and the state agricultural experiment station directors and researchers, who were accustomed to having some degree of local autonomy in designing their own projects and in determining the allocation of staff time at field stations. Before excavating archival evidence to illuminate some representative conflicts behind the scenes, it may be useful to first sketch out how Chilcott and other USDA leaders both mitigated and exacerbated these structural tensions. On the face of it, much of Chilcott's rhetoric in his published reports seems clearly chosen to preempt (or, in some later cases, perhaps react to) concerns about too much top-down hierarchy, with phrases such as "cooperation" and "cooperative" used routinely in all USDA publications, as indeed was the custom for all such work between USDA and the states. (The archival record reveals that tensions were endemic.) Such rhetoric was echoed by Chilcott's supervisor at the USDA, the head of the Bureau of Plant Industry, who described the USDA and state experiment stations as "cooperating in the agricultural development of the Great Plains."[32] Yet it was also undeniable that it seemed crucial, for the regionalization of knowledge as envisioned by Chilcott and his USDA colleagues, that experiments and observations be as uniform and standardized as possible throughout the Great Plains region. It was the USDA Office that was in the most powerful position, not only as the representative of the national government but also as the agency disbursing funds and personnel across the region, to make final decisions about research project development.

The key means by which Chilcott supervised the regionalization of knowledge production on the Great Plains were by negotiating agreements with the directors of state experiment stations, whose rival authority had to be reckoned with, and through his correspondence guiding his own USDA field agents, whom he besieged with written plans and requests that would standardize the office's work across the entire region and make it uniform enough to merit robust generalization. However, Chilcott foresaw from the very beginning of the office's work that coordination would require more than simply sending written instructions from the USDA headquarters to the states, and awaiting responses. Thus, he also organized face-to-face meetings where ideas—and trust—could flow in multiple directions, and among all the researchers, including those employed by both state and federal governments. To meet this perceived need, he established a "Cooperative Experiment Association of the Great Plains Area…in the fall of 1905 for the purpose of bringing about a closer relation between the officers of the various experiment stations of the Great Plains area and of the United States Department of Agriculture who are engaged in investigations upon the various problems of dry-land ag-

[32] Cooperative Experiment Association of the Great Plains Area (1908, p. 3).

riculture," with Chilcott himself serving as secretary.[33] The first full meeting was in Lincoln, Nebraska, in 1906 (although an organizing meeting had been held the previous November in Washington, D.C.), and the summer meeting location then moved around the region in subsequent years, with recurring winter meetings in the nation's capital. The proceedings of the Cooperative Experiment Association do not seem to be available for all meetings, but an extensive record exists of the second annual meeting in Manhattan, Kansas, where Chilcott reported an official agreed-upon emphasis: "coordination, systematization, and unification of all cooperative experimental work" conducted with the Office of Dryland Agriculture and other USDA agencies.[34]

Likewise, Chilcott's directives to his subordinates in the field, such as W. W. Burr, of North Platte, in western Nebraska, indicated a desire to maintain uniformity and precision not only in the agricultural practices themselves but also in the deployment of field instrumentation intended to measure the environmental variables that were so crucial to what I have been calling Chilcott's ecological approach to agricultural science. He therefore communicated the exact procedures needed for Briggs's physical investigations, especially the evaporation tank. While he allowed Burr, and presumably other local agents, some discretion in choosing the best location for the tank ("have it set in the ground at some suitable place out near the plots"), he admonished Burr to "guard against having it near enough to any obstruction that would prevent the free passage of wind over the tank at all times," he ordered him to enclose it with "a wire fence," and he carefully specified the dates on which measurements should be taken.[35] Of course, Burr may very well have welcomed such definite advice, especially given his likely unfamiliarity as a young agronomist with using evaporation tanks for scientific experiments in this way, as well as his own likely interest in helping to make the measurement of environmental variables as uniform as possible, as a matter of shared commitment to a collaborative research program.[36] In any event, the setup and use of scientific instruments, such as evaporation tanks, to measure environmental variables, does not seem to have generated any evident controversy, at least not in the correspondence available.

---

[33] Galloway (1908, pp. 319–20).

[34] Cooperative Experiment Association of the Great Plains Area (1908, pp. 7–8). The meetings of the organizing meeting in Washington, D.C., in November 1905, seem not to have been published, but a typescript labeled "Minutes of a Meeting Held at the Cosmos Club in Washington, November 15, 1905, for the Purpose Organizing a Cooperative Association for the Great Plains Area" can be found in University Archives, North Dakota State University, Institute for Regional Studies, Fargo, North Dakota Dickinson Experiment Station records (DES) 1/4.

[35] Love Library, University Archives, University of Nebraska, Lincoln, Agricultural Experiment Station records, North Platte Experiment Station 10/14/1 (NPES); E. C. Chilcott to W. W. Burr, 21 April 1906.

[36] The unfamiliarity of the branch field station personnel with the intricacies of evaporation tanks, not just as scientific instruments but more generally, is suggested by another letter indicating the problems posed by the field agent at Highmore, South Dakota, having received a disassembled tank even though the company that manufactured it had promised to "furnish the tanks complete," as noted in NPES; Chilcott to Burr, 14 May 1906.

At many branch stations, a typical arrangement was that a single state-appointed researcher was already present at the site to direct it locally, and the USDA field agent would represent a (typically, much desired) second scientific staff member to work under that branch station head—but also answering to Chilcott. Early on, state directors seem to have recognized and accepted Chilcott's authority to appoint personnel to serve as field agents, though often with the advice of the states. For instance, the director of the main North Dakota station wrote to his branch station director at Dickinson in western North Dakota that it was Chilcott's decision to make the appointment, but his "opinion" was that "if you have a good man to recommend,…he will consider your nominee favorably."[37] Yet in Chilcott's reply, after giving some very specific advice about how to mark field plots properly, he pushed hard to make sure that the nominee would meet his standards, saying that he "sincerely hope[d]" the North Dakotan recommendation was worthy. "Although these Special Agents will be entirely subordinate to the Superintendents of the substations," he reassured them, "we desire that they have sufficient training and education, so that their experience as special agents will fit them for larger and more important positions." If the nominee met Chilcott's standard of "well trained" and possessed the "necessary qualifications," then the USDA officer would "be glad to appoint him as soon as the work justifies it."[38] Such a hedged and carefully worded reply seems clearly to have been calculated to affirm the suggestions of the North Dakotans, but not at the expense of his own high standards.

The power relations were usually quite evident to the state agricultural scientists, and one North Dakotan noted privately to another that "where Mr. Chilcott furnishes us with certain money it becomes a delicate matter to dictate too largely to whom it shall go for services rendered." If they were dissatisfied with Chilcott's choice, they could "of course decline any assistance from his Department," but that was highly undesirable (and unlikely), and in their view, "that $1500," which was the salary of the additional field agent at the branch station, "is worth accepting thankfully."[39] The leverage that Chilcott possessed to bend state scientists to his terms, owing to the funding resources he controlled, meant that his plans for the regionalization of knowledge production on the Great Plains extended not just to personnel selection matters but also to the plans and specifications for the research. And collaborators at the state level seemed to tacitly understand, and even accept, this structural reality.

Where disagreement did erupt more openly, however, was typically in the allocation of the time of the field agents and in the competing lines of authority between the state stations and the USDA Office. Thus, when Chilcott wrote to the branch station scientist at Dickinson, he acknowledged that the local researcher's suggested plan for a research plot arrangement with only 67 plots was "a good idea," but at the same time he insisted on adopting his own system of 100 plots based on what was being done by "nearly all the other sub-stations" and asking that he "keep this

[37] DES 1/6; J. H. Worst to L. R. Waldron, 19 July 1906.

[38] DES 1/6; Chilcott to Waldron, 26 July 1906.

[39] DES 1/6; Worst to Waldron, 27 July 1906.

amount of land available...[including] sufficient land of uniform character."[40] Chilcott's demands were, no doubt, time consuming for his field agents as well, leaving them little time for additional work prescribed by the branch station and statewide directors. In late August of 1906, for example, Chilcott wrote to his agent at North Platte, Nebraska, giving very detailed instructions on the correcting of field plot diagrams, the keeping of proper records, various observations and data that needed to be reported, and regular reporting forms that would need to be filled out. The recurring use of the words "data" and "facts" in the instructions relates quite clearly to Chilcott's intention to bring together information collected from all the stations across the Great Plains region, in as uniform and consistent a fashion as possible. "We have taken some pains to work this system of note-taking," Chilcott noted, urging the submission of such data and facts to the USDA as the most important task for the field agents to accomplish.[41]

Yet sometimes the station directors, or their subordinates in charge of the branch stations, had other aspirations. The heavy workload of data and fact gathering, along with the precise monitoring of the agricultural field plots, were themselves time consuming activities, and they left precious little room for additional work prescribed at the state level. Thus, when Chilcott asked for his field agent in western Nebraska to undertake further "soil work" related to the measurement of the environmental variables that Briggs had specified as part of the USDA's physical investigations, Nebraska's experiment station director protested the directive. He requested that Nebraska's own soil specialist, along with the branch station director in North Platte, W. P. Snyder, "cooperate in the development of these plans." And he remonstrated, in particular, against the increased workload that would be involved, as well as the abrogation of what he saw as the proper line of authority. "Our agreement," he reminded Chilcott, "was that Mr. Burr should devote what time was necessary under Superintendent Snyder's instruction to the [USDA] experimental plats and the balance of his time should be given to our regular experimental work. ...If we should now double the amount of work which Mr. Burr is to do for you, he would have little or no time to assist us in other lines of work, and a new basis of agreement would be necessary."[42] To Snyder, he wrote more frankly and bitterly, questioning whether the collaboration "is a cooperative experiment as contemplated, or whether Professor Chilcott is assuming entire direction of the work."[43] The Nebraskans, like the North Dakotans, could see the emerging structure of power relations involved in the "cooperative" work,

[40] DES 1/6; Chilcott to Waldron, 6 August 1906.

[41] NPES 9; Chilcott to Burr, 29 August 1906.

[42] NPES 9; E. A. Burnett to Chilcott, 15 October 1905. At nearly the same time, as Nebraska station director, Burnett was also complaining directly to the USDA Bureau of Plant Industry chief, who was also Chilcott's supervisor, about collaboration requests that he viewed as too indefinite from another agency within the Bureau—though, interestingly, Chilcott was invoked there as a positive model, due to his more definite lines of communication about "details and plans of the work," where were deemed "satisfactory" as involving "mutual consideration," as noted in NPES 9; Burnett to B. T. Galloway, 9 October 1906.

[43] NPES 9; Burnett to Snyder, 31 October 1906.

but they also seemed to recognize that there was little they could do to change it, if they wanted to participate in this exciting and relatively well-funded regional Great Plains project at all.

Indeed, already by September of 1907, Chilcott and Briggs had together laid down the law, so to speak, by issuing a stern and uncompromising circular to all their regional collaborators. This document, littered with underlined imperatives and accusations, complained of rare, but deplorable, "gross violations of the spirit of the cooperative agreements." They proclaimed the "sole object" of choosing each field agent spread throughout the region to be finding "a man who is competent *to attend to all the technical details of the cooperative work and who does not have any other duties or responsibilities that in any way interfere with his giving to the cooperative work all the time and attention that it requires to obtain the very best results*." They objected, in particular, to the field agents being instructed to do other tasks that interfered with this work, including being "asked to perform common manual labor, such as team work," and stressing instead that "close, uninterrupted, personal attention" to the USDA's work would be "absolutely essential," as well as decrying that "such technical work as taking meteorological and physical observations" and "field notes" was being left to other staff members. This "substitution of observers except when absolutely unavoidable" was forbidden, since a "personal equation" was involved in the taking of measurements. "The success of the cooperative work is absolutely dependent," Chilcott and Briggs declared, upon having all the work done at *exactly the right time* and in *exactly the right manner*." Judging from the emphatic tone, and the specific tasks referenced, it seems fair to conclude that the broadly ecological approach followed by the USDA only heightened the potential for structural conflict and misunderstanding, given the imperative need for uniform methods of measurement and data collection for environmental variables such as soil moisture and evaporation in the field.

There is much more that could be said about the ongoing structural conflicts in this Great Plains project. To the extent that there was disagreement, or even occasional controversy, it often revolved around the resentment of state officials that they had so little control over the research design and the time allocation of the field agents. In Nebraska, for example, they were still complaining about this several years later, with the branch station head writing to his supervisor in Lincoln: "If Professor Chilcott's work is done properly, I do not see that it is any concern to him how it is accomplished." This same letter raised another issue about time allocation for field agents—a concern that sometimes surfaced in other correspondence—which was resistance to Chilcott's demand that all the field agents spend part of the winter season in Washington, D.C., with him, instead of in the states where they were based.[44] While Chilcott rhetorically invoked the importance of state authority, and did sometimes try to uphold that professed commitment when possible, in reality, a regionalized system of knowledge production, if it was to be uniform and coherent enough to merit the high standards of rigor that Chilcott advocated, seemed

[44] NPES 1; Snyder to Burnett, 9 March 1911. For another example expressing a similar sentiment, see NPES 1; Snyder to Burnett, 18 September 1911.

to require top-down directives from Washington, D.C., which would then have to be followed by each of the branch stations scattered across the Great Plains region, whichever state they happened to be in.

The disputes between Chilcott and his state collaborators also sometimes involved intellectual matters and the broader message of agricultural science for the dry-farming movement and the development of the Great Plains. Some of the most interesting controversies in this regard had erupted by the 1910s, when the USDA became widely known for advocating a "conservative" position against many of the practices associated with the popular dry-farming movement, such as summer tillage, or even the movement's tendency toward promotionalism and boosterism, which the USDA Office under Chilcott was especially cool about. (Following the dry-farming bust of the late 1910s, in which the specter of "broken fortunes, deserted farms, and ruined homes," which had been raised hypothetically in a USDA report, only a few years earlier during better times,[45] became an awful reality after a prolonged multiyear drought that devastated the northern Great Plains especially, Chilcott's attitude would be judged as "fully vindicated by the experiences of this severe extensive, and protracted drought," and as the "sane, conservative, and consistent attitude"!)[46] While many of the state experiment stations were mixed in their views, it is fair to say that most were somewhat less conservative than Chilcott, particularly in states such as Montana, which had close ties to railroad companies as research patrons, as well as exceptionally strong dry-farming booms during this period.

A full accounting of these disputes lies beyond the scope of this chapter, but it is perhaps worth noting that the divergent positioning of the USDA and state agricultural experiment stations on these issues follows quite predictably from their differing relationships to locality and region. If the state agricultural scientists were, as could be expected, more directly responsive—even if still somewhat cautious, compared to promoters like Campbell—to the boosterish agendas of their states, the USDA had a nationwide constituency, which included not only the Great Plains states themselves but also eastern (and even Midwestern) states whose farmers were in competition with the commodity grain produced in the newly opened land of the dry-farming frontier. Moreover, the closer proximity of the state agricultural scientists to their agricultural development-oriented supporters in state legislatures also likely shaped their worldviews, especially given the desperate search for status, patronage, and (ultimately) scarce funding. Finally, given the overall themes of this chapter, it is worth emphasizing the regional and ecological perspective of the USDA Office under Chilcott, which itself tipped the balance towards more nuanced conclusions about Great Plains agriculture, in which developmental possibilities are circumscribed more precisely by environmental variables.

[45] Taylor (1915, p. 150).

[46] Taylor (1920, pp. 186–87).

Conclusion

In the end, Chilcott's vision for rigorous regionalization of knowledge led to an emphatic emphasis on variability and diversity within the Great Plains environment, even as his office produced that knowledge through a ruthlessly uniform set of practices, carried out over a period of many years, and requiring a standardized research design imposed on a growing but still widely dispersed set of field stations across nearly a dozen states throughout the vast region. In its own reports published starting around 1915, by which time nearly a decade's worth of data were available from many field stations—Chilcott was reluctant to draw conclusions any earlier—this variability had been extended from geographical variability to include temporal variability as well. In Chilcott's view, to say that the Great Plains was "semiarid" was to mask a more complex ecological reality: "One season may have almost humid and another almost arid conditions."[47] In a sense, then, regionalizing knowledge on the Great Plains ultimately produced an ecological vision that, ironically, deconstructed the region as a stable object of analysis.

References

Billick, Ian, and Mary V. Price, eds. 2010. *The ecology of place: Contributions of place-based research to ecological understanding*. Chicago: University of Chicago Press.

Bowers, Janice E. 1990. A debt to the future: Scientific achievements of the Desert Laboratory, Tumamoc Hill, Tucson, Arizona. *Desert Plants* 10:9–12, 35–47.

Campbell, H. W. 1907a. Address delivered at Dry Farming Congress, Denver, Colorado, January 1907. *Proceedings of the Trans-Missouri Dry Farming Congress*, 111–122. Denver: Denver Chamber of Commerce.

Campbell, H. W. 1907b. *Campbell's 1907 soil culture manual: A complete guide to scientific agriculture as adapted to the semi-arid regions*. Lincoln: Campbell Soil Culture Co.

Chilcott, E. C. 1903. Some soil problems for practical farmers. *Yearbook of the United States Department of Agriculture*, 441–452. Washington: Government Printing Office.

Chilcott, E. C. 1911. Some misconceptions concerning dry farming. *Yearbook of the United States Department of Agriculture*, 247–256. Washington: Government Printing Office.

Chilcott, E. C., J. S. Cole, and W. W. Burr. 1915. *Crop production in the Great Plains Area: Relation of cultural methods to yields*. USDA Bulletin No. 268. Washington: Government Printing Office.

Cittadino, Eugene. 1993. A 'marvelous cosmopolitan preserve': The dunes, Chicago, and the dynamic ecology of Henry Cowles. *Perspectives on Science* 1:520–559.

Clements, Frederic E. 1916. *Plant succession: An analysis of the development of vegetation*. Washington: Carnegie Institution of Washington.

Cooperative Experiment Association of the Great Plains Area. 1908. *Dry-land agriculture: Papers read at the second annual meeting of the Cooperative Experiment Association of the Great Plains Area*, Held at Manhattan, Kanas, 26–27 June, 1907. USDA, Bureau of Plant Industry, Bulletin No. 130. Washington: Government Printing Office.

Cowan, John L. 1906. Dry farming-The hope of the West. *Century Magazine* 50:435–446.

[47] Chilcott et al. (1915, p. 3).

Dillman, Arthur C. 1910. *Breeding drought-resistant plants for the Great Plains Area*. USDA Bureau of Plant Industry, Bulletin No. 196. Washington: Government Printing Office.
Fleming, James Rodger. 1990. *Meteorology in America, 1800–1870*. Baltimore: Johns Hopkins University Press.
Fleming, James Rodger. 2000. Storms, strikes, and surveillance: The U.S. Army Signal Office, 1861–1891. *Historical Studies in the Physical and Biological Sciences* 30:315–332.
Galloway, B. T. 1906. Report of the Chief of the Bureau of Plant Industry. *Annual Reports of the Department of Agriculture*, 175–266. Washington: Government Printing Office.
Galloway, B. T. 1907. Report of the Chief of the Bureau of Plant Industry. *Annual reports of the Department of Agriculture*, 257–341. Washington: Government Printing Office.
Galloway, B. T. 1908. Report of the Chief of the Bureau of Plant Industry. *Annual reports of the Department of Agriculture*, 283–407. Washington: Government Printing Office.
Hargreaves, Mary W. M. 1948. Dry farming alias scientific farming. *Agricultural History* 22:39–56.
Hargreaves, Mary W. M. 1957. *Dry farming in the northern Great Plains, 1900–1925*. Cambridge: Harvard University Press.
Hargreaves, Mary W. M. 1958. Hardy Webster Campbell (1850–1937). *Agricultural History* 32:62–65.
Hargreaves, Mary W. M. 1977. The dry-farming movement in retrospect. *Agricultural History* 51:149–65.
Hargreaves, Mary W. M. 1993. *Dry farming in the northern Great Plains: Years of readjustment, 1920–1990*. Lawrence: University Press of Kansas.
Hersey, Mark. 2011. 'What we need is a crop ecologist': Ecology and agricultural science in Progressive-Era America. *Agricultural History* 85:297–321.
Jensen, C. A. 1910. *Seasonal nitrification as influenced by crops and tillage*. USDA Bureau of Plant Industry, Bulletin No. 173. Washington: Government Printing Office.
Kingsland, Sharon E. 1993. An elusive science: Ecological enterprise in the southwestern United States. In *Science and nature: Essays in the history of the environmental sciences*, ed. Michael Shortland, 151–179. Oxford: British Society for the History of Science.
Kingsland, Sharon. 2005. *The evolution of American ecology, 1890–2000*. Baltimore: Johns Hopkins University Press.
Klingle, Matthew W. 1998. Plying atomic waters: Lauren Donaldson and the 'Fern Lake concept' of fisheries management. *Journal of the History of Biology* 31:1–32.
Kohler, Robert E. 2002. *Landscapes and labscapes: Exploring the lab-field border in biology*. Chicago: University of Chicago Press.
McGinnies, William G. 1981. *Discovering the desert: Legacy of the Carnegie Desert Botanical Laboratory*. Tucson: University of Arizona Press.
Monmonier, Mark. 1988. Telegraphy, iconography, and the weather map: Cartographic weather reports by the United States Weather Bureau, 1870–1935. *Imago Mundi* 40:15–31.
Naylor, Simon. 2010. *Regionalizing science: Placing Knowledges in Victorian England*. London: Pickering and Chatto.
Phillips, Denise. 2003. Friends of nature: Urban sociability and regional natural history in Dresden, 1800–1850. *Osiris* 18:43–59.
Quisenberry, Karl. 1977. The dry land stations: Their mission and their men. *Agricultural History* 51:218–228.
Schneider, Daniel W. 2000. Local knowledge, environmental politics, and the founding of ecology in the United States: Stephen Forbes and 'The lake as microcosm' (1887). *Isis* 91:681–705.
Smith, Michael. 1987. *Pacific visions: California scientists and the environment, 1850–1915*. New Haven: Yale University Press.
Taylor, W. A. 1915. Report of the Chief of the Bureau of Plant Industry. *Annual Reports of the Department of Agriculture*, 143–158. Washington: Government Printing Office.
Taylor, W. A. 1920. Report of the Chief of the Bureau of Plant Industry. *Annual Reports of the Department of Agriculture*, 159–220. Washington: Government Printing Office.

Vetter, Jeremy. 2011. Lay observers, telegraph lines, and Kansas weather: The field network as a mode of knowledge production. *Science in Context* 24:259–280.
Vetter, Jeremy. 2012. Field life in the American West: Surveys, networks, stations, and quarries. In *Scientists and scholars in the field: Studies in the history of fieldwork and expeditions*, ed. Kristian H. Nielsen, Michael Harbsmeier, and Christopher J. Ries, 225–258. Aarhus: Aarhus University Press.
Way, Albert G. 2006. Burned to be wild: Herbert Stoddard and the roots of ecological conservation in the southern longleaf pine forest. *Environmental History* 11:500–526.
Worster, Donald. 1985. *Nature's economy: A history of ecological ideas, 2nd ed.* Cambridge: Cambridge University Press.

Jeremy Vetter is an assistant professor in the History Department at the University of Arizona, USA. His research is at the intersection of environmental history and the history of science and technology in the American West. His current projects focus on the Great Plains and Rocky Mountains between 1860 and 1920, including the history of scientific practices in the field and environmental sciences, and the relationship between the field sciences, economic development, and environmental transformation. He is also interested in agriculture and the history of the food system, the involvement of lay people in the field sciences, and the history of science in the national parks.

Chapter 15
Rexford F. Daubenmire and the Ecology of Place: Applied Ecology in the Mid-Twentieth-Century American West

Adam M. Sowards

Introduction

In 1952, Rexford F. Daubenmire, a botany professor at the State College of Washington in Pullman, concluded his *Ecological Monographs* article with this paragraph:

> The ideal management of forest lands involves a balanced consideration of their value in timber production, grazing capacity, wildlife production, and watershed protection. On account of the complexity of the problem and the fact that changing demands will undoubtedly call for frequent modifications of plans, it is difficult to see how such a multiple use policy can become effective until the fundamental potentialities of the major ecosystems are understood by all who are charged with the responsibility of planning land management.[1]

In two sentences, he captured much of his life's work, even though he was not quite at his career's midpoint. In sum, humans required much from natural communities; their demands would inevitably evolve and shift, so policies would fail unless managers could determine an ecosystem's ultimate potential. In the Pacific Northwest's interior grasslands and forests, Daubenmire spent more than four decades closely studying natural places to discern how ecosystems functioned and adjusting his ecological theories to fit what he found. With such knowledge, hard-won through meticulous fieldwork, Daubenmire's science could inform land management and reduce the likelihood of costly failures for farmers and foresters, ranchers and range managers.

Daubenmire worked as an ecologist from the 1930s through the 1970s, a time when plant ecology in the USA matured and evolved from its founding generation's

[1] Daubenmire (1952, p. 327).

A. M. Sowards (✉)
Department of History, University of Idaho, 875 Perimeter Drive, MS3175, Moscow, ID 83844-3175, USA
e-mail: asowards@uidaho.edu

D. Phillips, S. Kingsland (eds.), *New Perspectives on the History of Life Sciences and Agriculture*, Archimedes 40, DOI 10.1007/978-3-319-12185-7_15

roots.[2] Daubenmire serves as a window through which to examine how the discipline asked and answered questions and debated certain central concepts. Ecologists of his generation inherited dominant ideas that guided the field, but they did not accept all ideas uncritically or use them identically. Analyzing Daubenmire's research program and contemporary scientific debates show the ways a new generation of scholars accepted and extended, challenged and rejected, their teachers' ideas. Between the founders at the turn of the twentieth century and later ecologists who helped inspire the environmental movement of the 1960s and 1970s, the group Daubenmire represents has received comparatively little scholarly attention. This work, then, helps flesh out the history of ecology in that era. Recently, in *Measuring Plant Diversity,* ecologist Thomas J. Stohlgren claimed that "Daubenmire epitomized the science of vegetation ecology in the 1960s (and for many plant geographers and ecologists today)."[3] Daubenmire thus serves as an effective case study for a generation (or more) of applied ecologists solving problems related to land use in rural places.

The place where Daubenmire devoted his professional life also underwent important transformations while he worked there, changes shaped by economic activities and guided by science. Daubenmire focused most of his attention on the Columbia Plateau, the area between the Cascade Mountains and the Rocky Mountains that encompasses forested mountains and foothills, as well as an open plain that includes grazing rangeland, irrigated cropland, and dryland farmscapes. Hardly a pristine landscape in the 1930s when Daubenmire arrived, the region was poised for greater ecological disturbances with new agricultural possibilities and expanding timber production because of midcentury population growth and technological innovations. To best achieve these goals, conservationists promised that scientifically informed management would reduce wasteful, inefficient resource use.[4] Ecologists such as Daubenmire would provide the necessary understanding of nature to guide agricultural practices and resource development. Doing so required that Daubenmire and others like him reckon with past ecological disturbance to understand environmental impacts on regional landscapes and understand how an area's natural components fitted together so as to be able to predict the land's responses to various management possibilities. In looking at the land in the present, then, ecologists both looked backward and forward in time, accounting for change and forecasting the future. Daubenmire made this very point in 1953, writing, "a given condition of vegetation allows extrapolation into the past as well as prediction into the future."[5] His numerous publications offered relevant data and explanations useful to rural

[2] Overviews of ecology's founding and development include Bowler (1992); Golley (1993); Hagen (1992); Kingsland (2005), Tobey (1981); Worster (1994). Also, Real and Brown (1991).

[3] Stohlgren (2007, p. 34).

[4] The classic statement is Hays (1959). Although Hays' focus extended only to 1920, these priorities remained important much longer. For how these ideas evolved into the New Deal Era, see Fox (1981), esp. pp. 183–217; Maher (2008).

[5] Daubenmire (1953, p. 17).

land managers and users, especially his innovative approach to predicting a given habitat's *potential* vegetation.

However, Daubenmire was not just a lone scientist conducting case studies in an out-of-the-way part of the North American West. He trained with reputed ecologists and engaged widely in the profession's intellectual debates and institutions. Indeed, he led ecology's leading organization (the Ecological Society of America, ESA), earned national awards and recognition, and published textbooks on the fundamentals of plant ecology and geography for botany and ecology students.[6] Although not a widely recognizable name today in the history of ecology, Daubenmire was a substantial intellectual presence and can shine a light on central ideas and problems in ecology in the mid-twentieth century.

The approach taken here—studying one scientist in one place—helps develop the scholarly discussion of ecology of place. "Ecology of place," as characterized recently by scientists Ian Billick and Mary V. Price, describes a research approach that "pursues general understanding through…detailed understanding of a particular place."[7] Daubenmire illustrates this approach, for he mostly worked in the inland Northwest on both small-scale and landscape-scale research but consistently kept in mind larger questions about how vegetation units anywhere assembled and functioned and what factors affected them. He worked at intersections—the local and universal, the basic and applied—in understanding how his place fit within larger scientific and environmental frameworks. To interrogate the life sciences, agriculture, and the environment, we can investigate those who operated in the field trying to make sense of that very nexus. Daubenmire is an exemplar.

Vegetational Units in Early Ecological Theory

Ecologists inherited from biogeographers such as Alexander von Humboldt an interest in understanding how and why species were distributed across and interacted with the landscape.[8] They surveyed and mapped regions around the globe in ever-increasing detail with different methods and preferences developing over the course

[6] A brief biography that highlights Daubenmire's professional achievements is Hoffman (1996). The textbooks are Daubenmire (1947), *Plants and Environment;* Daubenmire (1959b), *Plants and Environment;* Daubenmire (1968b), *Plant Communities;* Daubenmire (1974), *Plants and Environment;* Daubenmire (1978), *Plant Geography.*

[7] Billick and Price (2010, p. 4). This approach also resembles Jeremy Vetter's discussion of field scientists "scaling up" their work from their local field sites to the regional scale (or even beyond) to reach broader claims of knowledge. See Vetter (2011) in his introduction, esp. pp. 2–3, as well as Chap. 14 in this volume.

[8] On Humboldt's work as a precursor to ecology, see Bowler (1992, pp. 205–208); Nicolson (1987); Worster (1994, pp. 133–137). Plant geography is discussed in various contexts in this volume in Chaps. 2 (Phillips), 3 (Güttler), 5 (Horan), and 16 (Lavelle).

of the nineteenth century and beyond.[9] In North America, and to a lesser extent in Great Britain, ecologists at the start of the twentieth century conceived of nature in discrete communities, forming through a process they eventually called succession.[10] After a disturbance like a fire or a landscape change like a receding glacier, new plant species would colonize an area, followed by another suite of species, and then another. Thus, nature was not static. As the theory's cofounder Henry Chandler Cowles of the University of Chicago put it in 1899, "Ecology is, therefore, a study in dynamics."[11]

Scientists differed in their theories about what caused succession and where it led. Some, like Cowles who had a background in geology and geography, saw the physical world as too dynamic to ever produce a stable array of species.[12] Succession was real enough, for he observed successive plant types as he walked from the sands along Lake Michigan through grasses and then into shrubs and inland toward forests.[13] Others, like Frederic E. Clements, a botanist from Nebraska, believed succession with any given locale's climate would lead to a single plant community that existed in self-replicating equilibrium, provided no disturbance or human interference disrupted the natural order of things. So coherent was this climax or monoclimax, he described it as an organism, or super-organism: "The unit of vegetation, the climax formation, is an organic entity. As an organism, the formation arises, grows, matures, and dies."[14]

The stability and predictability implied by Clements' vision failed to convince other ecologists, who proposed less deterministic alternatives. Rather than seeing succession as proceeding along community lines in an inherently progressive fashion, American ecologist Henry A. Gleason saw it driven by individual plants and plant species subject to unique environmental conditions and migration dynamics. As he concluded an influential 1926 article, "[I]t may be said that every species of plant is a law unto itself, the distribution of which in space depends upon its individual peculiarities of migration and environmental requirements.... A rigid definition of the scope or extent of the association is impossible, and a logical classification

[9] Ecologists have accounted for how scientists classified communities variously across the globe in Kendeigh (1954); Whittaker (1962).

[10] Robert Kohler contextualized American plant ecologists' classification activities from the 1890s to the 1930s within biology's other classification practices; see Kohler (2008).

[11] Cowles (1899, p. 95).

[12] For Cowles, see Cittadino (1993). Rumore (2009, pp. 84–86) describes Cowles' educational background.

[13] Cowles published his influential study in four successive 1899 journal issues: Cowles (1899).

[14] Clements' classic statement is in Clements (1916), quoted on p. 124. Daubenmire's characterization of Clements' theory is clear and helpful. In *Plant Communities,* he wrote: Clements hypothesized "that within a given area all differences among habitats due to soil and topography are eliminated with the passing of time, so that all the area is ultimately taken over by the same climax association, the nature of which reflects primarily the climate. His *monoclimax hypothesis,* as it later came to be known, therefore demanded that every piece of vegetation in a landscape be fitted into one or more seres, all of which converge in a common climax." Daubenmire (1968b, p. 240; original emphasis). Seres are transitory states of vegetation prior to reaching the climax state.

of associations into larger groups, or into successional series, has not yet been achieved."[15] If plant species had not assembled in a recognizable pattern but rather in a statistically random distribution caused by chance, then it made little sense to speak of plant communities at all. And if that were the case, little prediction was possible and efforts to control or improve nature would be difficult if not impossible for ecologists to recommend.[16] In the mid-twentieth century, ecologists centered at the University of Wisconsin-Madison picked up and extended this individualistic critique in a way that struck at the heart of Daubenmire's work (see below).[17]

Another alternative concept was the ecosystem. In a significant 1935 article in *Ecology,* Oxford botanist Arthur G. Tansley analyzed and criticized the "use and abuse" of various ecological terms and ideas within the Clementsian tradition but also proposed a novel way of thinking about the environment. To a greater extent than his predecessors, Tansley coupled the physical and biological: "Though the organisms may claim our primary interest, when we are trying to think fundamentally we cannot separate them from their special environment, with which they form one physical system." He continued: "It is the systems so formed which, from the point of view of the ecologist, are the basic units of nature on the face of the earth." He called these "basic units" ecosystems, a more diverse and integrated unit than the monoclimax. Qualifying Clementsian perspectives, Tansley allowed that ecosystems might organize into stable states determined by factors other than climate. Sometimes called polyclimax theory, it recognized that at times soil, topography, fire, or grazing created and maintained ecosystems in relative equilibrium, respectively known as edaphic, physiographic, fire, or biotic climaxes.[18] A critical distinction was that Tansley's system could incorporate disturbances and human activity as part of an ecological system, whereas Clements' approach excluded humans and saw disturbances as setbacks on a community's progress toward climax. It took ecologists two decades before the ecosystem became a widely adopted concept with methods devised to study the integrated system Tansley described. Nevertheless, Daubenmire and others such as his mentor William Skinner Cooper at the University of Minnesota saw this version of the natural world with biotic and abiotic factors "braided" together to more adequately represent complex nature.[19]

[15] Gleason (1926), quoted on p. 26; and Gleason (1939), where he is clearer and more insistent on his view's incompatibility with community ecology. Commentary is in Nicolson and McIntosh (2002).

[16] Kingsland (2005) also emphasizes how Gleason's concepts undermined ecologists' abilities to predict and thus be socially useful in *Evolution of American Ecology,* p. 160.

[17] On the resurgence of Gleason's influence in the 1950s, see Barbour (1995). Rumore has challenged the notion that there was as sharp a divergence as Barbour describes, because Barbour (and others) overemphasized the dominance of Clements. See Rumore (2009, pp. 10–11).

[18] Tansley (1935, p. 299).

[19] Daubenmire adopted "ecosystem" early and employed it throughout his career. His first use was in Daubenmire and Colwell (1942, p. 32). Cooper (1926, p. 397) famously compared dynamic vegetation communities to a "braided stream." Rumore (2009) has examined this effectively in "A natural laboratory."

Daubenmire immersed himself in these larger intellectual issues, and they shaped his scientific worldview and practice. Recognizing how these debates unfolded around him contextualizes his Northwest ecological fieldwork. Furthermore, it warrants emphasizing that his careful and data-rich approach followed that of Cooper.[20] In her study of Cooper and his work at Glacier Bay, Alaska, historian Gina Rumore characterized him as a careful ecologist working tirelessly to match theory with data. By contrast, Clements' organism framework and his notion of progressive change toward climax were dogmatic, much like the man himself.[21] Instead, Cooper practiced and instilled in his students care with ecological terms, avoidance of teleology, and careful tests of theories with field data. Cooper and his students still sought natural laws (unlike Gleason) to explain the constancy of ecological change, but their approaches closely integrated data with theory and readjusted them when data required (unlike Clements) and incorporated multicausal explanations and models (like Tansley) that changed over time.[22] Daubenmire seldom cited Cooper's influence, yet his undogmatic approach searching for underlying causes bears Cooper's intellectual imprint. For example, Daubenmire's dissertation study of Minnesota's Big Woods sought to understand the structure and physical limits to the biological community, incorporating climate, soils, and fire—this final factor being an innovative factor he later would develop further.[23] With this solid academic mooring, Daubenmire took his ecological practice west to the University of Idaho.

Besides the various personalities and schools of thought that have been recounted in the history of ecology, a central crux to the scientific debate at this time was whether plants existed in discrete objective units that could be described scientifically and what impelled their changes over time. For a half century and more, ecologists debated these tenets. Plant communities were real entities that could be delineated scientifically. Or not. Climax communities were homogenous states determined by climate in the absence of interference. Or not. These positions had practical and philosophical consequences. If real, plant communities could be classified scientifically—that is, objectively, or quantitatively. If not, they were merely

[20] Daubenmire studied with Stanley Cain as an undergraduate at Butler University. Cain worked at the University of Chicago at the same time Cowles taught there, although Cowles was not Cain's supervisor. Later, Cain was an assistant secretary of the Department of the Interior. Thomas (1995); Barbour (1995, p. 253). Daubenmire also took a master's degree at University of Colorado, working with Francis Ramaley. Hoffman (1996, pp. 143–144); Stout (1995, p. 85).

[21] Barbour relates a revealing story from Daubenmire about Clements' dogmatism. The two botanists were scouting plant communities in the Palouse when Clements misidentified a plant and announced it as a climax species. When Daubenmire corrected Clements and noted the plant was evidence of disturbance, Clements replied that "There's a negligible difference," suggesting how Clements might have overlooked details to fit his theories. Barbour (1995, p. 248).

[22] Rumore (2009, pp. 206–241) shows these ideas and influences in practice in Cooper's Glacier Bay fieldwork.

[23] Daubenmire (1936) was a pioneer researcher in fire ecology. His text, *Plants and Environment,* reportedly was the first plant ecology text to devote a chapter to fire as an ecological factor; see Hoffman (1980, p. 34). Also, in 1968, he published a review essay on fire in grasslands that remained a classic for a generation; see Daubenmire (1968a).

human contrivances and delimited by a given scientist's subjective interests. If they were real and predictable, then ecologists could diagnose problems and prescribe remedies based on natural laws. If they were not predictable but simply the result of various contingencies and historical accidents liable to move in any number of future directions, then scientists could make few relevant predictions and prescribe no effective policies. For ecologists such as Daubenmire who worked on applied questions in agriculture or forestry, the implications were tremendous. In 1936, when he relocated to the Northwest, he grappled with these questions and their applications in place.

Applied Ecology and Agriculture

Classifying landscapes was embedded within virtually all early ecologists' work. They sought to identify different confluences of biological and physical factors to capture their characteristics, especially those related to successional phases and climax states. Historian of science Robert E. Kohler has argued that the first generation of ecologists tried to create a vegetation type classification system, much as biologists had with species taxonomy, only to abandon the project by about 1940 when empirical data revealed that vegetation types were not like species.[24] However, Daubenmire engaged with questions surrounding classification throughout his career, never yielding the assumption that vegetation communities existed and therefore could be characterized, understood, and managed.

An early Daubenmire publication challenged one of North America's first classification systems, C. Hart Merriam's life zones.[25] Merriam built on a long tradition. Classifying vegetation groups began with Humboldt who, as the nineteenth century dawned, made plant geography modern by studying vegetation in relationships rather than just compiling individual lists of flora as followers of Carl Linnaeus had done.[26] By the end of the nineteenth century, such efforts expanded. As part of the US Department of Agriculture's Division of Ornithology and Mammalogy and later the Bureau of the Biological Survey, Merriam took up the Humboldtian mantle and examined the distribution of species, identifying six main life zones in North America. In perhaps his most famous study in 1890, he investigated Arizona's San Francisco Peak and saw life zones matching patterns based on altitude but determined mainly by temperature.[27] Such work was useful but lacked scientific rigor.

Writing in *The Quarterly Review of Biology* in 1938, Daubenmire summarized Merriam's biotic distribution and criticized it. Temperature, mapped onto latitude,

[24] To be sure, ecologists still named vegetation groups for pragmatic reasons, but, as Kohler wrote, "they no longer constructed *systems* of classification, nor inquired too deeply into biological meaning of their categories." Kohler (2008), quoted on p. 107 (original emphasis).

[25] Daubenmire (1938); Merriam (1898).

[26] Nicolson (1987).

[27] Merriam (1890). Worster linked Merriam to Humboldt in *Nature's Economy*, pp. 195–197.

was all Merriam used to explain patterns and included virtually no quantitative data. Testing Merriam's theory in the field, ecologists had found vegetation types to be evidence of coherent zones more than climatological data. In other words, plant communities indicated biological coherence better than climate readings. For instance, some places where instrumental data (e.g., temperature readings) suggested the existence of a new zone also contained the same biota and so "certain natural entities were artificially split," or the contrary, "very diverse vegetation types were at times lumped together," simply because they shared a common climate. Another problem with Merriam's perspective was that he relied on a single factor—temperature—while Daubenmire contended that "we now hold the environment to be such an intricate complex of interdependent factors that it is exceedingly difficult, if indeed not an impossibility, to attempt to evaluate the individual influences." This emphasis on myriad factors grew out of his ecosystem perspective and would remain consistent throughout Daubenmire's career. Ultimately, although he credited Merriam with stimulating new research and for being the first to use climatic data, he ultimately found the explanations "fallacious."[28] For his work, Daubenmire received an admiring letter from Joseph Grinnell, the eminent field biologist who directed the University of California, Berkeley's Museum of Vertebrate Zoology. Although Grinnell had used Merriam's schema and "feel a sort of responsibility for defending that concept," he found Daubenmire's "summation and appraisal… thought-provoking, hence worthy."[29] One can only surmise such praise would be gratifying to a young assistant professor.

Even as Daubenmire wrestled with continental-scale classification questions, he zeroed in on local landscapes. Daubenmire arrived in the inland Northwest during the Great Depression, which also corresponded with national concern over environmental problems and ambitious conservation programs. The Dust Bowl of the southern plains with its massive soil erosion brought to a national audience a concern about poor land-use decisions that resulted in both ecological and economic ruin. Clements used the opportunity to showcase ecology as an applied science, vocally criticizing agriculture's role in disturbing the plains' biological community and advocating ecologists' potential to advise conservation work.[30] Depression-era conservation focused on two arenas: One would ameliorate existing problems; the other would plan new projects scientifically to avoid repeating mistakes. Like ecologists and conservationists elsewhere, Daubenmire believed that science could and should guide human–land relations, and a large conservation project in the region offered him a timely opportunity to be useful.

---

[28] Daubenmire (1938), quotations on pp. 330–332. He developed his own assessment of zones in the Rocky Mountains not long after this publication; see Daubenmire (1943). Later, Daubenmire demonstrated shortcomings to other climate-based classifications in Daubenmire (1956a).

[29] Washington State University Manuscripts, Archives, and Special Collections; Rexford F. Daubenmire papers (unprocessed) MS-1997–05 (hereafter RFDP); J. Grinnell to Prof. Rexford F. Daubenmire, 11 November 1938.

[30] Worster (1979, 1994) accounts for Clements's work surrounding the Dust Bowl in *Nature's Economy,* pp. 221–253; and *Dust Bowl,* pp. 198–209.

The Columbia Basin Project was the impetus for much of Daubenmire's early Northwest work. Authorized in 1933, this project sought to transform the Columbia Plateau by, among other things, bringing water from the Columbia River to the plateau's rich, but dry, soil often hundreds of feet above the river. The western part of the Columbia Plateau's 63,000 square miles is flat, dry sagebrush plain where wildlife, then Native Americans' horses, and then Euro-American livestock, especially sheep, grazed. With Grand Coulee Dam as its centerpiece, the Columbia Basin Project promised to convert a million acres or more of that plain into irrigated cropland, bringing more intensive agriculture to the region. Yet this would displace some grazing lands.[31]

This environmental history set the stage for Daubenmire's research. As the Columbia Basin Project reconfigured the plateau's geography of agriculture, many were invested in doing it scientifically to avoid disaster. Daubenmire explained how this agricultural frontier would avoid the "misguided history" of the Dust Bowl and "follow a course dictated by the findings of scientific research. These findings must be the synthetic product of specialists: ecologists, soil scientists, agronomists, engineers, etc."[32] This multidisciplinary synthesis bespoke Daubenmire's ecological vision and exuded confidence in specialists, a faith representative of the Progressive-era conservation movement and its continuation in President Franklin D. Roosevelt's New Deal.[33] This work, as Daubenmire put it, offered an opportunity for "man to practice what conservation principles he has learned by his past mistakes."[34] As an applied ecologist, he would explain past impacts and advise on future use for project areas. Launching fieldwork in the region in the mid-1930s, Daubenmire initiated what became four decades of intensive research during which he became arguably the region's unrivaled botanical expert.

From the start, Daubenmire's ecology of place engaged with disturbed lands, larger ecological questions, and implications concerning land use and its impacts. Overgrazing had already produced on the Columbia Plateau "sorry conditions. So badly have they been overgrazed that no one knows just how much forage such lands are capable of producing under less injurious treatment."[35] Further study revealed four effects of overgrazing on plant communities. First, the characteristic climax plants—native bluebunch wheatgrasses (*Agropyron spicatum*)—declined, as grazing pressure destroyed perennial plants' capacity for photosynthesis, weakened their overall vigor, and prevented seed production in annuals all the while removing larger plants' protective coverage that helped grasses thrive. Second, a new set of plants that could withstand trampling and were generally hardier than native bunchgrasses thrived in overgrazed lands, but they were "woolly" or bristly

---

[31] Material on the region's history is synthesized well in Meinig (1995 pp. 3–25); Duffin (2007, pp. 16–31). For grazing, see Dwire et al. (1999); McGregor (1982). For the Columbia Basin Project, see Pitzer (1994).

[32] Daubenmire (1939, p. 33).

[33] Fox (1981); Maher (2008).

[34] Daubenmire (1940b, p. 8).

[35] Daubenmire (1939, p. 33).

or "otherwise distasteful," and thus seen not only as a biotic regression but also as economically worthless and undesirable. Third, a transitory plant community appeared as grazing removed competitors that allowed these plants to grow, but they were "not very well adapted" to the larger habitat and ultimately did not remain in significant numbers. Fourth, grazing did not affect some minor plant communities in frequency or distribution.[36] The upshot: A stable, productive grassland was being replaced by a disturbed, unpalatable one.

Daubenmire's joined other studies about grassland ecology but reached somewhat different conclusions, demonstrating the value of place-based inquiry. Native bunchgrasses on western rangelands had been a great boon to ranchers, but by the mid-twentieth century, overgrazing deteriorated prairies over much of western North America. Daubenmire reported that selected plants in Washington's bunchgrass prairies did not behave as expected based on observations elsewhere: Russian thistle (*Salsola kali* L.) was not present, despite its ubiquity in other regions; cheatgrass (*Bromus tectorum*) could dominate as it did elsewhere, but the relationship with grazing could not be drawn directly; and sagebrush (*Artemisia tridentata*) often invaded grazed lands, but in Washington it appeared complementary to, not competitive with, bunchgrasses. The conundrum facing range managers tasked with balancing livestock numbers and available forage was obvious. Ninety percent of the biological output in this ecosystem—measured by dry weight—came from just two plant groups, *Agropyron* and *Bromus,* but Daubenmire's study demonstrated that those plants declined markedly in overgrazed ecosystems. In fact, *Agropyron* was only negligibly present with the annuals that replaced it being "valueless as forage." Thus, grazing reduced and replaced over time the very grasses required or preferred by livestock. Daubenmire recommended cutting and curing grasses for hay later in the year, removing annuals to reduce competition and promote perennial vigor, and resting land from grazing, especially during spring growth. His experiments and observations in the field suggested that a haphazard grazing system would inevitably continue to destroy the range required to sustain livestock.[37]

Plants' successional responses to overgrazing were only one relevant element; to construct a more complete understanding of the ecosystem, Daubenmire also turned to the effects of overgrazing on soil. Such research was necessary, because while it was common knowledge that overgrazing caused "vegetational retrogression," few scientists investigated what it did to the soil. He examined two comparable plots only 50 m apart, one severely overgrazed, while the other protected from grazing for nearly three decades because of a railroad cut. In comparing these two virtually identical soil samples, Daubenmire found significant changes that could only be attributed to grazing. The annual plant communities that colonized heavily grazed

[36] Daubenmire (1940a), quoted on p. 60. He had presented these four stages in preliminary form in Daubenmire (1939, pp. 35–36).

[37] Daubenmire (1940a), quoted on p. 62. Earlier, he had recommended minimal spring grazing and relying on cured shoots in fall and winter for feedstock to ameliorate overgrazing's effects; see Daubenmire (1939, p. 36). Daubenmire's work related to cheatgrass invasion is contextualized in Young and Allen (1997).

land possessed shallower root structures, which in turn changed the way water accumulated and was absorbed in the soil, weakened soil aeration, and reduced soil aggregation. The evidence seemed clear: Grazing worsened soil functioning. Combined with his earlier study of grazing's botanical effects, Daubenmire recognized a causal chain from grazing that extended beyond obvious biotic reconfigurations to "secondary, or even more remote, effects of grazing."[38] His research had begun capturing in detailed scientific terms the negative consequences of the region's prevailing agricultural practice of maximizing production.

Meanwhile, Daubenmire sought to bring ecology's insights to other agricultural problems. To do so, he identified natural plant communities, seeing in them ecological clues to what the best crops for a habitat might be. This research constituted an outgrowth of his life zones work and ecology's general classification project, and he spent the bulk of his career classifying plant communities as a prerequisite to understanding an environment's subtle "potentialities."[39] The Columbia Basin Project's lands might appear uniform: "Apparently all that is needed is to grid the area into tracts of 40 acres as is now planned, supply irrigation water, and let the success of the project rest entirely upon the diligence of the farmers." But Daubenmire warned of greater complexity, "But nature has not endowed this area with uniform soil conditions, and farmers who settle tracts of good soil will prosper while their nabors [sic] may have a difficult time of finding subsistence on a tract of equal size and with the same amount of irrigation water."[40] In two separate studies—one brief and impressionistic, the other lengthy and statistical—he used "virgin and near-virgin relics" often found in cemeteries that had been protected from disturbances like plowing and grazing to determine natural plant communities and which environmental factors controlled their structure. Most important on the Columbia Plateau were soil types, which closely corresponded with the observed vegetation communities.[41] From this information, Daubenmire offered practical agricultural advice. For instance, the saltgrass-type community would be ideal for sugar beets or alfalfa, while sagebrush and rabbitbrush indicated good habitat for orchards. Understanding these botanical communities paid practical dividends for farmers, since the same environmental conditions affected any plant, even crops. "Native vegetation represents the final outcome of the operation of ecologic factors which have influenced plants throughout centuries and which are operating today not only on the remnants of the original flora, but on our crop plants as well," Daubenmire reasoned.[42] Knowing and implementing this ecological information, farmers might experience greater success and avoid expensive failures.

---

[38] Daubenmire and Colwell (1942), both quotations on p. 32.

[39] Daubenmire (1940b, p. 8. He used "potentiality" in various forms, including "biotic potentiality" or "crop potentiality" in many publications.

[40] Daubenmire (1940b, p. 8).

[41] Daubenmire (1940b); Daubenmire (1942), quoted on p. 60.

[42] Daubenmire (1940b, pp. 9–10, quotation on p. 10). He made a similar statement in Daubenmire (1942, p. 75).

To a large degree, that was the larger point: Ecologists sought practical applications for their work. This first spate of Daubenmire's Northwest work, rooted in questions surrounding the reclamation project's potential, found those outlets. Seeking to understand past impacts and future potentialities, the ecologist supported the region's agricultural interests. At the very least, Daubenmire believed in putting farming on an ecologically secure foundation that served farmers, although any suggestion of reducing grazing or questioning the inevitable success of irrigation across the project may have irritated farmers. Inland Northwest agriculture was in transition with expanding irrigation and increased mechanization meeting a landscape already showing signs of significant wear and tear. The science also was in transition, finally giving the region attention. In his major study of plateau vegetation in *Ecological Monographs* in 1942, Daubenmire noted that only two other scientists had examined the region's plants.[43] His study of grazing's impact on soil similarly brought attention to a question that had received little scientific investigation.[44] That Daubenmire was among the first to describe the region's ecosystems scientifically indicates just how recent was the conjunction of science and agriculture to this place. These detailed studies summarized field research but also engaged with ecology's larger questions of plant communities, illustrating Daubenmire's emerging ecology of place.

Habitat Types

Moving into the post–World War II era when Daubenmire relocated to the State College of Washington (renamed Washington State University in 1959), his growing research program found him still in agricultural fields, but also increasingly in the forested foothills, trying to make sense of timbered ecosystems. Diverse and disturbed landscapes challenged botanists, for succession's fundamental dynamism made classification difficult with constantly shifting biotic communities.[45] "The delimitation of natural sociologic entities in a complex and largely disturbed vegetation is by no means an easy task that can be resolved to simplicity in a short time"; Daubenmire explained in 1952, "even a small area of vegetation may contain thousands of species which, at first seem to form a chaotic pattern."[46] Finding the patterns in the chaos became his task. Paradoxically, by figuring out what nature might be like without human activities (i.e., disturbances), ecologists such as Daubenmire believed they could bring to bear scientific insight on environmental questions and guide natural resource development.

Daubenmire maintained his focus on natural vegetation communities and included larger landscapes from which he added more data and refinement to his earlier

[43] Daubenmire (1942, p. 55).

[44] Daubenmire and Colwell (1942, p. 32).

[45] Daubenmire (1946, p. 33).

[46] Daubenmire (1952, p. 321).

observations. Physiography and climate were insufficiently accurate indicators, he found; only biotic distribution worked, and since animals ultimately depended on plants, vegetation was the best criterion. The plants used to characterize the communities needed to be climax species, for otherwise the community's character would change with each successional stage. To Daubenmire, these "vegetation zones are fundamental natural entities." He acknowledged that individual species might be found in other zones, for a species' presence or absence was not what constituted a unique community. Their groupings and interactions, as well as their relative abundance, created "highly distinctive" communities that could be discerned and classified. Knowing these zones' characteristics allowed foresters, range experts, and game managers to understand potential biota and the "possibilities for controlling vegetational change" in each unique zone for various management goals.[47] Focusing on *potential* vegetation became a hallmark of Daubenmire's ecology, because planning to use a landscape over time required managers to know what could grow in a given type rather than what occupied the ground at the moment, which could be merely a transitory product of disturbance.[48] An early seral association, for instance, would be comparatively short-lived with the climax association ultimately dominating the area. Using existing cover type for classification might show foresters where commercially valuable trees were, but it would be subject to frequent change and was not an ecologically sound method.[49] Ecologists and managers required more basic and permanent classification schemes.

Daubenmire expanded and clarified these perspectives in a major study of northern Rockies forests, published in 1952 in *Ecological Monographs*.[50] It extended the geographic range of his earlier work from the west to the east, the ecological range from steppe vegetation to forests, and the economic focus from agriculture to natural resources more broadly. He had large ambitions for this project—nothing less than an exemplar of a universal scheme for vegetation classification. At the outset, Daubenmire labeled different natural units. *Unions* (also termed *synusias*) included a species or closely related species with similar environmental requirements; these were the smallest structural components. *Associations* were the basic units in classifying vegetation and included all unions in the same area characterized by climax species. He named associations binomially with the dominant and subordinate union identified, such as the *Pinus ponderosa/A. spicatum* association where ponderosa pines dominated with bunchgrasses as subordinates. *Zones* included areas of

[47] He explains his reasoning clearly in Daubenmire (1946), quotations on pp. 37, 36, 37, respectively.

[48] Daubenmire's approach to classification is contextualized historically for managers in Bailey et al. (1978); Franklin (1980); O'Hara et al. (1996). Both Franklin and Pfister (a coauthor in Bailey et al.) worked with Daubenmire for their doctorates. Daubenmire's approach is an antecedent to what is sometimes called potential natural vegetation (PNV). An explanation and application is found in Henderson et al. (2011), esp. pp. 2–5.

[49] Daubenmire (1952), "Forest vegetation of northern Idaho and adjacent Washington," p. 324.

[50] Ibid., pp. 301–30.

closely related associations, such as grasslands.[51] Using this nested system, while paying attention to climax and seral stages, Daubenmire could effectively describe ecosystems. This framework was foundational to field ecology, he concluded: "We should look upon complex ecosystems as the only natural units, and that macroscopic vegetation in its entirety comprises the best criterion of ecosystems."[52]

Daubenmire proposed using the *habitat type* as the basis of classifying land.[53] Habitat types included various environmental factors of a given place, including climate and soil. In effect, they provided the basic ecological context for the unions, associations, and zones. These would not fundamentally change because of natural disturbances like fire or human disturbances like logging. Habitat types were practically permanent and thus strong indicators for long-range planning; thus, they could be a valuable and welcome tool for land managers. He provided an example of mismanagement that could have minimized by using the habitat-type method. After fires moved through one Idaho stand, foresters planted ponderosa pines, which grew quickly but then stalled and declined. Meanwhile, natural regrowth of western white pine (*Pinus monticola*) started slowly but then far surpassed the ponderosas. Had managers understood the locale's true habitat type, they could have saved time and money by not planting trees likely to be supplanted. As he had done when advising for the Columbia Basin Project, Daubenmire searched for explanations to predict likelihoods and avoid costly efforts. "The trial and error method of ascertaining habitat potentialities of forestlands is very costly because of the many years that are needed to determine the ultimate effect of different practices as the tree crop matures," Daubenmire explained, "so that the habitat type concept has much to offer by indicating the degree to which each experiment can be extended throughout the mosaic of forest associations." In short, scientists could map habitat types that reflected ecological qualities so that wherever a certain habitat type was found—whether disturbed or pristine—managers could look into the future to see how that forest or rangeland would likely develop.[54]

---

[51] His system is described in ibid., pp. 302–303. Similar summaries are found in Daubenmire (1953), "Classification of the conifer forests," pp. 17–19; Daubenmire (1954), "Vegetation classification."

[52] Daubenmire (1952), "Forest vegetation of northern Idaho and adjacent Washington," quotation pp. 324–35. This definition differed from contemporaneous work that took a systems approach toward how and what moved through ecosystems; see Golley, *History of the Ecosystem Concept,* esp. pp. 35–108; Hagen, *Entangled Bank,* esp. pp. 78–145; Kingsland (2005), esp. pp. 185–99, 206–19; Worster, *Nature's Economy,* esp. pp. 301–15.

[53] Daubenmire described the origins of this idea in a paper given in 1987, see Rexford Daubenmire, "The roots of a concept," a paper presented at the Symposium Land Classifications Based on Vegetation: Applications for Resource Management," pp. 17–19 November 1987 (found in RFDP).

[54] Daubenmire (1952), "Forest vegetation of northern Idaho and adjacent Washington," quotation from p. 326. More on mapping habitat types is found in Daubenmire (1973), "A comparison of approaches to the mapping." Other publications also show Daubenmire attempting to predict ecological trends for managers, see Daubenmire (1956b), "The use of vegetation to indicate grazing potentials." Daubenmire (1976), "The use of vegetation in assessing the productivity."

Habitat typing would become an important practical tool.[55] Daubenmire believed that "land units defined on the basis of their potential or actual climax can and will play an increasingly more important part in the ecologic sciences as the use of uncultivated lands becomes intensified."[56] He was correct. In the postwar era, Northwest forests experienced significant harvest increases and intensified management, especially on federal lands. By 1960, Congress codified that national forest timber harvests be conducted on a sustained yield basis, while those lands would be managed for multiple uses, including timber, wildlife, watershed protection, and recreation. These competing goals, as well as pressures caused by a growing population's consumer and amenity demands, required the best and most informed management possible.[57] Scientists like Daubenmire tested and refined inherited ideas and formulated new approaches to assist this work.

Working foresters welcomed Daubenmire's efforts. Noted ecologist Frank E. Egler, then working at the research site Aton Forest in Connecticut, and often a critic of community ecology, praised Daubenmire's engrossing article in *Ecological Monographs* as a "mile-stone paper" and assured Daubenmire that he would be quoting it in the future.[58] Northwestern private foresters praised it, as did researchers abroad.[59] Wildlife experts also recognized the impact the classification system would have on their work.[60] Given the high proportion of Forest Service lands in the Northwest (today more than 20 million acres in Idaho alone),[61] perhaps the most significant praise came from a federal forester, Fred W. Johnson, who wanted 150 reprints of the article to distribute to all regional field officers. In particular, Johnson appreciated the applied ecological approach: "Your ecological interpretation of vegetative associations will form the basis for much of the silvicultural, range and wildlife habitat management which will be accomplished on the national forests of northern Idaho and eastern Washington in the future. Such a basis has long been needed." He continued by suggesting "your paper will go a long way toward selling

[55] Daubenmire, "Roots of a concept," (unpublished); Bailey et al. (1978); O'Hara et al. (1996);; Pfister and Arno (1980); Stout (1995); Hill Williams, "Shrubs, Herbs Used in Classing Forests," unnamed and undated newspaper article contained in RFDP; Hinz (1975).

[56] Daubenmire (1953, p. 17).

[57] Daubenmire made this very point about competing demands in Daubenmire (1973, pp. 87–91). An overview of these trends in the region is in Sowards (2007, pp. 176–82).

[58] RFDP; Frank E. Egler to Daubie, 2 December 1952. For Egler's position as a Clementsian critic in favor of the individualist school, see Whittaker (1962, pp. 82 and 124).

[59] RFDP; John H. Fagan (to Rexford Daubenmire, undated). Although it is not specified, Fagan was likely employed by Potlatch Corporation. Another Potlatch forester inquired about reprints to distribute to the company's foresters; see RFDP; Royce G. Cox to Dr. R. F. Daubenmire, 6 February 1953. RFDP; M. E. Solomon to Dr. R. Daubenmire (undated). Solomon worked in the Department of Scientific and Industrial Research, Pest Infestation Laboratory, Slough, England. RFDP; Lucy B. Moore to Dr. R. F. Daubenmire, 7 April 1953. Moore worked in the Botany Division of the Department of Scientific and Industrial Research in Wellington, New Zealand.

[60] RFDP; Paul D. Dalke to Dr. R. F. Daubenmire, 19 December 1952.

[61] Statistics derived from information on US Forest Service website, (http://www.fs.fed.us/) accessed 30 July 2013.

an ecological approach to forest land management of the area described."[62] Together, these comments demonstrate that Daubenmire offered generalized knowledge, useful to those in Pennsylvania, New Zealand, or England, where plant species and communities were quite distinct from the inland Northwest. But they also show practical, local applications from an ecological perspective on public and private forestlands. Daubenmire clearly conducted work that bridged, or at least appealed to, both sides of the basic and applied divide.

The Continuum Theory Challenge

Yet, while Daubenmire and coworkers traipsed through the forests finding climax or near-climax communities, other ecologists devised distinct approaches. Since its founding, the discipline struggled for acceptance, and one way it sought to enhance credibility was to develop greater rigor. Moving beyond what some saw as descriptive and subjective methods, a new school of ecology used statistical methods to create supposedly objective descriptions of plant ecology. Their innovations were part of a general quantitative turn in ecology, moving it more in line with so-called hard sciences, a self-conscious desire that seems to run throughout ecology's history.[63] Led by University of Wisconsin-Madison professor John T. Curtis, this school helped revive Gleason's individualist perspective. Working first in the Midwest, field-workers selected random plots and collected data on the vegetation and then arranged it along several axes tied to various environmental gradients.[64] The data revealed that distinct plant communities did not exist, but rather that vegetation grew in continuous variation—a continuum—whereby as one moved through a landscape, a species would appear, increase in quantity, then decline, and disappear, but in no particular pattern.[65] The continuum school constituted a significant shift in the 1950s, amounting to a paradigm shift according to plant biologist Michael

[62] RFDP; Fred W. Johnson to Dr. R. F. Daubenmire, 21 November 1952. The principal silviculturist from the Northeastern Forest Experiment Station in Pennsylvania concurred with the need to tie plant sociology with forest management, see RFDP; M. Westveld to Dr. R. Daubenmire, 8 December 1952. Perhaps too much can be made of the supportive statements by foresters, since historian Paul W. Hirt has shown that timber management in the region at this time initiated a disastrous set of unsustainable practices, see Hirt (1999).

[63] Kohler (2002) explored various efforts in biology to bring statistical and other methods into fieldwork around the turn of the twentieth century in *Landscapes and Labscapes*. Other histories of ecology note the quantitative shift. Kingsland (2005) focuses on how ecologists adopted systems perspectives to study ecosystems in *Evolution of American Ecology*, pp. 206–231. McIntosh explores a range of quantitative topics in *The Background of Ecology*, pp. 107–145. Also, Bowler, *Earth Encompassed*, pp. 535–46.

[64] The classic methodological paper is Bray and Curtis (1957).

[65] Explanations and context for Curtis' work can be found in McIntosh (1985), esp. pp. 137–45; Nicolson (2001).

G. Barbour.[66] It was a Gleasonian world without distinct communities behaving in predictable ways.

For the most part, Daubenmire's disagreement with the continuum school remained implicit in his own conclusions. Indeed, he frequently noted how there were good things to take from competing schools of thought.[67] However, his article in the prestigious journal *Science* directly criticized the continuum school. Describing Curtis' approach, he sardonically noted that the statistical methodology "makes the results more satisfying to a mathematician than to a botanist."[68] Daubenmire recognized what was at stake; without an organizing principle, vegetation science would be unable to predict and thus furnish useful information. As he once put it, "Without classification there can be no science of vegetation."[69] Daubenmire reported on his own Columbia Plateau research which revealed marked discontinuities among four vegetation zones. Rather than resting his case there, he provided a contrary reading of evidence. Random sampling, as Curtis advocated, in the same region could well have yielded islands of atypical plants—those growing on steep slopes, for instance. This was why Daubenmire advocated sampling, subjectively, from representative areas that were relatively homogenous in climax or near-climax states. This approach did not produce random objectivity but did generate an accurate characterization within the broader landscape.[70]

Furthermore, the continuum school focused on tabulating species' distribution and abundance, but Daubenmire pointed out that such a method was too simple, "as much a part of taxonomy as of synecology. In synecology we must come to grips with matters of more fundamental biologic importance, especially population structure and dynamics." The continuum school quantified plants as they existed in one moment of time, while the community-based approach examined how they interrelated with each other across space and time. Doing so required ecologists to pay closer attention to such factors as age structure and competition within stands, factors that revealed succession patterns post-disturbance. Curtis' statistical methods might have been ecologically innovative and mathematically sound. But the results, according to Daubenmire, merely showed "that continuum advocates have used disturbed vegetation mosaics in which seral mixtures can provide frequent bridging between otherwise reasonable distinct stable types, or in which degradation has

---

[66] Barbour (1995). The degree to which the shift truly represented a paradigm change is debatable, depending on one's comparative framework. Nonetheless, a revival of Gleason's influence was indisputable.

[67] For instance, Daubenmire (1952, p. 302); Daubenmire (1968b, p. x.)

[68] Daubenmire (1966, quoted on p. 291).

[69] Daubenmire (1960, p. 24). This paper was based on remarks at a Symposium on Forest Types and Forest Ecosystems during the IX International Botanical Congress in Montreal, 24 August 1959. It includes some of his most direct criticisms of the continuum school.

[70] Daubenmire (1966), esp. pp. 291–95. In 1959, Daubenmire published his own methodology, which explained his field approach in detail. Daubenmire (1959a). This article was widely cited (according to Google Scholar, nearly 2000 citations) and earned status as a "citation classic" for ecology; in fact, it was the 13th most cited article in ecology between 1947 and 1977. See McIntosh (1989). With modifications, his method continues to be used; see Bonham et al. (2004).

proceeded to a relatively stable network of variation that is infinitely simpler than the mosaic which replaced it." Daubenmire conceded *flora*—that is, the individual plants—represented a continua, for surely plants changed imperceptibly as one moved through the landscape; however, *vegetation*—that is, the cumulative plants in relationship with each other and the environment—was something arranged in distinct units.[71]

The *Science* article was a strong critique, and judging from the responses Daubenmire received privately, ecologists cared deeply about its implications. Although scientists have reputations for being rational, objective researchers, these letters of support exuded a combination of bellicosity and acclamation. Some described how Daubenmire "struck a blow" for the community perspective, characterized the article as a "rallying point" for community ecologists, and gave him the proverbial "Good show!" as if this were a schoolyard contest and not a set of scientific questions.[72] These reactions support one report that at least some of this rancorous debate was "maybe due to the delight in fighting each other that some people have."[73] Indeed, comments against the continuum school were often uncharitable, calling it "nonsense" or saying the method included "little of ecological value, mostly a maze of statistics." A zoologist at Curtis' university who had not been "entirely indoctrinated" sided with Daubenmire that "the study of ecology should not be reduced to numerical abstraction, in spite of the temptations our high speed machines offer in terms of data analysis."[74] The inimitable Frank Egler was sure that "This paper will go down in history as the long-overdue come-uppance for the continuumophilists."[75] Longtime Yale forester Harold Lutz enthusiastically "endorsed" Daubenmire's views but could not share his ideas about the continuum perspective "[w]ithout resorting to campfire language." More importantly, Lutz offered what was no doubt a common, though unscientific, point of view: "Plant communities are very real things to me; I have seen them, felt them, walked in them and *know* they are real and meaningful. The same is true for the climax concept and for succession."[76] Despite efforts to be objective, then, some ecologists still relied on a

[71] Daubenmire (1966), esp. pp. 292–96, quotations from pp. 295, 296, 298. His focus on space and time, as well as the emphasis on landscapes' mosaics, are all legacies of Cooper's teaching. Nicolson showed that Humboldt first distinguished between floral vegetation in "Humboldt, Humboldtian Science."

[72] All in RFDP; Francis C. Evans to Dr. R. F. Daubenmire, 26 January 1966 (struck a blow); John [no last name] to Dauby, 15 February 1966 (rallying point); Dr. Robert Linn to Dr. Rexford Daubenmire, 21 January 1966 (good show).

[73] Helen Buell, quoted in Barbour (1995, p. 242).

[74] All in RFDP; Ronald O. Kapp to Dr. R. Daubenmire, 25 January 1966 ("nonsense"); Philip V. Wells to Dr. R. F. Daubenmire, 15 February 1966 ("abstract nonsense"); Lawrence C. Bliss to Dr. Rexford Daubenmire, 28 February 1966 (little ecological value); James W. Drescher to Dr. Rexford Daubenmire, 19 April 1966 ("entirely indoctrinated").

[75] RFDP; Frank (Egler) to Daubie, 26 February 1966.

[76] RFDP; Harold Lutz to Dr. Daubenmire, 26 January 1966 (original emphasis). Lutz continued, "It may be smugness, but sometimes I wonder about the field experience of those who have trouble with these concepts."

felt sense of the way things were in nature. Too many scientists to list from throughout the USA and as far away as Costa Rica and India attested to Daubenmire's eloquence and discipline in publishing this important scholarship.[77]

A final comment from Richard S. Driscoll, a principal plant ecologist for the US Forest Service at its Rocky Mountain Forest and Range Experiment Station, made a critical point. Even though he knew others disagreed, classification could be done and in fact was essential: "I feel vegetation grouping is very necessary if we are to provide a rational scientific and factual basis for land use and management."[78] When managing vegetation, whether in forests or farms, the plant community concept offered something useful and necessary. For some at the time, that proved the essence of the debate. Jerry Franklin, a Daubenmire graduate student who finished his doctorate the year before the *Science* article appeared and who later became a central figure for the Forest Service in the spotted owl controversy of the 1980s and 1990s, recalled a continuum partisan telling him at the time, "This community stuff is OK for you managers, but I'm interested in the truth."[79] Even if it was not "truth," a continuum ecologist just might allow that it was useful for management.

The Ecology of Place

By the late 1960s and into the 1970s, after more than three decades in the inland Northwest, Daubenmire had witnessed much change in the region, not to mention in the science of ecology. Questions about power (hydroelectric and nuclear), the intensifying use of natural resources, and the preservation of wilderness made the Northwest a politically and economically contentious region centered on questions related to environmental quality.[80] Nationally, the environmental movement had emerged, and federal legislation like the National Environmental Policy Act (1970) and the Endangered Species Act (1973) reshaped Americans' legal and ethical relationship with nature.[81] Ecology played a role in this activism, providing data about the harm certain economic activities caused to natural systems. This context suffused the work Daubenmire conducted in the last stage of his career.

As well as anything, two major studies, *Forest Vegetation of Eastern Washington and Northern Idaho* in 1968 and *Steppe Vegetation of Washington* in 1970, exem-

[77] Others appreciated Daubenmire's account because it affirmed their own research findings and thus lent support against the wave of continuum studies. For instance, all in RFDP; Henry S. Conrad to Dr. Daubenmire, 22 January 1966; Donald Caplenor to Dr. R. F. Daubenmire, 23 February 1966. The Daubenmire Papers include dozens of supportive letters. *Science* published two letters from Curtis students, explaining their disagreements with Daubenmire's methods and interpretation of continuum perspectives; Vogl et al. (1966).

[78] Richard S. Driscoll to Dr. R. Daubenmire, 14 November 1966.

[79] Quoted in Barbour (1995, p. 241). A list of Daubenmire's graduate students is available in RFDP.

[80] A regional overview is found in Sowards (2007), esp. pp. 167–209.

[81] There are numerous studies that trace the contours of the environmental movement; for a representative introduction, see Rothman (1998).

plified Daubenmire's ecology of place and demonstrate how he had become more assured so that he could comment more openly, if briefly, on these broader environmental concerns.[82] The bulk of each study defined the myriad forest and steppe vegetation habitat types on the Columbia Plateau, the culmination of more than three decades of fieldwork. There were few surprises in these studies, as Daubenmire rehearsed his typical methods, his basic assumptions about the reality of plant associations, and his preference for sampling representative climax communities. Indeed, he noted that this work only strengthened his earlier conclusions. However, suggesting the era's zeitgeist, he included an impassioned rationale: "Remnants of primeval forest representing most of the associations are still to be found. However, as more and more of the land is brought under management, these stands are the first to suffer, for in terms of timber production they are 'overmature' and 'decadent.' Simple economics dictate their replacement by young and vigorously growing trees. Thus the possibility of making such a study as this is rapidly dwindling and another useful purpose, the historical, is served by recording the character of the primeval forest."[83] One readily senses Daubenmire's sense of urgency and passion for the place, as well as his impatience for "simple economics."

In *Steppe Vegetation of Washington,* Daubenmire offered less dramatic prose, but his criticism may have been more subversive. Ostensibly, the study could inform range management for maximum sustained yield.[84] Daubenmire proposed an ecologically informed grazing regime that differed from the "narrow view" typical of North American range managers who focused on just a few species, arguing that "plants of low economic value can have very high indicator significance."[85] Focusing on the entire ecosystem and not just economically valuable species improved management. For example, ranchers and range managers wanted to remove sagebrush to promote grasses. However, such so-called range restoration really was about increasing the productivity of specific grasses favored by livestock and was rooted in an ethos of maximized production and intensive agriculture. Ecological considerations were different. Sagebrush protected perennial grasses; thus, eradicating it would make grasses even more vulnerable to overgrazing. Little evidence existed that removing sagebrush did anything more than increase grass productivity

[82] Daubenmire and Daubenmire (1968); Daubenmire (1970). The forest vegetation study included Daubenmire's wife, who had earned an MSc degree, as a coauthor. She accompanied him on much of his fieldwork, and he faithfully acknowledged her assistance in numerous publications. This was their only coauthored piece. Unfortunately, the dynamics of their scientific partnership remain elusive in the extant record. Letters from his students contained in his papers frequently mention Jean, suggesting that she was an active and visible partner.

[83] Daubenmire and Daubenmire (1968, pp. 1–2). Much as anthropologists practice salvage anthropology where artifacts or communities are imminently threatened, what Daubenmire is describing here can be likened to salvage ecology: gathering as much ecological data as possible before the natural community was destroyed.

[84] Maximum sustained yield was a common managerial goal, although environmental historians have criticized its actual practice; see, for example, Hirt (1994); Langston (1995, pp. 157–200); McEvoy (1986, p. 6).

[85] Daubenmire (1970, p. 1). For further criticism of narrow management, see Daubenmire (1984).

in the short term; long-term effects were still unknown. Nor was there evidence about what ramifications there may be for soils, although Daubenmire hypothesized several negative consequences (e.g., declining mineral cycling). Sagebrush also furnished excellent bird habitat, which in turn aided in keeping insects in check. It also held snowpack longer in spring, which increased soil moisture during hot summers in the interior Northwest. He also warned against using powerful new herbicides, because these chemicals killed broadleaf plants indiscriminately, eliminating other plants that were economically unimportant but which could be ecologically significant. In short, Daubenmire challenged an article of faith—that removing sagebrush was beneficial because it enhanced economically valuable and palatable plants—by shifting the economic criteria to ecological values. More and more such conclusions permeated some ecologists' work, showing how a long career and widening perspectives promoted in Daubenmire a strong ethic of place.[86]

Daubenmire's decades of work earned recognition and accolades from his colleagues on a national level. He presided over the ESA in 1967, joining such other notable American ecologists as Aldo Leopold and Eugene Odum, both of whom had also served as ESA president. The Northwest Scientific Association honored him as their "Outstanding Scientist" in 1970. Daubenmire enjoyed national awards from the ESA who named him as the Eminent Ecologist for 1979, from the Society of American Foresters who awarded him the 1980 Barrington Moore Award, the Society for Range Management who gave him a "Special Award" in 1986 recognizing his "extraordinary contributions to the Society for Range Management and the range profession," and the Nature Conservancy granted him honorary lifetime membership. This recognition indicated both the esteem in which fellow scientists held Daubenmire and his diverse interests and expertise spread across forests and rangelands. Significant scientists also walked through his classroom and recognized Daubenmire's teaching influence, including F. Herbert Bormann (the longtime Yale researcher who worked on the notable Hubbard Brook Ecosystem Study), Tom Tidwell (the current US Forest Service chief), and Jerry Franklin (the erstwhile chief plant ecologist for the Forest Service's Pacific Northwest Research Station). A legacy of students and a full curriculum vitae meant that by traditional academic standards Daubenmire finished his career as a great success.[87]

After more than four decades examining the Columbia Plateau's varied and changing landscapes, Daubenmire had accomplished much. He applied ecological thinking to a place theretofore barely examined with modern scientific methods. He determined habitat types throughout the interior Northwest with an eye toward potential vegetation. He did this all caring how local results fit within broader schemes

[86] Daubenmire (1970, pp. 79–80), quotation on pp. 80. Knobloch points out that restoring overgrazed ranges was always about increasing economic productivity, not any ecologically based goal; see Knobloch (1996, pp. 99). Knobloch explores the chemical focus of weed eradication on pp. 136–142; see also Duffin (2007, pp. 102–26).

[87] Burgess and Ellstrand (1983); Hoffman (1980, pp. 34–35); Hoffman reviews his awards in Hoffman (1996). See also Bormann (1996, p. 3); Anonymous (n.d.). On his retirement, many students wrote letters of appreciation that revealed the deep admiration they felt for their mentor. This correspondence is bound and contained in the RFDP, which also contains the actual awards.

for organizing the world's vegetation. This meant he was engaged deeply in the local while simultaneously contributing to larger ecological projects, exemplifying the ecology of place approach to the discipline.

Conclusion

Daubenmire represents those many scientists seeking connections between generalized theories, local conditions, and practical problems. These contexts are important when investigating the intersection of life sciences, environment, and agriculture. In this case, ecology formed the scientific framework for Daubenmire's work. Yet the discipline changed during his career from a relatively immature science with few competing theories to one where changing methodologies and philosophies added nuances, challenges, and intellectual competition. Tracing Daubenmire's engagement with ecological debates during the transitional era between the 1930s and 1970s reveals some of these contours. Meanwhile, when lands opened to intensive agriculture or when forests opened to increased harvests, the regional environment transformed. The desire to avoid expensive trial-and-error approaches to growing plants and the hope to harvest nature's products sustainably held managers' and scientists' attention and drove Daubenmire and others to conceive of ways ecology could promote greater environmental quality. To ignore the policy or management dilemmas facing natural resource systems or to neglect the environmental changes and pressures in a landscape is to miss a prime motivating factor for many working ecologists. It is essential that historians of science keep in mind the practical and material contexts in which ecologists worked in addition to the ideas they developed and debated.

In *Plant Communities,* Daubenmire contextualized ecological work like his. "A major objective in any science is to predict and control," he claimed. "Since vegetation is dynamic, it is only through careful study of successional processes that man gains an ability to predict natural trends and to develop feasible objectives in modifying them, both of which are essential for success in managing vegetation."[88] Here, he summed up his work's raison d'être. Ecologists were not modern natural historians describing landscapes and listing species. Nor were they just discerning biological mechanics to determine plants' functioning. For ecologists such as Daubenmire, the necessary work they did served broader society by allowing natural resource decisions to be scientifically informed. For him, this grew organically out of his practice of the ecology of place, developed in the field, over time, and with deep engagement.

Acknowledgments The participants at the workshop provided a helpful, supportive, and engaging environment. Sharon Kingsland challenged me and forced me to clarify concepts and prose in a way that vastly improved this essay. I am grateful to Laura Kross for her research assistance

[88] Daubenmire (1968b, p. 25).

and many conversations about Daubenmire and ecological methods. Ian Chambers proved to be an excellent reader of multiple drafts. Andrew Duffin provided both invaluable research assistance and intellectual companionship; in an earlier iteration of this essay, he served as my coauthor, an experience I relished. Finally, to Kelley Sowards who listened to my findings and ideas and read countless drafts, I am grateful as we develop our own ecology of place here.

References

Anonymous. n.d. U.S. Forest Chief Tom Tidwell: 'WSU education taught me how to think'. Washington State University, Alumni and Friends website. http://cahnrs.wsu.edu/alumni/profile/u-s-forest-chief-tom-tidwell-wsu-education-taught-me-how-to-think/. Accessed 20 July 2013.

Bailey, Robert G., Robert D. Pfister, and Jan A. Henderson. 1978. Nature of land and resource classification—A review. *Journal of Forestry* 76:650–655.

Barbour, Michael G. 1995. Ecological fragmentation in the fifties. In *Uncommon ground: Toward reinventing nature,* ed. William Cronon, pp. 233–255. New York: W. W. Norton.

Billick, Ian, and Mary V. Price, eds. 2010. *The ecology of place: Contributions of place-based research to ecological understanding*. Chicago: University of Chicago Press.

Bonham, Charles D., Daryl E. Mergen, and Sam Montoya, 2004. Plant cover estimation: A contiguous Daubenmire frame. *Rangelands* 26:17–22.

Bormann, F. H. 1996. Ecology: A personal history. *Annual Review of Energy and the Environment* 21:1–29.

Bowler, Peter J. 1992. *The earth encompassed: A history of the environmental sciences*. New York: W. W. Norton.

Bray, J. R., and J. T. Curtis. 1957. An ordination of the upland forest communities of southern Wisconsin. *Ecological Monographs* 27:325–349.

Burgess, Robert L., and Norman C. Ellstrand. 1983. Of interest to ecologists. *Bulletin of the Ecological Society of America* 64:17–18.

Cittadino, Eugene. 1993. A 'marvelous cosmopolitan preserve': The dunes, Chicago, and the dynamic ecology of Henry Cowles. *Perspectives on Science* 1:520–559.

Clements, Frederic E. 1916. *Plant succession: An analysis of the development of vegetation*. Washington, D.C.: Carnegie Institution of Washington.

Cooper, W. S. 1926. Fundamentals of vegetational change. *Ecology* 7:391–413.

Cowles, Henry C. 1899. The ecological relations of the vegetation on the sand dunes of Lake Michigan. *Botanical Gazette* 27:95–117, 167–202, 281–308, 361–391.

Daubenmire, R. F. 1936. The 'big woods' of Minnesota: Its structure, and relation to climate, fire, and soils. *Ecological Monographs* 6:233–268.

Daubenmire, R. F. 1938. Merriam's life zones of North America. *The Quarterly Review of Biology* 13:327–332.

Daubenmire, R. F. 1939. Contributions to the ecology of the Big Bend area of Washington: Forage resources of the Scabland Prairies and the effects of grazing upon them. *Northwest Science* 13:33–37.

Daubenmire, R. F. 1940a. Plant succession due to overgrazing in the *Agropyron* bunchgrass prairie of southeastern Washington. *Ecology* 21:55–64.

Daubenmire, R. F. 1940b. Some forest-soil relationships: Contributions to the ecology of the Big Bend Area of Washington: II. Indicator significance of the natural plant communities in the region of the Grand Coulee irrigation project. *Northwest Science* 14:8–10.

Daubenmire, R. F. 1942. An ecological study of the vegetation of southeastern Washington and adjacent Idaho. *Ecological Monographs* 12:53–79.

Daubenmire, R. F. 1943. Vegetational zonation in the Rocky Mountains. *Botanical Review* 9:325–393.

Daubenmire, R. F. 1946. The life zone problem in the northern intermountain region. *Northwest Science* 20:28–38.

Daubenmire, R. F. 1947. *Plants and environment: A textbook of plant autecology*. New York: Wiley.

Daubenmire, R. F. 1952. Forest vegetation of Northern Idaho and adjacent Washington, and its bearing on concepts of vegetation classification. *Ecological Monographs* 22:301–330.

Daubenmire, R. F. 1953. Classification of the conifer forests of eastern Washington and northern Idaho. *Northwest Science* 27:17–24.

Daubenmire, R. F. 1954. Vegetation classification. *Veröffentlichungen des Geobotanischen Institutes Rübel in Zürich* 29:29–34.

Daubenmire, R. F. 1956a. Climate as a determinant of vegetation distribution in eastern Washington and Northern Idaho. *Ecological Monographs* 26:131–154.

Daubenmire, R. F. 1956b. The use of vegetation to indicate grazing potentials of forest land. In Converting the old-growth forest: Proceedings of the Society of American Foresters Meeting, October 16–21, 1955, 35–36. Washington: Society of American Foresters.

Daubenmire, R. F. 1959a. A canopy-coverage method of vegetational analysis. *Northwest Science* 33:43–64.

Daubenmire, R. F. 1959b. *Plants and environment: A textbook of plant autecology*. 2nd ed. New York: Wiley.

Daubenmire, R. F. 1960. Some major problems in vegetation classification. *Silva Fennica* 105:22–25.

Daubenmire, R. F. 1966. Vegetation: Identification of typal communities. *Science* 151:291–298.

Daubenmire, R. F. 1968a. Ecology of fire in grasslands. In *Advances in ecological research*, ed. J. B. Cragg, vol. 5, pp. 209–266. London: Academic.

Daubenmire, R. F. 1968b. *Plant communities: A textbook of plant synecology*. New York: Harper and Row.

Daubenmire, R. F. 1970. Steppe vegetation of Washington, Technical Bulletin 62. Pullman: Washington Agricultural Experiment Station, Washington State University.

Daubenmire, R. F. 1973. A comparison of approaches to the mapping of forest land for intensive management. *The Forestry Chronicle* 49:87–91.

Daubenmire, R. F. 1974. *Plants and environment*. 3rd ed. New York: Wiley.

Daubenmire, R. F. 1976. The use of vegetation in assessing the productivity of forest lands. *Botanical Review* 42:115–143.

Daubenmire, R. F. 1978. *Plant geography with special reference to North America*. New York: Academic.

Daubenmire, R. F. 1984. Viewpoint: Ecological site/range site/habitat type. *Rangelands* 6:263–264.

Daubenmire, R. F., and W. E. Colwell. 1942. Some edaphic changes due to overgrazing in the Agropyron-Poa prairie of southeastern Washington. *Ecology* 23:32–40.

Daubenmire, R. F., and Jean B. Daubenmire. 1968. *Forest vegetation of eastern Washington and Northern Idaho*. Pullman: Washington State University Extension.

Duffin, Andrew P. 2007. *Plowed under: Agriculture and environment in the Palouse*. Seattle: University of Washington Press.

Dwire, Kathleen A., Bruce A. McIntosh, and J. Boone Kauffman. 1999. Ecological influences of the introduction of livestock on Pacific Northwest ecosystems. In *Northwest lands, northwest peoples: Readings in environmental history*, ed. Dale D. Goble and Paul W. Hirt, pp. 313–335. Seattle: University of Washington Press.

Fox, Stephen. 1981. *The American conservation movement: John Muir and his legacy*. Madison: University of Wisconsin Press.

Franklin, Jerry F. 1980. Ecological site classification activities in Oregon and Washington. *The Forestry Chronicle* 56:68–70.

Gleason, H. A. 1926. The individualistic concept of the plant association. *Bulletin of the Torrey Botanical Club* 53:7–26.

Gleason, H. A. 1939. The individualistic concept of the plant association. *American Midland Naturalist* 21:92–110.

Golley, Frank B. 1993. *A history of the ecosystem concept in ecology: More than the sum of the parts*. New Haven: Yale University Press.

Hagen, Joel B. 1992. *An entangled bank: The origins of ecosystem ecology*. New Brunswick: Rutgers University Press.

Hays, Samuel P. 1959. *Conservation and the gospel of efficiency: The progressive conservation movement, 1890–1920*. Cambridge: Harvard University Press.

Henderson, Jan A., Robin D. Lesher, David H. Peter, and Chris D. Ringo. 2011. A landscape model for predicting potential natural vegetation of the Olympic Peninsula USA using boundary equations and newly developed environmental variables, General Technical Report PNW-GRT 841. Portland, Oregon: United States Department of Agriculture, Forest Service, Pacific Northwest Research Station.

Hinz, Sue. 1975. Daubenmire: A botanist 365 Days a Year. Hilltopics (October), 14.

Hirt, Paul W. 1994. *A conspiracy of optimism: Management of the national forests since World War Two*. Lincoln: University of Nebraska Press.

Hirt, Paul W. 1999. Getting out the cut: A history of national forest management in the Northern Rockies. In *Northwest lands, northwest peoples: Readings in environmental history*, ed. Dale D. Goble and Paul W. Hirt, pp. 437–461. Seattle: University of Washington Press.

Hoffman, George R. 1980. Eminent ecologist for 1979, Rexford Daubenmire. *Bulletin of the Ecological Society of America* 61:34.

Hoffman, George R. 1996. Rexford Daubenmire, 1909–1995. *Bulletin of the Ecological Society of America* 77:143–144.

Kendeigh, S. Charles. 1954. History and evaluation of various concepts of plant and animal communities in North America. *Ecology* 35:152–171.

Kingsland, Sharon E. 2005. *The evolution of American ecology, 1890–2000*. Baltimore: Johns Hopkins University Press.

Knobloch, Frieda. 1996. *The culture of wilderness: Agriculture as colonization in the American West*. Chapel Hill: University of North Carolina Press.

Kohler, Robert E. 2002. *Landscapes and labscapes: Exploring the lab-field border in biology*. Chicago: University of Chicago Press.

Kohler, Robert E. 2008. Plants and pigeonholes: Classification as a practice in American ecology. *Historical Studies in the Natural Sciences* 38:77–108.

Langston, Nancy. 1995. *Forest dreams, forest nightmares: The paradox of old growth in the inland west*. Seattle: University of Washington Press.

Maher, Neil M. 2008. *Nature's new deal: The Civilian Conservation Corps and the roots of the American environmental movement*. New York: Oxford University Press.

McEvoy, Arthur F. 1986. *The fisherman's problem: Ecology and law in the California fisheries, 1850–1980*. Cambridge: Cambridge University Press.

McGregor, Alexander Campbell. 1982. *Counting sheep: From open range to agribusiness on the Columbia Plateau*. Seattle: University of Washington Press.

McIntosh, Robert P. 1985. *The background of ecology: Concept and theory*. Cambridge: Cambridge University Press.

McIntosh, Robert P. 1989. Citation classics of ecology. *Quarterly Review of Biology* 64:31–49.

Meinig, D. W. 1995. *The Great Columbia Plain: A historical geography, 1805–1910. Reprint 1968 edition*. Seattle: University of Washington Press.

Merriam, C. Hart. 1890. Results of a biological survey of the San Francisco mountain region and desert of the Little Colorado in Arizona. *North American Fauna* 3:1–136.

Merriam, C. Hart. 1898. *Life zones and crop zones of the United States*. Washington, D.C.: Government Printing Office.

Nicolson, Malcolm. 1987. Alexander von Humboldt, Humboldtian science and the origins of the study of vegetation. *History of Science* 25:167–194.

Nicolson, Malcolm. 2001. 'Towards establishing ecology as a science instead of an art': The work of John T. Curtis on the plant community continuum. Web Ecology 2:1–6. http://www.web-ecol.net/2/1/2001/we-2-1-2001.html. Accessed 30 July 2013.

Nicolson, Malcolm, and Robert P. McIntosh. 2002. H. A. Gleason and the individualistic hypothesis revisited. *Bulletin of the Ecological Society of America* 83:133–42.

O'Hara, Kevin L., Penelope A. Latham, Paul Hessburg, and Bradley G. Smith. 1996. A structural classification for inland northwest forest vegetation. *Western Journal of Applied Forestry* 11:97–102.

Pfister, Robert D., and Stephen F. Arno. 1980. Classifying forest habitat types based on potential climax vegetation. *Forest Science* 26:52–70.

Pitzer, Paul C. 1994. *Grand Coulee: Harnessing a dream*. Pullman: Washington State University Press.

Real, Leslie A., and James H. Brown, eds. 1991. *Foundations of ecology: Classic papers with commentaries*. Chicago: University of Chicago Press.

Rothman, Hal K. 1998. *The greening of a nation? Environmentalism in the United States since 1945*. Fort Worth: Harcourt Brace College Publishers.

Rumore, Gina M. 2009. A natural laboratory, a national monument: Carving out a place for science in Glacier Bay, Alaska, 1879–1958. Ph.D. dissertation, University of Minnesota.

Sowards, Adam M. 2007. *United States west coast: An environmental history*. Santa Barbara: ABC–CLIO.

Stohlgren, Thomas J. 2007. *Measuring plant diversity: Lessons from the field*. New York: Oxford University Press.

Stout, David. 1995. R. F. Daubenmire, 85, botanist whose books inspired ecologists. *New York Times*, September 8.

Tansley, A. G. 1935. The use and abuse of vegetational concepts and terms. *Ecology* 16:284–307.

Thomas, Robert M Jr. 1995. Stanley Cain, ecologist 92, led the field. *New York Times*, April 2.

Tobey, Ronald C. 1981. *Saving the prairies: The life cycle of the founding school of American plant ecology, 1895–1955*. Berkeley: University of California Press.

Vetter, Jeremy, ed. 2011. *Knowing global environments: New historical perspectives on the field sciences*. New Brunswick: Rutgers University Press.

Vogl, Richard J., Grant Cottam, and Robert P. McIntosh. 1966. Vegetational continuum (letters to the editor). *Science* 152:546–547.

Whittaker, Robert H. 1962. Classification of natural communities. *The Botanical Review* 28:1–239.

Worster, Donald. 1979. *Dust bowl: The southern plains in the 1930s*. New York: Oxford University Press.

Worster, Donald. 1994. *Nature's economy: A history of ecological ideas*. 2nd ed. Cambridge: Cambridge University Press.

Young, James A., and Fay L. Allen. 1997. Cheatgrass and range science: 1930–1950. *Journal of Range Management* 50:530–535.

Adam M. Sowards is associate professor of history at the University of Idaho. He concentrates on environmental history in the North American West. He has written or edited several books, including *Idaho's Place: A New History of the Gem State* (2014) and *The Environmental Justice: William O. Douglas and American Conservation* (2009). Currently, his research centers on how scientists have come to understand particular places across changing scales.

Chapter 16
Agricultural Improvement at China's First Agricultural Experiment Stations

Peter Lavelle

Introduction

Experiment stations (*shiyan chang*) for field research in agriculture, sericulture, and forestry were established in China beginning in the first decade of the twentieth century. This chapter provides an analysis of the background and research agendas of China's first agricultural experiment stations. It argues that improving the country's stock of cultivars and its use of fertilizers was of prime concern to agricultural reformers and specialists who oversaw new agricultural institutions from the turn of the century to the late 1910s, prior to the institutionalization of agricultural research at Chinese universities. At the experiment stations, agricultural specialists focused much of their field research and experimental activity on plant propagation methods, crop acclimatization, and soil composition and fertility. Reformers who aimed to boost the output of China's agriculture to serve the needs of industrial and commercial development commonly asserted the urgency of disseminating better, more productive varieties of crops throughout the country and of identifying the best fertilizers. The experiment stations were the primary testing and demonstration grounds for these agricultural inputs.

Agricultural experiment stations emerged during the decade-long era of political reforms carried out by the central and provincial governments of the Qing dynasty (1644–1912) just before its end.[1] Historians in large part have ignored experiment stations in histories of Chinese agriculture and agronomy, the late Qing reforms, or the early years of the Republican era (1912–1949).[2] These new institutions were not

[1] On the reforms, see Reynolds (1993) and Reynolds (1995). In the main text throughout this chapter, Chinese and Japanese names are written according to the standard name order, with family name followed by given name, but the reference list uses English-language name order.

[2] One recent exception is Yuan (2012), pp. 110–129.

P. Lavelle (✉)
Department of History, Temple University, 1115 West Berks St., Philadelphia, PA 19122, USA
e-mail: plavelle@temple.edu

D. Phillips, S. Kingsland (eds.), *New Perspectives on the History of Life Sciences and Agriculture,* Archimedes 40, DOI 10.1007/978-3-319-12185-7_16

merely products of governmental reform in the last years of the empire. They were manifestations of a movement for agricultural reform and improvement (*nongshi gailiang*) which sought to leverage China's potential for economic growth through the technical development of farming and through the expansion of cultivation in the country's internal and peripheral frontiers. The turmoil in this era, including military and diplomatic confrontations with foreign powers, the downfall of the imperial court, and the ensuing territorial divisions and chaos of the warlord period (1916–1928), made the work of agricultural research at the experiment stations all the more urgent for reformers.

While the reformers who led the movement for agricultural improvement acknowledged the importance of drawing upon existing Chinese methods of production, they also made experiment stations the proving grounds for the "new logics" (*xin li*) and "new methods" (*xin fa*) of agriculture that had been pioneered in other countries. These new ideas and techniques came to China through multiple channels, including translated texts containing information about modern inputs like farm machinery and Peruvian guano.[3] But the actual condition of Chinese agriculture around the turn of the century, which was labor-intensive and capital-poor,[4] made ideas like the adoption of agricultural machinery or the large-scale importation of fertilizers far-fetched. Reformers nonetheless adopted new but accessible inputs such as cultivars from foreign countries and new knowledge about things like soil chemistry as they developed and implemented research agendas at the experiment stations.

The prime inspirations for agricultural improvement and the main sources of new agronomic knowledge in China at this time were Japan and the USA. The first agricultural experiment stations in both countries were established in the 1870s, and by the turn of the twentieth century, both had burgeoning networks of agricultural research at the stations and the educational institutions to which they were often attached.[5] In the nineteenth century, both countries had each borrowed from the other different tools for their own agricultural development. While the USA had appropriated seeds and plants from Japan, Japan had adopted organisms and institutional models for agricultural education and research from the USA.[6] China turned to both countries (and, to a lesser extent, European countries) for obtaining valuable seeds, for acquiring new ideas about soil management, for hiring experts for schools and

[3] The text about agricultural machinery is *Nongshi lun lüe* (Brief discussion of agriculture), in Liang (1897). The text about guano, *Pilu guo que fen lun* (Guano in Peru), was translated from Japanese and published in the *Nongxue bao* (*Journal of agriculture*) in 1902. See Zhong (1996), p. 157.

[4] Eastman (1988), p. 66. China was, however, rich in landesque capital, the human-made land formations that are part of agricultural infrastructure and technology, such as irrigation works or terraced fields.

[5] For the USA, see True (1970), pp. 82–106, 118–64. For Japan, see Hayami (1975), pp. 49–52; Ogura (2000), pp. 318–324; and Francks (1984), p. 79.

[6] On Japan's appropriation of American models for education and research, see Kargon (2008), esp. 65–66, and True (1970), p. 44, 46–47. On the American appropriation of germplasm and other biological materials from Japan and East Asia, see Kloppenburg (2004), p. 55, 60, 78.

research stations, and for providing advanced education to Chinese students. Both offered similar models of institutionalized research at experiment stations.[7] But in the 20 years after the first Sino-Japanese War of 1894–1895, owing to their recognition of the successes of reforms in Japan carried out after the Meiji Restoration of 1868 as well as to their close proximity to Japan, Chinese reformers sought experts, education, and texts from Japan more readily than from the USA.[8]

In the Qing dynasty's last decade of existence, the Chinese men who oversaw the stations adopted new ideas and technologies, but their analyses of the environmental conditions for production sometimes retained pre-twentieth-century logic about the relationship between plants and their terrestrial surroundings. That logic stressed the need to understand the natural characteristics of organisms and soils in order to determine a plant's viability in a particular environment in a way that seemed to assume the fixity of those characteristics. One reason for the continuation of earlier discourses of agriculture, plants, and soils alongside the new vocabularies of disciplines like agricultural chemistry was the diversity of educational backgrounds of the people who played a role in the experiment stations. In this transitional generation, only some of them had received special training at agricultural schools in China, Japan, or the USA. Others had grown up preparing for the imperial civil service examinations, which were abolished in 1905, and so they had a broad knowledge of classical texts and their discourses. This era of agricultural research thus saw the commingling of older and newer ways of thinking about plants, soils, and their potential for change via acclimatization research and agricultural chemistry. In this era, Chinese researchers concentrated on changing the varieties of crops grown in a particular location and on transforming soil fertility with fertilizers, rather than on changing the plants themselves with the techniques of plant breeding which were becoming commonplace in agricultural research elsewhere in the world.[9]

As it describes the origins and development of China's earliest agricultural experiment stations, this chapter pays particular attention to the research and researchers at the experiment station established in Beijing. It is one of the best documented of the earliest experiment stations in China, and its reports and related publications are important sources for understanding the agendas of Chinese agricultural research prior to the 1920s. Because of its location and affiliation with central governments during the late Qing and early Republican eras, it sat at the center of a countrywide network of exchange in ideas and organisms, and it played a key role in undertaking field research in support of the movement for agricultural improvement.

[7] In some instances, Chinese officials directly referenced American and Japanese experiment stations as inspirations for Chinese agricultural research (Ye 1909, p. 1a; Nong shang bu 1914c, p. 1a).

[8] One historian has called Japan "China's model and active partner" in the political and institutional reforms of the era. See Reynolds (1993), p. 5.

[9] Kingsbury (2009), pp. 144–166, and Kloppenburg (2004), pp. 66–84.

The Origins of Agricultural Experiment Stations in China

After the defeat of Qing forces by Japan in 1895, reform-minded scholars began to make vociferous calls for the imperial court to establish new institutions to strengthen the country, including institutions to spread new knowledge about agriculture. Among these scholars was Luo Zhenyu, who took a special interest in Japan's experience with agricultural and educational reforms.[10] Along with other scholars, Luo organized the Society for Agriculture (*Wunong hui* or *Nongxue hui*) in Shanghai in 1896. According to its regulations, the society was to pursue activities related to agricultural reform, such as establishing an agricultural school; experimenting with cultivation; distributing superior seed varieties; manufacturing fertilizers, pesticides, and tools; and holding competitive exhibitions of farm products.[11] The society is best known for its translation and publication of agricultural texts. The *Journal of Agriculture* (*Nongxue bao*), which it published from 1897 to 1906, as well as its voluminous *Collected Works on Agriculture* (*Nongxue congshu*) contained articles and treatises written originally in Chinese, Japanese, English, French, and German on a wide range of topics, from agricultural reform and management to new tools and technologies, to crop and soil sciences.[12] The society also made several attempts to set up testing grounds for foreign and domestic crops and silkworms, but none of these attempts seem to have reached fruition.[13]

In the summer of 1898, scholars pressed the Qing court to undertake a series of political and institutional reforms, including new initiatives in agricultural education, to develop China's economy and strengthen the state. Responding to these calls, the Guangxu emperor issued edicts to create a new government ministry to oversee agricultural, industrial, and commercial development in the empire, as well as to establish new agricultural schools in the provinces.[14] Although the planned reforms were scuttled several months later by the empress dowager Cixi, there was a surge of interest in agricultural reform throughout the country among scholars and high-ranking provincial officials in the following years. It was in this context that the first experiment stations and modern schools for agriculture, sericulture, and forestry in China were established. In Zhili Province, Governor-General Yuan Shikai set up bureaus for agriculture and land reclamation, an agricultural school, and an experiment station in 1902, followed by an agricultural association and an

[10] Luo published "Riben nongzheng weixin ji" (Japan's agricultural policy reforms) as part of a longer treatise devoted to ideas for agricultural reform in China. See Luo (2010).

[11] "Wu nong hui shiban zhangcheng nigao" (Draft of trial regulations of the Society for Agriculture), *Nongxue bao* (*Journal of agriculture*) 15 (November 1897), 2b, in Jiang and Jing (2009), 1:106, and also quoted in Li (1957), 1:866.

[12] Nearly two-thirds of the treatises in the *Collected Works* were translated from Japanese (Li 2008, p. 27). Lists of texts in the *Collected Works* appear in Dong and Fan (2000), pp. 837–840, and Li (1957), 1:868–70. For an overview of the Society's publications, see Ihara (2000), pp. 298–303.

[13] Li (2008), p. 29.

[14] Li (1956), p. 156, and Guo (1988), p. 463.

agricultural journal in 1905.[15] The experiment station sat on roughly 66 acres just west of the provincial capital, Baoding, and the school was located on its grounds. The two institutions shared the same leadership and a mission of agricultural research and education. The first crop of students at the school, who were admitted based solely on their literary aptitude, were rushed through the training program in 1 year so that they could be "made to do experiments on their own at the experiment station in order to be ready to serve the needs of agricultural education."[16] Although the school in Zhili was not China's first modern agricultural school, the experiment station seems to have been the first instance in China of a provincial-level institution created and funded for the sake of field research and experimentation in agriculture. Within a few years, officials in at least seven other provinces also established academies and training institutes in agriculture, sericulture, forestry, and animal husbandry with attached or nearby experimental stations so that students could be afforded the pedagogical benefits of putting into practice what they had learned in lecture halls.[17]

Efforts to create institutions devoted to agricultural research and education were bolstered by the actions of the Qing court after the turn of the twentieth century. Under Cixi's purview, the court reversed its opposition to institutional reforms, and in 1903 it took the initial step of establishing a Ministry of Commerce (*Shang bu*). The new ministry called upon provincial authorities to set up experiment stations to undertake "the analysis of soil quality, the examination and testing of seeds, the manufacture of fertilizers, and the observation and measurement of climate" and to allow farmers to observe field tests so that they would be convinced to adopt new crops and new methods of cultivation.[18] Three years later, the court established a Ministry of Agriculture, Industry, and Commerce (*Nong gong shang bu*).[19] Its direct contribution to research was to set up and oversee an agricultural experiment station in the imperial capital. Established in 1906, the station began operations 2 years later on roughly 70 ha of land just outside the Xizhi Gate, near the northwest corner of Beijing's city walls.[20]

The creation of the experiment station in Beijing came amidst a wave of new experiment stations in the provinces. From the turn of the twentieth century to the end of the Qing state, officials established agricultural experiment stations in all of the provinces; stations in Zhili (1902), Shanxi (1903), Shandong (1903), Fujian (1906),

---

[15] Tao et al. (1973), p. 4, 5–6 (Jia Shumo, "Zhili sheng nongye qingxing" (Agricultural conditions in Zhili Province)).

[16] Nong gong shang bu (2007a), 2:4a, 13a–b.

[17] These provinces were Shanxi, Hubei, Sichuan, Shandong, Henan, Gansu, and Guizhou. (Nong gong shang bu 2007a, 2:4b, 5a, 5b; Nong gong shang bu 2007b, 2:5a, 5b, 6a, 7a). At an agricultural school established in Gansu Province in 1907, an experiment station with space for a forestry section and mulberry garden was set up to give students a chance to conduct field tests on crops with "the expectation of bringing the academic principles of the lecture halls and practical experimentation (*shidi shiyan*) into concordance" (Nong gong shang bu 2007b, 2:6a).

[18] Quoted in Li (2008), p. 55.

[19] Li (1956), p. 210, 216.

[20] Ye (1909), pp. 1b–2a, and Nong gong shang bu (2007a), 2:9a.

and Fengtian (1906) were among the earliest.[21] Experiment stations also proliferated within individual provinces. For example, Guangdong Province had seven stations by 1911.[22] Many provinces had additional stations for sericulture or forestry. Beijing became the site of a second experiment station when the central government established a separate forestry experiment station in 1912 on the grounds around the Temple of Heaven in the southern section of the city.[23]

The agricultural experiment station in Beijing was among the largest and most active stations in the country and, as I discuss, it played a central role in the emerging network of experiment stations across China into the 1910s. When it began operations in 1908 under the purview of the Ministry of Agriculture, Industry, and Commerce, the station was divided into three research divisions: plant propagation, sericulture, and animal husbandry.[24] Of these three divisions, the plant propagation division had the largest staff with 31 members, including 2 division heads, 14 regular staffers, 1 foreign technician, and 4 foreign assistants. By contrast, the sericulture division had only 12 staff members, and the animal husbandry division had only 7 staff members. Together, the research staff shared ten different areas of the station, one each devoted to experimentation in rice, grains, mulberry trees, vegetables, fruit trees, flowers, fodder grasses, industrial crops, fish, and tree seedlings.[25] After 1912, the central government reorganized the research staff at the Beijing station into five divisions: plant propagation; horticulture; sericulture; chemical experiments; and diseases, pests, and disasters.[26]

Field experiments were the mainstay of the Beijing station's research work. Prior to the twentieth century, Qing scholars with an interest in promoting agriculture had sometimes conducted simple experiments with crops and cultivation techniques in which they compared the harvest yields of trial fields with the output of their regular fields.[27] What distinguished field research at the first experiment stations in China from its early modern antecedents was the large number of controlled field trials, the great variety of plants tested, and the emphasis on chemical analyses of soils and fertilizers. Researchers at the stations conducted experiments by testing single inputs or variables of production, such as seeds, fertilizers, and specific methods of cultivation. In 1909, for example, researchers in Beijing conducted three experiments on rice designed to determine the best method of seed selection, the best variety of seed, and the best mixture of fertilizers, with the primary measure of success being the total grain output per plot. They performed similar tests with dryland rice, sorghum, maize, buckwheat, soybeans, mung beans, and millet.[28] Within a

[21] Nong gong shang bu (2007a), 2:12a-15b; Nong gong shang bu (2007b), 2:10a–11a; Bai et al. (1995), p. 13; and Yuan (2012), pp. 111–112.

[22] Zhang (1979), p. 141.

[23] Nong shang bu (1914a), p. 1.

[24] Ye (1909), p. 17a.

[25] Ye (1909), p. 24a, 25b.

[26] Nong shang bu (1914c), p. 1b.

[27] Elvin (1975), p. 95, 101.

[28] Ye (1909), pp. 58a–59b.

few years of its opening, the Beijing station was undertaking dozens of field experiments per year with a wide range of crops. In 1913, the station carried out 52 different experiments on cereals, legumes, industrial crops, fiber crops, vegetables, root crops, and melons and gourds, and worked with 64 types of plants in total.[29]

Like agricultural experiment stations elsewhere in China, the station in Beijing served multiple purposes. Aside from conducting scientific research, the station also showcased new methods of agriculture to the Chinese public. In 1906, Qing officials had raised the idea of an experiment station in the capital to "increase and broaden the knowledge of the nation's farmers," whom they deemed ignorant of the best agricultural organisms, tools, and techniques.[30] Established on the former grounds of an imperial property called Leshan Garden (*Leshan yuan*), the station was intended to be a place of research and spectacle, experimentation and enjoyment, as it enabled researchers to conduct field tests of "all new and old logics and methods" while attracting the attention of city residents and farmers, who could come to have a look around.[31] Indeed, putting things on display for the public became a major trope of early twentieth-century Chinese attempts to propel economic development. Reformers organized special exhibitions of agricultural and industrial products, of which the Nanyang Industrial Exhibition (*Nanyang quanye hui*) of 1910 in Nanjing was the largest, while experiment stations functioned as active, ongoing exhibitions.[32] At some stations, displays of vegetable and animal specimens were erected for visitors to view, and photographs of specimens and test fields were sometimes reproduced in annual reports put out by the stations. In addition to its experiment fields, the Beijing station had a zoo and an exhibition hall where the public could observe samples of agricultural organisms and products. Although government officials were not always satisfied with the public's level of interest in the station, they strove to make it an interesting place of leisure and learning that would attract visitors.[33]

Before examining the research work of the experiment stations in greater detail, it is worthwhile to consider the sources of influence on those who directed programs of agricultural experimentation and education in China in this era. Among the Beijing station's 7 directors between 1906 and 1914, 6 had been trained overseas: 2 at Cornell University's College of Agriculture, 2 at Tokyo Imperial University's Faculty of Agriculture, 1 at Tohoku Imperial University's College of Agriculture (later part of Hokkaido Imperial University), and 1 at the University of California's

[29] Nong shang bu (1914c), *muci*:1–6, *fulu*:43–48.

[30] Ye (1909), p. 10a.

[31] Ye (1909), p. 1b, 4a, 9a, 17a–b.

[32] Zhu Ying, "On late Qing economic laws and regulations, (ca. 1901–1911)," in Reynolds (1995), p. 124.

[33] In 1914, officials lamented the public's general disregard for research taking place at the station. They decided to sell tickets for admission to the station, the zoo, and the exhibition halls to raise money for research, while continuing to encourage spectators to learn about new trends in farming and new agricultural products (Nong shang bu 1914c, pp. 1a–b; Nong shang bu 1914b, p. 2, 31–34).

College of Agriculture.[34] Despite the fact that Japan and the USA had each educated 3 future directors of the station, Japan was the more popular option for agricultural training among the station's other staff. Of the 59 other Han Chinese and Manchu staff members at the station in the same period, 7 studied at Japanese institutions but only 1 studied in the USA.[35] Moreover, foreign staff members at Chinese experiment stations were almost exclusively Japanese. Between 1907 and 1914, the Beijing station employed at least 2 Japanese specialists, 1 each in the sericulture and horticulture divisions.[36] At the agriculture and forestry stations in Shandong Province, a Japanese graduate of Sapporo Agricultural College (later part of Hokkaido Imperial University) served for a number of years as an advisor; the agricultural experiment station in Shanxi Province also hired Japanese advisors.[37] In the first decade of the twentieth century, Japanese presence and influence was even more apparent in Chinese agricultural schools insofar as much of the pedagogical material and most of the foreign teachers were Japanese.[38]

New Plants and Acclimatization

One of the main objectives of Chinese agricultural researchers in this era was to test the productivity and potential acclimatization of foreign and domestic crops on land at the experiment stations. This research agenda developed from the idea, commonplace in the first decade of the twentieth century, that importing valuable seeds and plants from overseas or from neighboring provinces was one of the most expedient means to raise the quality and quantity of output per unit of land and labor. Gathering and redistributing biological material in the form of seeds was a common practice of agricultural improvement in both Japan and the USA, and Chinese reformers hoped to emulate the apparent successes of both countries. Luo Zhenyu looked to Japan's experience with importing plants over the previous 30 years when he wrote in 1900 that China should promote "improvement through transplanting" (*yizhi gailiang*), and he called upon the imperial government to import crops and animals from abroad, including cotton and wheat from the USA, wheat, lucerne, cattle, and chickens from Europe, and horses from Central Asia.[39] A decade later, after the experiment stations had begun their work, scholars who gathered at the

[34] Nong shang bu (1914c), *fulu*:1, 4.

[35] Nong shang bu (1914b), pp. 91–99.

[36] Nong shang bu (1914b), p. 92, 96. Perhaps because of these Japanese connections, a catalog of dozens of photographed scenes from the experiment station in Beijing was printed in Tokyo after its 1st year of operation. See Anonymous (1909).

[37] Nong gong shang bu (2007a), 2:14a-14b, 15a.

[38] Governor-General Zhang Zhidong's decision in 1900 to switch from American to Japanese agricultural specialists, whom he employed as advisors, exemplified the spirit in which Chinese reformers more readily looked toward Japan in this period. See Stross (1986), pp. 42–49; Reynolds (1993), p. 108; and Zhang (1979), p. 140.

[39] Luo (2010), 11:304–306; Li (1957), 1:859–860; and Li (2008), p. 29.

Nanyang Industrial Exhibition in the summer of 1910 to exchange ideas about agriculture, industry, education, and other matters reiterated the significance of the importation and circulation of seeds and organisms for agricultural development. Tao Changshan, a graduate of Tohoku Imperial University's College of Agriculture who later served as a director at the Beijing station from 1913 to 1914, asserted that importing new varieties of crops was singularly the best method available for improving China's agriculture. Based on his impression of Japan's wild successes with crop acclimatization, he urged his fellow agronomists and reformers to import with urgency varieties of wheat and fodder grasses from abroad.[40] Zhu Zurong, who had helped Luo Zhenyu organize the Society for Agriculture in Shanghai and who had written about adopting foreign cotton varieties in China, likewise spoke of collecting and exchanging superior-quality seeds and of using chemical fertilizers, as the best means to raise agricultural output. According to him, these superior varieties were not only located in the USA, Japan, Russia, European countries, and South Asia. They, and especially dryland grain crops and American cottons, could now be found at experiment stations in China.[41]

The agricultural experiment stations played a major role in conducting propagation and acclimatization studies of crop varieties from overseas and from other parts of China. In field tests with a large number of crops, researchers compared the productivity of foreign and domestic cultivars and assessed the impact of the timing of sowing, the methods of cultivation, and the patterns of fertilization on harvest yields. In its early years, the Beijing station organized trials of grains from France, the USA, Italy, Japan, as well as from many regions of China. For its experiments with cotton and jute, the station obtained seeds from China, the USA, and Japan. Among the vegetables, legumes, and other plants that technicians propagated in field experiments were sugar beets from Holland; muskmelons, tomatoes, and hot peppers from Russia; and kidney beans, hollyhocks, radishes, and olives from the USA.[42] The Beijing station also made plans to purchase more than ten varieties of apple trees from the agricultural college in Hokkaido, Japan, for trial cultivation in an experimental orchard for apple cultivars slated to open in 1915.[43] Other stations also experimented with a variety of agricultural organisms. At the experiment station in Shandong Province, researchers tested local varieties of grains and vegetables; cotton and bean varieties from the USA; and grain, gourd, and vegetable varieties from Japan.[44] In Zhili Province, experiment station staff propagated many varieties of foreign organisms, such as Japanese maize, American cotton, American and German barleys, Japanese fruit and mulberry trees, and three kinds of Japanese

[40] Nanyang quanye hui yanjiu hui (2002), 2:272 (Tao Changshan, "Mai lei zhi xin zhong shuru shuo" (Discussion of the importation of new varieties of wheat)).

[41] Nanyang quanye hui yanjiu hui (2002), 2:171–172 (Zhu Zurong, "Jia zhong lu" (Record of superior seeds)). His earlier work is Zhu (1995).

[42] Ye (1909), pp. 5a–b.

[43] Nong shang bu (1914c), *fulu*:5.

[44] Nong gong shang bu (2007a), 2:14a.

silkworms.[45] The experiment station attached to the Guangdong Provincial Agricultural Association gathered and planted the seeds of various varieties of foreign cotton.[46] In Fengtian Province, among the 27 varieties of wheat cultivated for experiments in 1908, there were 3 from Japan, 6 from northeast China, 8 from Russia, and 10 from the USA.[47]

Experiment stations acquired agricultural organisms for use in field tests from many different sources. The station in Fengtian may have been typical in collecting local specimens by calling upon area farmers to send their best seeds to the station.[48] But foreign varieties were somewhat harder to obtain and required coordinated action by people inside and outside the country. In 1907, by the request of the Ministry of Agriculture, Industry, and Commerce, Qing diplomats abroad purchased seeds of foreign varieties of grain plants, fruit trees, and flowers and sent them to Beijing for experimentation and comparison with Chinese varieties.[49] (In addition, during a fact-finding mission abroad, Qing diplomat Duanfang acquired in Germany many animals from around the world which he gave to the Qing court and which were later transferred to the zoo at the experiment station in Beijing.[50]) In 1910, Zhu Zurong also urged overseas Chinese to participate in the acquisition of plant material from abroad by sending seeds and specimens to China as they had done for at least a dozen different organisms, including some from Southeast Asia.[51]

Experiment stations also exchanged and disseminated agricultural organisms to expand the geographic scope of cultivation of the most productive varieties. Occupying a premiere position in the network of experiment stations in China, the station in Beijing cooperated with provincial stations on gathering and circulating seeds and other biological material.[52] For one thing, the station received the seeds of plants which had undergone testing at other stations. Researchers at the Beijing station cultivated specimens of black millet, white peas, and white radishes named for the experiment station in Zhili, as well as a white winter wheat and a red winter wheat named for the experiment station in Fengtian.[53] By 1914, station officials had developed regulations governing the distribution of seeds, silkworm eggs, and tree seedlings from Beijing to agricultural organizations and private individuals. The regulations restricted distribution to only those organisms which had been shown through experimentation in Beijing to be especially productive, and each party could receive only a limited amount of biological material from the station. Interested parties were required to submit formal requests to Beijing during

[45] Nong gong shang bu (2007a), 2:13a–b.

[46] Tao et al. (1973), pp. 133–134 (Huang Qian, "Guangdong sheng nongye qingxing" (Agricultural conditions in Guangdong Province)).

[47] Fengtian nongye shiyan chang (1909a), 4:1:*nongyibu*:1a–2b.

[48] Nong gong shang bu (2007a), 2:12a.

[49] Ye (1909), pp. 4a–4b.

[50] Nong gong shang bu (2007a), 2:9a, and Ye (1909), p. 6b.

[51] Nanyang quanye hui yanjiu hui (2002), 2:196 (Zhu, "Jia zhong lu").

[52] Ye (1909), p. 17b.

[53] Nong shang bu (1914c), *shuyike*:12, 45, 56, 59, *yuanyike*:20.

specific periods of the year. Those who received specimens were required to report back to the station about growing conditions and harvests so that researchers could accumulate information about the climatic and terrestrial conditions of cultivation across China and refine their assessments of the productivity of specific varieties of organisms.[54]

Another organization, the National Federation of Agricultural Associations (*Quanguo nonghui lianhehui*), was also involved in the circulation of plant material. Established in 1910 at the Nanyang Industrial Exhibition, the Federation brought together like-minded reformers and agronomists from agricultural associations across the country.[55] Its creation marked a culmination of efforts to establish a network of provincial agricultural associations responsible for research and education. These efforts had begun with the Society for Agriculture in Shanghai and had been promoted by official action in 1907 when the Qing government issued a set of regulations for provinces to establish their own networks of agricultural associations, with the provincial association and its experiment station in the provincial capital and branch associations in sub-provincial administrative districts.[56] The Federation's first meeting in Beijing in the spring of 1913 enabled delegates from many provinces to exchange information about agriculture in different regions of the country, to visit the agricultural and forestry experiment stations in the city, and to present proposals, one of which was to establish a seed exchange bureau (*zhongzi jiaohuan suo*).[57] The Ministry of Agriculture, Industry, and Commerce also used the Federation to coordinate the dissemination of agricultural organisms to the provinces. At the meeting, it distributed seeds of a German variety of sugar beet to 16 provincial representatives as a means to jump-start the development of a national sugar beet industry. Provinces were supposed to send samples of grown sugar beets

[54] Regulations limited institutions to 2 l (*sheng*) of seed, 20 rings (*quan*) of silkworm eggs, and 8 tree saplings, and restricted individuals to receiving no more than one-fourth the allowances for institutions. Recipients of seeds were required to report to Beijing the sowing period, the transplant period, the harvest period, the average harvest per land area, the results of quantitative comparisons with previous harvests, and the existence of any insects, pathogens, or climatic irregularities or calamities (Nong shang bu 1914b, pp. 21–22).

[55] The National Federation of Agricultural Associations was organized by the Nanyang Industrial Exposition's Research Group during meetings in Nanjing in the summer of 1910 (Nanyang quanye hui yanjiu hui 2002, 1:3, 5).

[56] Hubei nonghui (1910), vol. 1, no. 1 *zhangzou*:3b-6a. ("Nong gong shang bu zou ding nonghui jianming zhangcheng er shi san tiao" (The Ministry of Agriculture, Industry, and Commerce memorializes on setting concise regulations for agricultural associations in 23 articles), 20 October 1907). Zhu, "On late Qing economic laws and regulations, (ca. 1901–1911)," in Reynolds (1995), pp. 114–115.

[57] Three years earlier, Zhu Zurong had suggested establishing a seed association (*miao zhong hui*) with two branches, one each for northern and southern China, for the purpose of facilitating seed exchanges between different parts of the country. See Nanyang quanye hui yanjiu hui (2002), 2:171–172 (Zhu, "Jia zhong lu"). On the activities of the Federation, see Tao et al. (1973), p. 60 ("Ben bu tiyi tongguo ge an" (Proposals made and passed by this Ministry)).

back to the experiment station in Beijing for chemical analyses of their sugar content, but by 1914, only 3 provinces had returned their samples.[58]

The abundance of foreign and domestic cultivars circulating in the country at this time and the growing prevalence of field experiments for agricultural research were both unprecedented for China. But when it came to explaining their research, agricultural specialists in Beijing regularly drew upon the long-standing and familiar vocabularies of Chinese agricultural thought. In particular, they often resorted to the idea of "suitability" (*yi*) to describe why, in certain cases, cultivated plants managed to grow and thrive in new settings. Suitability referred to the perceived concordance of the natural characteristics of a plant with the local climate and the quality of local soils. As in times past, agricultural specialists primarily paid attention to the plant's growth and its productivity at harvest time to determine the degree of concordance, and thus the degree of suitability. As they conducted experiments, researchers in Beijing worked to ascertain the natural and seemingly stable characteristics of plants—how they developed, what kinds of fertilizers and soils they liked, how much water they needed, what temperatures were optimal for their growth, etc.—so that they could choose the most suitable variety for a given soil and climate within a short period of time. They then could use the results of their experiments to distinguish among suitable and unsuitable varieties and counter any skepticism among farmers that certain exogenous crops could be cultivated profitably in the local area.[59]

At the experiment station in Beijing, researchers regularly spoke about the suitability of plants in local soils. They used seeds from other provinces and other countries in experiments designed to identify which plants could "match what is suitable to the land."[60] In 1913, Beijing researchers experimented with Chinese and foreign fruit trees "to observe whether or not they are suitable" and "to examine their condition and quality."[61] When experiments with 3 domestic varieties and 1 American variety of cucumber demonstrated the superior productivity of the Beijing variety, researchers attributed the results to the "mutual suitability of the local production variety and the local climate and soil."[62]

Yet researchers also acknowledged the potential of plants to adapt to local conditions. They conducted field experiments with dozens of varieties of foreign and domestic plants to see not only which cultivars could flourish immediately but also which cultivars evinced a potential to thrive after several years of cultivation and seed collection. Even as they analyzed what they called suitability and the natural characteristics of plants, researchers sometimes also recognized the plasticity of plants in their environments: "Plants change their natural qualities based upon differences in the climate and soil quality of each location. Thus, there are constant

[58] These three provinces were Shanxi, Henan, and Shandong (Nong shang bu 1914c, *huayanke*:33).

[59] Nanyang quanye hui yanjiu hui (2002), 2:172 (Zhu, "Jia zhong lu").

[60] Nong gong shang bu (2007a), 2:9a-9b, and Nong gong shang bu (2007b), 2:8a–b.

[61] Nong shang bu (1914c), *yuanyike*:56.

[62] Nong shang bu (1914c), *yuanyike*:40.

expenditures on research into methods of cultivation."[63] They knew that additional, multiyear research could yield useful results. Researchers in Beijing admitted that some cultivars would be unsuited to the North China Plain, which was colder and drier compared to southern China. But in conducting experiments year after year, they also accumulated knowledge about which varieties could be adapted to Beijing's conditions. Their 1913 report about experiments with 18 varieties of wet rice concluded that the growth of some strains was inevitably hampered by climate. But even varieties from southern regions could be made suitable over time. Although early-season and late-season rice varieties from southern China's Guangdong Province had failed to be productive, yellow japonica rice (*huang yingdao*) from Xiamen, in southeastern China's Fujian Province, had yielded a good crop with high-quality grains. The report stated that, after having been cultivated at the Beijing station for a number of years, the rice was "gradually becoming assimilated (*tonghua*) to the climate and quality of the soil."[64]

Agricultural Chemistry and Fertilizers

In their 1906 proposal for the experiment station in Beijing, Qing officials at the Ministry of Agriculture, Industry, and Commerce had asserted that "if we use chemistry (*huaxue*) to research the characteristics of matter, then the quality of the soil can be made uniform."[65] Chinese agronomists in earlier centuries had employed terms for the quality or nature of the soil (*tu xing, tu zhi*) alongside descriptions of specific soil characteristics as they categorized and analyzed the distinguishing features of soils in different topographic settings and geographical locations. Researchers in the first two decades of the twentieth century continued to use these existing terms to describe soil. They also drew upon China's extensive historical experience with fertilizers.[66] But with the advent of agricultural chemistry in China, they began to employ not only an additional vocabulary of chemicals but also a whole new range of fertilizers and new methods for testing soils, fertilizers, and biological material. Agricultural chemistry at the experiment stations encompassed a number of different practices, including assaying the chemical composition of agricultural products like sugar beets and soybeans as well as formulating pesticides for use in the fields.[67] But researchers were most concerned to use their knowledge of chemicals to improve soil management with fertilizers.

Agricultural chemistry captured the attention of Chinese scholars and agricultural reformers beginning in the last decade of the nineteenth century when a flurry of

[63] Nong shang bu (1914c), *yuanyike*:69.

[64] Nong shang bu (1914c), *shuyike*:5.

[65] Ye (1909), p. 9a.

[66] On the history of fertilizers in China, see Liang (1989), pp. 197–201, 409–412, 503–511, and Bray (1984), pp. 289–298.

[67] Nong shang bu (1914c), *fulu*:7, *bingchonghaike*:87–97.

newly translated texts related to soils and chemical fertilizers first appeared. One of the forerunners of agricultural chemistry in translation was the 1894 publication of *New Methods for Agriculture* (*Nongxue xin fa*), authored by American missionary William Preston Bentley. Addressed to agricultural reformers in China, Bentley's short tract explained how chemical analyses of soils and chemical fertilizers could make farmland fertile across the country.[68] Translations of much longer works related to agricultural chemistry were also produced in Shanghai by the Chinese and foreign men who worked in the translation department of the Jiangnan Arsenal. From the 1860s to the first decade of the twentieth century, the Arsenal produced well over 170 translations of Western-language works, many in the natural sciences.[69] Around the turn of the century, the Arsenal translated and published at least three texts about agricultural chemistry and soil sciences that had previously appeared in English: British chemist and agricultural scholar James F. W. Johnston's 1845 book *Catechism of Agricultural Chemistry and Geology* (*Nongwu huaxue wenda,* 1899); professor of agricultural physics at the University of Wisconsin Franklin Hiram King's 1895 work *The Soil: Its Nature, Relations, and Fundamental Principles of Management* (*Nongwu tu zhi lun,* 1900); and German-American author Tuisco Greiner's 1892 book *Practical Farm Chemistry: A Handbook for Profitable Crop Feeding* (*Nongwu huaxue jianfa,* 1903).[70] Luo Zhenyu's Society for Agriculture was also involved in the work of translating texts about agricultural chemistry through its journal and compendium. Around the turn of the century, the Society's large corpus of works included a handful of translated Japanese, American, and French texts about fertilizers and soil chemistry. Among the Japanese works in translation were Hara Hiroshi's 1892 work *Fertilizers* (*Feiliao pian*), Ikeda Masakichi's 1894 book *Soil Studies* (*Turang xue*), and Sawamura Makoto's 1900 treatise *Methods of Agricultural Chemistry Experimentation* (*Nongyi huaxue shiyan fa*).[71]

By the turn of the twentieth century, it was becoming common for researchers and reformers to consider agricultural chemistry necessary for farm production and for the development of Chinese agriculture. Reformers of the era regularly alluded to Bentley's work by referring to the "new methods for agriculture" as they touted the potential of chemical fertilizers to boost yields.[72] In this period, the focus of agricultural chemistry at the experiment stations was on chemical analyses of soils as well as tests which compared the efficacy of different combinations of fertilizers. The authors of the Beijing station's inaugural report of 1909 described the qualities

[68] Bentley (Bentley 1894). The text was republished several times thereafter, in Liang (1897) and in Yuan (1901). Several years later, Bentley wrote another short tract suggesting that China establish a national department of agriculture along the lines of the US Department of Agriculture (Bentley 1903). See Stross (1986), pp. 18–20.

[69] Reardon-Anderson (1991), p. 36, and Wright (2000), pp. 238–240.

[70] Johnston (1899); Johnston (1845); King (1900); King (1895); Greiner (1903); and Greiner (1892). See also Liu (2002), pp. 210–211.

[71] The original publications are Hara (1892), Ikeda (1894), and Sawamura (1900). See Li (1957), 1:869; Dong and Fan (2000), pp. 838–839; and Zhong (1996), pp. 157–158.

[72] Li (1957), 1:858 (Liang Qichao, "Nonghui bao xu," (Preface to the journal of the Society for Agriculture)); Nanyang quanye hui yanjiu hui (2002), 2:172 (Zhu, "Jia zhong lu").

and components of the topsoil and subsoil at the experiment station using "chemical analysis" (*huaxue fenxi*). The report analyzed the soil's composition of water, humus, nitrogen, and insoluble inorganic material. It also provided a tabulation of the chemical components of the topsoil and subsoil with measurements of elements and compounds like silicon, alumina, iron oxide, manganese, lime, magnesium, potassium, sodium, phosphoric acid, and sulfur oxide.[73] In 1913, soil analyses at the station were even more thorough insofar as they included assays for more elements and soil components.[74] Officials also made plans to carry out a complete analysis of soils at the station in 1914 in order to produce a chart of soils and their corresponding fertilizers.[75] Other stations, like those in Zhili and Fengtian, conducted chemical analyses of soils with the aid of chemical specialists.[76]

Chinese agronomists also brought their knowledge of soils and chemistry to bear on studies of agricultural conditions and inputs outside of the stations. Provincial soil surveys were increasingly common in this era and experiment station staff often conducted them. In 1909, researchers in Fengtian carried out a detailed survey of soils in the province by tabulating the topographic character, color, and quality of soils in over 300 villages.[77] The Zhili experiment station also made a full provincial survey of soils in the first decade of its existence.[78] The Beijing station did the opposite: Rather than dispatching people to survey land, laboratory technicians offered to accept samples of soil, fertilizers, and other materials from the public for chemical analysis at the station and to report the results for the mutual edification of researchers and those who owned the samples. Fees for this service were to be waived if the tests were considered to be in the public interest.[79]

Experiments with fertilizers were equally important to researchers who wanted to leverage their knowledge of chemistry to boost yields. Although the advent of agricultural chemistry brought with it the prospect of using new, powerful inorganic fertilizers on farmland, Chinese researchers did not limit their work to only these new concoctions. On the contrary, they approached the problem of soil fertility by assessing a wide assortment of potential fertilizer materials and ingredients, especially those already in use among Chinese farmers. Among the most common in China were human and animal manures, composted organic material, crop stalks and grain chaff, dregs and refuse from the production of vegetable oils and alcohol, and bone meal. In 1909, technicians at the Beijing station carried out an exhaustive analysis of dozens of fertilizers and soil additives and tallied their respective contents of nitrogen, phosphoric acid, and potassium, the triumvirate of chemicals that occupied the attention of agricultural chemists in this era. The list of fertilizers in the study included the manures of humans and 9 other animals; 13 organic

[73] Ye (1909), p. 27a.

[74] Nong shang bu (1914c), *huayanke*:5–6.

[75] Nong shang bu (1914c), *fulu*:6.

[76] Nong gong shang bu (2007a), 2:12a, 13a.

[77] Fengtian nongye shiyan chang (1909b), 1:15a–29b.

[78] Nong gong shang bu (2007a), 2:13a.

[79] Nong shang bu (1914c), *fulu*:7–8.

materials, such as blood meal, bone meal, crab shells, silkworm pupae, human hair, and algae; and 21 other substances, including the ash of fallen leaves, the ash of rice stalks, coal ash, potassium chloride, kitchen wastewater, and "mud along the road."[80] Several years later, researchers made plans to conduct more extensive analyses of the chemical components of all types of natural and man-made fertilizers.[81] Other experiment stations also tested specific fertilizers. At the station in Gansu Province, results of some experiments had encouraged local people to gather bone meal and ash to fertilize their fields.[82] Researchers at other stations assessed the value of using green manure, the cover crops grown to improve soil nutrients. These included Chinese milk vetch (*ziyunying*), a crop touted for fixing nitrogen in the soil. By 1910, a number of stations had planted Chinese milk vetch using seeds purchased from Japan.[83]

New knowledge about chemical elements and compounds in soils and fertilizers led the way for field tests comparing the efficacy of different combinations of organic and inorganic fertilizers. The Beijing station conducted a range of tests on fertilizer combinations in 1913. To paddy fields containing 14 varieties of rice, technicians applied a mixture of 3 fertilizers: dried human manure accounting for roughly two-thirds of the mixture by weight, along with equal proportions by weight of plant ash and a special fertilizer from the Japanese fertilizer company Taki.[84] In another experiment, technicians used 3 different combinations of compost, human manure, sesame seed dregs, and superphosphate of lime in varying proportions to fertilize fields growing a strain of rice from Beijing.[85]

Although researchers and reformers both took interest in the inorganic fertilizers that were new to China in this era, they quickly became concerned that the expenditures to purchase them reduced money for research and development in China and undermined their efforts to improve the country's agriculture and economy. They raised and responded to questions about how agronomists could develop domestic sources of fertilizers rather than buying them from foreign firms.[86] On the issue of formulating fertilizers, Zhu Zurong argued that researchers could simply replicate the ratios of phosphorous and nitrogen in special Japanese fertilizers for rice and wheat as they manufactured their own domestic fertilizers.[87] The experiment station in Guangdong Province took a more concrete step when it developed a booklet of

[80] Ye (1909), pp. 28a–30b.

[81] Nong shang bu (1914c), *fulu*:7.

[82] Tao et al. (1973), p. 105 (Wu Jun et al., "Gansu sheng nongye qingxing" (Agricultural conditions in Gansu Province)).

[83] Nanyang quanye hui yanjiu hui (2002), 2:232 (Sun Yue, "Yanjiu nongye yijian" (Ideas about researching agriculture)).

[84] According to the station's technicians, such a heavy reliance upon human manure was due to the shortage of phosphoric acid (Nong shang bu 1914c, *shuyike*:2).

[85] Nong shang bu (1914c), *shuyike*:5–6.

[86] Nong shang bu (1914c), *shuyike*:2, 66.

[87] Note that Zhu used the Japanese weight measure *kanme* (*guanmu,* approximately 3.75 kg) to discuss these fertilizers. See Nanyang quanye hui yanjiu hui (2002), 2:175, 178 (Zhu, "Jia zhong lu").

basic formulas for 16 fertilizers for crops like rice, cotton, sugarcane, and wheat.[88] But formulas and fertilizer recipes were one thing, and sources of the chemical ingredients for fertilizers were another. In this period, reformers and researchers at experiment stations began to seek the raw material ingredients for novel fertilizers throughout China and its borderlands. Perhaps most striking were the plans of officials at the agricultural experiment station in Guangdong to source phosphates for 8 of its fertilizer formulas from the Pratas Islands (*Dongsha dao*), located more than 300 km southeast of Hong Kong in the South China Sea, which held a phosphorus mine and guano deposits.[89] Guangdong agronomists also looked to the Paracel Islands (*Xisha dao*) to mine phosphorus deposits and to harvest calcium and phosphorus from coral.[90] Other agricultural reformers looked eastward and northward, to Jiangsu and Shandong provinces for deposits of saltpeter, to Nanjing for the production of ammonium sulfate, to Shanxi Province for sulfur and iron sulfide mines, and to the borderlands stretching across the north and west of the country from Manchuria and Mongolia to Xinjiang and Tibet, for potential sources of soybean cake fertilizers.[91] In other words, the search for fertilizers tied the work of the experiment stations and the movement for agricultural improvement to future plans for finding raw materials in the country's peripheries.

Conclusion

As this chapter demonstrates, China's earliest agricultural experiment stations arose from the desires of reformers and agricultural specialists to find the most expedient means of increasing the country's agricultural productivity and to put them on display for the Chinese public. For researchers at the stations, this primarily involved field experiments in plant propagation and acclimatization to identify the cultivars which best suited local soils and climates, as well as tests to determine which fertilizers could maximize yields. This research agenda suggests that Chinese agricultural science in the first two decades of the twentieth century was most interested in

[88] Nanyang quanye hui yanjiu hui (2002), 3:60–81 ("Guangdong nongshi shiyan chang ren zao feiliao shuoming shu" (The Guangdong Agricultural Experiment Station's explanatory booklet on manmade fertilizers)).

[89] The Pratas Islands were brought under Qing control in February 1910 after having been settled and mined for guano by Taiwan-based Japanese entrepreneurs several years earlier. See Rhoads (1975), p. 141.

[90] See the following two sources in Nanyang quanye hui yanjiu hui (2002): "Guangdong quansheng huafen kuangzhi suo kuang chanpin shuoming shu" (Explanatory booklet of the chemical ore products mined in Guangdong Province), 3:25–28; "Guangdong nongshi shiyan chang ren zao feiliao shuoming shu," 3:60–81; see also Tao et al. (1973), p. 115 (Huang, "Guangdong sheng nongye qingxing").

[91] Nanyang quanye hui yanjiu hui (2002), 2:180 (Zhu, "jia zhong lu") and 4:95–96 (Lu An, "Ouzhou zhongzhi zhitang luobo xinfa" (Europe's new methods for cultivating the sugar beet)).

the environmental and chemical conditions of production in the fields, and was not yet so concerned with developing better organisms.

By the 1920s, however, the institutional setting and focus of the most important experimental research for agriculture in China had shifted. The most active research agendas were to be found at departments and colleges of agriculture at several universities in China. Experiment stations continued their work, and stations devoted to cotton research proliferated after 1915.[92] But the quality of research at China's first agricultural experiment stations seems to have declined owing to the political chaos and territorial disunity of the warlord era. On his visit to the experiment station in Beijing in 1925, H. L. Russell, dean of the College of Agriculture at the University of Wisconsin, declared that the station and its experiments had become victims of political turmoil and had "no scientific value."[93]

By contrast, universities had gathered significant human and financial resources for agricultural research by that time. Two universities in the city of Nanjing, University of Nanjing and Southeastern University, were the most active in conducting agricultural research.[94] Both universities hired Chinese graduates of American universities with training in agriculture and related fields, and both participated in the surge of experimental work to acclimatize American cotton varieties and to improve domestic cotton varieties to create a better and more abundant supply of raw material for cotton manufacturing interests in China, which supported research efforts with funding.[95] Moreover, University of Nanjing, a private missionary school, received funding from the International Education Board for a cooperative program of crop improvement with Cornell University which lasted from 1925 to 1931.[96]

The agenda of Chinese agricultural research also changed as the focus of experimental work shifted from field tests of a huge variety of plants and fertilizers toward experiments for breeding better varieties of a few major industrial and food crops. It was the Nanjing–Cornell Crop Improvement Program that most clearly exemplified this shift. Researchers in the program aimed to improve crops like wheat and soybeans through such methods as rod row tests, with thousands of cultivated rows, for creating pure-line varieties.[97] The 1920s also saw the creation of China's first university programs in genetics, at Southeastern University and at Yanjing University in Beijing, alongside the program in plant genetics at University of Nanjing.[98] To be sure, some agricultural specialists at the first experiment stations had known

[92] From 1915 to 1918, political authorities established experiment stations for cotton in Hubei, Jiangsu, Hebei, and Henan provinces. By 1920, there were more than 20 experiment stations devoted to cotton in the city of Tianjin and the surrounding region. See *Zhengli mianye choubei chu* (1921), p. 1; Shen (1970), p. 211; and Pomeranz (1993), p. 97.

[93] H. L. Russell, Log Two, China and Philippines, 7, Rockefeller Foundation Archives, International Education Board, quoted in Stross (1986), p. 146.

[94] Shen (1970), pp. 212–213.

[95] Stross (1986), pp. 116–142, and Dongnan daxue nongshi shiyan chang mian zuo gailiang weiyuanhui (1925), Mei mian yuzhong baogao: shi er nian, shi san nian, 1.

[96] Schneider (2003), pp. 78–85; Shen (1970), pp. 214–220; and Stross (1986), pp. 143–160.

[97] Tan and Zhao (2002), p. 233, and Shen (1970), p. 216.

[98] Schneider (1988) and Schneider (2003), pp. 33–91).

about Mendelian principles of inheritance. In 1910, Tao Changshan had looked ahead to the time when scientists would apply these principles to "create new varieties through breeding" (*yu zao xin zhong*), and in the 1910s, there was growing discussion in Chinese journals of Mendel and the basics of genetics.[99] But unlike in the USA, where work in crop breeding and hybridization at agricultural colleges and experiment stations gave rise to the institutionalization of genetics as a field of research at American universities, genetics research and teaching in China did not develop in earnest before more Chinese graduates of American universities returned to work at universities in China and more funding became available in the 1920s.[100]

Finally, the 1920s witnessed a third shift in Chinese agricultural research: the waning influence in China of Japanese agronomists and agricultural education. At the turn of the century, Chinese had turned to Japan more than any other country for models of reform, for overseas education, and for new ideas about agriculture. But waves of returning graduates from American universities, institutional connections between universities in the USA and China, and funding from American philanthropic organizations all increased the relative importance of the USA to Chinese agricultural research. Moreover, Japan's demands for increased economic and political prerogatives in China in 1915 and its occupation of former German concessions in China's Shandong Province during and after World War I likely tarnished its reputation in the eyes of Chinese researchers who were working on behalf of agricultural improvement for the nation. While Japanese influence did not completely recede,[101] the era of Japanese influence at China's earliest agricultural experiment stations had passed.

References

Anonymous. 1909. *Nongshi shiyan chang quanjing (Panorama of the agricultural experiment station)*. Tokyo: Hironobu dō.

Bai, Hewen, Fuquan Du, and Zongdian Min. 1995. *Zhongguo jindai nongye keji shigao (Draft history of modern Chinese agricultural* technology). Beijing: Zhongguo nongye keji chubanshe.

Bentley, William Preston (Bei-de-li). 1894. *Nongxue xin fa (New methods for agriculture). Trans. Timothy Richard (Li Ti-mo-tai) and Tiesheng Zhu*. Shanghai: Guangxue hui.

Bentley, William Preston. 1903. *Guojia zhuan she nongbu yi (Proposal for a national department of agriculture)*. Shanghai: Commercial Press.

Bray, Francesca. 1984. *Science and civilisation in China*. Vol. 6, pt. 2, Agriculture. Cambridge: Cambridge University Press.

Dong, Kaichen, and Chuyu Fan, eds. 2000. *Zhongguo kexue jishu shi: nongxue juan. (The history of science and technology in China: Agriculture)*. Beijing: Kexue chubanshe.

[99] Nanyang quanye hui yanjiu hui (2002), 2:272 (Tao, "Mai lei zhi xin zhong shuru shuo"), and Tan and Zhao (2002), pp. 5–7.

[100] On the experience in the USA, see Kimmelman (1987).

[101] Even after 1915, Japan continued to have a measured degree of influence on the ideas of Chinese scholars and politicians. See Lu (2004), p. 253.

Dongnan daxue nongshi shiyan chang mian zuo gailiang weiyuanhui (Committee for the improvement of cotton at the agricultural experiment station of Southeastern University). 1925. Mei mian yuzhong baogao: shi er nian, shi san nian (Report on the breeding of American cotton: 1923, 1924). N.p.

Eastman, Lloyd E. 1988. *Family, fields, and ancestors: Constancy and change in China's social and economic history, 1550–1949*. Oxford: Oxford University Press.

Elvin, Mark. 1975. Skills and resources in late traditional China. In *China's modern economy in historical perspective*, ed. Dwight H. Perkins, pp. 85–113. Stanford: Stanford University Press.

Fengtian nongye shiyan chang (Fengtian Agricultural Experiment Station). 1909a. Fengtian nongye shiyan chang baogao shu (Report of the Fengtian Agricultural Experiment Station). Fengtian nongye shiyan zongchang.

Fengtian nongye shiyan chang (Fengtian Agricultural Experiment Station). 1909b. Fengtian quansheng nongye diaocha shu (Report on a survey of agriculture in Fengtian Province). 4 vols. Fengtian nongshi shiyan chang.

Francks, Penelope. 1984. *Technology and agricultural development in pre-war Japan*. New Haven: Yale University Press.

Greiner, Tuisco. 1892. *Practical farm chemistry: A handbook of profitable crop feeding*. New York: The Rural Publishing Co.

Greiner, Tuisco (Gu-lai-na). 1903. Nongwu huaxue jianfa (Practical farm chemistry: A handbook of profitable crop feeding). Trans. John Fryer (Fu-lan-ya) and Shushan Wang. Shanghai.

Guo, Wentao, ed. 1988. *Zhongguo nongye keji fazhan shilüe (A brief history of the development of agricultural science and technology in China)*. Beijing: Zhongguo kexue jishu chubanshe.

Hara, Hiroshi. 1892. *Hiryō hen (Fertilizers)*. Tokyo: Hakubunkan.

Hayami, Yūjirō. 1975. *A century of agricultural growth in Japan: Its relevance to Asian development*. Tokyo: University of Tokyo Press.

Hubei nonghui (Hubei Agricultural Association). 1910. Hubei nonghui bao (Journal of the Hubei Agricultural Association).

Ihara, Takushū. 2000. Wu nong hui zai wuxu bianfa yundong shi shang de diwei (The place of the Society for Agriculture in the history of the 1898 reform movement). In *Wuxu weixin yu jindai Zhongguo de gaige: wuxu weixin yi bai zhounian guoji xueshu taolunhui lunwen ji. (The hundred days reforms and reform in modern China: Collected essays from the International Conference on the Hundredth Anniversary of the Hundred Days Reforms)*, ed. Xiaoqiu Wang, pp. 293–310. Beijing: Shehui kexue wenxian chubanshe.

Ikeda, Masakichi. 1894. *Dojōgaku* (Soil Studies). Tokyo: Yūrindō.

Jiang, Yasha, and Li Jing, eds. 2009. *Zhongguo zaoqi nongxue qikan huibian (Compilation of early Chinese periodicals about agriculture)*. 40 vols. Beijing: Quanguo tushuguan wenxian suowei fuzhi zhongxin.

Johnston, James F. W. 1845. *Catechism of agricultural chemistry and geology*. 8th ed. Albany: Erastus H. Pease.

Johnston, James F. W. (Zhong-si-dun). 1899. Nongwu huaxue wenda (Catechism of agricultural chemistry and geology). Trans. Francis Huberty James (Xiu Yaochun) and Xiyong Fan. Shanghai.

Kargon, Robert. 2008. The U.S.-Japan nexus: Roots of the Japanese scientific enterprise. *Historia Scientiarum* 18:62–75.

Kimmelman, Barbara A. 1987. A Progressive Era discipline: Genetics at American agricultural colleges and experiment stations, 1900–1920. Ph.D. Diss., University of Pennsylvania, USA.

King, Franklin H. 1895. *The soil: Its nature, relations, and fundamental principles of management*. New York: Macmillan.

King, Franklin H. (Jin Fu-lan-ge-ling Xi-lan). 1900. Nongwu tu zhi lun (The soil: Its nature, relations, and fundamental principles of management). Trans. Edward Thomas Williams (Wei-li) and Xiyong Fan. Shanghai.

Kingsbury, Noel. 2009. *Hybrid: The history and science of plant breeding*. Chicago: University of Chicago Press.

Kloppenburg, Jack R., Jr. 2004. *First the seed: The political economy of plant biotechnology, 1492–2000.* 2nd ed. Madison: University of Wisconsin Press.

Li, Chien-nung. 1956. *The political history of China, 1840–1928*. Trans. and ed. Ssu-yu Teng and Jeremy Ingalls. New York: D. Van Nostrand Company.

Li, Wenzhi, ed. 1957. *Zhongguo jindai nongye shi ziliao (Historical materials about modern Chinese agriculture).* 3 vols. Beijing: Sanlian shudian.

Li, Yongfang. 2008. *Jindai Zhongguo nonghui yanjiu (Research on agricultural associations in modern China)*. Beijing: Shehui kexue wenxian chubanshe.

Liang, Jiamian, ed. 1989. *Zhongguo nongye kexue jishu shigao (Draft history of Chinese agricultural science and technology)*. Beijing: Nongye chubanshe.

Liang, Qichao, ed. 1897. *Xi zheng congshu (Compendium of books about Western governance)*. 32 vols. Shanghai: Shenji shuzhuang.

Liu, Guangding. 2002. Qingdai huaxue shuji mulu gao (Draft list of chemistry books from the Qing era). In *Zhongguo kexue shi lunji* (Collected essays about the history of Chinese science), Liu Guangding, pp. 201–222. Taipei: Guoli Taiwan daxue chuban zhongxin.

Lu, Yan. 2004. *Re-understanding Japan: Chinese perspectives, 1895–1945*. Honolulu: Association for Asian Studies and University of Hawai'i Press.

Luo, Zhenyu. 2010. Nongshi siyi (Personal ideas on agriculture). In *Luo Zhenyu xueshu lunzhu ji. (Collection of Luo Zhenyu's academic works)*, ed. Jizu Luo and Tongce Wang, vol. 11, pp. 294–345. Shanghai: Shanghai guji chubanshe.

Nanyang quanye hui yanjiu hui (The Research Group of the Nanyang Industrial Exposition), ed. 2002 (reprint 1913). *Nanyang quanye hui yanjiu hui baogao shu. (Report of the Research Group of the Nanyang Industrial Exposition).* 4 vols. Beijing: Quanguo tushuguan wenxian suowei fuzhi zhongxin.

Nong gong shang bu (Ministry of Agriculture, Industry, and Commerce). 2007a (reprint 1908). *Nong gong shang bu tongji biao: di yi ci (Statistical tables prepared by the Ministry of Agriculture, Industry, and Commerce: The first collection)*. Beijing: Beijing Yanshan chubanshe.

Nong gong shang bu (Ministry of Agriculture, Industry, and Commerce). 2007b (reprint 1909). *Nong gong shang bu tongji biao: di er ci (Statistical tables prepared by the Ministry of Agriculture, Industry, and Commerce: The second collection)*. Beijing: Beijing Yanshan chubanshe.

Nong shang bu (Ministry of Agriculture and Commerce). 1914a. Nong shang bu linye shiyan chang yaolan (Overview of the forestry experiment station of the Ministry of Agriculture and Commerce). Beijing.

Nong shang bu (Ministry of Agriculture and Commerce). 1914b. Nong shang bu nongshi shiyan chang yilan (Overview of the agricultural experiment station of the Ministry of Agriculture and Commerce). Beijing.

Nong shang bu (Ministry of Agriculture and Commerce). 1914c. Nongshi shiyan chang chengji baogao (Report of results from the agricultural experiment station). Beijing.

Ogura, Takekazu, ed. 2000 (reprint 1963). *Agricultural development in modern Japan*. London: Routledge.

Pomeranz, Kenneth. 1993. *The making of a hinterland: State, society, and economy in inland north China, 1853–1937*. Berkeley: University of California Press.

Reardon-Anderson, James. 1991. *The study of change: Chemistry in China, 1840–1949*. Cambridge: Cambridge University Press.

Reynolds, Douglas R. 1993. *China, 1898–1912: The Xinzheng revolution and Japan*. Cambridge: Council on East Asian Studies, Harvard University.

Reynolds, Douglas R., ed. 1995. *China, 1895–1912: State-sponsored reforms and China's late-Qing revolution: Selected essays from Zhongguo Jindai Shi (Modern Chinese History, 1840–1919)*. Trans. D. R. Reynolds. Armonk: M.E. Sharpe.

Rhoads, Edward J. M. 1975. *China's republican revolution: The case of Kwangtung, 1895–1913*. Cambridge: Harvard University Press.

Sawamura, Makoto. 1900. *Nōgei kagaku jikkenhō (Methods of agricultural chemistry experimentation)*. Tokyo: Kōbunsha.

Schneider, Laurence A. 1988. Genetics in republican China. In *Science and medicine in twentieth-century China: Research and education,* ed. John Z. Bowers, J. William Hess, and Nathan Sivin, pp. 3–29. Ann Arbor: The University of Michigan Center for Chinese Studies.
Schneider, Laurence A. 2003. *Biology and revolution in twentieth-century China*. Lanham: Rowman & Littlefield.
Shen, Tsung-han. 1970. First attempts to transform Chinese agriculture, 1927–1937. In *The strenuous decade: China's nation-building efforts, 1927–1937,* ed. Paul K. T. Sih, pp. 205–232. New York: St. John's University Press.
Stross, Randall E. 1986. *The stubborn earth: American agriculturalists on Chinese soil, 1898–1937*. Berkeley: University of California Press.
Tan, Jiazhen, and Gongmin Zhao, eds. 2002. *Zhongguo yichuan xue shi (A history of genetics in China)*. Shanghai: Shanghai keji jiaoyu chubanshe.
Tao, Changshan, et al., eds. 1973 (reprint 1913). *Quanguo nonghui lianhehui di yi ci jishi. (First record of the National Federation of Agricultural Associations)*. Taipei: Wenhai chubanshe.
True, Alfred Charles. 1970 (reprint 1937). *A history of agricultural experimentation and research in the United States, 1607–1925. United States Department of Agriculture Miscellaneous Publication No. 251.* New York: Johnson Reprint Corporation.
Wright, David. 2000. *Translating science: The transmission of western chemistry into late imperial China, 1840–1900*. Leiden: Brill.
Ye, Jizhen, ed. 1909. *Nongshi shiyan chang baogao shu (Report of the agricultural experiment station)*. Beijing: Nong gong shang bu.
Yuan, Junde, ed. 1901. *Fu qiang zhai congshu xu quanji (Collected works from the Studio of Wealth and Power, the continued and complete collection)*. 64 vols. Shanghai: Xiaocang shanfang.
Yuan, Pengxin. 2012. *Qingmo nongye xinzheng yanjiu (Research on new policies in agriculture at the end of the Qing)*. Jinan: Shandong renmin chubanshe.
Zhang, Yufa (Chang Yu-fa). 1979. Ershi shiji chuqi de Zhongguo nongye gailiang (1901–1916): yanhai yanjiang shi san ge shengqu de bijiao yanjiu (China's agricultural improvement, 1901–1916: Regional studies on thirteen provinces). *Shixue pinglun* (Chinese Historical Review) 1:119–159.
Zhengli mianye choubei chu (Preparatory Office for Organizing the National Cotton Industry), ed. 1921. Preface to Geng shen zhi mian shiyan lu (Record of experiments with cultivating cotton in 1920). Tianjin.
Zhong, Shaohua. 1996. Xifang keji dongliu shukan mulu (1607–1911). (Index of books and periodicals about western science and technology in the East, 1607–1911). *Zhongguo kexue shi tongxun.* (Newsletter for the History of Chinese Science). 11:138–168.
Zhu, Zurong. 1995 (reprint n.d.). *Quan zhong yang mian shuo (On urging the cultivation of foreign cotton)*. Shanghai: Shanghai guji chubanshe.

Peter Lavelle is assistant professor of history at Temple University, USA. His research interests include environmental history and the history of agricultural science and technology in China during the Qing period. He is currently writing a book about agricultural and geographical knowledge, technologies of land use, and imperialism in nineteenth-century Qing China.

Chapter 17
Did Mendelism Transform Plant Breeding? Genetic Theory and Breeding Practice, 1900–1945

Jonathan Harwood

Introduction

During the 1960s and 1970s, historians of technology, especially in the USA, wrote extensively on the relations between "science" and "technology" as forms of knowledge and practice. A recurring concern in that literature was to demonstrate that technology did not arise solely through the application of scientific knowledge, and by the 1980s one authoritative analysis of the literature reckoned that this argument had been won.[1] To demonstrate that technology was not "applied science"—thus not merely derivative of science—was certainly an important achievement, but it left a lot of unanswered questions. We still do not know much about the particular ways in which science may (or may not) have played a role in the development of various technologies. And that terrain remains largely unexplored because since the 1980s many historians of technology have abandoned study of the relations between science and technology,[2]probably reflecting the more general shift of interest in the field away from the genesis of technology toward its use. To be sure, it is very unlikely that there is any *single* answer to this question; inventors in different

[1] Staudenmaier (1985).

[2] From a study of the articles published in *Technology & Culture* between 1959 and 1980, for example, Staudenmaier (1985) found about seven articles per year which dealt with the relations between science and technology. From a full-text search via JSTOR of this journal between 1980 and 2010, by contrast, I found less than one paper per year on this topic, and most of these appeared during the 1980s.

J. Harwood (✉)
Centre for the History of Science, Technology & Medicine, University of Manchester, Manchester, UK
e-mail: jonathan.harwood@manchester.ac.uk

J. Harwood
c/o Lincoln, 36 Manor Rd., Salisbury SP1 1JS, UK

D. Phillips, S. Kingsland (eds.), *New Perspectives on the History of Life Sciences and Agriculture,* Archimedes 40, DOI 10.1007/978-3-319-12185-7_17

contexts seem to have made use of science (or not) in different ways and to varying extents.[3] My aim here, therefore, is not to try to find some kind of all-embracing formula but merely to make a plausible case for how science is likely to have had an impact upon technology in one particularly high-profile case.

Geneticists and some plant breeders have often made very far-reaching claims for the importance of Mendelism. The following statements are representative: Every advance in breeding "rests exclusively upon experimental genetic research"; "[genetic research] forms the basis on which practical breeding methods are formulated and is solely responsible for the development of present corn-breeding methods"; and after 1900, "plant breeding was applied Mendelism".[4] Some historians of plant breeding have also subscribed to this view. Gottfried Zirnstein, for example, argued that during the period 1895–1905 plant breeding underwent a major shift: from an empirically based practice to a scientific one. Nils Roll-Hansen has argued that, by prompting a series of small methodological improvements, genetic theory has led to a "revolutionary" change in breeding. And J. R. Kloppenburg has made a similar case for breeding in the USA.[5]

For a long time, claims of the mythical power of Mendelism remained unchallenged, partly because unlike the history of genetics, the history of plant breeding was largely uncharted territory. Over the last 20 years or so, however, the volume of historical work on this subject has grown rapidly, making it much easier to investigate the relations between biological theory and breeding practice. One point which has emerged from this new literature, for example, is that right from the start opinion was divided, among both geneticists and breeders, as to the practical value of the new Mendelism.[6] Although this literature has raised doubt as to whether Mendelism transformed breeding, few attempts have been made so far to look at this issue systematically, and the few which have done so tend, in my view, to exaggerate Mendelism's significance.[7] In what follows, therefore, I draw upon this literature as well as my own work on German-speaking Europe in order to unpack

---

[3] Mayr (1976) and Staudenmaier (1985, p. 83 ff.).

[4] The quotes are from Baur (1928, p. 52); Jenkins (1936, p. 493); and Becker (1947, p. 81). In a similar vein, see Riede (1927, p. 58); Babcock and Clausen (1927, p. 476); Stanton (1936, p. 355); Clark (1936, p. 240; Tschermak (1940, p. 13); Müntzing (1951, p. 473); and cf. Goldschmidt and Wilkes, both cited in Kloppenburg (1988, pp. 69–70).

[5] Zirnstein (1977, p. 149 ff.); Roll-Hansen (2000, p. 1109); and cf. Roll-Hansen (1997) and Kloppenburg (1988, pp. 88, cf. 66).

[6] Fitzgerald (1990); Palladino (1993); Harwood (1997); Wieland (2006); and Bonneuil (2006).

[7] E.g., Roll-Hansen (1997, 2000). For the last 20 years, my friend and colleague, Nils Roll-Hansen, and I have from time to time debated the importance of Mendelism for breeding. Although we tend to approach historical problems from different perspectives—his work is more informed by philosophy of science and mine by the sociology of knowledge—I have gradually come to agree with him that Mendelism's provision of a conceptual framework did have potentially significant consequences for practice. Nonetheless, an important difference of emphasis remains. As I show below, the improvements made possible by Mendelian theory constituted *incremental* changes to *pre-Mendelian* breeding methods. For that reason, I believe that it is misleading to assert—as many geneticists have since 1900—that Mendelism "revolutionised" the practice of breeding.

the specific ways in which Mendelian genetics may, in principle at least, have had an impact upon the development of breeding by the 1940s.

Before getting underway, two caveats are in order. First, what exactly is meant by "genetics"? Some elements of Mendelian genetics, for example, are widely agreed upon as having been important for breeders: e.g., evidence on the modes of inheritance of agriculturally important traits, linkage patterns, laboratory/field practices relevant for artificial hybridisation, etc. These are examples of what is often called "fundamental", "strategic", or "mission-oriented basic" research—research on basic processes which are potentially of practical relevance though of no immediate applicability—which the industry wishes to see carried out in public-sector institutions so that it is available as a resource to those inventing and developing technology.[8] In this chapter, however, I focus upon a much more controversial claim: that *concepts and theory* from the new Mendelism suggested new or improved practices.

The second caveat is a methodological one: that anyone wishing to demonstrate "impact" has to cope with a limited range of sources. That is, the nature of the problem requires that we find out what knowledge practitioners possessed and how they used it (if at all) in solving technical problems. For the sociologist, some form of participant observation offers a way to get at this process, but for the historian the necessary kinds of sources are difficult to come by. Occasionally, to be sure, one finds indirect sources, such as writings in which practitioners have reflected upon the nature of their work. While I have been able to draw upon a few sources of that kind, for the most part I approach the problem from a different angle. Rather than attempting to analyse the thought and practice of breeders *in action,* I focus instead upon what Mendelism offered breeders *in principle*. That is, I look at what proponents of Mendelism claimed were the practical implications of the theory and then assess the plausibility of those claims in two ways: (a) How well do they accord with what is known about nineteenth-century commercial breeders' practice, and (b) how well do they stand up against the judgments of early-twentieth-century public-sector breeders?

The result, as we shall see, is that some Mendelian claims were far more persuasive than others. I argue that, while Mendelism certainly did not "revolutionise" breeding during this period, it did affect breeding in three more limited ways. It provided a scientific explanation for breeding practice; it allowed breeders to reflect upon the adequacy of existing practices, and it served as a heuristic to open up the possibility of improved methods. By about 1945, therefore, breeding practice was hardly "applied genetics", but it had been affected nonetheless by the new body of theory.[9]

---

[8] Stokes (1997).

[9] It goes without saying that this conclusion applies only to the period in question. There are indications, for example, that in the late nineteenth century the reproductive biology of plants played a role in the breeding work of Wilhelm Rimpau (Wieland 2006; Meinel 2008).

Big Claims of Little Substance

Before the First World War, numerous advocates of the new theory made grand claims for its importance. Mendelism, it was often said, would "revolutionise" the practice of plant breeding. As William Bateson put it in 1905, "The science of heredity will soon provide power on a stupendous scale...".[10] Cornell University's Herbert J. Webber was equally enthusiastic: "From a condition of ignorance and largely of chaos [before 1900], where all advance was taken as a lucky chance, we have developed to a position where practically each step may be taken intelligently".[11] The theory was said to have replaced "crude empiricism" and "groping around in the dark" with precise prediction.[12]

Some of this excitement undoubtedly owed something to Mendelism's apparent simplicity. E. M. East seems to have perceived the theory in this way, as did R. H. Biffen who initially assumed that quality and yield in wheat would be inherited as simply as shape and colour in peas.[13] Perceiving a resemblance between the new theory and the atomic model in chemistry, several early Mendelians fantasised about the prospects of "synthesising" varieties from constituent genes.[14] "A science of stoichiometry", William Bateson proclaimed in 1902, "will now be created for living things, a science which will provide an analysis, and an exact determination of their constituents".[15]

But just how was genetics going to transform breeding? The breeding method championed by virtually all Mendelians was hybridisation. And although this method had been fairly widely used in the late nineteenth century, Mendelism's proponents often argued that the new theory demonstrated how hybridisation could be conducted in a "rational" manner. Their claim was that nineteenth-century breeders had crossed varieties largely at random merely in order to generate diversity ("breaking the type") from which the breeder would then select. Mendelism, by contrast, would allow for a "planned" approach to hybridisation since the breeder would be able to design crosses which would combine desired traits from each of the parents.[16]

---

[10] Cited in Harvey (1995, p. 116).

[11] Webber (1912, p. 29).

[12] Baur (1927, p. 722). Others agreed; see Paul and Kimmelman (1988, p. 295); Bonneuil (2006); and Charnley (2011, pp. 1–2).

[13] East (1907, p. 38); Charnley (2011, p. 44); and cf. Rimpau (1912, p. 128). Other breeders were more cautious, objecting that such simplification turned a blind eye to the actual complexity of organisms and breeding (Engledow 1931, p. 91; Beaven 1947, p. 4; Harlan 1957, p. 114).

[14] East (1907, p. 91); Babcock and Clausen (1927, p. 428); Zirnstein (1977, p. 185); and Bonneuil (2006, pp. 296–297).

[15] Royal Society (1902, p. 159) and cf. Radick (2013).

[16] E.g., Tschermak (1901); Lochow (1913); Baur (1913); Percival (1925); and Charnley (2011, pp. 140–41).

The claim was not entirely false; some nineteenth-century hybridisers do appear to have used crosses simply in order to generate diversity.[17] But this was not the whole story, for Mendelian enthusiasts were ignoring the fact that many nineteenth-century hybridisers had used the method in a thoroughly "rational" manner. In Germany, for example, Wilhelm Rimpau was one of several breeders constructing planned crosses from the 1870s.[18] And in the first German textbook of plant breeding, Kurt von Rümker distinguished between two existing approaches to hybridisation. "Unplanned" hybridisation simply carried out lots of different crosses without a particular goal in mind, but "planned" (*zielbewusste*) hybridisation chose particular parents in order to combine some of their traits in the progeny.[19] Nor was "rational" hybridisation confined to German-speaking lands. It was also practiced in the USA and in Britain.[20] Moreover, animal breeders in several countries were already conducting rational crosses in the late eighteenth century. Thus, rational crossing did not have to wait until the rediscovery of Mendelism.[21]

In other respects, too, the early champions of Mendelism seem either to have known little about nineteenth-century breeding practices or to have chosen to turn a blind eye to it. For example, some Mendelians took nineteenth-century breeders to task for relying upon mass selection on the grounds that external appearance was no guide to a plant's breeding value.[22] This, one geneticist suggested, was because early breeders lacked the distinction between genotype and phenotype.[23] But nineteenth-century breeders hardly needed a formal genotype–phenotype distinction. It was already well known to them by mid-century that some variations arose through the peculiar conditions of a plant's field location (so-called *Standortsmodifikationen*) and thus would not be inherited. Indeed, it was this recognition that prompted the development in mid-nineteenth-century France of "pedigree selection": a form of individual selection combined with progeny testing so that a plant's breeding value could be judged from its offspring rather than from its own appearance.[24] Finally, it is worth noting that the Mendelian critique of mass selection had little impact. Several public-sector breeders realised that mass selection was less efficient than

[17] Palladino (1994); Roemer (1914a); and Nilsson-Ehle (1913, pp. 71–72).

[18] Thiel (1904); Meinel (2008, pp. 35–36); cf. Gerland (1885); and Tschermak (1908).

[19] Rümker (1889, pp. 122–123).

[20] On the USA, see Webber (1912) and Fruwirth (1887). On the UK, see Roberts (1929) p. 113; Palladino (1993); Zirnstein (1977, p. 119); and Berris Charnley, pers.comm.

[21] On eighteenth-century animal breeders, see Wood and Orel (2001). The geneticist-breeder Erwin Baur made a further claim: that breeders using hybridisation sometimes abandoned a cross when they found nothing new in the *F1* generation because they did not realise that most new combinations would not appear until the F2 (Baur 1921, p. 91). Some French breeders also shared this misconception (Bonneuil 2006, p. 293). But it is not clear that many breeders were in the dark on this point. Any experienced nineteenth-century hybridiser would have known that the first generation of plants after a cross was generally uniform; only following a further generation of self-fertilisation would new combinations begin to emerge.

[22] Wacker (1923–1924, p. 41) and Charnley (2011, p. 141).

[23] Clark (1936, p. 219).

[24] Gayon and Zallen (1998).

individual selection, but they continued to recommend it anyway simply because it was quick and cheap.[25] And that is why mass selection is still used today.

Before the First World War, therefore, the Mendelian "revolution" in plant breeding was almost entirely a rhetorical one. None of the Mendelians' most ambitious claims stood up to scrutiny. As a result, a number of public-sector breeders in Germany doubted whether Mendelism was going to fundamentally transform practice.[26] Elsewhere, too, a few geneticist/breeders acknowledged that the theory had been oversold. In the USA, Liberty Hyde Bailey complained of wild exaggeration while Raymond Pearl worried that the failure of the new Mendelism to live up to its advocates' "perfervid oratory" would undermine the status of science in the minds of some practical breeders.[27] The same issue seems to have been troubling Hermann Nilsson-Ehle who wrote to a colleague in 1909 that "What breeding now needs above all is to abandon all the marketing and concentrate on getting precise results".[28] As we shall see, Nilsson-Ehle's prescription is roughly what happened between the wars.

Specific Claims Derived from Mendelian Theory

Understandably, the proponents of a new theory (or technology) are inclined to exaggerate its power. But not all of Mendelism's boosters made vast and ill-founded claims. Several advanced much more limited and precise claims for improvements to breeding practice which were grounded in specific features of Mendelian theory. What advantages, if any, did they offer to the breeder?

One such claim made by Theodor Roemer was that Mendelism demonstrated the importance of choosing the right parent lines in hybridisation:

> It must be emphasised that for *rational* hybridisation one must use only starting material which breeds true. Without this, hybridisation is just groping around [*Herumtappen*] or searching [*Herumsuchen*], and occasional successes are due to the *accidental* discovery of valuable combinations.[29]

This is a rather puzzling claim. Starting with "pure" parental varieties—by which he meant those which had bred true over several generations—is indeed important if one's aim is to establish the breeding ratios so characteristic of Mendelian research. But why should "impure" parents (i.e., which have not bred true, presumably because they are heterozygous at relevant loci) undermine rational breeding? If a breeder wants to achieve a particular trait combination (e.g., aabb), there is

[25] Kiessling (1912); Fruwirth (1911, p. 3); and Lang (1910).

[26] Harwood (1997, 2005).

[27] Paul and Kimmelman (1988, p. 295) and Pearl (1915, p. 159).

[28] "Was die Züchtung jetzt vor allem bedarf, ist sich von der Reklame zu entfernen und alles exakt zu behandeln"; letter to Carl Fruwirth, 31 August 1909, Hermann Nilsson-Ehle letters, University Library, Lund, Sweden. In a similar vein, see Engledow (1931, p. 90).

[29] Roemer (1914a, pp. 83–84), emphases in original.

nothing "irrational" about crossing "impure" parents such as AaBb and aaBb. On the contrary, such a procedure is rational in the sense that it saves the breeder the time and effort of getting the parents to breed true. During the 1950s, for example, breeders seeking to gain the advantage of hybrid vigour found they got better results when they crossed not inbred homozygotic lines but heterozygotic ones.[30]

Another advantage which Mendelism was said to offer was that it made "transgression breeding" possible.[31] In the nineteenth century, it was argued, breeders planning a cross believed that each parent should display a desired trait in a pronounced form.[32] For example, unless one parent was high yielding and the other displayed high quality, they assumed that the offspring would not be superior in both traits. But Mendel's work, Roemer claimed, showed that this was not necessary; even mediocre parent varieties could still produce hybrid offspring which displayed the desired traits in extreme form. One of the problems with this claim, however, is that breeders knew about the phenomenon of transgression before Mendelism was rediscovered. Both Wilhelm Rimpau and Erich Tschermak, for example, had found it in crosses conducted in the 1890s.[33] Another fact about transgression also proved rather inconvenient for Roemer. For although he had claimed that Mendelism made crossing more "rational", he himself later admitted that one could not predict in advance whether or not two mediocre parents would *in fact* spawn extreme progeny. This could only be established by carrying out the cross in question.[34] This meant that the method itself had to rely on trial and error—just as in the nineteenth century—to find the best parents.[35]

Probably the best-known argument that genetics had implications for breeding practice, however, derived from Wilhelm Johannsen's work on pure-line theory. In 1903, he argued that selection in pure lines was powerless to produce genetic change. Since a population of self-fertilising plants all descended from the same individual was supposedly genetically uniform, the variation within such a population would be merely phenotypic in character. Therefore, selection acting upon this population could not shift its genetic makeup.[36] The claim attracted considerable attention in Germany, perhaps because repeated individual selection upon self-fertilising plants was so common among commercial breeders that it was known as the "German method" (*deutsches Ausleseverfahren*). Accordingly, a substantial number

[30] Becker (1960, p. 108).

[31] Roemer (1914a) and Kappert (1931, p. 103). Transgression breeding is a form of hybridisation in which parent lines are chosen even though a desired trait is not very strongly expressed in either of them, but the breeder has reason nonetheless to expect that the trait will be very strong in the hybrid. The hybrid's phenotype thus "transgresses" that of either parent.

[32] See, for example, Bailey (1896, p. 109).

[33] Zirnstein (1977, p. 120) and Tschermak (1900, p. 531, 1898, p. 15).

[34] Roemer (1933, p. 171, 1940, p. 271).

[35] In France, some breeders became disillusioned with Mendelism for this reason (Bonneuil 2006, pp. 297–298).

[36] Roll-Hansen (2009).

of public-sector breeders—in German-speaking Europe as elsewhere—argued that Johannsen's argument was important for the reform of breeding.[37]

Nevertheless, other public-sector breeders in several countries were not so sure that Johannsen was correct.[38] A recurring objection was that the concept of "pure line" was a theoretical construct; in the field, pure lines simply did not exist.[39] One reason for this was recombination. Interestingly, Johannsen does not seem to have realised that cross-fertilisation occurs even in nominally self-fertilising plants.[40] Already in the late nineteenth century, however, commercial breeders like Wilhelm Rimpau had shown that natural hybridisation did take place occasionally in self-fertilising plants like wheat or barley,[41] and by the First World War this was widely recognised among breeders. As some Mendelians pointed out, this meant that one could not assume genetic homogeneity even in self-fertilising plants which had been inbred for several generations.[42] And even within a line apparently homozygous for the desired trait, there were almost certainly "adjacent segregations" going on in other traits.[43] The other source of variation in "pure" lines, of course, was mutation. By about 1930, for example, the public-sector breeder, Hans Kappert, reckoned that the high frequency of micromutations in natural populations demonstrated that the gene was much more labile than early Mendelians had thought. This meant that breeders should not abandon continued selection in "pure" lines too quickly.[44]

Some supporters of Johannsen responded to this criticism by conceding that natural hybridisation did generate new variation in "pure" lines but argued nonetheless that all that repeated selection could achieve was to maintain purity; it could not produce improvement.[45] But Johannsen's critics insisted that selection could go farther than this because "pure" lines were rarely pure even in the traits of interest. Important traits like yield or quality, they argued, were so genetically complex that obtaining homozygosity at all of their many loci would take a long time.[46] The German academic breeder George Sessous, for example, concluded from his own experiments and others' that continued selection in pure lines was effective in improving them because lines conventionally regarded as pure in respect of relatively easily measured traits (e.g., yield, morphology) still contained a lot of heterozygosity

[37] E.g., Tschermak (1904); Roemer (1910); Dix (1914); Webber (1912); Wieland (2006, p. 330 ff.); and Charnley (2011, p. 142).

[38] Percival (1925, pp. 68–69); Harwood (1997); Bonneuil and Thomas (2009, pp. 45–48); and Saraiva (2010, p. 483–484).

[39] E.g., Kiessling (1915, pp. 110–111) and Tschermak (1915).

[40] Roll-Hansen (2005, p. 47).

[41] Meinel (2008, pp. 165–166).

[42] East (1907, p. 55) and Nilsson-Ehle (1909, pp. 15–16).

[43] Nilsson-Ehle (1911, pp. 8–9).

[44] Kappert (1931).

[45] E.g., East (1907, p. 56); Webber (1912, pp. 126–127); and Tschermak (1915).

[46] Beaven (1947, p. 244 ff., 252–253) and Bonneuil (2006, p. 296).

in more subtle physiological traits.[47] This variability was much harder to detect and would only become noticeable if breeders tested each line in a variety of quite different growing conditions, but continued selection could improve them.[48] Finally, several public-sector breeders argued that repeated selection would lead to improvement in self-fertilising lines because a continuous stream of variation was generated by the inheritance of acquired characteristics.[49] This view was not peculiar to Germany and continued to be voiced by reputable biologists into the 1930s.[50] Although Mendelians liked to claim before the First World War that genetics had refuted neo-Lamarckian mechanisms, it is important to remember that such mechanisms were still regarded as credible by major evolutionary theorists in the 1930s.[51]

It is not clear when the debate over "pure lines" and selection finally subsided among public-sector breeders. It is perhaps significant that as late as 1940 Theodor Roemer still thought it necessary to declare the inefficacy of selection in pure lines,[52] but whom he felt needed persuading—public or private sector breeders—is hard to tell. The lesson of this episode, however, is that the inefficacy-of-selection argument could not, on its own, be persuasive in 1903 or even 20 years later. And that is because it relied upon background assumptions which themselves remained contestable through the 1930s: (a) that the inheritance of acquired characters was *not* a source of variation and (b) that complex polygenic traits *could* be obtained in homozygous form within a few generations of self-fertilisation. Only once those background assumptions were no longer contested—perhaps undermined by a growing number of studies showing that continued selection had proven fruitless?—could Johannsen's argument become convincing.[53]

---

[47] Sessous (1929). Beaven reckoned that in barley there were as many as 20 such non-observable characters which could affect both yield and quality (Beaven 1947, p. 333).

[48] One set of experiments to which Sessous may have been referring were those at the Bavarian Plant Breeding Station where it was found in the mid-1920s that continued selection in oats, though not in other crops, was effective (Anonymous 1927, pp. 10–11); see also Kalben (1923). A similarly qualified endorsement of Johannsen's thesis seems to have been maintained by Babcock and Clausen (1927, p. 388), who argued that continued selection would only be of value in those self-fertilising crop plants which displayed relatively high levels of natural crossing.

[49] Lang (1910, p. 48); Rümker (1911); Wittmack (1911); Ziegler (1930); and cf. Fruwirth (1925, p. 608).

[50] On the UK and France: Beaven (1947, p. 5) and Bonneuil (2006, p. 295). On the 1930s: Sessous (1929, p. 50) and Kappert (1978, p. 55).

[51] E.g., Pearl (1915) and Mayr and Provine (1980).

[52] Roemer (1940, p. 269).

[53] It has sometimes been argued that mutagenesis is another area of genetics which had an impact upon breeding practice (Zirnstein 1977, pp. 231–240). Evaluating this claim is too large a task to undertake here, but it is worth noting what two historians of mutagenesis have recently had to say on the subject. According to Helen Curry (personal communication), the expansion of mutation breeding which occurred in the 1950s was usually justified by reference to only a few successful earlier studies. And while many more new varieties were produced using mutagenesis from the 1950s, it remains to be seen whether this method was more cost-effective or efficient than other approaches. Some breeders thought not. About 1930, the mutation geneticist Lewis Stadler, for example, held out little hope that the method would soon have much practical impact (Zirnstein 1977), and others took the same view in the 1950s (Becker 1960, p. 109). As Hamblin (2009) has

What Did Theory Actually Contribute?

Much of the Mendelians' promise to transform breeding practice, therefore, did not amount to much. No radically new method was introduced, nor did a series of more specific claims to improve breeding stand up to scrutiny. That does not mean, however, that the new science had *no* impact upon breeders. On the contrary, by providing a theoretical foundation for breeding, Mendelism gave practitioners a framework for talking and thinking about their practice in a new way: no longer just at the phenomenological level but also at an underlying causal level. And this framework was consequential in three respects.

First, an obvious function of theory is to explain. Observable phenomena are accounted for in terms of the behaviour of hypothetical unobserved entities. Several of Mendelism's defenders regarded this in fact as the theory's major contribution. Acknowledging that Mendelism had not fundamentally altered breeding practice, they emphasised instead that it had put breeding on a scientific foundation: "The knowledge of breeding has developed into the science of genetics and is fast assuming… the form of an exact science. Yet with all this advance in our understanding, the methods of breeding…have changed but little in the last twenty years…".[54] Twenty years later, some continued to take this view: "Genetic principles have stimulated the breeder and altered his outlook but have not as yet given him any substantially new method".[55] And again in the 1950s: "Doubtless Mendel's papers have given us an assurance in our breeding program, but if one honestly tries to show how any genetic study since Mendel has modified his breeding procedure, he is hard put to it to find a case that is more than argument".[56]

Among the explanations which some thought important was that provided by the genotype—phenotype distinction. While acknowledging that a similar notion had existed in the nineteenth century, Theodor Roemer emphasised that Mendel had provided "the complete biological explanation" for it.[57] Another phenomenon

---

shown, in the 1960s, the International Atomic Energy Agency invested a good deal of effort in promoting the use of radiation in breeding and other areas of agriculture in the developing world. But a number of plant breeders and other agricultural scientists in the UN's Food and Agricultural Organization complained that the value of radiation was being overhyped and did not represent the best use of funds for Third-World plant breeding.

[54] Webber (1912, p. 29). In a similar vein, see Castle (1907, p. 34). Breeding method, Raymond Pearl noted, had made a great deal of progress without requiring the aid of science; Mendel's main contribution had been an explanatory one (Pearl 1915; cf. Nilsson-Ehle 1913, p. 69). It is perhaps significant that, in a 12-page paper on the implications of Mendelism for breeding practice, Erwin Baur devoted but a single page to Mendelism's consequences for *method,* spending the rest of the paper showing what Mendelian theory could *explain* (Baur 1913). That theory—at least initially—more often explains practice than transforms it has been recognised before, for example, by historians of medicine (e.g., Geison 1979).

[55] Engledow (1931, p. 87).

[56] Harlan (1957, pp. 164, cf. 212).

[57] Roemer (1914b, p. 266).

which had puzzled nineteenth-century breeders was transgression. Its explanation in Mendelian terms was also welcomed by some early-twentieth-century breeders.[58]

Explanations of puzzling phenomena, of course, are intellectually satisfying. And as noted above, they, no doubt, also lent a certain scientific respectability to plant breeding, a non-trivial consideration for the public sector before the First World War. But explanation also gave breeders something else: a tool for reflecting upon the nature of their *existing* practices. And as various writers at the time recognised, this provided an additional way to judge their efficacy.[59] Several argued, for example, that Mendelian analysis demonstrated *why* mass selection would be less effective than individual selection. One of these was Hans Kappert who argued that the genotype–phenotype distinction was very important because it provided a scientific confirmation of the validity of pedigree selection.[60] Animal breeders made much the same point. Mendelism had revealed that the long-standing emphasis upon a breed's ancestry—as an indicator of its capacity to breed true—was misplaced. Breeders should focus instead upon an animal's genotype as inferred from its *progeny*.[61] Whether these arguments in fact prompted breeders to alter their practice, of course, remains an open question.

Thirdly, as is well known, theory plays a heuristic role in inquiry. By metaphorically redescribing the process it accounts for, theory provides a new way of seeing a familiar phenomenon.[62] Instead of seeing hybridisation merely in phenomenological terms—i.e., noting the proportions of plants with particular trait combinations which are found in different generations after crossing—breeders after 1900 were able to see hybridisation as a process in which something like paired beads for each trait separated at meiosis and combined randomly with corresponding "beads" in fertilisation. This altered way of seeing then opened the possibility of *improving* existing practices as well as inventing new methods. This is what H. J. Webber

---

[58] Radick (2013) and Charnley and Radick (2013). The explanation is often credited to Nilsson-Ehle (1909).

[59] East (1907, p. 73) and Brieger (1927, p. 467).

[60] Kappert (1931, p. 105) and cf. Tschermak (1951, p. 262). On the other hand, not every breeder needed Mendelism to realise this; Wilhelm Rimpau had come to the same conclusion in the 1890s (Meinel 2008, p. 182).

[61] Pearl (1913, pp. 545–546); Theunissen, "Connecting genetics, evolutionary theory and practical animal breeding: Arend L. Hagedoorn (1885–1953)", unpublished manuscript, 2012; cf. (Bonneuil 2006, pp. 293–294). Many years ago, Gottfried Zirnstein suggested another respect in which Mendelism's ability to provide explanations may have changed breeders' perceptions (Zirnstein 1977, pp. 181, cf. 257). One of the things which a theoretical explanation does is to unify and simplify our understanding of the world. It draws together a range of phenomena, previously thought to be unrelated, under a common denominator. It makes "alike" what was earlier thought to be "unlike". In the case of plant breeding, Zirnstein remarked, many of the phenomena which Mendelism explained had been known to breeders in the nineteenth century, but since no one could then account for them, it may have been all too easy to dismiss them as one-off anomalies. Only with the arrival of a theoretical explanation for these apparently disparate phenomena would it have become apparent to breeders that they were all instances of a *general* process and should thus be taken seriously.

[62] Hesse (1966).

evidently meant when he wrote that the conception of unit characters and Mendelian segregation was necessary not only to clarify our understanding of hybridisation but also "to bring out the latent possibilities of the material presented by nature for the use of the breeder".[63]

Numerous breeders, for example, drew attention to the Mendelian vision of the organism as an "aggregate". During the 1890s, for example, breeders in Germany generally felt it important when using selection to take into account the properties of the whole plant.[64] The new Mendelians, however, argued that the plant should be seen less as an integrated whole than as an assemblage of independent traits.[65] And this altered focus had several important consequences. For one thing, as various breeders noted, it meant that desirable traits previously thought to be mutually exclusive (such as yield and quality)—due to some kind of physiological mechanism which made for an inverse relation between them[66]—might actually be combinable after all.[67] It is quite possible, therefore, that breeders would thus have been encouraged to attempt things previously thought unattainable.

Another implication of the "aggregate" model was that combined factors would maintain their integrity after hybridisation. Thus, if a parental trait seemed to disappear in the F1, the model predicted that it would reappear in the F2 or a later generation. This meant, for example, that when breeding for disease resistance, there was no need to be concerned about crossing a disease-resistant wild variety with a high-yielding domestic variety because there was no danger that disease resistance would be "diluted out" in the hybrid.[68]

The Mendelian framework also suggested ways in which hybridisation could be improved (an issue of particular concern to Erich von Tschermak[69]). One was based on the concept of dominance. In the late nineteenth century, for example, there was no known rule on how long one should select, following hybridisation, before a recombinant could be expected to breed true.[70] As a result, Tschermak complained, many hybridisers carried on selecting plants beyond the F2 generation unnecessarily because they treated all of the plants in the F2 alike.[71] Mendel, he argued, had shown that some F2 plants displaying desired dominant traits would be heterozygous and would thus indeed require further inbreeding before they bred true. But plants bearing desired recessive traits would immediately breed true so that there was no need to waste any time inbreeding them further. This is why he

[63] Webber (1912, p. 30, cf. 130).

[64] Anonymous (1896); Stoll (1905); and Wohltmann (1907).

[65] Fruwirth (1902, p. 229); Nilsson-Ehle (1909), p. 12; Pearl (1913, p. 544; and Kiessling (1914, p. 10).

[66] Harwood (2004).

[67] Tschermak (1905, p. 332); Roll-Hansen (1997, p. 203); Holdefleiss (1908, pp. 106–109); Engledow (1925, pp. 32–33); and Bonneuil and Thomas (2009), p. 42).

[68] Kappert (1931, p. 103).

[69] Gliboff, "Breeding better peas, pumpkins and peasants…" Chap. 19 in this volume.

[70] Fruwirth (1887).

[71] Tschermak (1903, 1905 p. 330); and cf. Fruwirth (1902, p. 229).

felt it crucially important to establish the dominance/recessiveness of agriculturally important traits.[72] The other Mendelian argument for the reform of hybridisation was that breeders too often failed to sow enough plants in the F2 generation so that they stood some chance of spotting rare recombinants. By allowing them to calculate the frequency at which a given recombinant ought to appear in the F2, however, Mendelism gave breeders an indication of how many plants they needed to sow.[73]

Finally, we must ask whether the Mendelian perspective helped breeders to develop *new* practices. The method which has probably attracted the most attention in this respect is heterosis breeding (also known as the "inbred-hybrid" method), the method which enabled the development of hybrid maize whose high-yielding varieties were quickly adopted by US maize growers from the 1940s.[74] Here, too, the importance of genetic theory has been heavily emphasised: "Hybrid corn has been developed as a result of researches in genetics...and is an outstanding example... of the influence of theoretical scientific research in revolutionizing the production practices of an agricultural crop".[75]

But what precisely was Mendelism's contribution to the development of this method? Several historians have claimed that it was important without actually demonstrating this in detail. Zirnstein, for example, says it was a new method "which was developed as a result of scientific research" but does not specify exactly what Mendelism contributed. Indeed, a few lines later he admits that the method was not predicted from science but rather "the unexpected result of studies directed at very different goals". Kloppenburg, too, states that genetic theory had a "tremendous practical impact" but does not actually show this. Deborah Fitzgerald's book provides the fullest historical account so far of the development of hybrid maize but does not address the issue of its relation to Mendelism.[76]

---

[72] Tschermak (1904, p. 42). In the UK, A. D. Darbishire later made the same point (Charnley 2011, pp. 134–137). According to Berris Charnley (pers. comm.), R. H. Biffen used the same principle to infer which plants in the F2 with desired *dominant* traits were homozygous and would thus breed true.

[73] E.g., Tschermak (1901, 1905, p. 330) and Webber (1912, p. 131); see also Bonneuil (2006, p. 293).

[74] Heterosis breeding was and is just one way in which breeders sought to take advantage of "hybrid vigour", a phenomenon known since the nineteenth century in which the offspring of a cross are often larger or higher yielding than their parents. Used with crops which normally reproduce by outbreeding (e.g., maize or rye), the method consists, first, of inbreeding single plants for seven or eight generations until each has produced a uniform line (because the plant is largely homozygotic at most loci). Then two such inbred lines are crossed to produce hybrid seed which produces not only high-yielding plants but also a uniform crop (since all of the plants possess the same heterozygotic genotype).

[75] Jenkins (1936, p. 471) and cf. Mangelsdorf (1951, p. 555).

[76] Respectively, Zirnstein (1977, p. 205); Kloppenburg (1988, p. 77); and Fitzgerald (1990, pp. 29–35). Historians are not the only ones who have failed to demonstrate this connection. One prominent maize breeder has claimed that the heterosis method was derived "through the application of the principles of genetics" (Mangelsdorf (1951, p. 555), but his attempt to show this is brief and unconvincing.

The role of genetic theory in the development of the heterosis method is too large a topic to be discussed thoroughly here, but it is worth noting that much of the literature on hybrid maize suggests that theory was rather less useful than has often been claimed. As one prominent maize breeder remarked in the 1950s, "…corn-breeding today is nearly as empirical as when Shull (1909) outlined the development and utilisation of inbred lines…. [Despite important advances] the basic problems then and now remain much the same".[77] One reason for this restrained assessment is that, despite the claims of early enthusiasts, Mendelism most certainly did not eliminate trial and error from maize breeding. As numerous breeders pointed out, there was no correlation between the characteristics of an inbred line (e.g., yield) and those of a hybrid constructed from it. This meant that the only way breeders could find hybrids which were high yielding was simply to examine all of the possible combinations which could be constructed from the available inbred lines. This was, of course, nothing less than trial and error of a very laborious kind.[78]

The literature on hybrid maize, however, allows us to ask more precisely what, if anything, Mendelism may have contributed. Was it, for example, *necessary* for the development of the new method? Apparently not. One maize breeder who was involved in the field around 1920 later played down the importance of Mendelism:

> In retrospect it appears that most of the principles now applied to the production of hybrid corn were known prior to the rediscovery of Mendel's laws in 1900. …These included most of the facts discovered later and presented in greater detail by Shull and East.[79]

Even if not necessary, was Mendelism perhaps *sufficient* to specify the heterosis method? Again, the answer seems to be "no". Looking back from the 1960s, one maize breeder concluded that "Basic genetic theory was inadequate to serve as a guide".[80] In the back of his mind was perhaps the fact that before the First World War breeders who had embraced the Mendelian model of heredity also advocated two *other* breeding methods designed to capture hybrid vigour. One of these was to cross varieties rather than inbred lines. E. M. East, for example, was one of many public-sector breeders who felt that the evidence from studies of inbreeding and cross-breeding suggested that varietal crosses could also be a useful method.[81] Indeed, some late-nineteenth-century varietal crosses gave yields which later turned out to compare favourably with those of hybrid maize.[82] Moreover, had this method been pursued, it would have avoided the huge costs involved with obtaining inbred lines and testing them in combination.

As Paul and Kimmelman demonstrated several years ago,[83] the other alternative to the inbred-hybrid method which was then being proposed by Mendelians was

---

[77] Sprague (1955, p. 283).

[78] E.g., Brieger (1927, p. 469); Hayes (1963, p. 49ff.); and Sprague (1975, p. 9).

[79] Hayes (1963, pp. 24–25).

[80] Sprague (1962, p. 106).

[81] Hayes (1963, p. 20) and Sprague (1955, p. 234).

[82] Sprague (1983, p. 48).

[83] Paul and Kimmelman (1988).

selection (i.e., without crossing). In order to understand why two such different methods were being proposed, one needs to look at the debate at that time over the causes of hybrid vigour. The main advocate of the inbred-hybrid method George H. Shull, explained the phenomenon of hybrid vigour in terms of the stimulation arising from heterozygosity as such (i.e., a synergy based on some kind of non-additive effect). This meant that, to secure the advantages of hybrid vigour, it would be necessary to construct varieties which were heterozygous at the important loci. Other Mendelians, however, argued that the data on hybrid vigour could just as easily be explained in terms of multiple loci with *additive* effects.[84] This meant that heterozygosity at key loci was not necessary for hybrid vigour; instead, it should be possible to obtain within a single line an assemblage of genes favourable to high yield in *homozygous* form by using selection alone.[85] In fact, by the 1920s, most geneticists favoured the additive interpretation, and it has been the dominant one since the 1960s.[86] Thus, one cannot claim that Mendelian theory "paved the way" for the inbred-hybrid method because most geneticists, while accepting Mendelian theory, rejected the explanation of hybrid vigour from which Shull had derived this method. All one can say is that the inbred-hybrid method was one of several methods which were *consistent* with Mendelian theory.

So did Mendelian theory have no impact at all upon the development of heterosis breeding? Not quite. Arguably, the theory opened up the possibility of improving breeding practice by changing the way in which breeders thought about an *older* method: inbreeding. As Fitzgerald has shown, in the late nineteenth century, inbreeding was widely perceived to weaken those crops which normally cross-fertilise.[87] But from about 1907, she suggests, E. M. East began to think that if Mendelism were correct, what inbreeding was doing was merely to increase the frequency of a trait in the population by bringing about homozygosity. In that case, most traits fixed by inbreeding might indeed be undesirable, but some might not be, in which case inbreeding could actually be useful. Although she does not argue the point in detail, Fitzgerald's suggestion is entirely plausible. Consider, for example, the classic paper by Shull (East's contemporary) which is commonly cited as a conceptual breakthrough in the development of heterosis breeding.[88] In the paper, he points out that inbreeding is not deleterious in *self*-fertilising plants and sometimes not even in *cross*-fertilising ones. Drawing upon a comparative study of in- versus out-breeding

[84] Keeble and Pellew (1910).

[85] Berlan and Lewontin (1986) and Lewontin and Berlan (1990).

[86] Sprague (1983); Jinks (1983); and Crow (1987). East, for example, while initially endorsing Shull's interpretation, soon changed his mind in favour of the additive model (East and Jones 1919, p. 164 ff.). One prominent breeder argued several years ago that this approach had a great deal of potential (Simmonds 1979, pp. 160–62), and some inbred lines of maize developed using selection have indeed displayed high yields, sometimes as high as F1 hybrid varieties (Babcock and Clausen 1927, p. 404; Crow 1987). Despite its credibility among geneticists and breeders, however, selection's potential was never systematically explored in the USA because the Department of Agriculture favoured F1 hybrids (Kloppenburg 1988; Fitzgerald 1990).

[87] Fitzgerald (1990, p. 35).

[88] Shull (1908).

in maize, Shull argues that the most interesting result is not that the process of inbreeding plants from an original stock population proved damaging but rather that it gave rise to a number of distinctive types—he calls them "races" or "elementary species", differing in a wide range of morphological traits—which then bred true.

Although he makes no mention of homozygosity/heterozygosity nor of Mendelism in an attempt to make sense of these results,[89] Shull does nonetheless draw upon genetic theory: He invokes Johannsen's concept of "biotype" as a genetically distinct variant within a heterogeneous population. Although the original stock population used for the experiment had appeared to be fairly uniform, Shull concludes, "an ordinary cornfield is a series of very complex hybrids produced by the combination of numerous elementary species". Inbreeding, therefore, "…eliminates the hybrid elements and reduces the strain to its elementary components".[90]

What, then, were the practical implications of this interpretation? As we saw above, Keeble and Pellew were among those who reckoned that hybrid vigour could be increased by selection alone. Shull, however, drew a different conclusion. Late-nineteenth-century maize breeders, he noted, initially sought to improve varieties by using inbreeding, just as breeders of self-fertilising cereals like wheat or oats had done so successfully. But they soon found that the lines thus produced were almost always severely weakened. As a result, they abandoned inbreeding and chose instead to use mass selection on existing (heterogeneous) populations in the hopes of capturing the most vigorous hybrid combinations. But this, Shull argued, was a slow, indirect, and inefficient way to isolate the best hybrids. What breeders ought to do instead, he argued, was to return to inbreeding, using it to produce a series of individual biotypes.[91] Crossing these in all possible combinations would then show *directly* which of these combinations (i.e., F1 hybrids) was most vigorous. Then these hybrids could be produced in bulk for sale. To my knowledge, this was the first sketch of what became known as the inbred-hybrid method.

If the argument above is correct, however, the usual portrayals of the invention of this method are highly misleading in two respects. First, to suggest that the inbred-hybrid method was "derived" from Mendelism is to downplay the fact that Mendelians other than Shull perceived the theory as perfectly consistent with the use of *other nineteenth-century methods*—i.e., mass selection, varietal crossing—which were also effective in capturing hybrid vigour. Thus, the theory was multivalent in its implications; it permitted a range of possible readings. And second, to describe the inbred-hybrid method as "new" is to take it out of historical context. What Mendelian theory did was to provide a framework for understanding what inbreeding meant at the genotypic level. This reinterpretation then enabled Shull to reassess its value as a method and put it together with other *pre-Mendelian* practices (i.e.,

[89] The following year, East (1909) did use this terminology in interpreting similar data.

[90] Shull (1908, p. 299).

[91] A year or two earlier, Hugo De Vries had come close to this view when he wrote that repeated selection (in maize) was done because breeders feared the detrimental effects of inbreeding, "but if experience should prove that one year's self-fertilisation is sufficiently harmless, the process of corn breeding could be shortened…" (Vries 1907, p. 151).

hybridisation). The resulting procedure, to be sure, was innovative, but it is better described as a novel combination of older methods than as anything radically new.

Conclusion

As I have shown, for the first few decades following the rediscovery of Mendelism, breeders and geneticists in several countries greatly exaggerated the implications of the new theory for breeding. Moreover, even several of the more circumscribed claims for its importance were unjustified. What Mendelism did provide, however, was a theoretical foundation which explained the breeding process. And that theory enabled breeders, not only to assess the efficacy of existing methods but also to think about how such methods might be improved, albeit through incremental rather than radical change. At least in principle, therefore, Mendelism offered something useful to breeders.

How they responded to it *in practice,* however, is another matter. One historian of plant breeding has suggested that the reason why some public-sector scientists exaggerated the benefits of Mendelism was that, in general, commercial breeders were simply not aware of its potential and needed to be persuaded.[92] Whether or not breeders were "aware" of Mendelism's potential is not easy to establish, and Zirnstein presents no evidence on this point. Moreover, he seems to overlook the fact that the major breeders, in Germany at least, would have been aware of the new Mendelism through lectures organised by the Seed Breeding Division of the German Agricultural Society and from 1908 by the commercial breeders' association (*Gesell. z. Förderung d. deutschen Pflanzenzucht*) as well as by articles in the agricultural press.

What the German evidence does suggest, on the other hand, is that interwar breeders were quite cautious about adopting hybridisation. Moreover, there would have been little point in trying to win them over through vigorous Mendelian propaganda since their caution was well founded. To begin with, German farmers in the 1920s were going through an agricultural crisis which forced several breeders into bankruptcy.[93] This was no time to be investing in an expensive new technology. And that is just what "the Mendelian method" was: Hybridisation was a very labour-intensive process. As one of its supporters conceded, "...as theoretically straightforward as hybridisation is, it is usually a lot more difficult in practice".[94] It took a long time, for example, to get stable combinations following hybridisation. Most experts estimated that it took 7–10 years to find promising combinations that were then worth field-testing.[95] With individual selection, on the other hand, breeders

[92] Zirnstein (1977. p. 259).

[93] Harwood (2012, p. 83 ff.).

[94] Baur (1921, p. 81; cf. Roemer (1927).

[95] Schindler (1909, p. 219); Baur (1921, pp. 80–82); Babcock and Clausen (1927, p. 425); Ackermann (1928); Engledow (1931, pp. 82–83); Stanton (1936, Table 2); and Harlan (1957, p. 107).

reckoned that it took between 4 and 7 years to get to the same stage.[96] But besides being slow, hybridisation required the breeder to evaluate a very large number of plants in the F2 and beyond. According to the Mendelian scheme, the frequency of a given recombinant falls logarithmically as the number of loci increases arithmetically. That meant that even if a trait were determined by only five loci—many economically important traits are much more complex—the breeder would have to screen about 1000 plants in the F2 generation in order to find a recombinant which bred true.[97]

Hybridisation, however, was not just labour intensive. The traits of agricultural interest were formidably complex. In order to know how many plants to screen, for example, the breeder needed some idea of how many loci were involved, and often such knowledge was lacking. One geneticist has claimed that the work of Hermann Nilsson-Ehle on the inheritance of physiological characters in wheat was "of fundamental importance to plant breeding".[98] Although this was certainly an important extension of Mendelian theory, it was not of much practical use.[99] In 1911, for example, Nilsson-Ehle was convinced that multiple Mendelian factors underlay the continuous distribution of yellow rust resistance in wheat, but he had no way of establishing the number of loci involved.[100] Similarly, by 1914, he believed that the time of germination in wheat was affected by at least four and perhaps five Mendelian factors, but he acknowledged that these were merely factors which inhibited germination; there were undoubtedly additional (unknown) factors which *promoted* the process.[101] A decade or two later, maize and wheat breeders still did not know the number of factors involved in yield and other important traits.[102] And one reason why it was so difficult for breeders to identify particular loci (through the segregation of a discrete phenotype) was that polygenic traits of economic interest tend to be very sensitive to environmental stimuli, such that segregation is masked by phenotypic variation.[103] All in all, as late as 1944, the quantitative geneticist Kenneth Mather admitted that genetics had not provided breeders with much help in dealing

[96] Holtmeier-Schomberg (1908, p. 366); Engledow (1931, pp. 80–81); and Ackermann (1929, p. 71).

[97] One breeder estimated that the number of plants one had to evaluate following hybridisation was 100 times the number necessary with pure-line selection (Engledow 1931). Another reckoned that finding the desired combination following a cross of two grape varieties—where the number of loci involved was thirty to forty—would require scanning several million plants (Baur 1927, p. 723; see also Babcock and Clausen 1927, pp. 423–424).

[98] Müntzing (1951, p. 475).

[99] Anonymous (1914, p. 19).

[100] Nilsson-Ehle (1911).

[101] Nilsson-Ehle (1914).

[102] Brunson (1926); Clark (1936); and Sprague (1955, p. 257). Sometimes, however, the number of loci can be estimated. Oil level in maize kernels, for example, is estimated to be affected by at least 20 genes, but one breeder reckoned that to find a plant homozygous for all of those loci would require planting 90 million acres (Sprague 1955, p. 257).

[103] E.g., Baur (1921, p. 81) and Hunter (1939, pp. 247–248.

with continuous variation.[104] Under the circumstances, one can hardly be surprised that interwar breeders did not rush to embrace Mendelism.

If we want to understand why Mendelism was so heavily oversold, therefore, we need to look elsewhere. The answer may well lie in status uncertainty. As several historians have shown, before the First World War, the new "genetics" lacked reliable sources of funding as well as institutional support in many countries.[105] Small wonder, then, that early Mendelians like Bateson or Hugo DeVries[106] made inflated claims for the theory's significance in breeding. A similar argument has been advanced to account for the early enthusiasm of academic plant breeders.[107] And there is little doubt that a "scientific foundation" for their field held considerable appeal for staff at some German agricultural colleges.[108] In the USA, too, "Mendelism seemed to take plant breeding from the arts and place it as a science overnight. It offered prestige".[109] One implication of this hypothesis, however, is that as both fields became more institutionally secure—as they did during the interwar period—the necessity for wild exaggeration would have begun to decline (apart from celebratory occasions or appeals for public funding). Interestingly, there are indeed signs that some Mendelians who had waxed lyrical in the early years began to tone down their claims later on. Though he had promised so much around 1902, for example, William Bateson was saying by 1911 that breeding practice was so far ahead of the science that the latter "can scarcely hope in finite time even to represent what has been done, still less to better the performance".[110] Two years later, Raymond Pearl said much the same, as did Erwin Baur two decades on. And in the late 1920s, Ernest Babcock and Roy Clausen recalled that in the early years of the century some Mendelians had been overly optimistic about the ease and speed with which the new laws of heredity could be applied to breeding.[111] Indeed, recent work in science and technology studies suggests that this may be a general phenomenon whereby "hype" characterises the early stages of new technologies, followed invariably by more sober assessments.[112] Whenever the "almighty power of science" is invoked, therefore, it would be well to look closely at the relevant institutional context.

Finally, to return to the question posed at the outset: What does this case study suggest about the relations between science and technology? During the 1920s, one academic breeder in the Netherlands remarked that "There is no way we can yet consider breeding as applied genetics".[113] A generation later, others were taking the

[104] Anonymous (1944, p. 781).

[105] Paul and Kimmelman (1988); Burian et al. (1988); and Harwood (1987).

[106] Theunissen (1912).

[107] Palladino (1994) and Wieland (2004).

[108] Harwood (2005).

[109] Harlan (1957, p. 96) and cf. Engledow (1931).

[110] Quoted in Radick (2013).

[111] Respectively, Pearl (1913), pp. 539–540); Baur (1932), p. 2; and Babcock and Clausen (1927), p. 337).

[112] Borup et al. (2006). I thank Thomas Wieland for alerting me to the existence of this literature.

[113] Cited in Maat (2001, p. 161).

same view. The geneticist A. L. Hagedoorn believed that geneticists had learned more from the best breeders than the other way round.[114] And at a British meeting in 1944 on the application of genetics to plant and animal breeding, the geneticist Cyril Darlington opened the meeting by declaring that the purpose of the meeting was to establish *whether* genetics would be able to contribute as much to breeding as the latter had to the development of genetics. At the same meeting, Kenneth Mather struck a similar note, remarking that "the progress of genetics has not yet led to the marked advances in…breeding which [have] been so confidently expected in the past".[115] In the case of plant breeding before 1945, therefore, the conclusion is clear: Although some breeders quite likely made use of Mendelian theory in devising new methods, they also drew upon a much wider range of resources, including successful, though empirically derived, practices from the nineteenth century. Thus, breeding was far more than "applied science" (a point which public-sector breeders have subsequently emphasised on numerous occasions[116]).

This conclusion will be no surprise to historians of technology, most of whom came to a similar conclusion 20 years ago. For a wider audience, however, it remains very important. On a general level, overinflated claims for the importance of Mendelism lend support to a widespread misconception that scientific theory plays the decisive role in technological invention. But more particularly, demonstrations that Mendelism's impact upon breeding was relatively modest may help to restrain the exuberance of those molecular biologists who never tire of declaring the "revolutionary" power of biotechnology.[117]

Acknowledgments This essay has been a long time in the making, starting with a paper to the conference "Ideas, Instruments and Innovation" at the California Institute of Technology in 1994. At that time, I gained much from discussions with Paolo Palladino and subsequently on numerous occasions with Nils Roll-Hansen. Recently, I have had helpful feedback from Thomas Wieland, Roger Wood, and Sander Gliboff. A first draft of this chapter was prepared in autumn 2012 at the Max-Planck Institute for History of Science in Berlin, thanks to an invitation from Veronika Lipphardt's research group.

References

Ackermann, Jakob. 1928. *25 Jahre Pflanzenzucht seit 1903*. Niederbayern: Saatzuchtwirtschaft Gut Irlbach.

Ackermann, Jakob. 1929. Bayerische Saatzucht. In *Die Landwirtschaft in Bayern: eine Sammlung von Darstellungen über Entwicklung und Stand der Bayerischen Landwirtschaft,* ed. Hans Dörfler, 70–74. Freising: Datterer.

[114] Theunissen, "Connecting genetics, evolutionary theory and practical animal breeding".

[115] Anonymous (1944, p. 781).

[116] E.g., Becker (1953, pp. 38–39); Schick (1958, p. 19); Dudley (1994, p. 163); Simmonds (1967, p. 15, 1979, p. viii); Riley (1983); and Duvick (2002, pp. 196–198).

[117] Brown (2003).

Anonymous. 1896. Versammlung der Saatgut-Abteilung am 12. Juni. *Mitteilungen der Deutschen Landwirtschafts-Gesellschaft* 11 (12): 123–125.

Anonymous, ed. 1914. *Bericht d. königlichen Bayerischen Saatzuchtanstalt in Weihenstephan,* Nr. 10 (1912 u. 1913). Freising: Datterer.

Anonymous, ed. 1927. *Bericht d. Bayerischen Landessaatzuchtanstalt,* 1922–1926, Nr. 14, 10–11. Munich: Carl Gerber.

Anonymous. 1944. Application of genetics to plant and animal breeding. *Nature* 153 (24): 780–783.

Babcock, E. B., and R. E. Clausen. 1927. *Genetics in relation to agriculture*. 2nd ed. New York: McGraw-Hill.

Bailey, L. H. 1896. *Plant breeding*. New York: Macmillan.

Baur, Erwin. 1913. Einige für die züchterische Praxis wichtige Ergebnisse d. neueren Bastardierungsforschung. *Beiträge zur Pflanzenzucht* 3:32–48.

Baur, Erwin. 1921. *Die wissenschaftlichen Grundlagen der Pflanzenzuchtung: ein Lehrbuch für Landwirte, Gartner und Forstleute*. 1st ed. Berlin: Borntraeger.

Baur, Erwin. 1927. Die experimentelle Erzeugung leistungsfähiger Rassen unserer Kulturpflanzen. *Die Naturwissenschaften* 15 (36): 721–725.

Baur, Erwin. 1928. Opening address to 5th International Congress of Genetics 1927. *Zeitschrift für induktive Abstammungs- und Vererbungslehre suppl.* I:51–53.

Baur, Erwin. 1932. Konsequenzen d. Vererbungslehre für d. Pflanzenzüchtung. *Handbuch der Vererbungswissenschaft* III D:1–30.

Beaven, E. S. 1947. *Barley: Fifty years of observation and experiment*. London: Duckworth.

Becker, G. 1947. Die wichtigsten Aufgaben der deutschen Pflanzenzucht. *Arbeiten d. Deutschen Landwirtschafts-Gesell., N.F.* 1:79–96.

Becker, G. 1953. Problematik der Pflanzenzüchtung. *Berichte u. Vorträge -Deutsche Akad. d. Landwirtschaftswissenschaften zu Berlin* 1:37–59.

Becker, G. 1960. Darwin u. die Pflanzenzüchtung. *Berichte u. Vorträge – Deutsche Akad. d. Landwirtschaftswissenschaften zu Berlin* 4:99–113.

Berlan, Jean-Pierre, and Richard Lewontin. 1986. The political economy of hybrid corn. *Monthly Review* 38 (3): 36–47.

Bonneuil, Christophe. 2006. Mendelism, plant-breeding and experimental cultures: Agriculture and the development of genetics in France. *Journal of the History of Biology* 39 (2): 281–308.

Bonneuil, C., and F. Thomas. 2009. *Gènes, pouvoirs et profits: Recherche publique et régimes de production des savoirs de Mendel aux OGM.* Versailles: Editions Quae and Fondation pour le progres de l'Homme.

Borup, M., et al. 2006. The sociology of expectations in science and technology. *Technology Analysis & Strategic Management* 18 (3): 285–298.

Brieger, Friedrich. 1927. Die genetische u. züchterische Bedeutung des Maises. *Illustrierte Landwirtschaftliche Zeitung* 9 (Sept): 467–469.

Brown, Nik. 2003. Hope against hype – accountability in biopasts, presents and future. *Science Studies* 16 (2): 3–21.

Brunson, A. M. 1926. The relation of inheritance studies to corn improvement. *Journal of the American Society of Agronomy* 18:308–314.

Burian, R., J. Gayon, and D. Zallen. 1988. The singular fate of genetics in the history of French biology, 1900–1940. *Journal of the History of Biology* 21:357–402.

Castle, W. E. 1907. The production and fixation of new breeds. *American Breeders' Association (Report of meeting on 15–18 Jan. 1907)* 3:34–41.

Charnley, Berris. 2011. Agricultural science, plant breeding and the emergence of a Mendelian system in Britain, 1880–1930. Ph.D. dissertation, University of Leeds.

Charnley, Berris, and Gregory Radick. 2013. Intellectual property, plant breeding and the making of Mendelian genetics. *Studies in History & Philosophy of Science* 44 (2): 222–233.

Clark, J. A. 1936. Improvement in wheat. In *Yearbook of Agriculture* 1936, 207–302. Washington: Government Printing Office.

Crow, James F. 1987. Muller, Dobzhansky and overdominance. *Journal of the History of Biology*, 20:351–380.

Dix, W. 1914. Die Anwendungen d. neueren Forschungsergebnisse auf d. Gebiete d. Pflanzenzüchtung in d. landwirt. Praxis. *Beiträge zur Pflanzenzucht* 4:122–139.

Dudley, John W. 1994. Plant breeding – a vital part of improvement in crop yields, quality and production efficiency. In *Historical perspectives in plant science,* ed. Kenneth J. Frey, 163–177. Ames: Iowa State University Press.

Duvick, Donald. 2002. Theory, empiricism and intuition in professional plant breeding. In *Farmers, scientists and plant breeding: Integrating knowledge and practice,* ed. D. A. Cleveland and D. Soleri, 189–211. Wallingford: CAB International.

East, E. M. 1907. The relation of certain biological principles to plant breeding. *Bulletin of the Connecticut Agric. Experiment Station* 158:1–93.

East, E. M. 1909. The distinction between development and heredity in inbreeding. *American Naturalist* 43:173–181.

East, E. M., and D. F. Jones. 1919. *Inbreeding and outbreeding: Their genetic and sociological significance*. Philadelphia: Lippincott.

Engledow, F. L. 1925. The economic possibilities of plant breeding. In *Imperial botanical conference: Report of proceedings,* ed. F. T. Brooks, 31–40. Cambridge: Cambridge University Press.

Engledow, F. L. 1931. Plant breeding: Its practices and scientific evolution. *Scientific Journal of the Royal College of Science* 1:74–95.

Fitzgerald, Deborah. 1990. *The business of breeding: Hybrid corn in Illinois, 1890–1940*. Ithaca: Cornell University Press.

Fruwirth, Carl. 1887. Züchtungsbestrebungen in den Vereinigten Staaten v. Nordamerika auf d. Gebiete d. landwirtschaftlichen Pflanzenbaues. *Fühlings Landwirtschaftliche Zeitung* 36:16–27.

Fruwirth, Carl. 1902. Neue Forschungen und ihre Verwertung bei der Pflanzenzüchtung. *Jahrbuch der Deutschen Landwirtschafts-Gesellschaft* 17:216–236.

Fruwirth, Carl. 1911. *Die Beziehungen der Pflanzenzüchtung zu den Wirtschaftsbetrieben*. Leipzig: Reichenbach'sche Verlagsbuchhandlung.

Fruwirth, Carl. 1925. Zur Frage erblicher Beeinflussung durch äussere Verhältnisse im Zuchtbetrieb. *Landwirtschaftliche Jahrbücher* 62:607–628.

Gayon, Jean, and Doris Zallen. 1998. The role of the Vilmorin Company in the promotion and diffusion of the experimental science of heredity in France, 1840–1920. *Journal of the History of Biology* 31:241–262.

Geison, Gerald. 1979. Divided we stand: Physiologists and clinicians in the American context. In *The therapeutic revolution: Essays in the social history of American medicine,* ed. M. J. Vogel and Charles Rosenberg, 67–90. Philadelphia: University of Pennsylvania Press.

Gerland, W. 1885. Die Hebung des Getreidebaues in Deutschland, namentlich in d. Provinz Sachsen. *Deutsche Landwirtschaftliche Presse* 12 (74): 489–490.

Hamblin, Jacob. 2009. Let there be light… and bread: The United Nations, the developing world, and atomic energy's Green Revolution. *History and Technology* 25:25–48.

Harlan, H. V. 1957. *One man's life with barley: The memories and observations of Harry V. Harlan*, New York: Exposition.

Harvey, R. D. 1995. Pioneers of genetics. *Notes & Records of the Royal Society* 49 (1): 105–117.

Harwood, Jonathan. 1987. National styles in science: Genetics in Germany and the U.S. between the world wars. *Isis* 78:390–414.

Harwood, Jonathan. 1997. The reception of genetic theory among academic plant-breeders in Germany, 1900–1930. *Sveriges Utsädesförenings Tidskrift (Journal of the Swedish Seed Association)* 107 (4): 187–195.

Harwood, Jonathan. 2004. Linkage before Mendelism? Plant-breeding research in central Europe, ca. 1880–1910. In *Classical genetic research and its legacy: The mapping cultures of twentieth century genetics,* ed. Hans-Joerg Rheinberger and Jean-Paul Gaudilliere, 9–20. London: Routledge.

Harwood, Jonathan. 2005. *Technology's dilemma: Agricultural colleges between science and practice in Germany, 1860–1934*. Frankfurt: Lang.

Harwood, Jonathan. 2012. *Europe's Green Revolution and others since: The rise and fall of peasant-friendly plant-breeding*. London: Routledge.

Hayes, H. K. 1963. *A professor's story of hybrid corn*. Minneapolis: Burgess.

Hesse, Mary. 1966. *Models and analogies in science*. South Bend: Notre Dame University Press.

Holdefleiss, Paul. 1908. *Landwirtschaftliche Pflanzenzüchtung*. Hannover: Verlagsbuchhandlung Dr. Max Jänecke.

Holtmeier-Schomberg, Hermann. 1908. Die Entwicklung und Organisation der Pflanzenzüchtung in Dänemark, Schweden und der Probstei. *Landwirtschaftliche Jahrbücher* 37:311–380.

Hunter, H. 1939. Developments in plant breeding. In *Agriculture in the twentieth century: Essays... to be presented to Sir Daniel Hall*, ed. Anon., 223–260. Oxford: Clarendon.

Jenkins, Merle T. 1936. Corn improvement. In *Yearbook of Agriculture 1936*, 455–494. Washington: Government Printing Office.

Jinks, J. L. 1983. Biometrical genetics of heterosis. In *Heterosis: Reappraisal of theory and practice*, ed. R. Frankel, 1–46. Berlin: Springer.

Kalben, Rudolf von. 1923. Erfahrungen aus Vienauer Pflanzenzucht unter wissenschaftlicher Beleuchtung. *Deutsche Landwirt. Presse* 50:154–155, 161–162, 170–171.

Kappert, Hans. 1931. Vererbungswissenschaft. In *Wissenschaft u.Landwirtschaft. Festschrift z. 50-jährigen Bestehen d. Landwirtschaftlichen Hochschule Berlin*, ed. Anon., 101–108. Berlin: Parey.

Kappert, Hans. 1978. *Vier Jahrzehnte miterlebte Genetik*. Berlin: Parey.

Keeble, Frederick, and C. Pellew. 1910. The mode of inheritance of stature and of time of flowering in peas (*Pisum sativum*). *Journal of Genetics* 1:47–56.

Kiessling, L. 1912. Die züchterische Bearbeitung der Landsorten in Bayern. *Beiträge zur Pflanzenzucht* 2:74–96.

Kiessling, L. 1914. Erfolge und Aufgaben der Pflanzenzüchtung mit besonderer Berücksichtigung der Bayerischen Verhältnisse. In *Bericht über die 48. Wanderversammlung Bayerischer Landwirte in Erlangen vom 23–27. Mai 1914* (reprint located in folder MK 40158, Bayerisches Hauptstaatsarchiv, Munich).

Kiessling, L. 1915. Untersuchungen über die Vererbung von Stickstoffgehalt u. Korngrösse d. zweizeiligen nickenden Gerste. *Zeitschrift für Pflanzenzüchtung* 3 (2): 81–147.

Kloppenburg, Jack. 1988. *First the seed: The political economy of plant biotechnology, 1492–2000*. Cambridge: Cambridge University Press.

Lang, Hans. 1910. *Theorie und Praxis der Pflanzenzüchtung*. Stuttgart: Eugen Ulmer.

Lewontin, Richard, and Jean-Pierre Berlan. 1990. The political economy of agricultural research: The case of hybrid corn. In *Agroecology*, eds. C. Ronald Carroll, J. H. Vandermeer, and Peter Rosset, 613–628. New York: McGraw-Hill.

Lochow, Ferdinand von. 1913. 25 Jahre deutsche Pflanzenzucht. *Illustrierte Landwirtschaftliche Zeitung*, 14 (June): 448–450.

Maat, Harro. 2001. *Science cultivating practice: A history of agricultural science in the Netherlands and its colonies, 1863–1986*. Wageningen: Ponsen and Looijen.

Mangelsdorf, P. 1951. Hybrid corn: Its genetic basis and its significance in human affairs. In *Genetics in the twentieth century*, ed. L. C. Dunn, 555–571. New York: Macmillan.

Mayr, Ernst, and William Provine, eds. 1980. *The evolutionary synthesis: Perspectives on the unification of biology*. Cambridge: Harvard University Press.

Mayr, Otto. 1976. The science-technology relationship as a historiographic problem. *Technology and Culture* 17 (4): 663–673.

Meinel, Albrecht. 2008. *Aufbruch in die wissenschaftliche Pflanzenzüchtung. Der Beitrag von Wilhelm Rimpau (1842–1903)*. Göttingen: Gesell. für Pflanzenzüchtung.

Müntzing, Arne. 1951. Genetics and plant breeding. In *Genetics in the 20th century*, ed. L. C. Dunn, 473–492. New York: Macmillan.

Nilsson-Ehle, Hermann. 1909. *Kreuzungsuntersuchungen an Hafer u. Weizen*. vol. 1. Lund: Häkan Ohlssons Buchdruckerei.

Nilsson-Ehle, Hermann. 1911. *Kreuzungsuntersuchungen an Hafer u. Weizen*. vol. 2. Lund: Häkan Ohlssons Buchdruckerei.

Nilsson-Ehle, Hermann. 1913. Ueber die Winterweizenarbeiten in Svalöf. *Beiträge zur Pflanzenzucht* 3:62–88.

Nilsson-Ehle, Hermann. 1914. Zur Kenntnis der mit der Keimungsphysiologie des Weizens in Zusammenhang stehenden inneren Faktoren. *Zeitschrift für Pflanzenzüchtung* 2:153–187.

Palladino, Paolo. 1993. Between craft and science: Plant breeding, Mendelian genetics, and British universities, 1900–1920. *Technology and Culture* 34:300–323.

Palladino, Paolo. 1994. Wizards and devotees: On the Mendelian theory of inheritance and the professionalization of agricultural science in Great Britain and the United States, 1880–1931. *History of Science* 32:409–444.

Paul, Diane, and Barbara Kimmelman. 1988. Mendel in America: Theory and practice, 1900–1919. In *The American development of biology*, ed. Ronald Rainger, Keith R. Benson, and Jane Maienschein, 281–310. Philadelphia: University of Pennsylvania Press.

Pearl, Raymond. 1913. Genetics and breeding. *Science* 37:539–546.

Pearl, Raymond. 1915. *Modes of research in genetics*. New York: Macmillan.

Percival, J. 1925. The value of selection work in the improvement of crop plants. In *Imperial botanical conference (London, July 7–16 1924): Report of proceedings,* ed. F. T. Brooks, 60–73. Cambridge: Cambridge University Press.

Radick, Gregory. 2013. The professor and the pea: Lives and afterlives of William Bateson's campaign for the utility of Mendelism. *Studies in History & Philosophy of Science* 44 (2): 280–291.

Riede, W. 1927. Beiträge zu den Grundlagen der Pflanzenzucht. *Beiträge zur Pflanzenzucht* 9:58–92.

Riley, Ralph. 1983. Plant breeding – an integrating technology. In *Plant breeding for low input conditions,* ed. K. ter Horst and E. R. Watts, 76–83. Lusaka: Ministry of Agriculture & Water Development.

Rimpau, W. 1912. Ueber Kreuzungsprodukte von Getreide. *Beiträge zur Pflanzenzucht* 2:115–129.

Roemer, Theodor. 1910. Variabilitätsstudien – Dissertation. *Archiv für Rassen- und Gesellschaftsbiologie* 7:397–469.

Roemer, Theodor. 1914a. *Mendelismus und Bastardzüchtung der landwirtschaftlichen Kulturpflanzen (Arbeiten der Deutschen Landwirt. Gesell., No. 266)*. Berlin: DLG Verlag.

Roemer, Theodor. 1914b. Vererbung von Leistungseigenschaften. *Fühlings Landwirtschaftliche Zeitung* 63:257–268.

Roemer, Theodor. 1927. Gedanken zur landwirtschaftlichen Pflanzenzüchtung. *Illustrierte Landwirtschaftliche Zeitung* 49 (9): 461–463.

Roemer, Theodor. 1933. Transgressionszüchtung. *Zeitschrift für Pflanzenzüchtung* 18:168–178.

Roemer, Theodor. 1940. Entwicklungslinien der Züchtungsmethoden. *Kühn-Archiv* 54:267–294.

Roberts, H. F. 1929. *Plant hybridization before Mendel*. Princeton: Princeton University Press.

Roll-Hansen, Nils. 1997. The role of genetic theory in the success of the Svalöf plant breeding program. *Sveriges Utsädesförenings Tidskrift (Journal of the Swedish Seed Association)* 107 (4): 196–207.

Roll-Hansen, Nils. 2000. Theory and practice: The impact of Mendelism on agriculture. *Comptes Rendus Acad. Sci. Paris, Sciences de la vie* 323 (12): 1107–1116.

Roll-Hansen, Nils. 2005. Sources of Johannsen's genotype theory. In *A cultural history of heredity III: 19th and early 20th centuries (Preprint 294)*, ed. Anon., 43–52. Berlin: Max-Planck Institut für Wissenschaftsgeschichte.

Roll-Hansen, Nils. 2009. Sources of Wilhelm Johannsen's genotype theory. *Journal of the History of Biology* 42:457–493.

Royal Society, ed. 1902. *Reports to the Evolution Committee. Report 1: Experiments undertaken by W. Bateson & Miss E.R. Saunders*. London: Harrison.

Rümker, Kurt von. 1889. *Anleitung zur Getreidezüchtung auf wissenschaftlicher und praktischer Grundlage*. Berlin: Parey.

Rümker, Kurt von. 1911. Ueber Bedeutung und Methoden der Saatgutzucht. *Mentzel und Lengerkes Schreibkalender* 64:64–90.

Saraiva, Tiago. 2010. Fascist labscapes: Geneticists, wheat, and the new landscapes of Fascism in Italy and Portugal. *Historical Studies in the Natural Sciences* 40 (4): 457–498.

Schick, Rudolf. 1958. 100 Jahre landwirtschaftliche Pflanzenzüchtung in Deutschland. *Sitzungsberichte der Deutschen Akademie der Landwirtschaftswissenschaften zu Berlin* 7 (2): 3–21.

Schindler, Franz. 1909. *Der Getreidebau auf wissenschaftlicher und praktischer Grundlage*. Berlin: Parey.

Sessous, George. 1929. Der Wert des deutschen Ausleseverfahrens für Selbstbefruchter. *Beiträge zur Pflanzenzucht* 10:44–57.

Shull, G. H. 1908. The composition of a field of maize. *American Breeders Association (Proceedings)* 4 (1): 296–301.

Simmonds, Norman. 1967. Report by the Director. In *Annual report of the Scottish plant breeding station,* ed. Anon., 5–20. Corstorphine: Scottish Plant Breeding Station.

Simmonds, Norman. 1979. *Principles of crop improvement.* New York: Longman.

Sprague, G. 1955. Corn breeding. In *Corn and corn improvement,* ed. G. Sprague, 221–292. New York: Academic.

Sprague, G. 1962. Hybrid corn. In *Yearbook of Agriculture 1962*, 106–107. Washington: Government Printing Office.

Sprague, G. 1975. *The development of hybrid corn technology in the United States and selected countries*. Washington, DC: U.S. Agency for International Development.

Sprague, G. 1983. Heterosis in maize: Theory and practice. In *Heterosis: Reappraisal of theory and practice,* ed. R. Frankel, 47–70. Berlin: Springer.

Stanton, T. R. 1936. Superior germ plasm in oats. *Yearbook of Agriculture 1936*, 347–413. Washington: Government Printing Office.

Staudenmaier, John. 1985. *Technology's storytellers: Reweaving the human fabric*. Cambridge: MIT Press.

Stoll, P. H. 1905. Entwicklung und Stand der Getreidezüchtung. *Badisches Landwirtschaftliches Wochenblatt* (no volume): 70–71, 84–85, 98–100.

Stokes, Donald. 1997. *Pasteur's quadrant: Basic science and technological innovation*. Washington, DC: Brookings.

Theunissen, Bert. 2012. Connecting genetics, evolutionary theory and practical animal breeding: Arend L. Hagedoorn (1885–1953). unpublished manuscript.

Thiel, Hugo. 1904. Zum Gedächtnis des Amtsrats Dr. Wilhelm Rimpau-Schlanstedt. *Jahrbuch der Deutschen Landwirtschafts-Gesellschaft* 19:6–15.

Tschermak, Erich. 1898. Ueber Veredelung u. Neuzüchtung landwirt. u. gärtnerischer Gewächse. *Zeitschrift für Naturwissenschaften* 71:1–16.

Tschermak, Erich. 1900. Ueber künstliche Kreuzung bei Pisum Sativum. *Zeitschrift für landwirt. Versuchswesen in Oesterreich* 3:465–555.

Tschermak, Erich. 1901. Ueber die Züchtung neuer Getreiderassen mittelst künstlicher Kreuzung I. Kritisch-historische Betrachtungen. *Zeitschrift für landwirt. Versuchswesen in Oesterreich* 4:1029–1060.

Tschermak, Erich. 1903. Die praktische Verwertung des Mendelschen Vererbungsgesetzes bei d. Züchtung neuer Getreiderassen. *Deutsche Landwirtschaftliche Presse* 30 (82): 712–713.

Tschermak, Erich. 1904. Die Lehre von den formbildenden Faktoren (Variation, Anpassung, Selektion, Mutation, Kreuzung) u. ihre Bedeutung für die rationelle Pflanzenzüchtung. *Jahrbuch der landwirt. Pflanzenzüchtung u. Tierzüchtung* 1:30–45.

Tschermak, Erich. 1905. Die Kreuzung im Dienste der Pflanzenzüchtung. *Jahrbuch der Deutschen Landwirtschafts-Gesellschaft* 20:325–340.

Tschermak, Erich. 1908. Die Kreuzungszüchtung des Getreides u. die Frage nach den Ursachen der Mutation. *Monatshefte für Landwirtschaft* 1:24–31.

Tschermak, Erich. 1915. Ueber die Notwendigkeit der Sammlung u. Erhaltung unserer bewährten, noch unveredelten Getreidelandrassen. *Wiener Landwirtschaftliche Zeitung* 65 (104): 759–761.

Tschermak, Erich. 1940. Die Einwirkung der Wiederentdeckung der Mendelschen Gesetze auf Pflanzenbau u. Pflanzenzucht. *Kühn-Archiv* 53:1–13.

Tschermak, Erich. 1951. Wien, der Ausgangsort des praktischen Mendelismus. *Zeitschrift für Pflanzenzüchtung* 29:262–275.
Vries, Hugo de. 1907. *Plant-breeding: Comments on the experiments of Nilsson and Burbank*. London: Kegan Paul, Trench, Trübner.
Wacker, Johann. 1923–1924. Die landwirtschaftliche Pflanzenzucht und ihre land- und volkswirtschaftliche Bedeutung. *Zeitschrift für Pflanzenzüchtung* 9:35–49.
Webber, Herbert J. 1912. The effect of research in genetics on the art of breeding. *American Breeders Magazine* 3 (1912): 29–36, 125–134.
Wieland, Thomas. 2004. *'Wir Beherrschen den pflanzlichen Organismus besser....' Wissenschaftliche Pflanzenzüchtung in Deutschland 1889–1945*. Munich: Deutsches Museum.
Wieland, Thomas. 2006. Scientific theory and agricultural practice: Plant breeding in Germany from the late 19th to the early 20th century. *Journal of the History of Biology* 39:309–343.
Wittmack, Ludwig. 1911. Die wissenschaftlichen Grundlagen der Saatzucht in Deutschland in den letzten 25 Jahren. *Jahrbuch der Deutschen Landwirtschafts-Gesellschaft* 26:101–119.
Wohltmann, Friedrich. 1907. *Saatgut-und Pflanzen-Züchtung in der Praxis*. Berlin: Bund der Landwirte.
Wood, Roger, and Vitezslav Orel. 2001. *Genetic prehistory in selective breeding: A prelude to Mendel*. Oxford: Oxford University Press.
Ziegler, Otto. 1930. Kritische Betrachtung zur Auslesezüchtung.... *Zeitschrift für Pflanzenzüchtung* 15:167–218.
Zirnstein, Gottfried. 1977. Zur Geschichte der Beziehungen von Pflanzenzüchtung und Biologie: von den Anfängen bis in die Dreissiger Jahre des 20. Jahrhunderts. Ph.D. dissertation, Universität Jena.

Prof. Jonathan Harwood is emeritus professor of the history of science and technology at the University of Manchester, UK. He has written extensively on the history of genetics, the emergence of laboratory biology, and the social history of the German scientific community. In recent years, his interests have turned to the history of agricultural higher education and plant breeding in German-speaking Europe as well as the comparative history of green revolutions. His latest book is *Europe's Green Revolution and Others Since: the Rise and Fall of Peasant-Friendly Plant Breeding* (2012).

Chapter 18
Chicken Breeding: The Complex Transition from Traditional to Genetic Methods in the USA

Margaret E. Derry

Introduction

"We can mix the blood of our birds as easily as we mix paints that give us different tints of color."[1] So wrote the renowned American chicken breeder, I. K. Felch, in the late nineteenth century. The breeding of chickens was a highly sophisticated endeavor by the mid-nineteenth century in the USA. By the mid-twentieth century, chicken breeding had undergone profound changes. Did the shift result primarily from the incorporation of new knowledge arising out of genetics? Or is the story more complicated? In this chapter, I look at practices in traditional chicken breeding in relation to the emergence of Mendelism and later developments in maize genetics and theoretical population genetics. My main concern is with the interface of genetics with chicken breeding, rather than the evolution of genetics itself. I focus on changes in the egg industry because it was here that the shifts which ultimately affected all chicken breeding initially took place.

Mendelism initially failed to influence poultry breeders in part because scientists lacked sophisticated understanding of breeding practices, and in part because geneticists preferred to pursue general biological problems rather than focus on practical problems. But the communication between scientists and breeders was affected also by changes within the poultry industry itself. Especially significant was the separation of the producing from the breeding arm of the industry. These changes played a role in creating an environment conducive to the introduction of

[1] Felch (1877, p. 47). For more on Felch's ideas on breeding, see as well Felch (1902).

M. E. Derry (✉)
Department of History, University of Guelph, Guelph, Canada
e-mail: mderry@poplarlane.net

20243 Heart Lake Road, Caledon, ON L7K 2A8, Canada

D. Phillips, S. Kingsland (eds.), *New Perspectives on the History of Life Sciences and Agriculture,* Archimedes 40, DOI 10.1007/978-3-319-12185-7_18

new approaches. Developments arising by the 1930s from both plant and livestock genetics might have laid the groundwork for a breeding revolution, but their impact must be seen in light of what effectively had become a breeding void. When new theoretical approaches to breeding arose out of population genetics and became linked with corporate involvement, genetics would reshape breeding for eggs in the 1940s. Similar patterns emerged in the meat or broiler industry over the 1950s.

My approach towards the subject matter differs from much of the scholarly literature which addresses agriculture and science generally, or chickens and genetics specifically. I come primarily from an agricultural point of view and therefore adopt a farm perspective, rather than the more common history of science perspective. My fundamental question is: what causes science-based innovation in breeding practices? My answer involves recognition that understanding agricultural innovation must consider not only the emergence of new science, or the communication channels between scientists and breeders, although these are important, but also changes in the organization of agricultural industry. I use the poultry industry to illustrate this complex relationship.

Excellent studies look at German and American orientation to genetic research for agricultural purposes,but these do not elucidate clearly how that research or resulting education related to or influenced breeding activities on farms.[2] A recent review of agricultural research in France in the early twentieth century follows the same pattern. The study focuses more on the French concern with developing education in American biology and attitudes to genetics than on the application of such education to farming.[3] Most material that does exist on the topic of genetics and agriculture in either Europe or North America addresses plant breeding.[4] Several articles on plant breeding and Mendelism in the USA, France, and Germany appeared recently in the *Journal of the History of Biology*'s special 2006 issue.[5] The relationship of genetics to animal breeding has received more limited attention, but an excellent set of articles on animal breeding, genetics, and technology in the twentieth century has been published in a 2007 issue of *Studies in History and Philosophy of Biological and Biomedical Sciences*.[6] Valuable information on the pre-genetics and genetics/animal breeding connection can be found in the writings of N. Russell, B. Theunissen, R. J. Wood, and V. Orel.[7]

Genetics and specifically chicken breeding have attracted the attention of historians of science and for logical reasons. The birds were commonly used in genetic experiments. Furthermore, business concerns were/are attached to the breeding of

---

[2] Harwood (1992, 2005a, b).

[3] Castonguay (2005).

[4] For example, Fitzgerald (1990, 1993), Kloppenburg (1988), Dreyer (1985), and Harwood (2005b).

[5] Kimmelman (2006), Bonneuil (2006), and Wieland (2006).

[6] Wilmot (2007a, b) and Grasseni (2007).

[7] Russell (1986) and Theunissen (2008, 2012b); "Connecting genetics, evolutionary theory and practical animal breeding: Arend L. Hagedoorn (1885–1953)" (unpublished manuscript, 2012), Wood and Orel (1981, 2000, 2001, 2005) and Orel (1977).

chickens under genetic principles. Control of intellectual property via a form of biological patenting is at the heart of breeding genetically.[8] My approach differs from this body of work. I stress the point that chicken production supported two separate industries, namely the egg and the meat industry, which reflected different cultures, breeding strategies, and industry structures. While most historical work concerned with chickens addresses the meat or broiler industry (often in conjunction with intellectual property issues), this material does not place the meat industry within its wider poultry context. Kathy Cooke's article (note 8) on the work of Raymond Pearl focuses on breeding for eggs but she does not deal with the meat industry. I augment information in the existing literature not only by separating egg from meat breeding but also by providing background on traditional chicken-breeding methods and exploring how changes in industry structure affected the continuity of breeding knowledge.

Chicken-Breeding Practices of the Nineteenth and Early Twentieth Century

To begin, I explore the chicken-breeding practices that had become well established before the emergence of the new science of genetics. Appreciating the sophistication of these practices is important for understanding how early efforts to apply Mendelian theory failed to influence breeders, not because the breeders were ignorant of science, but because the scientists were ignorant of breeding practices. These practices had developed since the late eighteenth century, and to understand them we must look beyond the formal establishment of the American Poultry Association in 1873, which will be discussed in a later section.

All of the breeding systems developed by master chicken breeders relied on general artificial selection principles established by livestock breeders since the late eighteenth century. Their concern was with the effects of inbreeding (that is the mating of related stock in order to achieve uniformity) and outcrossing (that is the breeding of unrelated stock in order to inject change and counterbalance the effects of inbreeding) on a population, and with ways of balancing the two against each other.[9] This fundamental method had been used to create all the known breeds of domestic animals developed over the nineteenth and early twentieth centuries. Concern with inbreeding and its effect on populations, rather than merely its effect on individuals, was basic to eighteenth-century Enlightenment thought about breeding farm animals.[10] Individual worth and parental or ancestral background might play a role in how selection for breeding worked, but neither approach took precedent over the idea of working with a group and using inbreeding to change that group.

[8] Cooke (1997), Shrader (1952), Horowitz (2004), Boyd (2001), and Bugos (1992).

[9] On this subject, see Russell (1986), Wood and Orel (2001), Theunissen (2012a), and Derry (2003).

[10] Wood and Orel (2001, p. 89).

Selection for inbreeding programs tended to be based on the progeny testing—namely, evaluating breeding potential on the basis of what an animal had already produced. Over the nineteenth century, this population approach was eroded by mediocre breeders through the widespread influence of purebred breeding, which tended to emphasize individuals and the ancestral background as a reflection of the importance of pedigrees to the system.[11] The best breeders, however, continued to adhere to eighteenth-century principles within complicated inbreeding and outcrossing programs.

Another early approach to breeding involved crossing of breeds, which was separate from the inbreeding/outcrossing systems described above. Cross-breeding was known to promote increased vigor (progeny that would be better than either parent), but the method resulted in stock that would not reproduce itself truly to the same good qualities or with any consistency. It was, therefore, rarely applied to the breeding of livestock strains which were meant to reproduce with consistency over generations. It was used for the terminal production of stock, namely animals destined for slaughter as meat for human consumption and therefore not intended for breeding. The inability to breed did not matter under these conditions.

A few examples of nineteenth-century chicken-breeding systems and the controversies surrounding breeding methods follow. All of these systems were designed to work within a single breed and to build up strains that could be used over generations. It should be noted as well that all of these methods focused on breeding from the point of view of populations, not individuals. This would be a hallmark of theoretical population genetics when it emerged in the 1920s.

H. H. Stoddard, a founding member of the American Poultry Association and an experienced breeder as well as a publisher of poultry journals, wrote numerous articles on an inbreeding/outcrossing system for egg-laying hens. In order to create a working commercial flock, he advised the breeder to start with 16 unrelated strains (but from within one breed) and interbreed the lines over 5 years, by selecting only males from certain lines for breeding and only females from other lines for breeding. Culling should be done at every level and no inbreeding incurred. The final cross of purely unrelated stock resulted in a vigorous flock that would be inbred, brother to sister, for at least 4 years. The inbred lines were to be used for table egg production. The stock would be weakened by the inbreeding after 4 years, but by that time the breeder would have new birds, arising from his original 16 strains, available for inbreeding and egg production.[12] Stoddard advised a well-planned balance between inbreeding and outcrossing, in order to harvest the advantages that inbreeding could bring (uniformity in the progeny) without incurring its dangers, namely a tendency to reduce fertility and vigor in the offspring.

Another method, developed about 1870, also focused on the development of a distinct male and female line.[13] One line was created by mating females producing

---

[11] Derry (2003, pp. 1–47).

[12] *American Poultry Journal*, April 1911, p. 749; May 1912, p. 971; July 1912, p. 1165; December 1913, pp. 1518, 1520–1521, 1542.

[13] See, for example, *American Poultry Journal*, May 1911, p. 1006.

especially good pullets (young hens) to males producing particularly good pullets, while the other was made by mating females producing good cockerels (young males) to males producing good cockerels. It was the cross of the male line on the female line that brought about the desired final results. The males and females that came from the male/female line cross, however, were useless as breeding pairs, because they would not reproduce truly. One had to access the parents or grandparents to achieve the same results.

This double mating system provoked controversy late in the nineteenth century and into the twentieth century.[14] Breeders who opposed double mating questioned the desirability of forcing farmers back to breeders for replacements and thereby making farmers relinquish any role in breeding, a situation which the double mating system enforced. The philosophy that all farmers should be breeders for the good of ongoing livestock improvement was at the heart of the matter. The breeding and producing of stock should be a seamless operation and improvement overall was deemed to arise from the concerted work of the many. To put breeding into the hands of the few would seriously undermine that situation. A great deal of discussion about elitism in breeding and restricting the occupation to a limited number of people took place in the poultry press, usually in relation to the double mating system as applied to a significant utility breed, the Barred Plymouth Rock.[15] Breeders using the double mating system must have been aware of the protection the biological lock provided them. Ideas concerning the control of intellectual property in breeding were as old as the Dishley Society, established in the eighteenth century by Robert Bakewell and other sheep breeders.[16] When the issue of biological locks and restriction of breeding to the hands of specialists emerged later in the 1930s, these ideas were anything but new. What was new was the fact that they became acceptable.

In the 1870s, the American breeder, I. K. Felch, developed a scheme which focused primarily on inbreeding, but controlled inbreeding. He knew, as did all breeders, that intense inbreeding could result in such undesirable characteristics as infertility, and therefore if one wanted to take advantage of the good results that inbreeding could generate—namely, the perpetuation of desirable qualities from generation to generation—one had to use the method in a restrained fashion. Felch developed a complicated chart which showed how the progeny over generations of a foundation male and female could be mated with each other in order to intensify the genetic input of one over the other. For example, in the third generation, a male could be mated with a female in order to intensify the inheritance of either member of the original pair. In the end, the breeder could produce stock whose genetic makeup showed a varying percentage between one half and seven eighths of each

[14] *American Poultry Journal*, April 1909, pp. 426, 428, 430; October 1915, pp. 1229, 1243; November 1915, pp. 1315–1316; July 1922, p. 737.

[15] *American Poultry Journal*, April 1909, pp. 426, 428, 430; October 1915, pp. 1229, 1243; November 1915, pp. 1315–1316; July 1922, p. 737.

[16] Bakewell to Culley, 15 December 1791 (p. 164); and Culley letter, 19 May 1792 (p. 9, 11–12). Part II in Pawson (1957).

foundation member. By recombining the blood of a selected foundation breeding pair through different mating combinations over generations of their descendents, one could inbreed forever without experiencing seriously reduced vigor.[17] Felch's percentage inbreeding system put forward a theoretically important idea that would be later developed in a more sophisticated way by the geneticist Sewall Wright.

At the beginning of the twentieth century, some egg-laying breeders communicated with trained biologists working at experiment stations about breeding matters, but it is unclear how much influence these scientists actually had on breeding programs. To the extent that the breeding methods of the biologists were similar to those of practical breeders, it is unlikely that breeders had much to learn and we cannot say how much breeders were affected by those discussions. Often breeders seemed to follow what experience, not science, had taught them. It is not evident, for example, how much (or even if) the large Texas breeder, M. Johnson, utilized biologist input when he began a huge breeding operation for table egg layers involving thousands of birds shortly after 1908.[18] D. Tancred discussed breeding with G. M. Gowell.[19] But whether his breeding program relied heavily on statistics as a result of Gowell's advice is not clear. Another breeder who conversed with a poultry scientist at an experiment station was the American J. A. Hanson.[20] His system resembled the future breeding of scientists, but it had an affinity as well to inbreeding and crossing systems designed by Stoddard.

Mendelism: Chicken-Breeding Experiments and Practical Breeder Reactions

The rise of Mendelism after 1900 brought escalating excitement into the scientific and practical breeding world.[21] Could farm breeding now really proceed "scientifically"?[22] One American breeder, E. Parmelee Prentice, hired qualified biologists to run his breeding operations. A Chicago lawyer, he bought a country estate in Massachusetts in 1910 and became interested in the potential of Mendelism for better farming.[23] Prentice hired H. D. Goodale, poultry specialist at the Massachusetts Experiment Station, to run cattle- and chicken-breeding operations (which concentrated on egg-laying Leghorns) at Mount Hope on a full-time basis.

---

[17] *American Poultry Journal*, August 1910, pp. 976–977; August 1911, p. 1268; Termohlen (1968, p. 12).

[18] *American Poultry Journal*, January 1927, pp. 11, 88, 90–92, 94–97; Fitzgerald (2003, pp. 106, 115).

[19] *American Poultry Journal*, December 1925, pp. 1032, 1038; Hanke et al. (1974).

[20] Hanke et al. (1974, p. 253).

[21] For example, Paul and Kimmelman (1988); Olby (1985).

[22] Palladino (1993), Kimmelman (1983), and Heape (1906).

[23] Van Riper (1932), Savage (1942), Prentice (1951, p. 483), and Hanke et al. (1974, p. 260).

Could a scientific approach, and especially Mendelian theory, introduce new methods that would improve upon many decades of practical knowledge? It was not obvious to breeders, even to those with an interest in the potential of Mendelism, that advice forthcoming in the early years of genetics was really helpful, or was even especially new. The case of Raymond Pearl, a biologist at the Maine Agricultural Experiment Station between 1907 and 1916, illustrates the controversy that such supposedly expert scientific advice could create. Initially trained in zoology, Pearl had become interested in statistics as applied to biology, a situation which drew him into the world of biometrics, or biometry.[24] He also started to utilize Mendelian theory when he attempted to understand how the transmission of certain characteristics worked.

When Pearl joined the Maine Agricultural Experiment Station, his first task was to analyze the extensive data that his predecessor, G. M. Gowell, had collected from results of breeding experiments to study an increase in the egg-laying capacity of chickens. Pearl's biometric approach to this data showed that there was no correlation between the production of hens and the capacity of their daughters to lay eggs. Selection of heavy layers as breeders over 10 years had not increased the average laying capacity of the flock.[25] Since it was assumed that farmers primarily followed this form of mass selection on their flocks, Pearl believed it was his duty first to point out that he had proved that this methodology did not work. Pearl apparently did not grasp the fact that much of the best chicken-breeding methodology, certainly those of Stoddard and Felch, rested on more complicated principles. Simple selection of the best egg layers was not the sole factor in their breeding programs, and did not explain the way a good breeder developed an egg-laying flock. Breeder opposition to egg-laying contests (which will be discussed in some depth later) was based on a similar conviction: namely that breeding for new generations of superior egg layers could not be done by simply selecting heavy laying hens.

Pearl believed his second duty was to learn how superior egg laying was inherited. Breeding experiments, Mendelian theory, and biometrics, when combined with his dissecting work led Pearl to argue (incorrectly, as it turned out) that superior egg-laying ability was mainly transmitted through males. Especially problematic was his apparent advocacy of the outright rejection of good hens in a breeding program, which the passage quoted below implied. Pearl outlined his theories for chicken breeders in a 1913 bulletin of the Maine Experiment Station, beginning his discussion with a synopsis of what artificial selection entailed. His synopsis explained breeding practices in the following way: like produces like, and breed the best to the best. This was the simplest system conceivable, and Pearl then proceeded to disparage it for its simplicity. As he argued, the success of this system depended upon the existence of equal simplicity in the phenomena of inheritance. If, for example, a breeder mates an individual that is larger than average to another individual larger than average and always gets offspring larger than average, then

[24] Cooke (1997, pp. 67–69).

[25] Cooke (1997, p. 73).

breeding "the best to the best" would, as Pearl put it, "offer a royal road to riches." But if, he continued:

> a character is not inherited in accordance with this beautifully and childishly simple scheme, but instead inherited in accordance with an absolutely different plan, which is of such a nature that the application of the simple selection system of breeding could not possibly have any direct effect, it would seem idle to continue to insist that the prolonged application of that system is bound to result in improvement.[26]

Pearl was suggesting that breeding methods might be seriously flawed if they were based on overly simple concepts of inheritance.

Pearl's concern with productivity in chickens drew the attention of breeders, but they could make little sense of his work, or of how it could be useful to them.[27] The idea that an emphasis should be put on males in mating systems, on the one hand, was not a novel concept to good breeders. They had always appreciated the value of the male in egg production (either via his female ancestry or his daughter progeny). But Pearl's further implication that good hens did not play a role in the breeding system struck breeders as nonsensical. In his articles for the *American Poultry Journal,* Stoddard challenged Pearl's conclusions, questioned the innovativeness of his suggestions, and took issue with some of the inflammatory language found in the Bulletin. Stoddard stated that:

> The fact is the bulletin is wrong. High fecundity and low, too, may descend from either sex to either sex or it may not descend at all directly from either to either. There will sometimes [be] great irregularity, and scattering every which way, and reversion to remote ancestor types, especially if there has been a cross of strains considerably diverse. Selection for the purpose of breeding from the best to get the best, even if it is 'childishly simple', will continue to be the only way to fix characteristics, and among many misses there will be some hits. Breed 'the best to the best' and though you may find that some of the progeny may not be as good as the average of their parents, yet some may be as good and some decidedly better.

Stoddard pointed out that the original egg production of wild fowl of six to eight eggs a year had been brought up to at least, and often more than 50 in domestic chickens.[28] He argued also that fecundity could be inherited through hens. "I do not deny the influence of the male bird in helping to build up a strain of great laying. Neither do I know of anyone who would….What I do deny is that dams have no finger in the pie of hereditary fecundity. They have a great deal to say about it."[29] Stoddard assumed that Pearl must mean, even if he did not articulate it, that the selection of males proceeds on the basis of their mothers. "What Dr. Pearl really teaches us is that fecundity is transmitted equally by both sexes, and by alteration," Stoddard argued, "but in his summary (misleading because incomplete) has laid a

[26] Bulletin 305, Maine Experiment Station, 1913, p. 388. Quoted in the *American Poultry Journal*, May 1913, p. 847.

[27] See, for example, *American Poultry Journal*, December 1913, p. 1517.

[28] *American Poultry Journal*, May 1913, p. 847.

[29] *American Poultry Journal*, April 1913, p. 672.

trap for his readers; for, although it tells what the daughters inherit, and from which parent, it is silent as to what the sons inherit and from whence."[30]

Stoddard concluded that Pearl had needlessly confused the existing situation by advocating what breeders had always done. Under "Pearlite" theory, as Stoddard put it, a heavy layer would be mated to a cock whose dam was a heavy layer. Is that not the same as the best to the best, he queried, that is, both males and females were selected from families in which the female members were good egg layers? Pearl seemed to be advocating something new and different, but Stoddard disagreed. "Mendelism, or the new genetics, or whatever it may be called," Stoddard stated, "offers at its present stage no new practical instructions for mating and breeding either the lower animals or humans. The professors who say that the old rule of 'breeding the best to the best', is no good; turn right around and prescribe methods that amount to the same thing."[31] He summarized his impressions concerning Pearl's approach to breeding in the following words: "If the 'childishly simple scheme' or 'breeding from the best to the best', which 'is the simplest system conceivable', was so totally and disgustingly fruitless in the past, will the identical practice result differently because of masquerading under a new name?"[32] In spite of his criticism of Pearl's comments, Stoddard was prepared to admit that genetics might ultimately be of revolutionary value to chicken breeding. He concluded: "The whole problem offered by Mendel's discovery, one of the most important as well as wonderful, in the annals of science, is such a complicated one that it will take generations to solve it, and at present the breeders of domestic animals…can derive little benefit or none at all from all that Mendelism can offer—in its present stage of development."[33]

Some breeders took Pearl's words simply to mean that egg-laying capacity was not an inherited factor at all, and rejected what he had to say on that basis. J. B. Morman, for example, concluded that Pearl's work implied egg laying was not even an inheritable trait. Morman, therefore, saw Pearlism as a dangerous trend, and one which had already convinced the famous English biometrician, Karl Pearson (with whom Pearl had studied), to "relegate the problem of inheritance of egg-laying power in fowls to oblivion." Since experiments done in 1912 by Morman himself on the problem had convinced him that egg laying was an inheritable characteristic, this breeder wondered what was the good of science, if the experimenters could be so misguided as to believe that chickens do not inherit egg laying?[34] Breeders' experiences could not easily be set aside, and the efforts to substitute "scientific" advice for breeding practices simply appeared confusing and ill-supported by evidence.

Pearl was unusual for his time in his attempt to address characteristics important to farmers. Many biologists, when they started to explore the implications of Mendelian theory for poultry, focused on understanding the general biological problems

[30] *American Poultry Journal*, May 1913, p. 847.

[31] *American Poultry Journal*, October 1913, p. 1278.

[32] *American Poultry Journal*, May 1913, p. 847.

[33] *American Poultry Journal*, October 1913, p. 1278.

[34] *American Poultry Journal*, December 1913, p. 1517.

raised by Mendel's laws rather than on aiding agriculture, even though much breeding research in North America was done at American agricultural experiment stations.[35] Biologists might have shown a keen and ongoing interest in working with chickens, but virtually none of their research, in either North America or Europe, was aimed at commercial productivity before 1930.[36] Biologists/geneticists who joined the poultry departments of experiment stations before 1930 in the USA hoped, for example, to illustrate dominance/recessive characteristics by exploring the way chickens inherited such features as feather coloring, shape of comb, and skeletal defects. Other features studied were flightlessness, crooked neck, feathering, silkiness, ragged wings, feathered shanks, multiple and double spurs, blindness, and dwarfism. Inheritance of characteristics on the basis of sex also interested early geneticists.[37] The studies explored the overall genetic constitution of poultry. They did not address the inheritance of traits of economic importance: egg-laying and/or meat-producing capacity, feed conversion, or resistance to disease; all features that were of economic value for commercial chicken breeders.

The work and thought of Leslie C. Dunn within this period serve as an example of what problems concerned geneticists when they used chickens in research. Dunn undertook several experiments for the Connecticut research station at Storrs between 1920 and 1928, in which he investigated, for example, the inheritance of plumage color patterns. Some of Dunn's work involved inbreeding and subsequent cross-breeding of inbred lines in poultry in order to study the process of fitness decline from inbreeding and its recovery with cross-breeding. These experiments were not undertaken for practical ends. Dunn saw the phenomenon of fitness decline and recovery as a process which might relate to speciation. Fitness decline and recovery would interest scientists working with chickens for the same reason for some years to come. Dunn ran experiments as well studying the relationship of hatchability to egg weight, which he believed to be, correctly as it turned out some 30 years later, of importance to evolutionary theory. He looked into skeletal variations, the presence of lethal genes, egg-laying patterns of different poultry breeds, and color of the leg shank.[38] Research of this nature undermined any sense of relevance that might have evolved between the developing science of genetics and the breeding of farm chickens.

Changes Within the Industry

Further complicating this story of the relationship between science and breeding was the role of the American Poultry Association as a regulating agency. Its position on breeding was vague and the problems it created were compounded when

[35] Kimmelman (1983, pp. 163–204).

[36] Warren (1958).

[37] Warren (1958, pp. 4–5).

[38] Lerner (1974).

government later took a more active role in encouraging egg breeding. The Association, formed in 1873, regulated how breeding should proceed through its Standard of Excellence (after 1888 called the Standard of Perfection), which described all breeds and varieties, and evaluated each under a point system that indicated what comprised good quality.[39] The standard made it possible to measure the quality of individuals against each other within each breed for exhibition conditions, but did not set rules for breeding methods. All breeding aimed simply at producing birds that matched the standard as closely as possible. The importance of the show ring raised a basic dichotomy from the beginning: were chickens to be bred for beauty or utility, and were these two things related? The same dichotomy arose over the way horse, cattle, and dog shows affected breeding in both Britain and North America.[40] But the beauty/utility conflict became most blatant within the poultry breeding/show system. By the late 1880s, the idea that beauty and utility could go together in chicken breeding was being criticized; for example, a writer in the *Mark Lane Express* in 1888 argued that "the fancier who minces the matter, preferring to allow the world to continue to believe that exhibitions instruct and improve the people in a particular direction, is insincere. In answer to the question, What has the poultry fancy done for profitable poultry? We must answer, clearly enough, nothing."[41]

The American Poultry Association reacted to such criticism by considering standards based on productivity as early as 1903.[42] At the 1907 meeting of the association, members resolved that "the American Standard of Perfection [gave] undue prominence to the beauty value of standard-bred fowls, to the detriment of the utility value of domestic poultry."[43] By this time, a specialized meat industry no longer existed. The egg industry was, quite simply, *the* commercial chicken industry, and therefore the association's focus on utility was aimed at egg-laying capacity of hens. The idea that competitions encouraged good breeding made the association consider running egg-laying contests.[44] Although breeders who opposed egg-laying contests were often accused of being only interested in beauty, this accusation was not accurate.[45] Rather, breeders viewed such competitions as serving no useful purpose because identifying heavy layers did not teach anyone how to breed them. Relying on one winner of an egg-laying contest to produce superior daughters was not, in their opinion, the proper way to breed.[46] The problem of breeding for beauty versus utility was not overcome by these efforts, and the American Poultry Association's only solution was to organize competitions for either beauty or utility purposes, but with no advice about how to breed. Thus the American Poultry Association was so

---

[39] Sawyer (1971, p. 18) and Hanke et al. (1974, pp. 35–36).

[40] Ritvo (1986, 1987), White (1992), and Lytton (1911).

[41] Quoted in *Farmer's Advocate*, June 1888, p. 178.

[42] *Advocate*, 15 June 1903, p. 559.

[43] *American Poultry Journal*, September 1907, p. 690.

[44] *Advocate*, 12 January 1905, p. 50; 8 February 1912, p. 226.

[45] *American Poultry Journal*, November 1915, pp. 1321–1322.

[46] *American Poultry Journal*, February 1912, p. 298; June 1921, p. 642.

fractured in its outlook as to breeding direction that it could not guide how breeding should proceed. Widespread knowledge of good breeding methods that existed (and which I described earlier) declined under these conditions.

When government entered the picture, its attempts to encourage better egg-producing hens compounded the problems that stemmed from the American Poultry Association's position on breeding. The idea of government involvement with poultry breeding for eggs was initiated first by William Graham of the Ontario Agricultural College in Canada before the First World War.[47] By the summer of 1919, regulations for a national Canadian Record of Performance, known as the ROP, had been established. The ROP registered hens that met the standards of the American Poultry Association and were capable of laying a determined number of eggs per year.[48] While the idea of competition was removed from ROP recording, the emphasis on individuals and meeting a standard remained in place. So did the undermining of a sense that breeding should be directed at populations. Because the ROP firmly linked beauty with utility, the state system fed as well into the beauty/use dichotomy by uniting beauty standards with production standards.

A move towards an American ROP developed fairly quickly after the Canadian structure was established. While the American Association of Instructors and Investigators in Poultry Husbandry authorized the initiation of national ROP as early as 1919, American ROP recording tended throughout the 1920s to stay more regionally or state oriented, and these operated under regulations that were similar to the Canadian ROP standards. The linkage of beauty via reliance on the Standard of Perfection to utility stayed in place. (It was not until 1930 that the US ROP Association was formed by 16 states.)[49] American farm breeders who had opposed egg-laying contests tended to oppose the ROP, and for the same reasons: namely, reliance on individual heavy laying hens for breeding purposes.[50] The ROP aggravated the inherent tensions in the traditional chicken-breeding world, by dividing utility farm breeders into two camps (particularly in the USA): those who entered the ROP and those who did not.

Increasingly after the beginning of the twentieth century, it was unclear to the average breeder how breeding should proceed. The hegemony of the American Poultry Association over general poultry affairs, combined with its ambiguous stand on the beauty/utility issue and its emphasis on competitions, seemed to play the initial role in weakening a general understanding of breeding methodologies promoted by men like Felch and Stoddard. Over the years, other factors played into the situation, such as the linkage of the ROP with beauty and individual-worth breeding.

---

[47] Derry (2001, pp. 73–83, 2003, pp. 36–44).

[48] *Advocate*, 7 August 1919, pp. 1429–1430; "Official record of performance for poultry", *Agricultural Gazette of Canada*, 1919, p. 796.

[49] See *American Poultry Journal*, June 1921, p. 642; June 1922, pp. 672–673, 674; April 1926, pp. 466, 468; Hagedoorn and Sykes (1953, pp. 217–220), and Hanke et al. (1974, pp. 702, 703, 704).

[50] Warren (1958, p. 13).

As early as 1914, there were signs of confusion over how breeding methodology worked. For example, C. D. Cleveland (secretary of the New York Poultry and Pigeon Club, breeder, judge, and writer of many articles on chicken husbandry) commented on the poverty of material concerning selective breeding methods of breeders. "I have never been able to understand why it was that the breeder was so loath to give away anything in regard to the essentials of the way he breeds his varieties…. Breeders ought not to hold back their so-called breeding secrets." The editor of the *American Poultry Journal* thoroughly agreed, stating "Mr. Cleveland's remarks summed up the situation very nicely. We have been trying for years to get articles [on breeding] we want, and believe should be published, on the how and why of mating and breeding, but we can't get the information…. Perhaps some breeders may not be able to tell how they get results."[51] Articles devoted to breeding methods appeared less frequently in the poultry press, a pattern quite evident by 1930. Virtually no articles on the subject were printed after 1930. The chicken-breeding industry presented a complex and indeed fractured face by 1930 to poultry producers. The situation was ripe for change.

Exacerbating the decline of traditional breeding knowledge was the ever-widening cleavage between the breeder and the producer, which was a result of growth of the hatchery industry. Superior artificial incubation methods in the early twentieth century, coupled with the fact that baby chicks did not need to be fed for 72 hours after hatching, meant that chicks could be shipped long distances. The hatchery industry as a result could act as a middleman between breeders and producers, encouraging a division between the breeder and producer/grower.[52] The work of running huge incubators meant that increasingly by the late 1920s men operating them were not involved in either the breeding or producing side of the chicken business.[53] The growing custom of buying day-old chicks from the farm and from nonbreeding hatcheries further reduced the involvement of producer/growers with breeding. Relying on hatcheries, not breeders, for birds started a trend that would be ongoing; namely, the reliance of producer/growers in breeding matters on outside bodies and the reduction of their control over what type of chicken they used. The breeding occupation no longer functioned in a seamless way with the producing occupation. By the 1930s, most poultry people raising chickens for table eggs took no part in breeding. Those interested in producing fowl tended to fall into two separate camps: true breeders who created distinct lines of stock and producers/growers who simply multiplied and/or used the birds. While the splintering and masking of breeding methodology promoted the removal of poultry people from the breeding activity, the hatchery industry played an even more critical role in that trend.

---

[51] *American Poultry Journal*, April 1915, p. 693.

[52] Sawyer (1971, p. 26); *American Poultry Journal*, December 1910, p. 1446.

[53] Derry (2012); *Art and Science in Breeding*, pp. 128–153.

Scientific Innovations: Hybrid Vigor in Plants and Animals

Specific developments in plant genetics, rather than in poultry genetics per se, laid the groundwork for a new genetic approach to agricultural breeding, ultimately affecting breeding of egg-laying chickens. As early as 1908, botanist George H. Shull began to wonder why improved fertility and vigor resulted from the crossing of inbred parental lines of corn. He had noted the year before that such progeny could not sustain that superiority when bred with each other for the next generation.[54] In 1914, he created a new word, "heterosis," to describe hybrid vigor, a phenomenon well recognized by practical breeders.[55] But Shull, much like his fellow Mendelians, was more concerned with understanding genetic laws, than he was with corn improvement. "For Shull corn was the window, not the landscape," as one historian put it.[56]

In 1917, D. F. Jones considered using heterosis to increase the farm productivity of corn.[57] He worked out a method that combined inbreeding and crossing of lines in order to produce superior hybrid vigor.[58] In order to maintain that level of hybrid vigor over succeeding generations, seeds from the commercial crop would not be used for breeding. New commercial generations would always be regenerated by stock belonging to the parent and grandparent generations. The commercial plant was seen as a terminal product.

The method interested American corn breeders who were prepared to fund research to study its feasibility. (It would be years, and after considerable effort and expense, before the method worked more effectively than traditional plant-breeding methods which relied on lines that reproduced truly, in spite of propaganda that suggested otherwise.[59]) The corn breeders recognized that by producing corn in this fashion, farmers would be forced back to them to buy next year's seeds. The breeders, therefore, would have a guaranteed market. The biological lock could be used to provide a form of biological patenting. The only way such a breeding system could work, however, was if the producers relinquished their part in breeding. That did not come readily, as studies on the acceptance of this innovation show.[60]

At about the same time, the development of theoretical population genetics in the 1920s and 1930s, which synthesized Mendelism and Darwin's theory of natural selection, initiated new directions in livestock breeding. Critical to the rise of population genetics, and more specifically genetics aimed at livestock breeding, was the work of British scientist R. A. Fisher (with his emphasis on the inheritance of quantitative characteristics, that is how much or how little a characteristic was

[54] Dunn (1965, p. 125).

[55] Shull (1948, p. 440).

[56] Fitzgerald (1990, p. 39).

[57] Jones (1917, p. 477).

[58] Fitzgerald (1990, p 55) and Kloppenburg (1988, p. 99).

[59] Fitzgerald (1990, p. 64).

[60] Rogers (2003, pp. 31–36, 53–55).

inherited), and that of American scientist Sewall Wright (who studied inheritance in populations via inbreeding and the subsequent bottlenecking of breeding groups). Wright played the more important role in the development of livestock genetics through his influence on the founder of livestock genetics, the American J. L. Lush.

Wright is best known as a theoretical population geneticist, and his primary interest throughout his life was evolution. But in his early professional years Wright studied livestock production (especially the historic breeding of Shorthorn cattle[61]) and also worked for the Bureau of Animal Industry. He wrote articles in the livestock journal, the *Breeder's Gazette,* although his language was so specialized that breeders would have found little of the information helpful. One of the most important things that Wright did for livestock genetics and subsequently for future chicken breeding was to quantify the effects of various inbreeding strategies.[62] His work in effect provided a more complicated method of controlling the level of inbreeding than, for example, Felch's chart. Perhaps one reason that future theoretical animal breeders came by the 1940s to see Wright as the true founder of genetics for farm animal breeding was the fact that his path coefficient calculations for inbreeding reflected, in a way that the work of other geneticists did not, the refined attitude to the interbreeding of blood-related individuals that historically all the great breeders of the past had adhered to. Between 1915 and 1922, Wright devised a way of calculating the level of shared genes that would result from different inbreeding systems—brother to sister, first cousins, double first cousins, half brother to half sister, and so on.[63]

Heterosis also interested Wright. He was aware of Shull's work with inbred corn and knew from his own experiments that crossing inbred lines often led to progeny superior to either parent.[64] Wright believed his path coefficient might work well in the production of synthetic lines within one breed, lines that would be crossed for heterosis. In other words, one could promote hybrid vigor (and therefore improvement) by crossing lines within one breed, not just by crossing breeds. He theorized that he could quantify the level of inbreeding and thereby reduce its intensity, thus avoiding some of the dangers it could incur. The path coefficient in the end would make it easier to create vigorous lines resulting from matings of related animals that could be used to cross for heterosis. (The coefficient could equally well be used to produce superior pure lines designed to breed truly.) Research geneticists began to explore how hybridizing could work within the controlled framework of Wright's path coefficient theory.[65]

J. L. Lush initiated the move to make Wright's theories applicable to the breeding of livestock. Trained at Kansas State Agricultural College in animal husbandry, Lush began corresponding with Wright in 1918. Between 1918 and 1922, while

[61] Wright (1923a, b).

[62] Crow (1990, pp. 58, 62–66).

[63] Provine (1986, p. 156).

[64] Provine (1986, pp. 138–139, 140, 141, 1971, pp. 160–161). See also Wright (1922, 1958), which contains reprints of (1921a, b, 1931, 1934).

[65] Babcock and Clausen (1918) and Wriedt (1930).

Wright developed his path coefficient theory of inbreeding and wrote about systems of mating, Lush kept up with the literature as it appeared, and quickly saw that much of it could be made applicable to livestock breeding strategies on farms.[66] At Iowa State University, Lush assessed inbreeding levels practiced on breeds of different livestock species. He synthesized the theories of Wright with statistics and what he could learn from Mendelian geneticists at the college into a comprehensive animal breeding theory that could be utilized on the farm.[67] His graduate and postdoctoral students would carry his theory literally all around the world.[68] Lush always felt indebted to Wright's work, even though Lush himself actually created a more usable and practically oriented set of theories designed to improve farm animals.[69]

Chicken breeding interested Lush. At Ames in 1945, he tried to produce chicks along the lines of hybrid corn breeding, by inbreeding various lines within a breed and crossing for heterosis.[70] He was not alone in this poultry breeding work by that time. A number of geneticists (e.g., D. C. Warren, a poultry geneticist at the Kansas State University) were highly focused on applying the hybrid corn-breeding method to chickens. So were the corn-breeding companies who hired geneticists to experiment with using the method for the production of commercial, egg-laying hybrid chicks. All inbreeding/line-crossing research of this nature was directed at working within the confines of a single breed, normally the Leghorn. Line crossing, not the crossing of breeds, lay at the heart of the matter. By the 1930s, innovations in corn-breeding methodology dovetailed with much of the outlook towards animal breeding that had evolved under the Lush school. Genetics would, as a result, now offer an entirely new way to approach chicken breeding.

Genetics, Egg-Laying Chicken Breeding, and Corporate Enterprise

By the 1930s, the American corn-breeding companies had succeeded in making the hybrid method work for corn, and had also convinced farmers to buy seeds from them every year rather than breed next year's crop. The success of the hybrid corn-breeding method made the companies want to explore the idea of using the same system for the production of egg-laying chickens. Under these conditions, Lush's (and his school's) inbreeding and more importantly heterosis work with poultry attracted their attention. It would be the beginning of important new developments in relation to chicken breeding; namely, the entrance of corporate involvement and an emphasis on breeding for lines that do not produce truly. Unlike many scientists in the USA working at research stations in the 1930s, the managers of the corn

[66] Wright (1923b, pp. 405–422).

[67] D. C. Warren, in Hanke et al. (1974, p. 273).

[68] Havenstein (2006, p. 30) and Theunissen (2008).

[69] Provine (1986, p. 321–326).

[70] McClary (1969, p. 121).

companies believed the hybrid corn-breeding system could be successfully applied to chickens. They were willing to finance expensive experimental programs in order to find a way to achieve this end. The Wallace family, that is Henry A. Wallace who developed the hybrid seed company Hi-Bred Corn Company in 1926 (renamed Pioneer Hi-Bred Corn Company in 1935) and his son Henry B. Wallace, initiated this effort to produce commercial hybrid chicks in 1936. By 1942, the Wallace family was selling hybrid egg-laying Leghorns under the name of Hy-Line.[71] How these hybrid chicks were generated was kept secret, even if inbreeding and line crossing were central to the process.

The hybrid corn-breeding method, as applied to chickens, needed a huge number of birds in order to operate properly, because many breeding experiments and the stock used in them would have to be discarded. No ordinary breeder had the numbers required to carry out any breeding system of this nature. Flocks (breeding or otherwise) rarely numbered above a few 100 on farms at this time, and the average per farm was in fact a great deal lower, and had not changed in size much from the nineteenth century. I. K. Felch had estimated that in the 1870s American flock size varied from 12 to 50. Larger ones were extremely rare.[72] The flocks on some American farms had risen by 1913 to between 100 and over 200 hens.[73] As late as 1930, the average flock in the USA remained about the same as Felch's estimate for the late nineteenth century. Over half of farms reporting chicken keeping in 1930 stated they had fewer than 50 hens, and the average number of hens per farm was estimated to be no higher than 23. The vast majority of commercial flocks until after the 1950s continued to number below 200.[74] The hybrid corn-breeding method, which was immensely expensive and also wasteful, needed the kind of corporate enterprise that the Wallace family could provide.

By the early 1940s, American geneticists working with breeding companies had succeeded in breeding hybrid hens with increased egg-laying capacity via the hybrid corn-breeding method, and companies of egg-laying birds began to franchise hatcheries, which were themselves independent, in the USA and Canada. The appeal of the new hybrid production method can be gauged especially by its great success in Canada, even though the hens had to be imported from the USA. Both Hy-Line and the DeKalb Hybrid Corn Company marketed chicks through franchised Canadian hatcheries by the 1950s.[75] Hatcheries advertized what they had been franchised to sell.[76] Canadian hatcherymen were forced to confront the fact that producer/growers liked egg-laying birds resulting from the crossing of breeds or of strains within a

[71] Sawyer (1971, p. 112) and Schapsmeier and Schapsmeier (1968, pp. 21, 27, 28).

[72] Felch (1877, pp. 20, 22).

[73] Hawthorne (1918, pp. 27, 29–31).

[74] Hanke et al. (1974, p. 218).

[75] *Advocate*, 11 January 1958, p. 37; 8 February 1958, p. 5; 14 February 1958, p. 3; 9 January 1960, p. 35.

[76] See *Advocate*, 11 January 1958, p. 37; 8 February 1958, p. 5; 14 February 1959, p. 3; 9 January 1960, p. 35.

breed, even if these came from the USA.[77] They grew faster and converted feed better than stock bred the traditional way. And they laid more eggs. As early as 1941, one Ontario hatcheryman running incubators in various locations stated that 65 % of his sales had been hybrids and he prophesized that it would be more in the next year. Once a producer had experienced hybrids, it was impossible to sell him or her anything else.[78] ROP birds were not wanted, and few if any hybrids were available from Canadian breeders. The result was increased importation of American hybrid chicks into Canada.

From Egg Laying to Meat Breeding: Expansion of the Industry

By the 1950s, after the revolution in the breeding of egg-laying chickens was in place, dramatic changes in the chicken meat industry were underway in the USA. By the late 1920s, a market for chicken meat had begun to exist along the eastern seaboard. This situation encouraged poultry people to raise chickens for meat, and not simply fatten birds no longer useful for egg laying (spent hens), or not useful at all (excess males) to supply the limited market that existed for chicken meat before that time. Since the meat industry was a by-product of the egg industry until the 1920s, virtually no single purpose breeding for meat existed before the 1930s. When a renewed interest in meat breeding evolved with rising broiler markets, traditional ideas on how to breed for meat—namely, cross-breeding breeds (not lines within a breed) in order to promote hybrid vigor—reemerged.[79] Breeders, such as the Hall brothers, began to supply the hatcheries with cross-bred chicks to be raised for the broiler market in the eastern seaboard US cities.[80]

The most important event in the twentieth century for the poultry meat-breeding industry was the Chicken-of-Tomorrow contest, run by A & P (Atlantic and Pacific Tea Company) Food Stores between 1948 and 1951 in the USA.[81] A & P sponsored and underwrote a long-range project designed to improve meat-type birds by teaching the breeders what the consumer wanted in a meat bird. The Chicken-of-Tomorrow contests rather showed breeders what to breed for. Several breeders unknown outside their local communities were skyrocketed into national fame, especially the two winners of the national contests.[82] The contests and an ever expanding market for chicken meat meant that breeding specifically for meat increasingly made economic sense. Meat breeding continued to follow traditional cross-breeding tech-

[77] E. S. Snyder, A *history of the poultry science department at the Ontario Agricultural College, 1894–1968* (unpublished manuscript, 1970), p. 289.

[78] *Advocate*, 26 June 1941, p. 426.

[79] *Advocate*, December 1868, p. 185.

[80] Sawyer (1971, pp. 113–114).

[81] Shrader (1952, pp. 7–8), Horowitz (2004), and Boyd (2001).

[82] Shrader (1952, pp. 7–8).

niques, although more complicated systems of breed crossing evolved over time. This hybridist approach to meat breeding in turn attracted corporate enterprise, and for the same reasons that had made corporate enterprise enter egg-laying chicken breeding: a guaranteed market for the hybrid chicks and protection of the investment needed to turn out those chicks in massive numbers. Classical approaches to meat breeding lent themselves to this new interest in biological patenting. Breeders who had been successful in the contests soon found themselves at the head of companies that functioned increasingly with the aid of geneticists who established many cross-bred lines involving a number of breeds.

Conclusion

Chicken breeders showed considerable understanding of the process of heredity, even if no detailed understanding of hereditary mechanisms existed, by at least the mid-nineteenth century. Many of the attitudes of the best breeders would match those of scientists working with artificial selection after 1900. Felch and Stoddard clearly comprehended the effects of inbreeding, for example, and appreciated the percentage inbreeding ideas put forward much later by Wright. Hybrid breeding via crossbreeding was fully understood by breeders (even if they had not focused on line crossing within a breed for heterosis) long before the phenomenon interested scientists. All the best chicken breeders worked theoretically with the idea of breeding from the point of view of populations. The two groups had much in common when it came to conceptual breeding strategies, namely an emphasis on groups, inbreeding to establish lines, and some effort at quantifying breeding results. It is, therefore, difficult to argue that genetics offered critically new knowledge on strategic approaches to breeding. Until at least 1930, scientists, in fact, made little effort to interact with farm breeders. When geneticists began to look specifically at productivity in egg-laying chickens by the late 1930s, it was apparent that they differed from practical breeders over one highly significant point: Should stock be bred to reproduce truly or be bred in such a way that the lines would not reproduce truly? The traditional breeders tended to uphold the first, while the geneticists came to support the latter. Why did the geneticist stance take precedence over the traditional breeder stance in relation to egg-laying chickens?

There were three fundamental reasons. First, North American chicken breeding was so hopelessly divided in outlook by the 1920s, regardless of the overall commercial emphasis on the egg industry, that effectively no clear idea as to how breeding should proceed could be discerned. A sort of intellectual vacuum resulted. Traditional organizations, which supported chicken breeding, had introduced a particularly serious and entrenched dichotomy: Should birds be bred for beauty or for utility, and did beauty mean utility? The adherence of the American Poultry Association to the Standard of Perfection and beauty breeding, as well as its promotion of competitions to encourage good breeding for egg laying, had forced a confrontation between commercial breeders and fancy breeders. The linkage of programs like the ROP with fancy did not help that situation. The fractured nature of the breeding world seemed to promote a loss of knowledge by traditional breeders of what breeding methodology actually meant. In contrast, a sense of clarity emerged from breeding efforts by geneticists and corporate breeding companies when they entered the egg-laying chicken-breeding world by 1940.

The second reason lies in the structure of the general North American table egg industry and the division that existed between the breeder and producer/grower, which was encouraged by such trends as the rise of the hatchery industry. The division provided the rationale for a critical change in chicken-breeding strategies for egg-laying hens. The idea that breeding should result in true producing lines was no longer important if the producer was not part of the breeding structure. Once the producer/grower abdicated any role in breeding, one of the main advantages that true breeding lines offered (namely the possible inclusion of producing farmers in the breeding process) was no longer critical to the breeding structure. The producer/grower began to demand hybrid stock when it became available. The producer/grower might consume the product of the breeder, but the opinions of the former directed what breeding would be acceptable and therefore how breeding would evolve.

And here we come to the third reason, namely the entrance of corporate enterprise, essential for the funding needed to create good hybrid lines. The cheapness of the individual birds and their fast reproductive life all lent itself to quantification, but quantification required capital investment to make it work. Success of the American corn companies with inbreeding and heterosis made them interested in financing expensive experiments on egg-laying hens. Corporate enterprise was attracted to the idea of undertaking such a project because the biological lock indigenous to hybrid breeding created a natural patent, thereby protecting the investment of the company. Corporate hybrid chicken breeding could only work if poultry farmers who produced table eggs agreed to buy the chicks and in the process abdicate any role in breeding, a situation in place by the late 1920s. The buyer of company hybrid stock could not use that stock for breeding and would, therefore, be a return customer for every new generation. Since the producer/growers were the main buyers, and because they did not care to breed, they had no problem accepting hybrid stock. With the growth of the broiler industry, cross-breeding methods traditionally used for meat production reemerged. When this form of crossing was utilized to control intellectual property in the biology of the birds (with the entrance of corporate enterprise in the 1950s), the structure of the meat-breeding industry took on patterns evident in the egg-breeding industry by the 1940s.

References

Babcock, Ernest Brown, and R. E. Clausen. 1918. *Genetics in relation to agriculture*. New York: McGraw-Hill.

Bonneuil, Christophe. 2006. Mendelism, plant breeding and experimental cultures: Agriculture and the development of genetics in France. *Journal of the History of Biology* 39:281–308.

Boyd, William. 2001. Making meat: Science, technology, and American poultry production. *Technology and Culture* 42:631–664.

Bugos, Glenn E. 1992. Intellectual property protection in the American chicken-breeding industry. *Business History Review* 66:127–168.

Castonguay, Stéphane. 2005. The transformation of agricultural research in France: The introduction of the American system. *Minerva* 43:265–287.

Cooke, Kathy J. 1997. From science to practice, or practice to science? Chickens and eggs in Raymond Pearl's agricultural breeding research, 1907–1916. *Isis* 88:62–86.

Crow, James F. 1990. Sewall Wright's place in twentieth-century biology. *Journal of the History of Biology* 23:57–89.

Derry, Margaret. 2001. *Ontario's cattle kingdom: Purebred breeders and their world, 1870–1920*. Toronto: University of Toronto Press.

Derry, Margaret E. 2003. *Bred for perfection: Shorthorn cattle, collies, and Arabian horses since 1800*. Baltimore: Johns Hopkins University Press.

Derry, Margaret E. 2012. *Art and science in breeding: Crating better chickens*. Toronto: University of Toronto Press.

Dreyer, Peter. 1985. *A gardener touched with genius: The life of Luther Burbank*. Berkeley: University of California Press.

Dunn, L. C. 1965. *A short history of genetics: The development of some of the main lines of thought, 1864–1939*. New York: McGraw-Hill.

Felch, I. K. 1877. *The breeding and management of poultry*. Hyde Park: Norfolk County.

Felch, I. K. 1902. *Poultry culture: How to raise, mate and judge thoroughbred fowl*. Chicago: Donohue, Henneberry.

Fitzgerald, Deborah. 1990. *The business of breeding: Hybrid corn in Illinois, 1890–1940*. Ithaca: Cornell University Press.

Fitzgerald, Deborah. 1993. Farmers deskilled: Hybrid corn and farmers' work. *Technology and Culture* 34:324–343.

Fitzgerald, Deborah. 2003. *Every farm a factory: The industrial ideal in American agriculture*. New Haven: Yale University Press.

Grasseni, C. 2007. Managing cows: An ethnography of breeding practices and uses of reproductive technology in contemporary dairy farming in Lombardy (Italy). *Studies in History and Philosophy of Biological and Biomedical Sciences* 38:488–510.

Hagedoorn, A. L., and Geoffrey Sykes. 1953. *Poultry breeding: Theory and practice*. London: Crosby Lockwood.

Hanke, Oscar A., John L. Skinner, and James L. Florea, eds. 1974. *American poultry history, 1823–1973*. Madison: American Poultry History Society.

Harwood, Jonathan. 1992. *Styles of scientific thought: The German genetics community, 1900–1933*. Chicago: University of Chicago Press.

Harwood, Jonathan. 2005a. On the genesis of technoscience: A case study of German agricultural education. *Perspectives on Science* 13:329–351.

Harwood, Jonathan. 2005b. *Technology's dilemma: Agricultural colleges between science and practice in Germany, 1860–1934*. New York: Peter Lang.

Havenstein, Gerald B. 2006. Performance changes in livestock following 50 years of genetic selection. *Lohmann Information* 41:30–37.

Hawthorne, H. W. 1918. *A five-year farm management survey in Palmer Township, Washington County, Ohio, 1912–1916*. United States Department of Agriculture Bulletin 716, Washington DC: Government Printing Office.

Heape, Walter. 1906. *The breeding industry: Its value to the country, and its needs*. Cambridge: Cambridge University Press.

Horowitz, Roger. 2004. Making the chicken of tomorrow: Reworking poultry as commodities and as creatures, 1945–1990. In *Industrializing organisms*, ed. Susan R. Schrepfer and Philip Scranton, 215–237. New York: Routledge.

Jones, Donald F. 1917. Dominance of linked factors as a means of accounting for heterosis. *Genetics* 2:466–479.

Kimmelman, Barbara A. 1983. The American Breeders' Association: Genetics and eugenics in an agricultural context, 1903–1913. *Social Studies of Science* 13:163–204.

Kimmelman, Barbara A. 2006. Mr. Blakeslee builds his dream house: Agricultural institutions, genetics, and careers 1900–1915. *Journal of the History of Biology* 39:240–280.

Kloppenburg, Jack Ralph, Jr. 1988. *First the seed: The political economy of plant technology, 1492–2000*. Cambridge: Cambridge University Press.

Lerner, I. Michael. 1974. L.C. Dunn (1893–1974) and poultry genetics: A brief memoir. *Journal of Heredity* 65:185–186.

Lytton, Hon. Mrs. Neville (Judith Anne Dorothea Wentworth Blunt-Lytton, Baroness Wentworth). 1911. *Toy dogs and their ancestors*. London: Duckworth.

McClary, C. F. 1969. Reciprocal recurrent selection response in poultry, other animals and plants. In *Proceedings, Eighteenth Annual Poultry Breeders' Roundtable*, 120–149.

Olby, Robert. 1985. *Origins of Mendelism*. 2nd ed. Chicago: University of Chicago Press.

Orel, Vítezslav. 1977. Selection practice and theory of heredity in Moravia before Mendel. *Folia Mendeliana* 12:179–199.

Palladino, Paolo. 1993. Between craft and science: Plant breeding, Mendelian genetics, and British Universities, 1900–1920. *Technology and Culture* 34:300–323.

Paul, Diane B., and Barbara A. Kimmelman. 1988. Mendel in America: Theory and practice, 1900–1919. In *The American development of biology*, ed. Ronald Rainger, Keith R. Benson, and Jane Maienschein, 281–310. Philadelphia: University of Pennsylvania Press.

Pawson, H. Cecil. 1957. *Robert Bakewell: Pioneer livestock breeder*. London: Crosby Lockwood.

Prentice, E. Parmalee. 1951. Food for America, 1980-The supply of animal proteins: The agricultural colleges. *Political Science Quarterly* 66:481–506.

Provine, William B. 1971. *The origins of theoretical population genetics*. Chicago: University of Chicago Press.

Provine, William B. 1986. *Sewall Wright and evolutionary biology*. Chicago: University of Chicago Press.

Ritvo, Harriet. 1986. Pride and pedigree: The evolution of the Victorian dog fancy. *Victorian Studies* 29:227–253.

Ritvo, Harriet. 1987. *The animal estate: The English and other creatures in the Victorian age*. Cambridge: Harvard University Press.

Rogers, Everett M. 2003. *Diffusion of innovations*. New York: Free Press.

Russell, Nicholas. 1986. *Like engend'ring like: Heredity and animal breeding in early modern England*. Cambridge: Cambridge University Press.

Savage, E. S. 1942. Review of E. Parmalee Prentice, American dairy cattle: Their past and future. *Journal of Economic History* 2:227–229.

Sawyer, Gordon. 1971. *The agribusiness poultry industry: A history of its development*. New York: Exposition.

Schapsmeier, Edward L., and Frederick H. Schapsmeier. 1968. *Henry A. Wallace of Iowa: The agrarian years, 1910–1940*. Ames: Iowa State University Press.

Shrader, H. L. 1952. The chicken-of-tomorrow program: Its influence on 'meat-type' poultry production. *Poultry Science* 31:3–10.

Shull, George Harrison. 1948. What is 'heterosis'? *Genetics* 33:439–446.

Termohlen, W. D. 1968. Past history and future developments. *Poultry Science* 47:6–21.

Theunissen, Bert. 2008. Breeding without Mendelism: Theory and practice of dairy cattle breeding in the Netherlands, 1900–1950. *Journal of the History of Biology* 41:637–676.

Theunissen, Bert. 2012a. Darwin and his pigeons: The analogy between artificial and natural selection revisited. *Journal of the History of Biology* 45:179–212.

Theunissen, Bert. 2012b. Breeding for nobility or production? Cultures of dairy cattle breeding in the Netherlands, 1945–1995. *Isis* 103:278–309.

Van Riper, Walter. 1932. Aesthetic notions in animal breeding. *Quarterly Review of Biology* 7:84–92.

Warren, Don C. 1958. A half century of advances in the genetics and breeding improvement of poultry. *Poultry Science* 37:3–20.
White, Kay. 1992. Victorian and Edwardian dogs. *Veterinary History n.s.* 7:72–78.
Wieland, Thomas. 2006. Scientific theory and agricultural practice: Plant breeding in Germany from the late 19th to the early 20th century. *Journal of the History of Biology* 39:309–343.
Wilmot, Sarah. 2007a. Between the farm and the clinic: Agricultural and reproductive technology in the twentieth century. *Studies in History and Philosophy of Biological and Biomedical Sciences* 38:303–315.
Wilmot, Sarah. 2007b. From 'public service' to artificial insemination: Animal breeding science and reproductive research in early twentieth-century Britain. *Studies in History and Philosophy of Biological and Biomedical Sciences* 38:411–441.
Wood, Roger J., and Vítězslav Orel. 1981. Early development in artificial selection as a background to Mendel's research. *History and Philosophy of the Life Sciences* 3:145–170.
Wood, Roger J., and Vítězslav Orel. 2000. Scientific animal breeding in Moravia before and after the discovery of Mendel's theory. *Quarterly Review of Biology* 75:149–57.
Wood, Roger J., and Vítězslav Orel. 2001. *Genetic prehistory in selective breeding: A prelude to Mendel*. Oxford: Oxford University Press.
Wood, Roger J., and Vítězslav Orel. 2005. Scientific breeding in central Europe during the early nineteenth century: Background to Mendel's later work. *Journal of the History of Biology* 38:239–272.
Wriedt, Christian. 1930. *Heredity in livestock*. London: Macmillan.
Wright, Sewall. 1921a. Correlation and causation. *Journal of Agricultural Research* 20:557–585.
Wright, Sewall. 1921b. Systems of mating. *Genetics* 6:111–178.
Wright, Sewall. 1922. *The effects of inbreeding and crossbreeding on guinea pigs*. United States Department of Agriculture Bulletin 1090. Washington: Government Printing Office.
Wright, Sewall. 1923a. Mendelian analysis of the pure bred breeds of livestock, part 1: The measurement of inbreeding and relationship. *Journal of Heredity* 14:339–348.
Wright, Sewall. 1923b. Mendelian analysis of the pure bred breeds of livestock, Part 2: The Duchess family of shorthorns as bred by Thomas Bates. *Journal of Heredity* 14:405–422.
Wright, Sewall. 1931. Evolution in Mendelian populations. *Genetics* 16:97–159.
Wright, Sewall. 1934. The method of path coefficients. *Annals of Mathematical Statistics* 5:161–215.
Wright, Sewall. 1958. *Systems of mating and other papers*. Ames: Iowa State College Press.

Margaret E. Derry is adjunct professor in the History Department, University of Guelph, Canada. An established artist as well as historian of genetics and breeding, she is the author of numerous books that explore two quite different themes: a series on "Life and Living" deals mainly with Canadian experience, especially in Ontario, and a series on "Animals and Culture" focuses on animal breeding. Her recent book, *Art and Science in Breeding: Creating Better Chickens* (2012), and her forthcoming book, *Masterminding Nature: The Breeding of Animals*, 1750–2010 (2015) are related to the subject of her chapter in this volume.

Chapter 19
Breeding Better Peas, Pumpkins, and Peasants: The Practical Mendelism of Erich Tschermak

Sander J. Gliboff

Introduction

The historiography of Mendelism and early genetics have taken a practical turn in the recent years. From the context of sheep breeding in pre-Mendelian Moravia and Mendel's own involvement in scientific agriculture and plant breeding[1] to the leading roles of breeders in Mendel's twentieth-century reception,[2] historians have come to see practical interests motivating and informing every stage of the theory's and the discipline's development.

With few exceptions,[3] however, historians of genetics have treated the pivotal event in the story—namely the "rediscovery" of Mendel's work in 1900—primarily as an intellectual breakthrough. The three co-rediscoverers are celebrated for recognizing the theoretical significance of Gregor Mendel's 1865 paper, particularly its conception of paired intracellular hereditary "elements" or "factors" and its manner of accounting for 3:1 segregation ratios in terms of such factors and their distribution.

Or at least two of the rediscoverers are so celebrated. There has been considerable doubt about how much credit is due Erich Tschermak (1871–1962, also known as Erich von Tschermak-Seysenegg). Tschermak defended his priority tirelessly. His papers at the Academy of Sciences in Vienna are full of apologetic notes from editors and publishers to whom he evidently had protested when authors referred to

---

[1] Orel (1998), Wood and Orel (2001), and Wood and Orel 2005. On the current state of the Mendel literature, see Gliboff (2013).

[2] Kimmelman (1997), Paul and Kimmelman (1988), Olby (2000), Allen (2000), and Onaga (2010).

[3] E.g., Harwood (2000).

S. J. Gliboff (✉)
Department of History and Philosophy of Science, Indiana University, 130 Goodbody Hall, 1011 East 3rd St., Bloomington, Indiana 47405, USA
e-mail: sgliboff@indiana.edu

D. Phillips, S. Kingsland (eds.), *New Perspectives on the History of Life Sciences and Agriculture,* Archimedes 40, DOI 10.1007/978-3-319-12185-7_19

Carl Correns and Hugo de Vries as Mendel's *two* co-rediscoverers. His 1958 memoir even has a special section combating such slights.[4]

Nonetheless, he was the last of the three to publish, and his initial explanation of segregation is indeed rather hard to follow or even to find in his 90-page paper of 1900, and there is no trace at all in that paper of any hypothetical intracellular elements or factors like Mendel's.[5] His detractors conclude from this that he must not have understood Mendel on these crucial points and should therefore not count among the rediscoverers.[6]

This is, however, an extremely uncharitable and selective reading of Tschermak. And it demands that any rediscoverer worthy of full credit accept a particular physical interpretation of the hereditary factors and how they paired up and separated again—essentially that given by Correns in his rediscovery paper. Whereas Mendel himself had avoided an explicit commitment to hereditary particles,[7] Correns ascribed to him a system of discrete hereditary/developmental *Anlagen* that paired up in the nucleus after fertilization, interacted directly so as to allow one of each pair to dominate the other, and parted company during meiosis.[8] Yet, one did not have to embrace such a model to count as a good Mendelian in the early years of the field. Even Thomas Hunt Morgan resisted as late as 1909.[9] It does not seem reasonable to make it the norm for 1900. It belongs rather to the twentieth-century reinterpretation than the rediscovery of Mendel.

His early agnosticism about segregating particles is not Tschermak's only posthumous problem. His practical orientation and relative disinterest in theory or in discipline building in genetics have also contributed to his lower status among the rediscoverers. But now, given the practical turn in the literature, perhaps the time has come to reevaluate Tschermak and better integrate the plant breeder into the rediscovery story and early genetics. In particular, I would like to put aside the question of whether he understood segregation properly (or what a "proper" understanding should even have been in 1900) and focus instead on his practical use of it as a tool in plant breeding, a justification for his favored breeding methods, and leverage for advancing his career.

After 1900, hybridization and Mendelian segregation served Tschermak as guiding principles in the design of practical breeding programs. For Mendel had made no explicit provision for his hereditary factors to vary. He explained variation (and, in a limited way, evolution, too) in terms of changes not in the factors themselves, but in their assortment. Mendelism implied, as it seemed to Tschermak and many others, that the key to breeding new varieties was to seek out existing ones with

[4] Tschermak-Seysenegg (1958, pp. 58–59).

[5] Tschermak (1900).

[6] Stern and Sherwood (1966, pp. xi–xii), Monaghan and Corcos (1986), and Monaghan and Corcos (1987).

[7] Olby (1979).

[8] Correns (1990/1965).

[9] Morgan (1909).

desirable traits, hybridize them, and get those traits to segregate and reassort in new combinations.[10]

In outline, that was Tschermak's main approach for the rest of his long career at the Agricultural College [*Hochschule für Bodenkultur*] in Vienna, and he used his status and authority as a rediscoverer of Mendel to promote it. He not only applied it himself but also campaigned to establish a series of new agricultural experimental stations in and around Vienna, Lower Austria, Bohemia, and Moravia, which would be devoted to its use.

In addition to reopening the question of Tschermak's role in the rediscovery and reception of Mendelism, I also wish to provide a brief overview of Tschermak's career, about which very little information has been available in the secondary literature. Least of all seems to be known about Tschermak's Nazi-era activities. I have found no mention of this subject in his own many memoirs. Secondary accounts tend to follow the pattern of Tschermak's memoirs and trail off by the time of the *Anschluss* of Austria in 1938.[11]

But Tschermak was still quite active even as he approached his 70th birthday and his emeritization. His correspondence shows him consulting with the German Ministry of Agriculture on selecting and crossing strains of crops, animals, and even the peasants to go with them to planned settlements in occupied Eastern Europe. He also applied the principles of Mendelian segregation in a very odd critique of the Nuremberg marriage-restriction laws.

Early Life and Education

Erich Tschermak was born into an academic family in Vienna on 15 November 1871. His father was the mineralogist Gustav Tschermak, his grandfather the botanist Eduard Fenzl, who had been one of Mendel's professors. His elder brother Armin would also have a successful academic career, as professor of physiology in Prague, and evidently was always considered the smart one in the family. His elder sister Silvia studied painting and chemistry and worked for a time as an unpaid university *Assistentin,* but did not have an academic career beyond that.[12] In 1906, Emperor Francis Joseph elevated Gustav into the hereditary nobility with the formal title of Gustav Tschermak, Edler von Seysenegg, which the family simplified to "von Tschermak-Seysenegg."

---

[10] On the nineteenth- and early-twentieth-century breeding methods and the pros and cons of hybridization, see Bailey (1892), Harwood (2000), Theunissen (2012), and Wieland (2006).

[11] An exception is Veronika Hofer, "Mendelism and eugenics in Vienna: Mendel's rediscoverer Erich Tschermak-Seysenegg and his active involvement with eugenics," paper presented at the annual meeting of the History of Science Society, Montreal, November 6, 2010.

[12] Unless otherwise noted, biographical information is from Tschermak-Seysenegg (1958), Ruckenbauer (2000), and the editors' introduction to Michal Simunek et al. (2011a).

Erich Tschermak began his studies in Vienna in 1891, simultaneously at the University and the Agricultural College. Possibly to differentiate himself from his brother, he soon devoted himself entirely to agriculture. After two semesters of study in Vienna, he went to Freiberg in Saxony, Germany, to do a kind of unpaid internship at a large agricultural estate. He was well received there, made the acquaintance of prominent farmers and foresters, and decided to stay in Saxony and finish his studies at the University of Halle. He still did not show much interest in an academic career, but asked to be assigned a quick dissertation topic, so that he could be done with his degree and go back to practical work as soon as possible. His dissertation project was to use dyes and salt solutions to trace the pathways of vascular tissues in woody and herbaceous plants. He seems to have needed his brother's help to get it done in 1896, and it is easy to get the impression from their correspondence, and even from Erich's own accounts, that Armin was the brains of the operation.[13]

Erich Tschermak's father supported him in his career turn to agriculture, allowing him to spend time volunteering at several different seed companies, learning various methods of plant breeding, and most important, making contacts with leading breeders, including Kurt von Rümker and Wilhem Rimpau.[14] Upon his return to Austria, he was able to speak with considerable authority about German breeding methods and their relative merits. His 1898 survey of current practices there already reveals his preferences and future directions.

Tschermak was looking for scientifically justifiable methods, not just successful rules of thumb. And in addition to improving existing varieties, he wanted to be able to produce new forms without simply waiting and hoping for spontaneous variations to appear or for selection to do its gradual work. He was skeptical of selection alone as a means of producing new forms, but thought it could be used to stabilize and improve varieties created by hybridization or a lucky variation.[15] He quoted Rimpau to the effect that German breeding methodology (particularly for grain) was still in its infancy, and lamented that the Germans were still far ahead of the Austrians, who had hardly even begun to use hybridization: "The method that underlies grain breeding everywhere in Austria is that of selection alone. No experiments with hybridization have been made at all, at least not in practical operations."[16] Tschermak's ambition was to change all that.

In 1898, one of his father's connections at the University of Ghent talked him into going there, but he did not find the expected opportunities for agricultural fieldwork in the nearby countryside. Instead, he asked to do some crossing experiments in Ghent at the city's botanical gardens. He had no particular research question in mind, but the director of the gardens suggested to him that he read Darwin's *Effects of Cross and Self Fertilisation in the Vegetable Kingdom* for inspiration. So, with

[13] For further details on their partnership, see Simunek et al. (2011b).

[14] For more on whom he worked with and what breeding methods they used, see Harwood (2000); also his own report, Tschermak (1898).

[15] Tschermak (1898, esp. pp. 3–8).

[16] Tschermak (1898, p. 22).

the intention of investigating inbreeding depression and hybrid vigor, he chose to work on peas, because they normally self-pollinate, but could also be crossed by hand. He chose pea varieties of different colors, shapes, plant heights, and flower colors, coincidentally much like Mendel's selections.

For Tschermak, however, experimental work was secondary in importance to making the rounds of the major plant breeders and seed companies of western Europe, so he left for an extended stay at the firm of Vilmorin and Andrieux in Paris,[17] without harvesting his results. He left instructions for a gardener to pack up the peas and mail them to Vienna.

Before returning to Vienna, Tschermak took the time to visit breeders in London and Amsterdam, too. In Amsterdam, he called on future co-rediscoverer Hugo de Vries, who showed him his *Oenothera* mutants. Tschermak later wrote that he was glad he had not brought up the subject of his pea crosses on this visit, because de Vries might have explained their significance to him and made it impossible for him to claim an independent role in the rediscovery of Mendel.[18] The story of this twist of fate might support his claim of independence from de Vries, but at the expense of making his approach to the pea experiments seem rather haphazard. That they replicated Mendel's work looks all the more like a case of chance, contra Pasteur's adage, favoring the *un*prepared mind.

Rediscovery Revisited

Luckily, the pea collection from Ghent reached Tschermak in Vienna intact. In 1899, while volunteering at an imperial farm in Esslingen, Lower Austria (now Essling, Vienna), he found time and space to continue the crossing experiments, raise a second generation from the hybrids, and perform backcrosses of hybrids to pure recessive parents, unwittingly replicating some of Mendel's trials. He was astounded, at first, to get 3:1 ratios repeatedly in the former and 1:1 in the latter, but upon returning to Vienna in winter of 1899, his literature search led him to Wilhelm Focke's compendium on hybridization and its discussion of Mendel's paper. He found Mendel's paper at the University library, and the rest is history. Or at least that is how he later liked to remember it.[19] The memoirs differ somewhat from the published account of his experiments from 1900, in which the Mendelian ratios were not quite so consistent and Tschermak was much more cautious in his embrace of Mendel.

He wrote up his results first as his *Habilitationsschrift*, the postdoctoral thesis required for teaching at the university level, which he handed in January 1900.

[17] On the Vilmorins and what Tschermak would have learned from them about the importance of hybridization, see Gayon and Zallen (1998).

[18] Tschermak-Seysenegg (1958, pp. 47–48).

[19] Tschermak-Seysenegg (1958, pp. 52–53); for other versions in English, see Roberts (1929, pp. 343–347) and Tschermak-Seysenegg (1951).

While he was waiting for it to be accepted, the other rediscovery papers (two by de Vries and one by Carl Correns) appeared, and he rushed to publish his own account and crossing results.

Tschermak's publication was not a focused explication of Mendel's laws, but his entire 90-page thesis.[20] It was framed as a test of Darwin's ideas about the effects of inbreeding—the original purpose of the Ghent experiments—and the main research question was whether a natural self-pollinator like the pea was immune to inbreeding depression or could be made more vigorous by outcrossing. The Mendel-style crosses and backcrosses seem to have been an afterthought, inspired by some of the literature on hybridization, possibly including Mendel. As Tschermak wrote in 1900, "When, in the course of my work, I became acquainted with additional relevant literature on crosses done with peas, I inserted another series of different experiments that were to study the inheritance of unequally valued [*ungleichwerthigen*], dominant or recessive (Mendel) traits…."[21] This suggests a change in direction in mid-project, and helps explain why the paper was not framed as an explication or test of Mendel's laws.

The explications were scattered throughout the paper and dwelt most of all on the rules of dominance. To Tschermak, dominance was a special kind of hereditary value or potency, which he usually called the *Werthigkeit* of the trait. But the trait's relative power to determine the appearance of a hybrid individual was only one aspect of its *Werthigkeit*. Segregation, though mentioned nowhere by name, was treated as a second aspect, which determined the numerical preponderance of the higher-valued trait among all the offspring of a hybrid. In Mendel's examples, the traits that were dominant also preponderated in 3:1 ratios, but Tschermak decoupled the two phenomena.

This treatment of segregation has been viewed by later Mendelians as a misunderstanding, but it had certain practical advantages for Tschermak. It unified the Mendelian phenomena under the single concept of *Werthigkeit,* and it did not commit him to a model of paired hereditary particles that physically segregated from each other and went into equal numbers of gametes. Such a model explained 3:1 ratios perfectly, but what if 3:1 turned out not to be the rule? Tschermak's approach was safer and could easily accommodate other ratios, because it left the physical model entirely out of the discussion.

Mendelism for Breeders

So, what did Tschermak get out of Mendel's paper if not a conception of hereditary elements or particles and their distribution among the gametes? Given that he had learned his practical breeding techniques before 1900, and continued to use the same ones, did he or any breeder really benefit from the new theory? There is reason to believe that Tschermak and other breeders were overstating the direct impact of

[20] Tschermak (1900).

[21] Tschermak (1900, p. 466).

Mendelian theory on their methods, and that it was actually more useful as a rhetorical tool or emblem of professional status.[22]

For Tschermak, it was a little of both. He valued Mendelism as a new, scientific justification for his favored hybridization methods, which he was determined to make the standard in Austria. It also gave him some guidance in planning particular crossing programs: How many individual plants to start with and how many of their seeds to sow, given the probability of a desired combination; or which intermediate crossing products would breed true and which harbored hidden recessives?

Certain post-rediscovery developments in hereditary theory further reinforced his case for hybridization over selection. For example, in a 1903 paper on heredity and evolution for breeders, he adopted Wilhelm Johannsen's pure-line theory, under which selection could only sort out preexisting types or lines from a mixed population, never produce anything really new. As Tschermak put it:

> The types within a species and race turn out to be *constant,* i.e., the constant centers of the range of variation, in spite of all selection. An origination of new types within a species, subspecies, or race is not to be achieved through selection, but occurs either in connection with hybridization (hybrid mutation [*Kreuzungsnova*]) or through spontaneous mutation.[23]

Rather than wait around for spontaneous mutations, as rival de Vriesians might advise, the best thing for breeders to do, according to Tschermak, was try to create novel hybrids. Most of Tschermak's immediate post-rediscovery publications use Mendelism to provide the theoretical underpinnings for such a practical approach to hybridization. In particular, he favored a method that he associated with the Vilmorins and attributed to Mendel as well: hybridization of varieties with different desirable traits, followed by isolation and inbreeding of individual lines of the hybrid offspring.[24] The goal was to discover novel combinations of traits among those isolated lines, or sometimes even novel traits, and to get favored lines to breed true. He developed this method further in his own work and in collaboration with his friend Herman Nilsson-Ehle at Svalöf, Sweden.[25]

Tschermak's status as a co-rediscoverer, along with his command of plant-breeding methodology and his international connections, opened unexpected opportunities for him in academia, and he was lured back into the family tradition. Following a series of assistantships and lectureships, he became professor in ordinary at the Agricultural College in Vienna in 1909, at the age of 38. He kept that position until his retirement in 1941, but even thereafter he continued to do his research at the College.

---

[22] Palladino (1994) and Harwood (1997).

[23] Tschermak (1903), on 36, emphasis original.

[24] On Philippe de Vilmorin and French applications of Mendelism that were similar to Tschermak's, see Bonneuil (2006); but cf. Gayon and Zallen, who do not see Mendelian theory playing such a direct role in French plant breeding.

[25] Contra Harwood and Palladino, who play down the influence of genetical theory on agricultural practice, Roll-Hansen argues that the development of this method, particularly at Svalöf, did indeed depend upon Mendelian and pure-line theory. Roll-Hansen (1997). Certainly, the scientists themselves insisted on the importance of Mendelian theory: Nilsson-Ehle (1924), Tschermak-Seysenegg (1940), and Stubbe (1942, p. 696).

Tschermak's post-rediscovery work was mostly along two lines: breeding new varieties and developing guidelines for breeders. Tschermak's message to the breeders was twofold: that they should use Tschermak's preferred methods of hybridization, and that Mendel's laws could help them design their crosses:

> Through the renewal of the Mendelian theory of the lawful valuation [*Werthigkeit*] of traits, the breeding of new, constant plant forms by means of artificial crossing appears placed on a new, rational basis. When these principles are observed, the intentional combination of certain traits from different parental varieties takes shape in a significantly surer and simpler way than with purely empirical methods of crossing and selection.[26]

But Mendel's laws were highly idealized, and it was not clear just how they could be translated into practice.[27] Tschermak's approach was to set the idealizations aside and investigate how particular traits really behaved in crosses.

To that end, Tschermak began compiling reference tables of important traits in important varieties—their degrees of dominance, segregation ratios, linkages, etc.—not as predicted by Mendel, but as observed by breeders. For he noted that the same trait might behave differently in different crosses. It might dominate over one alternative from one variety and be recessive to another. The complete dominance, independent assortment, and neat 3:1 segregation ratios observed by Mendel were not presumed to be the norm.[28]

From the rediscovery of Mendel to the *Anschluss,* Tschermak left a distinct mark on Austrian agriculture. Following a three-month trip to America with Rümker in 1909, to make the rounds of all the major agricultural and evolutionary research stations, from Cold Spring Harbor on Long Island to Luther Burbank's operations in California, he returned with new ideas for improving Austria's research-and-development infrastructure. He was involved in establishing several agricultural research stations. But most of his efforts went into breeding an astonishing number of new varieties of wheat, rye, wheat-rye hybrids, barley, oats, peas and beans, primroses, and many other flowers and vegetables. His winter ryes were especially successful.[29]

Of Pumpkins and Peasants

In the 1930s, Tschermak got interested in the pumpkins grown in the Austrian province of Styria for their thick, dark green seed oil, which is prized in Austria as a salad oil. What probably attracted Tschermak's attention to them, aside from their local economic importance, was a mutation that gave the Styrian variety "hull-less"

26 Tschermak (1901, p. 1029).

27 For more on the actual contributions of Mendelian theory to agricultural practices, see Jonathan Harwood, Chap. 17 in this volume.

28 Tschermak (1901, pp. 1039–1064).

29 For lists of both his plant-breeding stations and his crop plant varieties, see Ruckenbauer, Tables 1 and 2; and also the Appendix of Tschermak-Seysenegg (1958).

(actually very thin-hulled) seeds. In Tschermak's estimation, it must have arisen spontaneously around 1880 and spread throughout the region. It also was a good candidate for improvement by means of hybridization, because so many other squash varieties were available with traits that might be combined with their hull-lessness. Tschermak used a vegetable marrow with a bushy growth form, as opposed to the more common long vines and tendrils, to produce what later came to be known as the Tschermak oil pumpkin [*Tschermak Ölkürbis*]. It had the hull-less, oily seeds of the Styrian pumpkin without the vinous form, so it could be grown in neat rows and harvested more easily by machine. It never caught on in Styria, though, mainly because it also had smaller fruits and smaller seeds than its Styrian parent, but perhaps also because it produced a lighter-colored oil.[30]

After the *Anschluss*, interest in the Tschermak pumpkin increased, because of cooking-oil shortages and pressure to improve domestic production of seed oils.[31] In a Nazi-era popularization of his work, Tschermak described how he had added desirable traits to the pumpkin, and he made a pitch for the seeds as both an oil source and an almond ersatz.[32] The light color made the oil more attractive to markets outside of Austria and the seeds more plausible as a substitute for nuts. The home economics unit of the German Women's Welfare Organization developed recipes for pumpkin-seed macaroons and cakes and affirmed of the seeds that, "As an addition to cookies and cakes, and as a cake filling, they will be welcomed by every Hausfrau."[33]

Tschermak's pre-*Anschluss* views on Nazism are difficult to determine. He was never active politically and left little documentation of his opinions, but they probably fell in the center-right, conservative Catholic region of the Austrian spectrum. Certainly, his pro-Nazi colleagues did not treat him like one of their own when they took over the Agricultural College in 1938, but they did not remove him from his position, either, as they did their perceived opponents. At the time of the *Anschluss,* Tschermak was 66 years old, still four years short of the standard retirement age and eager to continue his work and to defend his academic perquisites, regardless of who was running his country or his institution. It did not take him long to figure out how to ingratiate himself with the new authorities.

[30] Tschermak touted it to Austrian and German farmers in Tschermak-Seysenegg (1934a, b). See also recent descriptions by Teppner (2000), Ruckenbauer (2000), Winkler (2000), and Teppner (2004).

[31] Oil seeds were a priority for Herbert Backe, state secretary (later minister) at the Ministry of Food and Agriculture, who also promoted plant-breeding work at the Kaiser-Wilhelm Society. He served on its senate from 1937 and as its first vice president from 1941 until the war's end (Heim 2008, pp. 15–27).

[32] Tschermak-Seysenegg, "Wien als Ausgangsort des praktischen Mendelismus," *Böhmen und Mähren: Blatt des Reichsprotektors in Böhmen und Mähren*, Juli/August 1942, from a copy in Nachlass Erich von Tschermak-Seysenegg, Archiv der Österreichischen Akademie der Wissenschaften, Vienna (henceforth cited as the Tschermak Papers), box 15, on 244.

[33] Deutsches Frauenwerk, Gau Niederdonau, Kreisstelle Brünn Hauswirtschaftliche Beratungsstelle, "Verwendung von Kürbiskernen im Haushalte," mimeographed flyer, n.d., Tschermak Papers, box 1, folder 77 (filed under "Brünn").

Nazi sympathizers and secret party members like Tschermak's colleague Franz Sekera had been secretly plotting their takeover of the Agricultural College for several years. Sekera took charge on March 12, 1938, the very day German troops arrived. He promptly had the *Rektor* arrested, dismissed the few Jewish or half-Jewish students, removed all the professors with known anti-Nazi sentiments, and reinstated any of his fellow "illegals" who had lost their positions in 1934, under the Schuschnigg government.[34] Within a few months, Sekera was short of rooms for his political appointees and reappointees, so he set his sights on Tschermak's workspace.

Sekera asked politely, but ominously, for Tschermak to comply with the wishes of his new colleagues and give it up voluntarily:

> What matters to me most especially in this whole affair: I would like for you, through an act of camaraderie, to show these gentlemen here, who are not quite well disposed toward you, that you stand entirely on our side and have fully gotten over bygone events. You see, honored Court Councilor, how hard I am trying to build a pleasant mode of coexistence at our College….[35]

Tschermak was wary, but did not rush to comply. He wrote back to Sekera to ask for more time, at least until the end of his spring growing season.[36] Simultaneously, he went over Sekera's head to appeal to State Secretary Anton Reinthaller at Austria's post-*Anschluss* Ministry of Agriculture, who eventually intervened on his behalf and enabled him to keep his room.

From then on, Tschermak made an effort to stay in the good graces of the Austrian and the Reich Ministries of Food and Agriculture. He made the personal acquaintance of *Reichsminister* R. Walther Darré, from whom he also received official congratulations and high praise on his 70th birthday and retirement in 1941.[37] He also received the "Goethe Medal" (*Goethe-Medaille für Kunst und Wissenschaft*) from Hitler that year.[38]

As a scientific plant breeder, Tschermak embodied an ideal of the Aryan researcher, learned, but anchored in the practical realm, and in this case in the soil as well. As a high Darré staff member put it in a newspaper article: "In Tschermak, there is a happy combination of knowledge gained at the scholar's desk with practical experiences that he was able to gather through agricultural and gardening

[34] Ebner (1997a, p. 112 ff.).

[35] Franz Sekera to Erich von Tschermak-Seysenegg, May 29, 1938, Tschermak Papers, box 4, folder 95.

[36] Erich von Tschermak-Seysenegg to Franz Sekera, May 30, 1938, Tschermak Papers, box 4, folder 95.

[37] M. A. Prinzessin Reuss zur Lippe, "Ein Pionier der Vererbungsforschung: Professor Dr. Tschermak von Seysenegg 70 Jahre alt," *Nationalsozialistische Landpost: Hauptblatt des Reichsnährstandes*, November 14, 1941, 46, from a clipping in the Tschermak Papers, box 12, folder 38.

[38] "Der Wiederentdecker der Mendelschen Gesetze: Der Führer verlieh Professor Dr. Tschermak zu seinem 70. Geburtstag die Goethe-Medaille," Captioned portrait of Erich von Tschermak-Seysenegg, *Nationalsozialistische Landpost: Hauptblatt des Reichsnährstandes*, November 14, 1941, 1, from a clipping in the Tschermak Papers, box 12, folder 38.

work. Thus all the results of his decades-long research serves practical daily life."[39] Tschermak tried to conform to this ideal. In a letter draft from around this time to an unnamed state secretary (apparently intended for Herbert Backe in Germany), he declared his support for key elements of Darré's *Blut und Boden* [blood and soil] ideology. He portrayed himself as just the sort of scholar-farmer that Germany needed, but no one appreciated properly.

Tschermak lamented the low levels of "recognition for plant breeding achievements, which are so important for the food industry, compared to industrial and commercial achievements," and he expressed his delight with the Ministry's conviction that "the balance between city and countryside, which had been shaken, to the detriment of the Volk and the economy, must be restored." He asserted that he deserved the Nobel Prize for his rediscovery of Mendel, but would not be considered for it because he was a farmer (*Landwirt*), and he hinted that he could use further support for his work. This letter draft is remarkable for the mercenary attitude revealed in handwritten annotations, of which some are in Armin's hand. Erich must have sent it to his big brother for approval and for advice on how to make it "bear even more fruit."[40]

But there are also indications that much of the admiration Tschermak professed for Darré was sincere. From the end of 1939, when his wife got Darré to inscribe a book for him as a Christmas present,[41] Tschermak cultivated a friendly relationship with Darré, continually sending him articles on Mendel and Mendelism, and reports on his wheats, ryes, and pumpkins. He joined the Nazi farmers' associations under Darré's aegis, serving on the board (*Kuratorium*) of the *Gesellschaft der Freunde des Deutschen Bauerntums* and the council (*Landesbauernrat*) of his regional *Landesbaurnschaft*.[42] When Darré resigned from the government in 1942, under pressure because of food-supply problems and differences with Heinrich Himmler over settlement plans for occupied Eastern Europe, Tschermak expressed his concern, thanked him for years of support, and asked for his home address to continue the correspondence,[43] which he kept up even after the end of the war.[44]

There is one letter draft in which Tschermak discussed Mendelian heredity in humans. It dates from 1942, after Darré had left office. In it, Tschermak first complained about how he still had to fight to keep his laboratory space, about how the younger generation was not being trained properly and had no respect any more for their elders, and about the failures of the recent university reforms. Then, he added that nobody understood the Jewish question properly, either, and he proceeded to

39 Reuss zur Lippe (n. 37), first column on 46.

40 Erich von Tschermak-Seysenegg to Staatssekretär, 1941? Tschermak Papers, box 5, folder 10 (filed with the Sekera letters).

41 I infer this from Darré's response to a thank-you note from Tschermak: R. Walther Darré to Erich von Tschermak-Seysenegg, January 9, 1940, Tschermak Papers, box 1, folder 105.

42 Based on various letters in the Darré folder.

43 Erich von Tschermak-Seysenegg to R. Walther Darré, 1942, Tschermak Papers, box 1, folder 105.

44 One last letter draft in the collection is from 1951 and still addresses Darré as Herr Minister.

explain the principles of Mendelian segregation to Darré, as they applied to Jewish–Aryan intermarriage:

> The Jewish question, too, which interests me very much, is dominated by multiple misunderstandings. I am against the inhuman tormenting and killing of the Jews, but surely for the *merciless* sterilization of all people in the Reich who have even a drop of Jewish blood in them. There are still a lot of 1/2, 1/4 and 1/8 Jews running around among us, including such as have swindled themselves an *Ahnenpass,* or were not recognized as Jews, because they knew how to disguise themselves and joined the Party very early.

Tschermak had no faith in bureaucratic methods of keeping track of ancestry, not only because of Jewish trickery but also because of their supposed promiscuity and adultery. They would frequently father illegitimate half-Jews who are born into unwitting Christian families and recorded as Christian births. It was better, he argued, to identify the part-Jews by their Jewish looks and behavioral traits than by their recorded parentage, even at the risk of misclassifying the occasional Christian.

Tschermak's real problem with part-Jews—and here he could speak with the authority of the great hybridizer—had to do with the rules of segregation, and what happened if you let them marry each other, as they were wont to do:

> The Jews really must be able to smell or feel each other out, because usually 1/2, 1/4, 1/8 Jew-mixtures marry among themselves all over again, for which reason absolutely pure Jews or more of the awful Jew-mixtures must Mendelize out of such marriages![45]

Ironically, if Tschermak had described the inheritance of pea coloration in this manner in 1900 that might have earned him more respect from critics who said he did not understand how segregation was supposed to work. But for 1942, and for complex human cultural characteristics, it is an astonishingly primitive view.[46] He had nothing to gain by feigning such a view to impress Darré, since the latter was out of office and out of favor. It would seem that he thought about intermarriage the way he thought about practical breeding, where hybridization was a means of breaking a type.

Tschermak left us one more Nazi-era discussion of human breeding, in an eight-page draft of "Remarks on the future task of German settlement in the East," coauthored with Armin. It is undated, and there is no indication of what, if anything, it was ever used for, but it seems likely to have been written for Darré or maybe his

[45] Original text: "*Auch in der mich sehr interessierenden Judenfrage herrscht noch vielfach Unverständnis. Ich bin gegen das unmenschliche Quälen und Töten der Juden, wohl aber für die* unbarmherzige *Sterilisierung aller Menschen im Reiche, die auch nur einen Tropfen jüdischen Blutes in sich haben. Es laufen bei uns noch eine Menge 1/2, 1/4 and 1/8 Juden herum, auch solche die sich ihren Ahnenpass erschwindelt haben, oder weil sie es verstanden hatten sich zu tarnen und sich sehr frühzeitig der Partei angeschlossen hatten, nicht als Juden erkannt wurden….Die Juden müssen sich gegenseitig wirklich riechen oder tasten, denn meistens heiraten wieder 1/2, 1/4, 1/8 Judenmischlinge untereinander, weshalb wieder absolut reine oder noch arge Judenmischlinge aus solchen Ehen wieder herausmendeln müssen!*" Erich von Tschermak-Seysenegg to R. Walther Darré, December 26, 1942, Tschermak Papers, box 1, folder 105, emphasis original.

[46] Not that there were not any American eugenicists left who still conceived of *feeblemindedness* or other ill-defined mental and moral traits as simple Mendelian recessives, but no one with Tschermak's experience and status as a founder of modern Mendelism.

more-ruthless successor, Backe, who is known for his eponymous plan to divert Soviet food supplies and starve most of Russia.[47] In this document, the brothers Tschermak focused on "questions of soil, water, and climate, further, questions of management and usage (*Wirtschaftsweise*), of mechanical and financial resources, of plant and animal production and breeding, and finally of settler selection and of course settler production."[48] They avoided military matters and wrote as if the land to the east, from Poland through the Ukraine to the Black Sea, were uninhabited. Their description of the territory skirted Russia proper, perhaps an indication that it dated from after June 1941 and the end of the Nazi–Soviet nonaggression pact, but before the siege of Leningrad, which began in September.

Concerning plant and animal breeding, they pointed out how varied the terrain and climate were in that large territory, so that the challenge, at least in the short run, would be to find appropriate existing breeds for each situation, rather than to breed new varieties. They took the opportunity to plug Erich's work and the likely usefulness of his ryes, wheats, and barleys, and even his oil pumpkins, for the eastern settlements. For the longer term, they made a pitch for establishing more agricultural research stations to test and breed new varieties. They praised the work of Nicolai Vavilov in Leningrad and his extensive seed collections as resources for new hybridization efforts, but stopped short of saying they should be raided (perhaps another indication that German troops had not yet attacked Leningrad). Instead, they opined that there would surely be some abandoned breeding stations in the occupied territories, where locally adapted varieties would be found that would be worth trying out.[49]

When they took up the subject of selecting and breeding settlers, all talk of hybridization came to an end. However, local adaptation was still a consideration. Their first preference was to recruit ethnic Germans already living in Ukrainian and Russian territories, whose availability should be provided for explicitly in the eventual peace agreement. Their second choices would have been ethnic Germans who had successfully migrated to other places and maintained their German-ness there. From among these, those most willing to migrate, and younger sons of farmers, should be recruited first. Further consideration should be given to identifying talented organizers and leaders among the settlers, sending enough skilled workers and professionals along with the farmers, keeping members of preexisting communities or coreligionists together, and maybe even sending convicts or other forced migrants (but to strictly segregated settlements).[50]

---

[47] For more on such plans and the technocratic logic behind them, see Aly and Heim (1991); English version: Aly and Heim (2002).

[48] Erich von Tschermak-Seysenegg and Armin von Tschermak-Seysenegg, "Bemerkungen zu der Zukunftsaufgabe Deutsche Ostsiedlung," typewritten manuscript with manual corrections, n.d. [ca. 1941], Tschermak Papers, box 10, folder 26, p. 1.

[49] Erich von Tschermak-Seysenegg and Armin von Tschermak-Seysenegg, "Bemerkungen zu der Zukunftsaufgabe Deutsche Ostsiedlung," typewritten manuscript with manual corrections, n.d. [ca. 1941], Tschermak Papers, box 10, folder 26, pp. 2–4.

[50] Erich von Tschermak-Seysenegg and Armin von Tschermak-Seysenegg, "Bemerkungen zu der Zukunftsaufgabe Deutsche Ostsiedlung," typewritten manuscript with manual corrections, n.d.

Finally, the authors turned from the recruiting of new settlers to the planning of their (re)production (*planmässige Siedlerproduktion*). They foresaw strict observation and classification of their achievements, from which would emerge a hierarchy of fitter and less-fit families. In order to encourage intermarriage within the top level, there could be specially planned social events, and travel stipends to send the fittest farmers' sons to visit or even work on the estates of selected families. Farmers' daughters could also be sent the other way to see how the prospective grooms' families and farms worked. It was all pretty conventional positive eugenics, except perhaps for an unusually strong pitch for organized childcare, both to lighten the workload of the parents and to protect the children from being taken out to the fields all day.[51]

Tschermak's collaboration with the Nazi authorities was probably not very consequential—there is no reason to think that his eugenical ideas had any effect on policy or that the Tschermak pumpkin made much of a dent in the cooking-oil shortage, for example—but it was also unprompted and unnecessary. The path of least resistance would have been to scale back his research program in 1938, as Sekera wanted, then lie low, and retire in 1942. No one in authority demanded anything more of him. Instead, he sought recognition and resources from the state. He was probably motivated mostly by egotism and opportunism, but he was also genuinely attracted to some aspects of the ideology, especially the value placed on agriculture and the ties of the German people to the soil. The Nazi concepts of racial purity also resonated with his views on the nature of plant varieties and the power of hybridization to modify them.

Postwar Career

On April 24, 1945, while fighting was still going on in the vicinity of Vienna, a small group of professors met to begin reorganizing and de-Nazifying the Agricultural College. Most of the group had lost their jobs for political reasons in 1938, under Sekera, and they spoke scornfully of their absent colleagues—the opportunists who had not minded working under the Nazis, but now abandoned ship when there might be consequences to face or danger from bombing. Many of them had headed for the relative safety of the mountain lakes of the Salzkammergut. As one professor later recalled, everyone with reason to fear a liberated Austria had "preferred to set up house on the Wolfgangsee in Upper Austria or some place in the west until the critical days of the occupation were history."[52] This caricature of the fair-weather faculty fit Tschermak perfectly.

[ca. 1941], Tschermak Papers, box 10, folder 26, pp. 6–7.

51 Erich von Tschermak-Seysenegg and Armin von Tschermak-Seysenegg, "Bemerkungen zu der Zukunftsaufgabe Deutsche Ostsiedlung," typewritten manuscript with manual corrections, n.d. [ca. 1941], Tschermak Papers, box 10, folder 26, pp. 7–8.

52 Josef Flatscher in his inaugural address as rector in 1948, translated from Ebner (1997b, p. 142).

After the group reinstated Alfred Till and elected him *Rektor,*[53] Tschermak wrote to him obliviously from his lakeside cottage on the Wolfgangsee,[54] in June, about getting his laboratory back. Despite age and arthritis, he wrote, he believed he would be "able to participate usefully, especially in the area of plant breeding and to serve as a scientific worker and possibly a silent advisor to the Agricultural College." He pointed to all the foreign scientific societies he still belonged to and claimed to be respected in all the occupying countries. He wanted to revive Austrian plant breeding and reconstitute the Society for Plant Breeding (which he had cofounded in 1912,[55] but had been closed down in 1939). All he needed was his ground-floor workspace, so he would not have to climb stairs, a small plot of land in the College garden, and maybe a technical assistant. Also, he was worried that someone might have appropriated his bomb-damaged house, and would appreciate it if the College could make official inquiries about that and tell his property manager he wanted to come back. And he wanted a job for his former assistant Franz Frimmel, who was stuck in Czechoslovakia.[56]

After two more letters to Till (with the additional demand to have his pension payments resumed) failed to elicit a response, he turned, in December, to Leopold Figl, Austria's first postwar federal chancellor, who happened to be a graduate of the Agricultural College. Tschermak reminded him that he was a former teacher and said he shared Figl's desire to rebuild his country: "Of course I follow with great interest all the things you went through under the Nazi regime[57] and how you now, in your position are doing everything you can to help our poor, shattered Austria onto its feet again." Despite seven job offers from foreign countries over the course of his career, he had always been faithful to his College and his fatherland, yet, he complained, his offer to help out at the College seemed not to resonate there. He claimed to be much in demand at other institutions but wanted to work on rebuilding plant breeding in Vienna and Lower Austria, to reconstitute the Society for Plant Breeding, which the Nazis had shut down, and to see that Frimmel was recruited back from Czechoslovakia.[58]

The response came from Anton Steden, another reinstated colleague at the Agricultural College (forced to resign under Sekera because of suspicion that his wife's grandmother had been Jewish), who was also serving in Figl's government as head of the provisional department of agriculture.[59] Steden assured Tschermak that his

[53] Ebner (1997a, p. 125) and Ebner (1997b, pp. 141–142).

[54] In St. Wolfgang, not the same town frequented by the Exners and the other academic families discussed by Deborah Coen, but on the same lake: Coen (2007).

[55] Tschermak-Seysenegg (1937).

[56] Erich von Tschermak-Seysenegg to Alfred Till, typewritten carbon copy, June 15, 1945, Tschermak Papers, box 4, folder 54 (misfiled under Titl).

[57] He had just spent 6 years in concentration camps and had a death sentence hanging over him when the war ended.

[58] Erich von Tschermak-Seysenegg to Leopold Figl, typewritten carbon copy, December 7, 1945, Tschermak Papers, box 1, folder 126.

[59] Ebner (1997a, p. 139).

offer of support really had resonated at the College, but added cryptically that: "The College of Agriculture, however, has a right, in this connection, to receive some clarifications, which would be best done through oral discussion after your return."[60]

Still at the lake in January, Tschermak was incensed. He complained to Steden again about the lack of response to his earlier letters, all the trouble he had repeatedly had to go to keep the ground-floor workspace that had been promised to him on his 60th birthday because of his arthritis, his pension, and—what Steden probably wanted to talk about—accusations of uncollegial behavior toward his former *Rektor* following the latter's dismissal in 1938. He claimed that he had been the only professor to pay a call on the ex-*Rektor*, once it was permitted, and had spoken in favor of him being paid his pension.[61]

Tschermak's postwar correspondence reinforces the impression of him as an egotist and opportunist, concerned only with securing resources and privileges, no matter from whom. Still, the speed with which he was able to change his political stances and switch patrons is dizzying.

Despite the friction with Steden, Tschermak got most of what he wanted. Frimmel never got his job in Austria, but was able to stay on in Brünn. Tschermak soon returned to Vienna and resumed work at the College and remained active in plant breeding for many more years. He lived to be almost 91.

Conclusion

This study was intended to serve two purposes: to use the case of Erich Tschermak to bring out the practical dimension of early Mendelism, and to fill out the biographical picture of Tschermak's post-rediscovery career.

Breeding practice both contributed to Mendel's rediscovery and benefited from it. Young Tschermak had already embarked on a career in plant breeding before 1900, and was led by practical questions about hybrid vigor to begin his hybridizing experiments with peas and read up on pea hybridizers like Mendel. After 1900, he opted to play a leading role in developing hereditary theory or building the discipline of genetics but preferred to promote the use of Mendelian principles in plant breeding. I believe his reputation as an independent co-rediscoverer suffered because of the low profile he maintained in the field of pure genetics, except perhaps for a few years after 1900.

The breeding method Tschermak promoted emphasized the use of hybridization and the principles of segregation to plan sequences of crosses that would bring together desired traits from preexisting varieties. For this, he needed his Mendelism to provide a theoretical justification, some practical guidance, and prestige. He had

[60] Anton Steden to Erich von Tschermak-Seysenegg, typewritten copy of letter, January 17, 1945, Tschermak Papers, box 1, folder 126.

[61] Erich von Tschermak-Seysenegg to Anton Steden, January 24, 1945, Tschermak Papers, box 1, folder 126.

only limited use for theory, and he did not care to play a leading role in the development of the gene concept, chromosome mapping, or even to say much about the physical basis of heredity. The breeding methods were extremely successful, both in creating new and useful crop varieties and in spreading to many experimental breeding stations within the Austro-Hungarian Empire, interwar Austria, and abroad. His status as a co-rediscoverer and founder of Mendelism certainly helped promote the method.

Tschermak's status, along with his devotion to practical breeding, paid off in his personal life as well as his research. It opened a niche for him within his academic family and in the Austrian academic world, as the scholar-practitioner. It was not a bad niche to occupy during the Nazi period, when it earned him favor with Darré at the Ministry of Food and Agriculture and kept him in his laboratory long after he might have been forced into retirement. And it let him make a case for being needed again at the end of the war, to rebuild the Austrian agriculture.

Acknowledgments Work on this paper was supported by a grant from the NSF (award no. 0843297). I thank Andy Fiss for research assistance and Jonathan Harwood, Sharon Kingsland, Renate Kasak, and participants in the Biology Studies Reading Group at Indiana University for comments on various drafts. Thanks also to the library staff at the Austrian Academy of Sciences for their friendly support.

References

Allen, Garland E. 2000. The reception of Mendelism in the United States, 1900–1930. *Comptes Rendus de l'Académie des Sciences—Series III—Sciences de la Vie* 323:1081–1088.

Aly, Götz, and Susanne Heim. 1991. *Vordenker der Vernichtung: Auschwitz und die deutschen Pläne für eine neue europäische Ordnung.* Hamburg: Hoffmann & Campe.

Aly, Götz, and Susanne Heim. 2002. *Architects of annihilation: Auschwitz and the logic of destruction.* Trans. A. G. Blunden. Princeton: Princeton University Press.

Bailey, Liberty Hyde. 1892. *Cross-breeding and hybridizing: The philosophy of the crossing of plants, considered with reference to their improvement under cultivation.* New York: Rural Publishing.

Bonneuil, Christophe. 2006. Mendelism, plant breeding and experimental cultures: Agriculture and the development of genetics in France. *Journal of the History of Biology* 39:281–308.

Coen, Deborah R. 2007. *Vienna in the age of uncertainty: Science, liberalism, and private life.* Chicago: University of Chicago Press.

Correns, Carl. 1900/1965. G. Mendel's Regel über das Verhalten der Nachkommenschaft der Rassenbastarde (1900). In *Fundamenta genetica: The revised edition of Mendel's classic paper, with a collection of 27 original papers published during the rediscovery era*, ed. Jaroslav Kříženecký, 108–109. Prague: Czechoslovak Academy of Sciences.

Ebner, Paulus. 1997a. Krise in Permanenz (1918–1945). In *Die Universität für Bodenkultur Wien: Von der Gründung in die Zukunft 1872–1997*, ed. Manfred Welan and Paulus Ebner, 65–139. Vienna: Böhlau.

Ebner, Paulus. 1997b. Wiederaufbau und Stabilisierung 1945–1971. In *Die Universität für Bodenkultur Wien: Von der Gründung in die Zukunft 1872–1997*, ed. Manfred Welan and Paulus Ebner, 140–170. Vienna: Böhlau.

Gayon, Jean, and Doris T. Zallen. 1998. The role of the Vilmorin Company in the promotion and diffusion of the experimental science of heredity in France, 1840–1920. *Journal of the History of Biology* 31:241–262.

Gliboff, Sander. 2013. The many sides of Gregor Mendel. In *Outsider scientists: Routes to innovation in biology*, ed. Oren Harman and Michael Dietrich, 27–44. Chicago: University of Chicago Press.

Harwood, Jonathan. 1997. The reception of genetic theory among academic plant-breeders in Germany, 1900–1930. *Sveriges Utsädesförenings Tidskrift/Journal of the Swedish Seed Association* 107:187–195.

Harwood, Jonathan. 2000. The rediscovery of Mendelism in agricultural context: Erich von Tschermak as plant-breeder. *Comptes Rendus de l'Académie des Sciences, Series III: Sciences de la Vie* 323:1061–1067.

Heim, Susanne. 2008. *Calories, caoutchouc, careers: Plant breeding and agrarian research in Kaiser-Wilhelm-Institutes,1933–1945*. Trans. Sorcha O'Hagan. New York: Springer.

Kimmelman, Barbara A. 1997. The influence of agricultural practice on the development of genetic theory. *Sveriges Utsädesförenings Tidskrift/Journal of the Swedish Seed Association* 107:178–185.

Monaghan, Floyd, and Alain Corcos. 1986. Tschermak: A non-discoverer of Mendelism, I: An historical note. *Journal of Heredity* 77:468–469.

Monaghan, Floyd, and Alain Corcos. 1987. Tschermak: A non-discoverer of Mendelism, II: A critique. *Journal of Heredity* 78:208–210.

Morgan, Thomas Hunt. 1909. What are 'factors' in Mendelian explanations? *American Breeders Association Reports* 5:365–368.

Nilsson-Ehle, Hermann. 1924. Einige Züge aus der Entwickling des Mendelismus. *Naturwissenschaften* 12:757–61.

Olby, Robert C. 1979. Mendel no Mendelian? *History of Science* 17:53–72.

Olby, Robert C. 2000. Mendelism: From hybrids and trade to a science. *Comptes Rendus de l'Académie des Sciences, Series III: Sciences de la Vie* 323:1043–1051.

Onaga, Lisa. 2010. Toyama Kametarō and Vernon Kellogg: Silkworm inheritance experiments in Japan, Siam, and the United States, 1900–1912. *Journal of the History of Biology* 43:215–264.

Orel, Vítězslav. 1998. Constant hybrids in Mendel's research. *History and Philosophy of the Life Sciences* 20:291–299.

Palladino, Paolo. 1994. Wizards and devotees: On the Mendelian theory of inheritance and the professionalization of agricultural science in Great Britain and the United States, 1880–1930. *History of Science* 32:409–444.

Paul, Diane B., and Barbara A. Kimmelman. 1988. Mendel in America: Theory and practice, 1900–1919. In *The American development of biology*, ed. Ronald Rainger, Keith R. Benson, and Jane Maienschein, 281–310. Philadelphia: University of Pennsylvania Press.

Roberts, Herbert F. 1929. *Plant hybridization before Mendel*. Princeton: Princeton University Press.

Roll-Hansen, Nils. 1997. The role of genetic theory in the success of the Svalöf plant breeding program. *Sveriges Utsädesforenings Tidskrift/Journal of the Swedish Seed Association* 107:196–206.

Ruckenbauer, Peter. 2000. E. von Tschermak-Seysenegg and the Austrian contribution to plant breeding. *Vorträge für Pflanzenzüchtung* 48:31–46.

Simunek, Michal, Uwe Hossfeld, Florian Thümmler, and Olaf Breidbach, eds. 2011a. *The Mendelian Dioskuri: Correspondence of Armin with Erich von Tschermak-Seysenegg, 1898–1951*. Prague: Institute of Contemporary History of the Academy of Sciences, Prague, and Department of Genetics/"Mendelianum" of the Moravian Museum, Brno.

Simunek, Michal, Uwe Hossfeld, and Volker Wissemann. 2011b. 'Rediscovery' revised: The cooperation of Erich and Armin von Tschermak-Seysenegg in the context of the 'rediscovery' of Mendel's laws in 1899–1901. *Plant Biology* 13:835–841.

Stern, Curt, and Eva R. Sherwood, eds. 1966. *The origin of genetics: A Mendel source book*. San Francisco: W. H. Freeman.

Stubbe, Hans. 1942. Erich von Tschermak-Seysenegg zum 70. Geburtstage. *Naturwissenschaften* 29:696.

Teppner, Herwig. 2000. *Cucurbita pepo*: History and thin coated seeds. *Cucurbit Genetics Cooperative Report* 23:126–127.

Teppner, Herwig. 2004. Notes on *Lagenaria* and *Cucurbita (Cucurbitaceae)*: Review and new contributions. *Phyton* 44:245–308.

Theunissen, Bert. 2012. Darwin and his pigeons: The analogy between artificial and natural selection revisited. *Journal of the History of Biology* 45:179–212.

Tschermak, Erich. 1898. Methoden der Veredelung und Neuzüchtung landwirthschaftlicher und gärtnerischer Gewächse in Deutschland. *Wiener landwirthschaftliche Zeitung: Illustrierte Zeitung für die gesamte Landwirtschaft*. March 30:216–217.

Tschermak, Erich. 1900. Ueber künstliche Kreuzung bei Pisum sativum. *Zeitschrift für das landwirtschaftliche Versuchswesen in Österreich* 3:465–555.

Tschermak, Erich. 1901. Ueber Züchtung neuer Getreiderassen mittelst künstlicher Kreuzung: Kritisch-historische Betrachtungen. *Zeitschrift für das landwirtschaftliche Versuchswesen in Österreich* 4:1029–1060

Tschermak, Erich. 1903. Die Lehre von den formbildenden Faktoren (Variation, Anpassung, Selektion, Mutation, Kreuzung) und ihre Bedeutung für die 'rationelle' Pflanzenzüchtung. *Jahrbuch der landwirtschaftlichen Pflanzen- und Tierzüchtung* 1:30–45.

Tschermak-Seysenegg, Erich. 1934a. Der Kürbis mit schalenlosen Samen: Eine beachtenswerte Oelfrucht. *Wiener landwirtschaftliche Zeitung*: *Illustrierte Zeitung für die gesamte Landwirtschaft*, February 17:41–42, 48–49.

Tschermak-Seysenegg, Erich. 1934b. Der schalenlose Kürbis als Ölfrucht. *Deutsche Landwirtschaftliche Presse*: *Vereinigt mit: Illustrierte Landwirtschaftliche Zeitung*, February 24:94.

Tschermak-Seysenegg, Erich. 1937. Zum 25 jährigen Bestehen der Gesellschaft für Pflanzenzüchtung 'Z' Wien. *Der Züchter* 9 (6–7): 137–144.

Tschermak-Seysenegg, Erich von. 1940. Die Einwirkung der Wiederentdeckung der Mendelschen Gesetze auf Pflanzenbau und Pflanzenzüchtung. *Kühn-Archiv: Arbeiten aus dem Landwirtschaftlichen Institut der Universität Halle* 53:1–13.

Tschermak-Seysenegg, Erich von. 1951. The rediscovery of Mendel's work: An historical retrospect. *Journal of Heredity* 42 (4): 163–171.

Tschermak-Seysenegg, Erich von. 1958. *Leben und Wirken eines Österreichischen Pflanzenzüchters: Beitrag zur Geschichte der Wiederentdeckung der Mendelschen Gesetze und ihre Anwendung in der Pflanzenzüchtung*. Berlin: Paul Parey.

Wieland, Thomas. 2006. Scientific theory and agricultural practice: Plant breeding in Germany from the late 19th to the early 20th century. *Journal of the History of Biology* 39:309–343.

Winkler, Joanna. 2000. The origin and breeding of the hull-less seeded Styrian oil- pumpkin varieties in Austria. *Cucurbit Genetics Cooperative Report* 23:101–104.

Wood, Roger J., and Vítězslav Orel. 2001. *Genetic prehistory in selective breeding: A prelude to Mendel*. Oxford: Oxford University Press.

Wood, Roger J., and Vítězslav Orel. 2005. Scientific breeding in central Europe during the early nineteenth century: Background to Mendel's later work. *Journal of the History of Biology* 38:239–272.

Sander J. Gliboff is associate professor in the History and Philosophy of Science Department at Indiana University, USA. His research interests are in the history of biology, especially evolution and genetics, and science in modern Germany and Austria. He is the author of *H. G. Bronn, Ernst Haeckel, and the Origins of German Darwinism: A Study in Translation and Transformation* (2008). He is currently writing a book on Darwinian evolution at the turn of the twentieth century, as well as a series of articles on Erich Tschermak and early Mendelism.

Chapter 20
More than Metamorphosis: The Silkworm Experiments of Toyama Kametarō and his Cultivation of Genetic Thought in Japan's Sericultural Practices, 1894–1918

Lisa Onaga

Introduction

The history of silk craft in Japan dates back hundreds of years. Between the late 1870s and the 1930s, the mass production of silk cocoons intensified in the islands and territories of Japan. These objects contained the metamorphosing chrysalises of the silkworm, the larva of the domesticated silk moth, *Bombyx mori*. Once harvested, the whole cocoons, with the chrysalis still snugly inside, were usually taken to filature factories to be degummed in boiling water so that single strands of silk could be coaxed away, unraveled, reeled, and twisted together with others to make raw silk necessary for textile industries. In the nineteenth century, the delicate raw silk fibers had variable qualities, often broke, or lacked in qualities desirable for industrial scales of production, such as long length. The need to address these issues, while also increasing the volume of annual harvest, laid the ground for various scientific investigations that were directed to the improvement of the silk industry. Heredity constituted an important area of research for generating cocoons that would suit factories using ever popular spinning machines.

One of the key scientific figures in the study of the inheritance of silkworm characteristics at the time was agricultural scientist Toyama Kametarō (1867–1918), trained at the Imperial University of Tokyo.[1] He served as the founding headmaster of the Fukushima Sericultural School between 1896 and 1900 before returning to the Imperial University to work on his doctorate in agricultural sciences. During a sojourn between 1902 and 1905 as a consultant for the kingdom of Siam (present-day Thailand), which had an interest in establishing a silk industry, Toyama carried

[1] Citations of Japanese names in the main text appear as they do in Japan, with surname preceding given name. Citations in the reference list use English-language name order. All translations, in quotes, are mine except for those from documents published in English.

L. Onaga (✉)
History Programme, School of Humanities and Social Sciences, Nanyang Technological University, 14 Nanyang Drive, Singapore 639798, Republic of Singapore
e-mail: lonaga@ntu.edu.sg

D. Phillips, S. Kingsland (eds.), *New Perspectives on the History of Life Sciences and Agriculture,* Archimedes 40, DOI 10.1007/978-3-319-12185-7_20

out key experiments that explained the generational transmission of cocoon color in the hybridization of silk moths from Siam, and Japan.[2] When he returned to Japan, his research findings were incorporated into his doctoral dissertation, which he wrote in English. Interestingly, this took place at the same time that botanists were "rediscovering" Gregor Mendel in Europe.[3]

Toyama first published his research on Mendelism as a chapter in his 1906 Ph.D. dissertation on what he described as the hybridology of insects, verifying the founding laws of modern genetics based on Mendel's garden pea experiments on the patterns of inheritance of traits.[4] To this day, scientists familiar with silkworms refer to Toyama as Japan's "pioneer of genetics." One might also be tempted to label Toyama as the pioneer of industrial agriculture, for upon his return to Japan, the Ministry of Agriculture hired him to apply his knowledge, especially of hybrid vigor, toward improving the production of silkworm cocoons.[5] His work hinted at the lucrative potential of scientific breeding that saw silkworms as biological entities constructed of both their natural endowment that we may today consider as genetics, and the effects of the environment in which they existed. By examining the range of Toyama's silkworm experiments, this essay illustrates how his investigations of Mendelism helped air other questions about biology beyond the genetic determinants of inheritance, and thus complicates more straightforward understandings of how Mendelism entered the everyday practices of silkworm husbandry.[6]

Toyama's studies led him to deeply consider the relationships between biology and productivity as he researched silkworms to improve the efficiency and profitability of Japan's raw silk industry. Toyama's career, spanning just over two decades, was cut short by his death in 1918 at age 51, but he left behind a rich trove of research, popular writings, and speeches that demonstrate how new biological ideas about Mendelism, and related genetic phenomena such as mutation, linkage, and breeding to create pure lines, had developed in Japanese thought and practice. Toyama circulated among different groups of people associated with affairs of the silkworm, from peasants to politicians. Many Japanese people of different social arenas were already familiar with the silkworm and were experts in their own right. This required Toyama to relate to each group with a tone appropriate to their backgrounds without appearing to insult their intelligence, in order to elicit changes in breeding practices. Thus, even as he discussed ideas about the relationships between

---

[2] Onaga (2010b).

[3] Tschermak (1900); Vries (1900); Correns (1900). See also Bateson (1900); Olby (1966).

[4] Toyama (1906b). Toyama's doctoral thesis was written completely in English and submitted in May 1905 before its publication in 1906. Additional citations in Toyama's literature review were added in May 1906 before publication in October 1906. A 15-page abstracted version of his paper was published in English in the back of the trade publication *Dainihon sanshikaihō* (Bulletin of the Great Japan Silk Association), May 1906.

[5] The term hybrid vigor serves here as the translation for *zasshu kyōsei,* the phenomenon of heterosis in which the offspring of a crossbreed exceeds either parental strain in size, yield, or resistance to disease or abiotic stresses.

[6] Yokoyama (1959); Matsubara (2004).

biological productivity and energy and labor expenditures with farmers, he engaged with business leaders and lawmakers about sericulture reforms in Japan.

The ways in which different members of Japanese society interacted with Toyama's sericulture-entrenched work gestures to how differently imagined futures tied to intellectual or popular thought, whether shared or in conflict, together informed the course of Japan's history of genetic knowledge. The changing knowledge of the silkworm's sexual reproduction serves as an illustration of how basic scientific understanding of genetics grew in Japan. Toyama's work makes clear why this was so, as the groundwork for modern genetics in Japan came more from the capital-cultivating silkworm nursery than the tranquil gardens of gentlemanly scholars or monks. I argue that attention to Toyama's pragmatic silkworm improvement efforts, however, ought not to overshadow the fuller historical picture of his biological investigations and what he left behind for other researchers in Japan to study.

The activity of silk production created a way for Toyama to envision a world that could be organized by the application of rational genetic principles. His scholarly research in genetics grew from his keen attention to the silkworm husbandry practices that he critiqued. Toyama was one of Japan's early interlocutors of Mendelism, but his experiments also contributed to broader explorations of silkworm heredity, such as gynandromorphism, exemplified by the co-appearance of male and female characteristics in the bodies of some silk moth specimens, and the non-Mendelian transmission of characteristics. As he sought to instill scientific thought in the craft of sericulture, Toyama thus gained insights into different kinds of hereditary phenomena, and he broadly referred to instances that did not seem obviously to reflect the predictive ratios of dominant to recessive traits as non-Mendelian. His research on this broad category particularly interrogated whether maternal bodies determined the appearance of particular silkworm characteristics, such as egg color, in what he called *bosei iden* or maternal inheritance.[7] I also analyze how Toyama communicated his experiment-based ideas to farmers and sericulturists in the years surrounding the formation of a set of national sericulture policies in 1911, which provides insight into the various efforts to convey new or unsettled scientific ideas to instill practical changes.[8] Finally, I discuss the significance of Toyama's research for designing new silkworm breeds used in national experimental stations, and how other scientists attentive to his work took on methodological and intellectual questions about inheritance that he left behind.

Toyama's Early Career

Toyama came from a landowning family not far from the port city of Yokohama. Supported by his family, he attended the schools best available to him and so gained admission to the Tokyo School of Forestry, which, before he graduated in 1893,

[7] Toyama (1913, p. 353). A translated title of this text refers to maternal inheritance as *bosei iden*.

[8] Further discussion on this point is in Onaga, (2012).

became the College of Agriculture of the Imperial University of Tokyo in 1890.[9] At first he was to work under the entomologist Sasaki Chujiro (1857–1938), but Toyama eventually found a more compatible intellectual relationship with zoologist Ishikawa Chiyomatsu (1860–1935). Although both senior biologists had trained under American zoologist Edward Sylvester Morse (1838–1925), they differed with respect to their research approaches and purposes. Sasaki had significantly fewer theoretical interests in biology and focused on the natural history of insects. He also wielded considerable power within the sericultural science community as an expert in the applied study of microbial diseases of silkworms.[10]

In contrast, Ishikawa, who would later write extensively on human improvement, focused on broader, internationally discussed questions about heredity at the level of cellular reproduction in eggs and sperm of marine organisms. Ishikawa's interests related to current discussions of the mechanism of hereditary transmission of traits, which grew from Darwin's evolutionary theories and which he would have encountered both from Morse's teachings, and from another mentor, German biologist August Weismann (1834–1914).[11] Toyama's tutelage under these two scholars, Ishikawa and Sasaki, likely colored his own outlook on and approach to research.

Toyama is known as much for his intimate understanding of *B. mori*, as he is for his intellectual concerns with inheritance and evolution. Although his own family had not been involved with the business of silk production, his initial decision to work under Sasaki suggests that it was an interest in the insect species that had drawn Toyama to study agricultural science. Toyama's affinity for basic questions led him to work on an undergraduate research thesis on silkworm spermatogenesis, a topic that was better handled under the supervision of Ishikawa.[12] In 1894, Toyama published his thesis, which offered a chromosome count for spermatogenic cells of *B. mori*. A paper on the anatomical structure of the sex organs of silkworms followed this work before he left the Imperial University to assume a position as the founding headmaster of the Fukushima Sericultural School.[13]

Machida Jirō (1885–1964), a student of Toyama who later joined the faculty at the Imperial University of Tokyo and simultaneously worked at the Sericultural Experiment Station, commented that his mentor's early work on spermatogenesis was "significantly distanced from any utility," and that the topic itself elicited some controversy among professors who thought it inappropriate for a student of agriculture. One argument was that Japanese cytological research materials and its

---

[9] Takeuchi (1940).

[10] Sasaki came from a family that commanded considerable authority when it came to sericulture, especially with respect to the control of contagion. Onaga (2013).

[11] Ishikawa had studied at the Stazione Zoologica in Naples, Italy, in 1888, where he encountered Weismann. He later worked with him in Germany in 1889. Edward Sylvester Morse Collection, Philips Library, Peabody Essex Museum; Ishikawa Chiyomatsu to Edward Sylvester Morse, 19 August 1889. For more on Ishikawa, see Godart (2009).

[12] Toyama (1894).

[13] Toyama (1895).

technical language had not advanced enough in the 1890s for the investigation of such problems.[14]

Meanwhile in 1896, amidst all his work on silkworms and sex, Toyama also wrote, illustrated, and published a seminal dissection manual with his colleague, Ishiwatari Shigetane (1858–1941), which students of the silkworm would continue to rely on for decades.[15] At the Fukushima Sericultural School, Toyama could pursue more research projects that engaged deliberately with sericulture. With ample labor available through students and technicians, he began investigations on the relationship between sericulture and temperature, silkworm larval molting, disease, silkworm varieties, and *jōzoku,* the moment of the silkworm's life cycle as it prepares to spin its cocoon.[16]

Of all of these experiments, the survey of silkworm varieties and their behaviors foreshadowed the kind of research Toyama would pursue in the larger part of his career as he gained interest in the biological phenomena of inheritance. Indeed, his enthusiasm for the silkworm seems deeply rooted in his attempts to help sericulturists improve their silk yields. The increasing scale of silk production in Japan during the early 1900s necessitated his particular type of expertise, and the need for new institutions and silkworm nurseries provided the young researcher with an ideal situation in which to find support for his research. Traveling to Fukushima, a historic cradle of silkworm egg production in Japan, allowed him to access and crossbreed different kinds of silkworms for experimental purposes.

By 1900, Toyama had approached the question of inheritance from multiple angles. For instance, he compared biometric measurements of silk cocoons preserved from the eighteenth century with those from the Meiji period. He used the data to critique the eroding wisdom of sericulturists who selected silkworms for larger cocoons instead of paying attention to the whole picture of what constituted high silk content. Another angle of analysis would relate to the inheritance of discrete traits.[17] Later in the spring of 1900, Toyama returned to Tokyo to begin his doctoral research. In a preliminary experiment, he crossbred white cocoon-spinning silkworms from Japan with yellow cocoon-spinning silkworm varieties that had originated from France. Interest in this crossbreeding event produced only yellow cocoon-spinning progeny, an observation that would fuel much of his subsequent research, starting with projects undertaken during his deployment to Siam in 1902.

Based on his experience in Fukushima, authorities at the Ministry of Foreign Affairs, consulting with the Ministry of Agriculture, had likely recognized Toyama's potential ability to set up a sericulture school and industry in Siam.[18] Although Siam was neither a common nor necessarily desirable assignment for most Japanese, it

[14] Machida (1940).

[15] Toyama and Ishiwatari (1896).

[16] Toyama oversaw and coauthored research on these topics, published in the first two years (1898–1899) of the *Fukushima-ken sangyō gakkō hōkoku* (Bulletin of the Fukushima Prefecture Sericulture School). For example, Toyama & Murakoshi (1898) and Toyama (1899).

[17] Toyama (1900).

[18] Onaga (2010b).

seems that Toyama converted his situation into an opportunity to conduct his research in relative freedom.

The research for Toyama's 1906 paper on Mendelian crosses used a *B. mori* strain common to Siam and one conventional Japanese strain. Toyama attempted to improve the Siamese silkworms for commercial use by mating them with Japanese varieties, and thus simultaneously collected data that would contribute to his doctoral dissertation. His hybridization studies have been discussed elsewhere,[19] and here, I would like merely to emphasize that the industrial potential of hybrid vigor suggested by some of Toyama's findings supported both the Siamese court and the Japanese government's belief that their independent silk industries would help resist colonization. The work that Toyama produced in Siam identified him as a key expert who would contribute to new scientific methods for reproducing silkworms in Japan.[20]

Long after Toyama completed his Ph.D. requirements, the mandate that he work for the state and help advance the mass production of silk cocoons evoked the same tensions he had faced earlier as an undergraduate. Indeed, the human capital generated through the imperial universities at the time was synonymous with the production of servants of the state. Toyama could never fully extricate his career from the state's demands of the silk industry even though he had broader interests in heredity, not only in silkworms. After completing his doctorate, he regularly wrote articles for various periodicals and journals, and finally, in 1909, he wrote *Sanshuron,* a thick, two-part tome directed at the literate sericulturist as well as the lay reader interested in silk cultivation, heredity in general, and the biology of silkworms. That he penned as much as he did on the topic of sericulture suggests that his focus on the silkworm was as much a choice as it was a result of his bond to the Ministry of Agriculture.

Maleness and Femaleness

Toyama's research on the biology of silkworms led him to many different questions, some of which appeared to depart from Mendelian rules of inheritance. His research most relevant to silk cultivation had grown alongside basic questions concerning embryological development and sex determination.[21] These cases provide additional scope for apprehending how the biology of the silkworm did not always "fit" Mendelian ratios, as well as Toyama's continued concerns with the development and implications of maleness and femaleness in silkworms. While he had shown that the silkworm provided a straightforward system for understanding Mendelism, within his lifetime, he could not fully explain what seemed like exceptions to the genetic principles.

[19] Ibid., Moriwaki (2010).

[20] Onaga (2010b); Onaga (2012).

[21] Toyama (1901, 1902, 1906a).

Gynandromorphs

In the spring of 1901, Toyama crossed two silk moths that he had reared from a "striped French yellow" female silkworm and a male of a "common Japanese white" silkworm.[22] Two very unusual larvae appeared among the offspring, which, like the female parent, exhibited stripes—but only on the left halves of their bodies. Their right sides were plain white, like the male parent. The larvae that survived to adulthood displayed male moth behavior and sexual dimorphism split down its dorsoventral axis. Toyama dissected it, showing that it was a definite gynandromorph, containing half a pair each of female and male sexual organs.[23] In the early 1900s, the biology of sex determination in moths was uncertain, and geneticists were keen to understand mosaics and gynandromorphs. Toyama viewed his findings as an addition to a number of other animals that exhibited maternal and paternal traits on very distinct locations on their bodies, as described previously by scientists such as Darwin.[24] Among entomologists, Max Standfuss and Georges Coutagne had attempted to breed gynandromorph moths and butterflies and found new occurrences of black-and-white silkworms, respectively.[25]

Toyama's encounter with spontaneously occurring gynandromorphs had illuminated the mosaic phenomenon in the juvenile and the adult forms of *B. mori*. His published findings inspired subsequent work by Thomas H. Morgan, who in 1905 had contested Theodor Boveri's 1888 hypothesis that delayed egg fertilization led to the generation of both haploid and diploid nuclei in the resulting offspring and that gynandromorphism would occur only when the male characteristics dominated.[26] Meanwhile, Toyama dedicated the greater part of his career to experiments to improve silkworms through breeding. The gynandromorph constituted part of Toyama's excitement about the prospect of creating completely new varieties of silkworms. In 1909, he wrote, "The creation of a new animal by crossing species seems to emerge from the idea of crossing varieties within species." He did not just stop at crossing *B. mori* varieties; he had also tried to cultivate a cross between the domesticated silk moth species and its closest wild relative, the *kuwako* (*Bombyx mandarina*), to generate what he dubbed the *kuwago*. This insect, he reported, bore the traits of both species.[27] The prospect of making new species must have motivated Toyama to further study the intricacies of genetics.

---

[22] Toyama (1906b, p. 354).

[23] Ibid., pp. 353–358.

[24] Darwin (1868/1998, vol. 2, 70).

[25] Standfuss (1900); Coutagne (1902).

[26] Morgan (1905); Boveri (1915); Morgan *et al.* (1919); Sander (1994).

[27] Toyama (1909b).

Egg Color

Toyama's observations about mosaics joined a broader study of inheritance connected to maleness or femaleness. His experiments on egg colors represented this question, and they underscored his belief in the importance of using precise genetic knowledge to breed for and to control silkworm qualities. Toyama had studied egg colors since 1906, when he published the finding that Siamese silkworm egg colors followed Mendel's law of heredity. In 1907, a very reputable silkworm breeder from Fukushima, K. Ishiwata, gave Toyama eggs laid by a white breed called Shinkawachi. The breeder had difficulty inbreeding the silk moths so that they would produce white or brown eggs exclusively. Egg color in silkworms seemed to depend on the shell, the yolk, and the serosa, a membrane tissue between the yolk and shell. Most egg color traits followed the general mathematical predictions of Mendelian laws, but in this case, the serosa pigment exhibited different colors anomalously, which concerned Toyama.[28] By 1908, Toyama had separated the white and brown eggs and reared their hatched larvae. The spring generation did not exhibit any serosa color, but they did in the summer that year. Through inbreeding experiments, he found that the brown and normal egg types could each breed true.[29]

Egg color, along with the problem of mosaics, fueled Toyama's keen interest in creating novel types, as exemplified by his 1911–1913 visits to scientists in Europe, such as Arend L. Hagedoorn and Louis Blaringhem, who worked on new breed development and mosaics, respectively.[30] The egg color study also gave Toyama a reason to visit William Bateson in England on that study trip, funded through a joint project with Ishikawa.[31] Bateson advised Toyama to include data on a wider variety of silkworms before the paper's publication. In February 1913, Toyama would be the first Japanese scientist to publish in the *Journal of Genetics,* then edited by Bateson and Reginald Punnett. In that paper, Toyama framed his study as one that would help equip breeders with better information to create pure lines.[32] Proponents of the Mendelian chromosome theory would selectively cite Toyama's 1913 study to support their position, but his paper, which did not deny maternal inheritance categorically, offers insight as to why Toyama may not have unequivocally endorsed the chromosome theory.[33]

One breed that Toyama additionally discussed in his 1913 paper included an example of what we now consider sex linkage, which he categorized at the time as a case of maternal inheritance. The breeding behaviors of the yellow and white forms of a Japanese silkworm breed called Onodahime showed that newly laid egg

[28] Toyama (1913). Given name of Ishiwata was not provided.

[29] Toyama (1913, pp. 351–405).

[30] Takeuchi (1940, pp. 14, 34); also Vilmorin (1913).

[31] Matsui (1967).

[32] Takeuchi (1940, p. 33); Toyama (1913).

[33] Sturtevant (1965, p. 122).

colors corresponded with the color of the egg that the mother had hatched from.[34] Indeed, Toyama's researches had shown that different instances of maternal inheritance, the passing of traits down the mother's side, could have different explanations. Toyama thought the serosa had resulted from the joined nucleic material of both male and female parents, and he expected to see the characteristics inherited from the male parent if it were carrying the dominant trait. Instead, he wrote, "[the inherited characteristic] is entirely maternal in certain colours, such as the reddish brown, blue, normal colour, etc.," regardless of which parental trait was the dominant one. Toyama then qualified that this example of maternal inheritance actually exemplified "dormancy" or "masking" of the paternal characteristics, even dominant ones, during egg development.[35]

Varying reasons for the inheritance of various egg shapes, sizes, or colors could compare with the heredity of "purely maternal" plant seed coats discussed in 1901 and 1905 by Carl Correns and Rowland Biffen, respectively, Toyama thought.[36] He also recognized similarities between the inheritance of some egg colors and shapes and of reciprocal crosses of "indented" (not wrinkled) and round peas that Bateson had described.[37] After various crossbreeding experiments, Toyama eventually confirmed that verification of the dominant or recessive egg colors necessitated a consideration of maternal inheritance. As his broad conceptualization of maternal inheritance might suggest, Toyama did not take any obvious stance in debates at the time about either cytoplasmic inheritance or the role of chromosomes.[38]

Toyama's acknowledgment of both maternal inheritance and Mendelism in moths likely caught the attention of German biologist Richard Goldschmidt, who contacted Toyama about spending a fellowship at the Tokyo Imperial University's zoological laboratory that summer.[39] Goldschmidt had been studying gypsy moths' sexual phenotypic variation, and his papers would later suggest that grades of maleness or femaleness resulted from the quantitative potency of sex determinants in the cytoplasm. He had great skepticism about the exclusive importance of chromosome structures in heredity that Morgan and others recognized.[40] While Toyama's paper had not mentioned Goldschmidt's research, his correspondence assured Goldschmidt of access to laboratory supplies and informed him that Machida Jirō had been experimenting and cultivating gynandromorphic moths bred from a "Japanese ordinary form with Hokkaido form," suggesting that he knew of Goldschmidt enough to welcome his research.[41] While many of Toyama's own studies would

[34] Toyama (1913, pp. 376–377). Tetravoltine, meaning breeding four times a year.

[35] Toyama (1913, pp. 351–405).

[36] Correns (1901); Biffen (1905).

[37] Bateson (1909); Lock (1906, p. 184); also Cock and Forsdyke (2008, p. 384).

[38] Sapp (1987).

[39] Goldschmidt Papers, Bancroft Library, University of California at Berkeley, 72/241z, Toyama to Goldschmidt, 12 November 1913.

[40] Goldschmidt (1916); Morgan et al. (1915); Allen (1980); Richmond (2007); Dietrich (2008).

[41] Goldschmidt Papers, Bancroft Library, University of California at Berkeley, 72/241z, Toyama to Goldschmidt, 12 November 1913.

seem to support the chromosome theory of gene transmission, his silkworm egg analyses also suggested that certain aspects of physiology remained maternally constituted.[42] These theoretical investigations of how maleness and femaleness worked in genetics also played out in Toyama's pragmatic concerns.

Discussions of egg characteristics represented both Toyama's interests in the somatic presentation of sex in silkworms as well as how maleness or femaleness seemed to inform heredity.[43] Given his primary aim to create new species or varieties, maternal inheritance may be seen within the framework of Toyama's interest in optimizing breeding decisions. Toyama had acknowledged since 1906 that some traits follow other laws than Mendel's, pointing not just to qualitative but plastic quantitative traits such as cocoon size or silk length.[44] Understanding the dynamics between genetics and external inputs and stresses in the production of phenotype were both part of Toyama's desire to understand the expression of traits. Toyama anticipated that understanding non-Mendelian inheritance, including cases of maternal inheritance that seemed to obscure straightforward Mendelian patterns, would help make breeding more effective. His work on silkworm varieties thus continued to survey for irregularities and then determine which traits could be inbred.[45] This pragmatic interest in distinguishing between different means of inheritance gave Toyama a way to delve more deeply into the scientific problems of sex and genetics.

The Responsibility for Sericulture

Toyama's research reflected simultaneously the vitality of his collegial relationship with breeders who helped him locate these irregularities, and his efforts to integrate scientific thinking into their work. His public speeches provide useful context for understanding how he conceptualized the improvement of farming and breeding practices in Japan. Toyama's interactions with silkworm egg producers before and after the passing of the Sericultural Law of 1911 show how he sought to empower producers of silkworms and plant crops with knowledge stemming from his or others' experiments.

In Toyama's lifetime, the Sericultural Law of 1911 created new institutions and procedures to encourage the systematic production of silkworms in Japan. The Ministry of Agriculture established the National Institute of Silkworm-Egg Production, which assigned Toyama to lead new experiments to develop parent stocks that would produce ideal hybrid offspring. This law regulated who was allowed to grow which silkworms and generated a chain of command whereby licensed egg producers could only use stocks approved or distributed by local experiment stations that

[42] Toyama (1913).

[43] Ibid., pp. 351–405.

[44] Toyama (1906b).

[45] Ibid. He also considered voltinism, the number of generations the insect could cycle through before overwintering, as an example of maternal inheritance.

received parent strains of silkworms from a regional experiment station, managed by the national institute. While Toyama saw a number of advantages in the support given to sericultural scientists, he had reservations about the degree to which scholars and the government ought to manage silkworm varieties and assume responsibility of Japan's silk craft. Although the government claimed to want to start the national breeding program gradually, Toyama remained skeptical and viewed uncritical acceptance of these developments as a mistake for the industry.[46]

Farmers still had much to contribute to the improvement of silk qualities and yields, Toyama felt, but how much their knowledge and expertise could improve, or embrace scientific thinking, remained a question. When Toyama spoke to farmers, he often shared insights and advice that urged cultivators to think about the multiple biological dimensions of their organisms' productivity. His frankness stood out compared to most scientific workers in the sericultural industry. Toyama seemed more interested in delivering ideas connected to his identity as a scholar than as a government lackey. In the years leading up to the 1911 law, Toyama gained a reputation as a much-sought-after public speaker of early Japanese genetics, and he keenly shared his high standards, hopes, and expectations for Japanese agriculture with all manner of farmers, and not just sericulturists. Toyama often met with people in rural Japan. For example, in 1909, he visited a group of farmers in the important agricultural region of Niigata Prefecture, on the northern side of the island of Honshu. His lecture centered on the topic of experimental evolution and focused on improvement by directing attention to the need for the farmer to first allow himself to improve by considering the limitations of an organism's performance.[47]

Toyama thought that he had a responsibility to help agriculturists use their heads to increase their earnings rather than sell things mindlessly. In the process of critiquing an overreliance on external inputs, such as fertilizer, Toyama called upon his listeners to consider what had to improve *within* their fruits of harvest, whether rice, pears, or potatoes. The means to understand those fundamental insights about what guided the development of crops depended on knowledge gained through what Toyama referred to as "experimental evolution."[48]

Experimental evolution would also be used to understand the definition of the metaphorical *nemoto,* or root, of plants (and animals), which embodied the problems that could be potentially improved. This concern of Toyama's referred to how each organism had a fixed range of ability that could be determined experimentally and then be brought out either naturally or by the cultivator's efforts. Desired features would never come within reach if the organism lacked something rooted in their *honshō*—what they were born with. In a whimsical example, he explained, "It's the same idea behind the fact that an eggplant can't grow from a watermelon seed, but a tadpole can become a frog." Strange as this explanation may sound, it illustrated Toyama's point that a species can exhibit a whole developmental range between immaturity and maturity. This point was less about metamorphosis or a life

[46] Onaga (2012).

[47] Toyama (1909c).

[48] Ibid., p. 14.

cycle, but more about the extent to which one could elicit change toward a desired phenotype. The impossibility of growing an eggplant from a watermelon seed, then, spoke to the impossibility for certain varieties to materialize a desired outcome, no matter what effort the farmer showered upon it.[49]

Questions about what exactly constituted *honshō* were central to Toyama's research. His use of this term in popular arenas usually went undefined, which meant listeners could flexibly interpret *honshō* as they liked. This may seem rather unscientific, yet Toyama often fed his audience a salad of stories and metaphors to describe scientific concepts. This strategy of conveying the relation between the non-speaking silkworms and their breeders with evocative language joined Toyama's straightforward biology lesson-like accounts of experiments. Together, they helped relate his ideas about Mendelian inheritance to everyday Japanese agricultural life, and depending on to whom Toyama had to direct his words, he allowed the contextual significance of *honshō* to change. Far from a rigid scientific term, *honshō* granted Toyama poetic license to explain biology.

Toyama rarely lectured farmers about Mendelism exclusively. Rather, he raised it along with discussion of broader activities and goals of cultivation and how better choices can be made if people use a more informed basis for judgment. Many of these speeches that he gave to the farmers recounting bicolored flower hybridization experiments that yielded Mendelian ratios of 9:3:3:1, summaries of mouse coat color breeding patterns, or the appearance of "waltzing" mice, served to familiarize them with the scientific language and activities used to describe Mendelian inheritance.[50] Toyama also used folklore to convey the mechanics of Mendelism to ordinary people. For instance, to the Niigata farmers, he said,

> varietal improvement today is not a desktop theory but an experiment from a college of science. Intelligent newspapers think that I must know how to make people rich and beautiful. I cannot do that… Applying this to humans requires more research. For plants and animals today, the exchange of *tamashi* (souls) is possible. For example, take the *kuwako* (wild silkworm moth). It is possible to combine one feature of this with the silkworm in your house. Varietal improvement means you can actualize your thoughts even more.[51]

This exchange of the *tamashi* is something Toyama used to illustrate the behaviors of Mendelian factors. This would be neither the first nor the last time he talked about the metaphysical to describe hereditary phenomena. Folk understandings of sleep in Japan during the premodern period suggested that *tamashi* disembodied themselves from their embodied states. In an undated article for a youth magazine, *Shōnen,* Toyama portrayed a story from the famous *Tales of Ise,* about a farmer, who upon inadvertently startling a monk and a pilgrim, napping and in the middle of dreaming, caused their *tamashi* to mistakenly jump into the wrong bodies.[52] He

[49] Ibid., pp. 15–16.

[50] Ibid., pp. 18–19.

[51] Ibid. pp. 19–20. Parenthetical translations are added to allow use of some Japanese terms.

[52] *Ise monogatari* is a collection of Japanese tanka poems and stories from the ninth century.

used this phantasm as an example to emphasize the possibility for various characteristics of living things to interchange freely.[53]

Farmers could willfully direct the *tamashi* (or trait) to appear in a body to produce an altogether new variety or organism. The fact that Americans had already begun to conduct such studies on various organisms added fuel to Toyama's emphasis on this point in Niigata. This story of exchanging spirits also gives us insight into his methodology for improving varieties through the creation of a "pure line." Acknowledging Wilhelm Johannsen, the Danish geneticist, though not citing him by name, Toyama explained, "This method was first developed in Svalöv. Today, they are doing this in every country with barley and wheat. They are selecting varieties."[54] Making new varieties by using this progressive method of experimental evolution would ensure a 10% increase in income without the use of fertilizer, Toyama suggested. Although he did not remark on exactly how the results of experimental evolution would translate into yield, the story of the exchanging spirits helped stress the feasibility of developing new strains and the benefits of thinking specifically about the internal constitution of the desired organism. Most of all, Toyama directed attention to the breeder's own self.

At that time, in 1909, Toyama felt that the understandings and execution of agricultural priorities in rural Japan were unclear and he thus supported state-level reform in order to enhance productivity. Following this rhetoric, according to Toyama, the farmer's ability to generate "good" seeds would require him to first improve himself. This meant disciplining oneself to negate any indulgent mind-set that could inhibit the embrace of scientifically grounded change to their practice.[55] Toyama, for example, had witnessed Japan's growing dependence on fertilizers as soybean lees from Manchuria gradually replaced fish fertilizer in the 1800s. By World War I, Japan had grown reliant on commercial chemical fertilizers.[56] The absence of critical thought about the intensification of agricultural practices concerned Toyama. Fertilizer available by the ton epitomized the kind of nonthinking that motivated him to espouse experiment-based evidence and critical thinking. In sericulture, the increased application of fertilizer on mulberry plants and feeding of mulberry leaves to silkworms represented a disregard of the relationship between external factors and the biology of the organism in question. Such disregard wasted labor and capital, which Toyama spoke about to sericulturists near the city of Hamamatsu in Shizuoka Prefecture in 1914.[57]

In his speech, "Story of Silkworm Seeds," in which seeds referred to both eggs and varieties of silkworms, Toyama imparted key knowledge about genetics and outlined a way to reduce the cost of fertilizer application. The publisher of the sericulture magazine *Sangyo zasshi,* Ishida Magotaro, transcribed and printed the speech's text, giving the scientist an audience beyond Shizuoka. After the magazine

---

[53] Toyama (n.d.).

[54] Toyama (1909c, p. 20).

[55] Ibid., pp. 20–21.

[56] Nakamura and Odaka (2003, vol. 3).

[57] Toyama (1914).

sold out, the transcript remained in demand, so the *Dainihon Sangyō Gakkai* (Great Japan Academic Society of Sericulture Industry) republished it.[58] By combining his understandings of the environment and genetics to critique sericulture at the time, Toyama called for a stop to believing in false promises of profit, and he asked people to instead pay attention to the biological limits of living things.

Neither farmers nor scholars could be faulted for believing that continuously increasing fertilizer or feed could elicit prosperity, but Toyama, though sympathetic, tried to wean people from the empty mantra of improvement.[59] "Until now," he explained, "there have been hopes and expectations that cocoons may become so big that a horse will only be able to carry two on its back…."[60] Ridiculing this hyperbolic image, Toyama emphasized that toying with food levels, soil fertility, or temperature did not produce limitless positive growth but required more income-draining costs.

The concept of *honshō* reprised in Toyama's communication to the Shizuoka sericulturists. He argued that making silkworms with a collection of characteristics comprising *honshō* depended not on those who reared the larvae but on the makers and sellers of silkworm varieties. As mentioned earlier, Toyama appropriated the term *honshō* to describe the sciences that governed an individual's inherited abilities. Similar to the biological arguments against the inheritance of acquired traits, *honshō* helped Toyama to emphasize that single improved parental qualities could not pass down immediately to the offspring's generation.[61] Yet, the collection of characteristics that constituted *honshō* resided within an organism and could be transmitted to redefine the range of an offspring's performance.

Honshō could change between generations. To explain how this occurred according to his version of Mendelian principles of inheritance, Toyama forcefully argued that fertilizer and mindless expenditures of labor and capital had little impact on *honshō*. It was not an organism's positive response to environmental inputs but its fundamental genetic constitution that could be potentially passed down to offspring. To make this clear, Toyama explained that a characteristic for disease susceptibility could be exchanged for strength. That day, he had brought with him an assortment of eggs that bred true for different colors, all purified from one highly heterogeneous line. To illustrate this purification process, Toyama showed how "normally purplish brown" silkworm eggs could be "disassembled" into red, yellow, brown, green, and white forms. The generation of different colors, even albinism, Toyama said, "is not something that can be made possible because of the human hand; it is the opposite, the source of this magic exists within the organism." He then clarified that this "magic" resides in both sexes and in all living things, including people.[62]

[58] Ibid., also Ishida (1908); Ishida (1913).

[59] Toyama (1914, pp. 16–20).

[60] Ibid., pp. 20–21.

[61] Ibid., p. 22.

[62] Ibid., pp.33–34; Toyama (1912).

Creating the Egg

The representation of pure lines through egg color allowed Toyama to show the Shizuoka sericulturists how *honshō* could change by exploiting disassembly to produce precise and desirable changes through its converse, assembly. Toyama's use of the figurative language of assembly and disassembly worked to expand his audience's consciousness to grasp the possibilities that could arise from analytical awareness of heredity. This freeing language may have also allowed Toyama to borrow certainty about the mechanics of Mendelian inheritance in spite of the debates that surrounded it at the time.[63]

Elsewhere, in 1909, Toyama wrote that Mendel's law of inheritance explained how offspring characteristics result from independently assorted characteristics of two parents. He also described a mechanism of partial dis-segregation, in which a new offspring trait would arise as an intermediary form of its two parents, so as to create a novel independent characteristic.[64] Toyama also used disassembly to explain when both male and female egg color characteristics appeared in successive generations of offspring, which would necessitate a breeder to continually select their desired color. Similarly, in the 1914 "Story of Silkworm Seeds," though not stated explicitly to his audience, Toyama conveyed a particulate understanding of inheritance, in which disassembly pointed to the segregation of traits, or factors, in an organism. Assembly hinted at the processes associated with a mating event, wherein the independently assorted traits would come together in the making of the offspring. Toyama suggested that a simple understanding of the range of movements that male and female factors could make in disassembly and assembly processes could allow breeders to concretely "fix" a trait in a given strain and create a pure line, at least to the extent possible for a sexually reproducing animal that could not self-fertilize like plants.[65] He also exercised benevolent shame through comparison with agricultural innovations in the USA. He explained that Americans rapidly arrived at fresh ideas for breed improvement due to their "laziness" and inherent dislike of work compared to the Japanese.[66] The ease of manipulating traits excited Toyama, who wished to engender a spirit of creativity, and likened this work to that of a chemist making new things by first analyzing the constitution of various compounds. If it held that novel chemicals could be created to make medicines or serve other new functions, surely this could be applied to agricultural situations, he mused.[67]

The implementation of assembly and disassembly was not just about novel species creation; Toyama encouraged breeders to imagine how they could use inbreeding and crossing to eventually replace weak silkworms with strong ones. He knew

[63] Sapp (1987, pp. 7–21); Bowler (1989, pp. 110–127).

[64] Toyama (1909a).

[65] Toyama (1914, pp. 38–39).

[66] Ibid., pp. 1–4.

[67] Ibid., p. 39.

that many characteristics were inherited together and could not be isolated by the process of disassembly, in which distinct traits would avail themselves for breeders to select to interbreed like kinds. Although Toyama did not use the term "linkage," this phenomenon fed his ideas about how to improve silkworms through a combination of interbreeding like kinds, crossbreeding, and selection, and in general to encourage close observation of the hereditary traits of different lineages of silkworms. One practical result, for example, would be a better understanding and control of disease in silkworms. Toyama pointed out that genetic knowledge, specifically, the ability to identify disease susceptible lines of silkworms from resistant lines, could enhance understandings that the removal of disease-causing agents alone would not ensure a good cocoon harvest. To understand why "bad" characteristics emerged, therefore, it was necessary to have detailed knowledge of the lineage going back several generations.[68]

Toyama's interest in silkworm lineages sat couched between his larger concern about improving silk harvests without needless expenditures and his thought that disease was a function of both genetics and environment. The degree of care in controlling environmental conditions, in silkworms' mulberry litter, for instance, could alter the spread of infections. While silkworm rearing methods, which involved controlling the environment in precise ways, had over time become extremely complicated, the reasons for these "improved" practices were unclear to Toyama, who argued that cultivators were operating more on the basis of conjecture than scholarly understanding of what made silkworms grow optimally. Moreover, such methods were very labor intensive. Toyama argued that more could be achieved with fewer human hands if appropriate silkworms were selected to be cultivated in given environments.[69] That is, one had to consider not just the environment of the silkworm but also its hereditary qualities.

Conclusion: Toyama's Legacy

This essay has shown how Toyama negotiated new ideas about inheritance in the early days of genetics from the perspectives of scientific research and agricultural practices. Much of the science imparted by Toyama came from his interactions with silkworm breeders in Japan, but he also drew upon examples of the research of other scientists in Europe and the USA, as well as insights from his own research on silkworms and his observations of breeding other organisms that were not agriculturally critical.[70] The discussions of new biological knowledge that Toyama brought to sericulture comprised part of his broader vision whereby he sought to bring attention to the value of universal scientific knowledge. By speaking to various audiences about how a silkworm's robustness would depend on understanding

[68] Ibid., pp. 44–45.

[69] Ibid., pp. 52–56.

[70] Takeuchi (1940).

the scientific limits of the organism's performance, Toyama underscored the importance of comprehending and embracing Mendelian genetics, even when he could not explain some silkworm traits that appeared to escape the explanatory bounds of mathematical predictions of heredity. Analysis of his lectures and scientific papers has shown that the growth of Mendelian thought in Japanese sericulture occurred through an interactive process, in which Toyama asked breeders and silkworm egg producers to share peculiar specimens with him. Breeders also actively brought perplexing varieties to Toyama, which helped push the categorical boundaries of his understandings of non-Mendelian inheritance.[71] Toyama maintained an open stance to these cases of exceptions and viewed them as topics that deserved dedicated study. As a result, his work aired a number of intellectual issues that others would examine later in insect genetics.

Toyama's obligations and interests toward helping sericulturists in a highly economic and national pursuit required theoretical and experimental biology to mingle in the silkworm nursery. Despite the relatively lower status that sericultural scientists and technicians generally occupied in Japanese life sciences, and in spite of Toyama's infrequent presence at the Imperial University of Tokyo, which replaced the silkworm geneticist's position after he died with a plant scientist, it is possible to piece together his intellectual lineage. In contrast to professors such as Sasaki, who had intellectual "descendants" who would carry forward the retiree's research program, Toyama's legacy did not hinge primarily upon his students. In fact, accounts within the field of sericulture suggest that Toyama could not lecture at the university with regularity because he was virtually bedridden with illness when he was promoted to chair the "No. 3 zoological science, entomology, and sericulture science teaching unit."[72] As the first researcher of heredity and genetics to have received the Imperial Academy Prize in 1915—the country's most prestigious research accolade—Toyama's expertise and role in clearing the field for a new generation of researchers was by that point unmistakable, and it remains unclear why the university did not do more to preserve his research lineage.

Ultimately, it was a small handful of Toyama's students and intellectual compatriots in different parts of Japan, and who had similar interests in silkworm heredity, who helped move his research agenda forward over the years. Toyama's legacy thus survived through the connections he had made through extensive interactions with the sericultural experiment stations, and during his tour of laboratories abroad. The working worm certainly informed Toyama's conceptions of heredity, specifically, that both genetics and environment played a role in the overall yield of its cocoons or the viability of offspring. Toyama's research on maternal inheritance and egg color had likely piqued the interest of Goldschmidt, who had maintained an active interest in Toyama's and his colleagues' research ever since his visit in 1914. Although Toyama spent more of his time at the Imperial Sericultural Experiment Station than his university, and Goldschmidt ultimately formed a lifelong relationship

[71] Toyama (1912).

[72] Fukuda (1990, p. 11); Goldschmidt Papers, Bancroft Library, University of California at Berkeley, 72/241z, Ishikawa Chiyomatsu to Goldschmidt, 26 February 1918.

instead with Ishikawa, Toyama, known for his scientific description of spontaneously occurring silkworm mosaics, served as a conduit for Goldschmidt to other researchers interested in moth sex determination.[73]

Goldschmidt's relationship with the Japanese genetics community is today symbolized through the books and reprints that he donated to the National Institute of Genetics library after World War II, but that relationship had formed initially around collaborations with Japanese silkworm researchers.[74] Katsuki Kitō, one of Toyama's students and intellectual successors, was a prominent teacher and researcher at the Ueda Sericulture Professional School in Nagano, who published on the cocoon shapes of first filial hybrids.[75] Katsuki, who also worked at the Sericultural Experiment Station, would coauthor papers on the genetic formation of silkworm gynandromorphs with Goldschmidt between 1927 and 1931.[76] Another exponent of Toyama who had caught Goldschmidt's interest was Machida, who lectured in the same department as Toyama at the university, and who had also managed to breed gynandromorphs.[77] Meanwhile, details of Toyama's gynandromorph experiments also produced ripples in American genetics, with the renowned Morgan later citing Toyama's studies in his work on *Drosophila* sex differentiation.[78]

Toyama's career shows that he was interested not only in breeding silkworms but in questions about what constitutes biological sex. His 1913 experiments sought to understand how sex determines phenotype, and silkworm scientists continued to study this for decades.[79] The key figure to push forward on the study of silkworm sex chromosomes within Japan was Tanaka Yoshimaro (1884–1972), who came of professional age while reading about silkworm hybridology as a student. Tanaka eventually taught the first university genetics course in Japan at the Tohoku Imperial University College of Agriculture and pursued the study of sex determination on a more general level.[80] His experiments critiqued Toyama's, and in 1916, he proposed that the silkworm's sex chromosome formulae were ZW for the female and ZZ for the male.[81] Furthermore, discussions of Tanaka's sex-determination research, which used cytological studies and induced mutation methods to map and

[73] Goldschmidt Papers, Bancroft Library, University of California at Berkeley, 72/241z, Toyama to Goldschmidt, 12 November 1913.

[74] Goldschmidt (1960, pp. 108–110).

[75] Katsuki (1917). See Goldschmidt Papers, Bancroft Library, University of California at Berkeley, 72/241z, K. Katsuki to Goldschmidt, 12 November 1934.

[76] Goldschmidt (1912); Goldschmidt and Katsuki (1927); Goldschmidt and Katsuki (1928); Goldschmidt and Katsuki (1931).

[77] Goldschmidt Papers, Bancroft Library, University of California at Berkeley, 72/241z, Toyama to Goldschmidt, 12 November 1913; Ibid., Ishikawa to Goldschmidt, 26 February 1918. Discussion of Goldschmidt's relationship to Japanese genetics is subject of a separate paper.

[78] Morgan (1914); Morgan (1916); Morgan et al. (1919).

[79] For example, Kikkawa (1943); Tsujita (1961).

[80] Tanaka (1967).

[81] Tanaka (1916).

verify the roles of the Z and W chromosomes in the 1920s and 1930s, continued in the postwar period.[82]

The observation of the mosaic silkworms described by Toyama and analyses of other rule-defying varieties that sericulturists brought to him have symbolized the importance of the interactions between geneticists and breeders in prompting the scientific investigations that generations of researchers have pursued since. Today, researchers at the University of Tokyo and elsewhere use genomic and other advanced technologies to investigate the genetics of various moths, including the cases of silkworms exhibiting traits with hereditary patterns that were previously inexplicable to Toyama.[83] The Laboratory of Insect Genetics and Bioscience (IGB) at the university has inherited the workspace of Sasaki and the line of silkworm scientists who followed him, but it seems as if it is the *tamashi* of Toyama that infuses the contemporary genetic research of the silkworm and its relatives. In the IGB office hangs one of two oil portrait paintings by Takashima Yajurō, made on the occasion of the 25th anniversary of Toyama's death, keeping watch over the work of future generations of insect geneticists. The other painting, housed at the National Institute of Agrobiological Sciences in Tsukuba, Japan, keeps company with entomological research laboratories and a library that includes documents collected from now defunct sericultural experiment stations.[84] In these settings, the history of Toyama's research continues to be retold and transformed as researchers now, in a time that hardly depends on selling silk to distant markets, pursue new kinds of projects and add their own chapters to the "story of silkworm seeds."

Japan's raw silk industry faces a denouement today, only to remind us of how Toyama's research took the path it did in relation to a history of an export commodity that thus involved particular relationships with people and places within and beyond the archipelago. Tied to the practical concerns of silk harvesting, Toyama's research on *B. mori* made eventual contributions to both Japanese sericulturists and to various scientists through a common framework of challenging the norms of silkworm nursery practices. Despite his enthusiasm for a particulate understanding of Mendelian transmission of traits, Toyama also maintained a complicated scientific understanding of inheritance that included a serious consideration of what we call environmental effects. His view of heredity balanced multiple ideas of what together explained phenotypes and shaped a variety's evolutionary course. Ultimately, through a mind committed to experimentation, he believed that one could wisely build upon multiple biological understandings of inheritance to make sericultural and agricultural production most efficacious.

[82] Tanaka (1922); Onaga (2010a).

[83] "Kenkyūshitsu no rekishi—Tōkyō daigaku—konchū idengaku kenkyūshitsu kōshiki uebusaito, *Laboratory of Insect Genetics and Bioscience Official Website,* http://papilio.ab.a.u-tokyo.ac.jp/igb/ja/profile2.html. See, e.g., Osanai-Futahashi *et al.* (2012).

[84] Kitamura and Nozaki (2004, p. 3).

References

Allen, Garland E. 1980. The historical development of time law of intersexuality and its philosophical implications. In *Experientia supplementum: Richard Goldschmidt: Controversial geneticist and creative biologist,* ed. Leonie K. Pieternick, 41–48. Basel: Birkhäuser.

Bateson, William. 1900. Hybridisation and cross-breeding as a method of scientific investigation. *Journal of the Royal Horticultural Society* 24:59–66.

Bateson, William. 1909. Miscellaneous, exceptional and unconformable phenomena. In *Mendel's Principles of Heredity*. Cambridge: Cambridge University Press.

Biffen, Rowland H. 1905. Mendel's laws of inheritance and wheat breeding. *Journal of Agricultural Science* 1:4–48.

Boveri, Theodor. 1915. Über die Entstehung der Eugsterschen Zwitterbienen. *Archiv für Entwicklungsmechanik der Organismen* 41:264–311.

Bowler, Peter J. 1989. *The Mendelian revolution: The emergence of hereditarian concepts in modern science and society*. Baltimore: Johns Hopkins University Press.

Cock, Alan G., and Donald R. Forsdyke. 2008. *Treasure your exceptions: The science and life of William Bateson*. New York: Springer.

Correns, Carl. 1900. G. Mendel's Regel über das Verhalten der Nachkommenschaft der Rassenbastarde. *Berichte der Deutschen Botanischen Gesellschaft* 18:158–168.

Correns, Carl. 1901. *Bastarde zwischen Maisrassen, mit besonderer Berücksichtigung der Xenien*. Stuttgart: E. Nägele.

Coutagne, Georges. 1902. Recherches expérimentales sur l'hérédité chez les vers à soie. *Bulletin scientifique de la France et de la Belgique* 37:1–193.

Darwin, Charles. 1868/1998. *The variation of animals and plants under domestication, Vol. 2,* introd. Harriet Ritvo. Baltimore: Johns Hopkins University Press.

Dietrich, Michael R. 2008. Striking the hornet's nest: Richard Goldschmidt's rejection of the particulate gene. In *Rebels, mavericks, and heretics in biology,* ed. Oren Solomon Harman and Michael R. Dietrich, 120–125. New Haven: Yale University Press.

Fukuda, Kibun. 1990. *Waga kuni no sanshigyō wo sasaeta omona sanshikagaku to gijutsu [Major sericultural science and technology that supported our country's silk industry]*. Tokyo: Dainihon Sanshi Shimbunsha.

Godart, Gerard Rainier Clinton. 2009. Darwin in Japan: Evolutionary theory and Japan's modernity, (1820–1970). Ph.D. dissertation, University of Chicago, 2009.

Goldschmidt, Richard. 1912. Erblichkeitsstudien an Schmetterlingen 1. *Zeitschrift für Induktive Abstammungs- und Vererbungslehre* 7 (1): 1–62.

Goldschmidt, Richard. 1916. Experimental intersexuality and the sex-problem. *The American Naturalist* 50:705–718.

Goldschmidt, Richard. 1960. *In and out of the ivory tower: The autobiography of Richard B. Goldschmidt*. Seattle: University of Washington Press.

Goldschmidt, Richard, and K. Katsuki. 1927. Erblicher Gynandromorphismus und somatische Mosaikbildung bei Bombyx mori L. *Biologisches Zentralblat* 47:45–54.

Goldschmidt, Richard, and K. Katsuki. 1928. Cytologie des erblichen Gynandromorphismus von Bombyx mori L. *Biologisches Zentralblat* 48:685–699.

Goldschmidt, Richard, and K. Katsuki. 1931. Vierte mitteilung uber erblichen Gynandromorphismus und somatische Mosaikbildung bei Bombyx mori L. *Biologisches Zentralblat* 51:58–74.

Ishida, Magotaro. 1908. *Kashusan shiikuho [Method of rearing summer-fall silkworms]*. Tokyo: Shotoku Taishi Sankyo Hojunkai.

Ishida, Magotaro. 1913. *Shunkashusan shaken hosakuho [Policy method for testing spring, summer-fall silkworms]*. Tokyo: Dainihon Sangyo Gakkai.

Katsuki, Kitō. 1917. Kōzatsuichidai ni okeru mayu no katachi ni tuite [On the shapes of F1 hybrid cocoons]. *Dainihon sanshikaihō*; 26:710–717.

Kikkawa, Hideo. 1943. The maternal inheritance in the egg-color of *Bombyx Mori. Japanese Journal of Genetics* 19:125–126.

Kitamura, Chikayoshi, and Minoru Nozaki. 2004. *Norin Suisansho ni okeru sanshi shiken kenkyu no rekishi* [History of sericultural experiments of the Ministry of Agriculture, Forestry, and Fisheries]. Tsukuba-shi: Nogyo Seibutsu Shigen Kenkyujo.

Lock, R. H. 1906. Studies in plant breeding in the tropics. III. Experiments with maize. *Annals of the Royal Botanic Gardens, Peradeniya* 3 (2): 96–184.

Machida, Jirō. 1940. Toyama sensei no gyōseki [The achievements of Toyama Kametarō]. In *Toyama Kametarō kinenroku* [Commemoration of Toyama Kametarō], ed. Nagamusa Takeuchi, 1–8. Koayumura, Kanagawa Prefecture: Toyama Naoyasu.

Matsubara, Yoko. 2004. The reception of Mendelism in Japan, 1900–1920. *Historia Scientiarum* 13:232–240.

Matsui, Keiichi. 1967. Kenkyūjo ni okeru hakase no inshō [Impressions of the Professor at the Research Institute], *Nihon sanshigaku zasshi* [Journal of Sericultural Science of Japan] 36:461–462.

Morgan, Thomas Hunt. 1905. An alternative interpretation of the origin of gynandromorphous insects. *Science* 21:632–634.

Morgan, T. H. 1914. Mosaics and gynandromorphs in Drosophila. *Proceedings of the Society for Experimental Biology and Medicine* 11:171–172.

Morgan, T. H. 1916. The Eugster gynandromorph bees. *The American Naturalist* 50:39–45.

Morgan, T. H., A. H. Sturtevant, H. J. Muller, and C. B. Bridges. 1915. *The mechanism of Mendelian heredity*. New York: Henry Holt.

Morgan, Thomas Hunt, Calvin Blackman Bridges, and Alfred Henry Sturtevant. 1919. *The origin of gynandromorphs*. Washington: Carnegie Institution.

Moriwaki, Yasuko. 2010. Toyama Kametarō to meijiki no sanshigyō ni okeru kaiko no 'shurui kairyō [K. Toyama and silkworm breeding in Japan, from 1891 to 1913]. *Kagakushi kenkyū*; [Japanese Journal of History of Science] II 49:163–173.

Nakamura, Takafusa, and Kōnosuke Odaka. 2003. *The economic history of Japan, 1600–1990: Vol. 3, Economic history of Japan 1914–1955: A dual structure*. Oxford: Oxford University Press.

Olby, Robert. 1966. *The origins of Mendelism*. Chicago: University of Chicago Press.

Onaga, Lisa. 2010a. Tracing the Totsuzen in Tanaka's silkworms: An exploration of the establishment of *Bombyx mori* mutant stocks. *Preprints of the Max-Planck Institute for the History of Science* 393:109–117.

Onaga, Lisa. 2010b. Toyama Kametarō and Vernon Kellogg: Silkworm inheritance experiments in Japan, Siam, and the United States, 1900–1912. *Journal of the History of Biology* 43:215–264.

Onaga, Lisa. 2012. Silkworms, science, and nation: A sericultural history of genetics in modern Japan. Ph.D. dissertation, Cornell University.

Onaga, Lisa. 2013. Bombyx and bugs in Meiji Japan: Toward a multispecies history? *Scholar & Feminist*, 11(3). http://sfonline.barnard.edu/life-un-ltd-feminism-bioscience-race/bombyx-and-bugs-in-meiji-japan-toward-a-multispecies-history/. Accessed 3 July 2014.

Osanai-Futahashi, M., K.-i. Tatematsu, K. Yamamoto, J. Narukawa, K. Uchino, T. Kayukawa, T. Shinoda, Y. Banno, T. Tamura, and H. Sezutsu. 2012. Identification of the Bombyx red egg gene reveals involvement of a novel transporter family gene in late steps of the insect ommochrome biosynthesis pathway. *Journal of Biological Chemistry* 287:17706–17714.

Richmond, Marsha. 2007. The cell as the basis for heredity, development, and evolution: Richard Goldschmidt's program of physiological genetics. In *From embryology to evo-devo: A history of evolutionary development*, ed. Manfred D. Laublichler and Jane Maienschein, 169–211. Cambridge: MIT Press.

Sander, Klaus. 1994. True alternatives: Boveri and Morgan arguing about the origins of honeybee gynanders. *Roux's Archives of Developmental Biology* 203:175–177.

Sapp, Jan. 1987. *Beyond the gene: Cytoplasmic inheritance and the struggle for authority in genetics*. Oxford: Oxford University Press.

Standfuss, Max. 1900. *Synopsis of experiments in hybridization and temperature made with lepidoptera up to the end of 1898*. London: West, Newman.

Sturtevant, Alfred H. 1965. *A history of genetics*. New York: Harper & Row.

Takeuchi, Nagamasa, ed. 1940. *Toyama Kametarō kinenroku* [Commemoration of Toyama Kametarō]. Koayumura, Kanagawa Prefecture: Toyama Naoyasu.

Tanaka, Yoshimaro. 1916. Genetic studies of the silkworm. *The Journal of the College of Agriculture, Tohoku Imperial University, Sapporo, Japan* 7:129–255.

Tanaka, Yoshimaro. 1922. Sex-linkage in the silkworm. *Journal of Genetics* 12:163–178.

Tanaka, Yoshimaro. 1967. Nihon idengaku no yoake [Dawn of genetics in Japan]. In *Idengaku no Ayumi*: *Menderu idenhōsoku 100nen kinen* [Upon the steps of Mendel: In commemoration of the Mendel centennial anniversary], ed. Menderu idenhōsoku 100nen kinen shuppan iinkai [Publication Committee of Mendel Centennial Anniversary in Japan], 317–318. Tokyo: Shokabo Co.

Toyama, Kametarō. 1894. On the spermatogenesis of the silk-worm. *Bulletin of the College of Agriculture, Tokyo Imperial University* 2:125–157.

Toyama, Kametarō. 1895. Sanji no seishokuki [The sex organs of silkworm larvae]. *Dōbutsugaku zasshi* [Journal of Zoology] 8:339–344.

Toyama, Kametarō. 1899. Kaiko no shuruihikaku shiken [Silkworm variety comparison experiments]. *Fukushima-ken sangyō gakkō hōkuku* 3:1–70.

Toyama, Kametarō. 1900. Hyakunen izen ni okeru honpō kaiko no shurui [Varieties of silkworms in Japan 100 years ago]. *Dainihon sanshikaihō*; 9:1–9.

Toyama, Kametarō. 1901. Santainai no shin kikan [New organ inside silkworm body]. *Dainihon sanshikaihō*; [Bulletin of the Great Japan Silk Association] 108:1–5.

Toyama, Kametarō. 1902. Contributions to the study of silk-worms. I. On the embryology of the silk-worm. *Bulletin of the College of Agriculture, Tokyo Imperial University* 5:73–118.

Toyama, Kametarō. 1906a. Sanran no kouzoku narabi ni haishi no hattatsu [Structure of silkworm eggs and development of embryos]. *Nōgakukai kaihō*; [Bulletin of the Agricultural Association] 28:28–40.

Toyama, Kametarō. 1906b. Studies on the hybridology of insects, I. On some silkworm crosses, with special reference to Mendel's law of heredity. *Bulletin of the College of Agriculture, Tokyo Imperial University* 7:259–393.

Toyama, Kametarō. 1909a. Futatabi shurui kairyō to kaiko no iden ni tuite ichigensu [Another brief comment on breed improvement and silkworm genetics]. *Dainihon sanshikaihō*; 200:1–6.

Toyama, Kametarō. 1909b. Gutaiteki shurui kairyō hō" [Definite breed improvement method]. *Dainihon sanshikaihō*; 204:3–6.

Toyama, Kametaro. 1909c. Jikken shinkaron to nōgyō tono kankei [The relationship between theory of experimental evolution and agriculture]. In *Kōenshū*; [Collected lectures]. Niigata: Niigata-ken nōyōkai.

Toyama, Kametarō. 1912. On the varying dominance of certain white breeds of the silkworm, Bombyx mori, L. *Zeitschrift für Induktive Abstammungs- und Vererbungslehre* 7:252–88.

Toyama, Kametarō. 1913. Maternal inheritance and Mendelism (first contribution). *Journal of Genetics* 2:351–405.

Toyama, Kametarō. 1914. *Sanshu no hanashi [Story of silkworm seeds]*. Tokyo: Dainihon Sangyō Gakkai.

Toyama, Kametarō. n.d. Ikimono no seishitu wo irekaeru hō [Method to exchange the nature of living things]. *Shōnen* [The Boy] 76:138–46.

Toyama, Kametarō, and Shirō Murakoshi. 1898. Furanki saisei shiken [Silkworm egg incubation experiment]. *Fukushima-ken sangyō gakkō hōkuku* [Bulletin of the Fukushima Prefecture Sericulture School] 1:1–11.

Toyama, Kametarō, and Shigetane Ishiwatari. 1896. *Jikken santai kaibō: Fu sanbyo narabini kenbikyo shiyoho [Experiments in silkworm dissection: How to use microscope with silkworm disease, supplement]*. Tokyo: Fuzanbo.

Tschermak, Erich. 1900. Über künstliche Kreuzung bei Pisum sativum. *Berichte der Deutschen Botanischen Gesellschaft* 18:232–239.

Tsujita, Mitsuo. 1961. Maternal effect of + lem gene on pterine reductase of Bombyx mori. *Idengaku Zasshi* [Japanese Journal of Genetics] 36:337–346.

Vilmorin, Philippe de, ed. 1913. *IVe conférence internationale de génétique, Paris, 1911: Comptes rendus et rapports*. Paris: Masson et Cie, 1913.
Vries, Hugo de. 1900. Sur la loi de disjonction des hybrides. *Comptes rendus de l'Académie des Sciences Paris* 130:845–847.
Yokoyama, Tadao. 1959. *Silkworm genetics illustrated.* Tokyo: Japan Society for the Promotion of Science.

Lisa Onaga is an assistant professor in the History Programme at Nanyang Technological University, Singapore. She is finishing a book that examines the intertwined histories of Japan's pursuits of the ideal silk cocoon and the mastering of heredity. Her research interests include agricultural craft, biological experimentation, and engineering in modern Japan.

Chapter 21
Genetics and "Breeding as a Science": Kihara Hitoshi and the Development of Genetics in Japan in the First Half of the Twentieth Century

Kaori Iida

Introduction[1]

Japanese genetics as a discipline developed rapidly in the early twentieth century after Japanese breeders and biologists learned of Mendelian principles. As Matsubara Yoko has pointed out in 2000, Mendelism was quickly disseminated and began to be examined in the agricultural context of Japan, especially under the urgent need to develop agricultural industry for the expanding nation.[2] However, there are no detailed studies about how Japanese genetics subsequently developed in relation to breeding studies. In this chapter, I show that the development of genetics in Japan was deeply embedded in the agricultural context, and academic (university) geneticists did not clearly demarcate their area of study from breeding studies for a long time until around 1950; this close connection reflected a socioeconomic context that valued practical science.

Barbara Kimmelman's study of the early history of genetics in the USA has argued that genetics arose in agricultural institutions; her analysis is suggestive for the interpretation of Japanese genetics.[3] According to her, American breeders became interested in Mendel's results very early, believing that they might have relevance to hybridization work. By 1915, researchers working at agricultural institutions transformed into geneticists and created the foundation for a new discipline of genetics.

[1] The Japanese convention of placing surname first, followed by the given name, has been adopted for all Japanese names in the main text. The reference list uses English-language order for all Japanese names.

[2] Matsubara (2000, 2004).

[3] Kimmelman (1987, 2006) and Paul and Kimmelman (1988).

K. Iida (✉)
Department of Evolutionary Studies of Biosystems, The Graduate University for Advanced Studies SOKENDAI Hayama, Kanagawa, 240-0193 Japan
e-mail: iida_kaori@soken.ac.jp

D. Phillips, S. Kingsland (eds.), *New Perspectives on the History of Life Sciences and Agriculture*, Archimedes 40, DOI 10.1007/978-3-319-12185-7_21

During the early period, genetics had earned a place "as both an applied and a basic science," adapting to the aim of those institutions.[4]

Similarly in Japan, genetics developed in the agricultural context and was promoted as both basic and applied science. Genetics as a discipline was shaped by its agricultural setting at least until the early 1950s. The Genetics Society of Japan emerged in 1920 through the reformation of the Breeding Society of Japan (*nihon ikushu gakkai,* established in 1915). It was 30 years later when a new Japanese Society of Breeding (with the same Japanese name, *nihon ikushu gakkai*) again branched off from the Genetics Society. Though this divergence in 1951 by no means signifies the end of the agricultural association with genetics, the event marks the emergence of newer genetic research that no longer had agricultural relevance. In turn, it also implies that such divergence was felt unnecessary until around 1950.

In the academic environment in the first half of the twentieth century, Japanese geneticists often chose agricultural organisms as their research materials under strong pressure to conduct applied science, and in turn capitalized on practical implications of genetics to attract public support. This appeal to practical applications was not just rhetorical and researchers often contributed directly to breeding studies. As a result, geneticists were able to increase the importance of their discipline in relation to agriculture. At the same time, geneticists working at universities had the desire to conduct fundamental work as academic scientists (they conceptually distinguished between "basic" and "applied" science). Consequently, they created scientific research that could flexibly be considered as "applied" or "basic." I discuss how this type of research was influential in shaping Japanese approaches and concepts in genetics, in particular their preference for physiological genetics and for a multidisciplinary approach to understanding the whole organism.

I discuss this early history of genetics in Japan as seen through the career of Kihara Hitoshi (1893–1986), a plant cytogeneticist who was a leader in the development of the discipline of genetics in Japan.[5] A graduate of Hokkaido Imperial University, where he first became interested in wheat genetics, Kihara went to Germany in 1925 and studied under Carl Correns. He returned to Japan in 1927 as professor at the Faculty of Agriculture of Kyoto Imperial University, where he headed up a Laboratory of Experimental Genetics that became famous as Japan's "Mecca" of genetics. In 1942, he established the Kihara Institute for Biological Research in Kyoto, where a wide variety of problems in genetics, physiology, and cytology were studied in many species of plants (mostly crops). He also held important positions including president of the Genetics Society of Japan (1944, 1949–1952) and director of the National Institute of Genetics (NIG) in Japan (1955–1969), and perhaps it is fair to say that he was internationally the best-known Japanese geneticist in his generation.[6] Thus Kihara is an exceptionally good lens through which to view the

[4] Kimmelman (2006, p. 273). Also Paul and Kimmelman (1988, p. 285).

[5] Crow (1994) and Iida (2010).

[6] According to the American geneticist James Crow, the best-known Japanese geneticists in the 1950s were Kihara and a younger geneticist Kikkawa Hideo (personal communication 29 July 2009).

early history of Japanese genetics. Mainly through his career, I analyze how genetics in Japan developed and how their approach to genetic studies was shaped, with particular attention to agriculture and the wartime context of Japan in the first half of the twentieth century.

In the following, I first illustrate how some of the important institutions for the discipline of genetics emerged and developed in close connection with agriculture. Kihara's own laboratory in Kyoto, the first government-funded genetics laboratory in Japan, was established under a new department where biological research was to be conducted to solve agricultural problems. Since the 1930s, at Kihara's laboratory and at the Institute for Biological Research, the practical significance of his group's work was made explicit, and consequently Kihara was able to expand their projects. There they created a type of research in which there was no boundary between basic and applied work. I argue that as geneticists were committed to work with applied goals, it inevitably led the researchers to take a multidisciplinary approach beyond genetics to understand the full range of biological processes of the organism. The approach was very similar to what the Russian geneticist Nikolai Vavilov had proposed, "breeding as a science." Kihara was very interested in and influenced by Vavilov's work; the Japanese approach to genetic studies was similarly placed within the larger project of "breeding as a science." Japanese geneticists were generally also interested in the approach of German geneticist Richard Goldschmidt, who argued that the future of genetics lay in "physiological genetics" or the pursuit of genetics alongside the study of the physiology and development of organisms. I end with a brief discussion of how such a holistic approach to organisms remained at the newly established NIG in the 1950s.

Rise of Genetics in the Agricultural Context

Mendelism was introduced into Japan soon after the rediscovery of Mendel's work in 1900, and began to be examined at agricultural institutions. The earliest record to associate the Japanese with Mendelism is known to be in 1901, when Hoshino Yūzō, an agronomist at the Sapporo Agricultural College in Hokkaido, mentioned in his paper on corn the rediscovery papers by Hugo de Vries and Carl Correns. In the following year, Hoshino introduced Mendelian laws, based on Correns's paper, to readers of a Japanese agricultural journal.[7] More publications on Mendelism subsequently appeared, including two well-known writings in 1906. One was a section in a book, *Phylogeny of Plants* (*Shokubutsu keitō gaku*), written by Ikeno Seiichirō, who was associate professor at the Agricultural College of Tokyo Imperial University. It is said that this book introduced Mendelian concepts widely in Japan. The

[7] Moriwaki (2010), "Hoshino Yūzo no kisenia kenkyū to 1902-nen no menderizumu shōkai" [Xenia study by Hoshino Yūzo and his introduction of Mendelism in 1902], presented on 26 May 2012 at the 59th annual meeting of the History of Science Society of Japan, Tsu, Mie, Japan; Noguchi (1978, p. 244). Hoshino's 1902 paper: Hoshino (1902).

other was an article of Toyama Kametarō who was also associate professor at the same college (see Chap. 20 by Lisa Onaga for discussion of Toyama's career and legacy). This reported new Mendelian data from his silkworm study conducted in Siam and was one of the first Mendelian results shown in animals in the international scientific community.[8] At experimental stations, too, breeders soon began examining Mendelism. In 1904, researchers at one of the experiment stations initiated large-scale breeding studies of crops, including Mendelian studies of various characters in rice, and created a new rice hybrid by 1909.[9]

This early interest in Mendelism (as part of broader studies of inheritance) by both scholars and practitioners was partly due to the long practice of breeding that had existed since the pre-Meiji era and to the pressing need for the Meiji government to develop agricultural industry. Around 1900 when Mendelism was rediscovered, for example, Japan was under pressure to improve the quality of silk. Silk, which had been an important industry for a long time, was considered to be a particularly important export product to obtain foreign currency because by rearing silkworms there was no need to import raw materials (such as in the case of cotton). Toyama examined various characters of silkworms in his crossing experiments and reported both Mendelian and non-Mendelian patterns of inheritance in his paper of 1906. Soon he began promoting the use of hybrids (from a cross between two different pure strains) in the Japanese silk industry, which drastically changed the silkworm business.[10]

Agriculture had additional importance for the nation's imperial expansion. In 1895, as a result of the Sino-Japanese War (1894–1895), Taiwan was ceded to Japan from China. After the victory in the Russo-Japanese War (1904–1905), Japan in 1905 declared Korea to be its protectorate and officially annexed it in 1910. As the Japanese Empire expanded, the Japanese government also established experimental stations in colonies (Taiwan, Korea, Sakhalin, and Manchuria). To train people who would manage agricultural projects in the expanding nation, imperial universities gained importance. Japan's first department related to colonization strategy (the Department of the Study of Colonization and Agricultural Administration) was established in the former Sapporo Agricultural College in 1907, when this college was integrated into one of the imperial universities (Tohoku Imperial University).[11] In fact, many graduates of this college led breeding projects in the colonies (especially in Taiwan and Manchuria).[12]

The relevance of agricultural studies to Mendelism led this college in Hokkaido to become one of the centers of genetic studies in Japan. Following the first known reference to Mendelism by Hoshino, the first known cases in Japan of a journal club

[8] Matsubara (2000). On Toyama, see Onaga (2010); see also her Chap. 20 in this volume.

[9] Fujihara (2012, pp. 74–75).

[10] Toyama (1906), Moriwaki (2010), and Matsubara (2000).

[11] The college became independent again as Hokkaido Imperial University in 1918. About the department, Inoue (2006).

[12] Tanaka and Imai (2006, pp. 108–111) and Yamamoto (2011). For case studies of such breeders, see Fujihara (2012).

to read genetics journals and of a course specialized in genetics also occurred in this college in the early 1910s.[13] Kihara's association with this college was enormously important for his career. During his college years (1915–1918), he majored in plant physiology but was also exposed to genetics. In his graduation year, he encountered his lifelong research material, wheat, because this college had a good collection of wheat varieties. Moreover, being a graduate of the college, he would later expand his projects over the large map of the empire using the college network.

By the time of Kihara's graduation in 1918, two institutions that were critical for the development of the discipline of genetics in Japan were established. One was a new society related to genetic studies. In 1915, seven scholars (including Toyama), all working at agricultural institutions, founded the Breeding Society of Japan, the forerunner of the Genetics Society of Japan. At the founding meeting of the Breeding Society held in Tokyo, scholars gave ten lectures on the heredity or breeding of organisms familiar to the Japanese people: morning glories, *medaka* (small fish common in rice paddies), white snake (considered as good luck), rice, wheat, chestnuts, *shiso* (perilla, used as herb), bellflower, and garden peas. In the following year, the society published a journal, which discussed similar subjects.[14] The journal published only two issues, in 1916 and in 1918, and with Toyama's death in 1918 and the move of another central member to Taiwan the society became less active. By this time, however, about 200 people (many from agricultural experiment stations) were registered as members of the society. This membership then became the basis for the Genetics Society of Japan.

The Breeding Society was reformed into the Genetics Society of Japan in 1920. While the main office was placed in the headquarters of the agricultural experimental station in Tokyo, the new society was not just about breeding studies. One of the reasons for the reformation of the society was, according to the silkworm geneticist Tanaka Yoshimaro (1884–1972), to accommodate recent discoveries in genetics beyond agricultural breeding studies, particularly the fruit-fly genetics led by Thomas Hunt Morgan (1866–1945) in the USA (using silkworm, Tanaka had been conducting genetic analysis similar to the studies done in fly)[15]. In the reformation, scholars from nonagricultural institutions also joined as board members. One was Fujii Kenjirō (1866–1952), a German-trained plant cytologist, who established the first genetics laboratory in Japan in 1918.[16]

The development of genetics and breeding studies was facilitated by and in turn further encouraged the development of cytology in Japan. Since 1905, in the laboratory of "morphological studies" at the Department of Botany of Tokyo Imperial University, Fujii had been training students in cytology. For example, the cytologist and a former student of Fujii, Tahara Masato (1884–1969), published in 1915 one of the earliest reports of polyploidy (i.e., having more than the normal two sets of chromosomes). He identified the chromosome numbers in different varieties of the

[13] See for example, Tanaka (1961a).

[14] See *Nihon ikushu gakkai kaihō* (*Japanese Journal of the Breeding Society*) 1, no. 1 (1916).

[15] Tanaka (1916).

[16] Tanaka (1961b).

chrysanthemum family and discovered that the numbers were multiples of nine (i.e., 18, 36, 54, 72, 90).[17] Others, also trained under Fujii, determined for the first time the correct numbers of chromosomes in rice (1910) and in wheat (1918).[18] Thus cytogenetics was developing fast in Japan particularly under Fujii's influence.

Fujii played a central role in establishing in 1918 Japan's first Genetics Laboratory (*idengaku kōza*) under the Department of Botany of Tokyo Imperial University. The importance of balancing basic and applied goals for sustaining the growth of genetics is underscored by comparing this laboratory to Kihara's own genetics laboratory, established nearly a decade later. While Kihara's laboratory would be placed under the Faculty of Agriculture with the aim to conduct research to solve agricultural problems, Fujii's laboratory was in the Department of Botany of the College of Science, and Fujii's aim in establishing the laboratory was to develop "genetics based on cell biology" as a "purely independent field."[19] Clearly, Fujii was hoping to conduct more basic research in the new laboratory.[20] One consequence was that unlike Kihara's laboratory, the establishment of Fujii's laboratory was made through a contribution from wealthy entrepreneurs, not through the government's support. These entrepreneurs were "commended" in an academic journal for their "praiseworthy act" of supporting "basic" science especially when it was rare that men of wealth would make contributions to "pure academic work that could not easily lead into direct benefits."[21] Although Fujii explained, in an official letter to the university requesting approval of the establishment of the new laboratory, that genetics was an "important" field with two potential applications, agriculture and eugenics, the "basic" nature of Fujii's laboratory might have made the establishment and support of this laboratory low priority for the government.[22] Because the funding for the laboratory came exclusively from an outside source, the imperial university did not allocate official faculty positions for the new laboratory. Fujii, who was the head of the existing laboratory for morphological studies, concurrently took a position as the head of the new genetics laboratory without abandoning the morphological laboratory.[23]

While conducting cytological analysis under the microscope, however, Fujii tried to maintain a connection with agricultural goals. Fujii recommended that students choose a "practically important plant," and they investigated cells of rice, corn, mulberry, chrysanthemum, and so forth.[24] The use of practical materials worked advantageously for geneticists in Japan because it could be used to demonstrate that their research was needed for the progress of the nation through betterment

[17] Tahara (1914, 1915).

[18] For rice, Kuwada (1910). For wheat, Sakamura (1918).

[19] Fujii (1918). Also see Fujii's letter, reprinted in: Nihon kagakushi gakkai (1965, pp. 277–278).

[20] See for example, Fujii (1920).

[21] Fujii (1918).

[22] The letter is in Nihon kagakushi gakkai (1965).

[23] Shinotō (1967, pp. 382–384).

[24] Tahara (1941). *N.I.G. danwashitsu* was a newsletter that came with the *Japanese Journal of Genetics* (or *idengaku zasshi*); bound with the journal in some libraries in Japan.

of biological resources. This practical benefit was, however, not just rhetoric. The Ministry of Agriculture and Commerce was promoting the improvement of silkworm and crops, and more technical staff were needed at experimental stations of each prefecture. Therefore, students must have been encouraged to be trained with agricultural organisms to be able to fit into the job market of the time.[25]

With this overall demand for agricultural research, it may not be a coincidence that the first *government-funded* genetics laboratory in Japan—although the second genetics laboratory in the country—was established under the Faculty of Agriculture. Kihara started directing the newly established Laboratory of Experimental Genetics at the Faculty of Agriculture of Kyoto Imperial University in 1927. This university founded departments for physics, chemistry, and mathematics when it was established in 1897 as the second national university after the first one in Tokyo. Only after 22 years was the Department of Biology established (1919), followed by the Faculty of Agriculture (1923). The university acquired experimental forests in Taiwan, Korea, and Sakhalin over the years as the nation expanded. Because of this expansion, it became necessary to establish the new Faculty of Agriculture.[26] One of the six departments under the Faculty was the Department of Biology for Agriculture and Forestry, which conducted biological research closely linked to agricultural problems. An author who had written for the compiled "70-year history of the Faculty of Agriculture of Kyoto University" notes that the establishment of this new department was viewed as "unorthodox" by agriculturalists.[27] It is said that Kihara's undergraduate adviser, Kōriba Kan (1882–1957) who was a plant physiologist and had just moved from Hokkaido to Kyoto, played a central role in designing the new department by modeling it on the departmental structure in Hokkaido (where basic biology had been conducted within the agricultural department). As a result of Kōriba's integration of biological research in the agricultural department, Kihara and another Hokkaido graduate were able to secure positions under the new department in Kyoto.

Under the Department of Biology for Agriculture and Forestry, three laboratories (*kōza*) were established initially: plant pathology, entomology, and experimental genetics. Kihara, who had just submitted his doctoral dissertation on cytogenetic studies of wheat, was promoted to associate professor (*jokyōju*) at the Faculty of Agriculture and was named as a future director of the new Laboratory of Experimental Genetics in 1924. In 1925, the Ministry of Education sent Kihara to Germany. This was because the Japanese government had made it mandatory for all professors to have the experience of studying in Europe or the USA. Kihara went to Germany to work with Carl Correns for 2 years. Upon returning to Japan in 1927 as professor, Kihara started running the new laboratory. In the same year, Fujii retired. Though his disciples continued running the laboratory, the university did not allocate official faculty positions until 1951.[28] Therefore, within academic institutions

[25] Abe (1939, p. 27).

[26] Kyoto Imperial University (1943, p. 1069).

[27] Kyoto daigaku nōgakubu 70-nenshi henshū iinkai (1993, p. 256).

[28] Shinotō (1967, p. 384).

Kihara's laboratory remained for a long time the only independent and government-supported genetics laboratory in Japan. This laboratory would later be known as Japan's "Mecca" of genetics.[29]

After returning to Japan in 1927, Kihara rapidly developed his research program and he became a recognized authority in cytogenetics, particularly for his superb analysis of polyploidy in wheat. As will be discussed in the next section, the devoted assistance of his collaborator, Flora Lilienfeld, ensured that he was able to communicate internationally. The imperial context was also critical for the development of his group. Kihara's laboratory was expected to contribute to the state's agricultural needs through biological research since its inception. In wartime, Kihara was isolated from the international community but was able to expand his group by successfully integrating genetics into agricultural projects. The role of genetic research within the agricultural department might have been seen with skepticism by agriculturists in the mid-1920s; however, by 1940, various agricultural industries asked Kihara's group for advice and collaboration for their breeding projects.

Evolution of Kihara's Group through Basic and Applied Studies of Polyploidy

In the new laboratory in Kyoto, Kihara conducted cytogenetic studies and published a series of important articles in wheat studies and plant genetics. In 1929, Flora A. Lilienfeld (1886–1977), Kihara's most important collaborator in his lifetime, joined his laboratory.[30] Lilienfeld had been a student of Correns from Poland and had shared an office with Kihara at the Kaiser Wilhelm Institute. She came from a well-educated Jewish family in Lvov and had studied botany at Lvov University, obtaining her Ph.D. in 1914. After leaving Correns's laboratory, she had hoped to find a permanent position in Poland. However, she was unable to do so, and in 1929 accepted Kihara's invitation to come to Japan. Kihara was trying to train students in presenting and discussing research results in German and was hoping to get some help from foreign scholars. Fortunately, Lilienfeld agreed to move all the way to Japan. She was hired as "lecturer" (a rank between *joshu,* "assistant," and associate professor) but functioned more as "associate professor" (*jokyōju*) who would assist the professor in both research and education.[31]

As soon as Lilienfeld arrived, they started to formalize a method that Kihara had developed and gave it a name: *Genomanalyse* or genome analysis.[32] The purpose

---

[29] Genetic research was conducted and geneticists were trained in many places in Japan at that time in laboratories under older disciplines. A notable geneticist was Komai Taku (1886–1972) at the Department of Zoology under the Faculty of Science of Kyoto Imperial University. Komai was at Morgan's laboratory for 2 years, returned to Japan with fruit flies in 1925 and then started the first-fly group in Japan. On the reference to "Mecca," see Kimura (1986, p. 726).

[30] Majewski (1989). Also see Kihara (1951, pp. 74, 130).

[31] Kyoto daigaku nōgakubu 70-nenshi henshū iinkai (1993, p. 259).

[32] Kihara (1951, p. 134).

of this analysis was to identify types of chromosome sets. It was known that wheat varieties were polyploids with the base chromosome number, seven (i.e., 14, 28, 42). Based on his previous study, Kihara concluded that seven chromosomes should be treated as one unit, and referred to this unit as "Genom." Seeking to analyze the chromosomal composition of the entire genus, Kihara along with members of his laboratory identified what kind of a genome each polyploid plant had in all species and varieties of wheat (genus *Triticum*) and a closely related genus (*Aegilops*) over many years. According to the American botanist and evolutionary biologist George Ledyard Stebbins (1906–2000), "In the history of polyploidy, Kihara was the first person to analyze a whole genus," and his polyploid analysis was quickly followed by other botanists in the 1930s.[33]

Lilienfeld's role in this enterprise as a researcher and translator was vitally important to its success. In the year following her arrival in Japan, Kihara published the first paper on "genome analysis" in the series.[34] She rewrote Kihara's German draft into more formal-style German, which even Kihara later had some difficulty understanding.[35] (This paper was published as Kihara's single-authored article.) Among a total of ten articles of the series, she coauthored three articles, the fourth (1932) to the sixth (1935), until she left Kihara's laboratory for the USA just before the outbreak of war, and single-authored the final tenth article (which was a review of the past articles of the series) in 1951 after she returned to Japan. Lilienfeld not only helped edit Kihara's (and often his laboratory members') publications but also wrote letters for Kihara's international correspondence in German or English.[36] When the official network among wheat scholars (called the Wheat Information Service) was later established in 1953, Kihara played a central role in its organization, which suggests that he had successfully become an important part of an unofficial network of wheat exchange in the prewar period. This network was, however, mediated effectively by Lilienfeld. She essentially connected Kihara and his laboratory with the research community outside Japan.

One of the goals of their comprehensive genome analysis was to understand the evolution of bread wheat and to identify a close ancestral species of the wheat. (He and the American team of Edgar McFadden and Ernest Sears would independently identify the species in 1944.[37]) Thus polyploidy was an important tool for investigating evolution and cytogenetics of plant species, and Japanese work contributed significantly to the development of such fields.

However, polyploidy was also important for applied work. Expectations about the use of polyploidy for plant breeding were high because it was at the time "the most powerful tool yet available to a geneticist for molding living matter into new shapes."[38] During the war years in Japan (especially from the Manchurian Incident

[33] Stebbins (1980, p. 145).

[34] Kihara (1930).

[35] Kihara (1951, p. 147).

[36] Lilienfeld to Brandes, 26 February 1940, reprinted in *Zihō* 1 (1941): 13. A journal published by Kihara's group; currently held at the library of the Kihara Institute for Biological Research, Yokohama, Japan.

[37] Kihara (1944) and McFadden and Sears (1944).

[38] Dobzhansky (1937, p. 207).

in 1931 to the end of the Second World War in 1945), the need to mobilize science for the war effort further encouraged an emphasis on applied research, and geneticists came under stronger pressure to present the practical value of their research. Polyploidy thus became a critical tool for Kihara's survival and success as a scientist during wartime.

Lilienfeld left Japan in 1936 before Japan entered a full-scale war against China. Starting in 1937, Japanese scientists became increasingly isolated from the rest of the world. In isolation, Kihara would develop a research center based on the connections with former students, friends, and college alumni and on accumulated experiences in cytogenetics in his laboratory. There he would expand practical projects aiming at the production of new polyploid crops in order to adapt to the wartime climate. At the same time, their approach to genetics would diverge even more from the study of nuclear chromosomes to the whole cell and whole organisms.

Just at that time, a new method was discovered that made the production of polyploidy easier and more reliable. In 1937, two American researchers, Albert Blakeslee and Amos Avery, reported that they had succeeded in inducing polyploidy using a chemical inducer, colchicine.[39] Colchicine brought a significant change in plant breeding and horticulture. As one cytologist described, beginning in 1938, a "Colchicine fad" took hold among plant breeders.[40] Kihara's group was not an exception.

Using polyploidy and colchicine, Kihara shifted his research direction toward more practical work beyond wheat. Kihara expanded his research projects, materials, and the geographical range of research fields. He formed ties with industry and developed new research programs on plants that were valuable *and* polyploid (such as sugar beet, sugarcane, barley, and cotton). One of the products of this expansion of research was the creation of the seedless watermelon. Kihara's group created the fruit by converting the normal diploid (having two sets of chromosomes) into a triploid (having three sets) because there had been several examples of triploid plants being seedless or sterile. (After the war they would successfully harvest seedless watermelon.)

In the middle of the war in 1942, with financial support from industries, Kihara established the Kihara Institute for Biological Research in Kyoto to accommodate the new projects.[41] Kihara commented at the opening ceremony that people should appreciate that there was also a "science war" in agriculture, not just in engineering. The research at the Institute was their way to fight a quieter yet important "science war," which would contribute to making "the best use of Greater East Asian (*daitōa*) resources."[42]

[39] Blakeslee and Avery (1937) and Curry (2010).

[40] Eigsti (1957, p. 273).

[41] About Kihara Institute, also see Iida (2010, pp. 544–546); in that article, all three references to *sweet potato* on pp. 544–546 are errors and should be corrected to *sugarcane*.

[42] Kihara (1942a, p. 104). *Seiken zihō* was a journal published by Kihara's group, currently held at the library of the Kihara Institute for Biological Research, Yokohama, Japan.

Since the beginning of the Pacific War in December 1941, Japan quickly occupied a vast area, including Burma, Malay, Indonesia (Dutch), the Philippines, some Pacific islands, and part of New Guinea. The Kihara Institute's research zone was in fact as vast as the Japanese Empire: from Hokkaido and Manchuria (sugar beets and barley) to Saipan (sugarcane and watermelon). The Institute in Kyoto was headquarters for all projects. Kihara envisioned that his Institute was a type of "dojo" (school for martial arts) to train biologist warriors who could eventually help fight the "sacred war" in Greater East Asia.[43] He had already been sending his students to experimental stations and industry in various places in the empire such as Manchuria, Taiwan, and Saipan. In isolation during wartime, it became necessary for Japan to train all technicians and scholars within its own geographical boundary, and Kihara began to train technical staff for industry. For example, five employees at the fiber industry Tōyō bōseki (Toyo cotton spinning company) entered the Kihara Institute in 1942 for training in theory and practice.[44] These botanist warriors were to be sent to the South for cotton agriculture.

A Holistic Approach to Organisms, Cells, and Genes

Plant breeding required a multidisciplinary approach to achieve better growth, flowering, and fruition of plants. I argue that because of the applied interests of Japanese scientists, they pursued a holistic approach to biological study. In Kihara's work, and elsewhere in Japanese biology, there was no clear distinction between "basic" and "applied" science in actual research. Fundamental research was pursued with applications in mind, in particular the need to control and improve species that had economic value. Such an approach included genetics but also a variety of other disciplinary approaches that had to be deployed to understand the full range of biological processes within the organism.

The research program at the Kihara Institute was similar to what the Russian geneticist Nikolai I. Vavilov (1887–1943) had proposed in advocating the development of "breeding as a science." Vavilov, director of the Institute of Applied Botany and Plant Breeding in Leningrad, had visited Kihara in 1929. Earlier during his European sojourn, Kihara had been particularly excited to visit Vavilov's Institute, considered to be the world's leading research institute on crops and the world's first seed bank. During Vavilov's subsequent visit to Japan, he lectured on the origin of cultivated plants at Kyoto Imperial University. Kihara took great interest in Vavilov's theory of the origin of cultivated plants and was in general sympathetic to his approach to biology and to the science of breeding.

According to Vavilov, breeding was a complex "scientific system" that borrowed various methods from other "fundamental sciences" for the production of a new variety. He wrote: "In controlling heredity it relies wholly on the findings

[43] Kihara (1942b, p. 108).

[44] Yamashita (1942).

of genetics, cytology, and embryology, while in the study of breeding technique it depends upon the biology of flowering, physiology, chemistry, technology, phytopathology, entomology."[45] These all contributed necessary knowledge for plant breeding. Vavilov had also noted the centrality of including the environment as part of this scientific enterprise. As he wrote, "The question of environment and the interaction of the organism and the environment is one of the most important branches of breeding." The environment represented by the various climate zones of the empire became one of the biggest concerns for plant breeders of Japan. To develop crops adapted to various climates of the expanding empire, it was inevitable for the researchers to examine interactions between organisms and the environment. As a result, Japanese breeders were compelled to extend their approach in a multidisciplinary direction, particularly physiological and developmental studies that went beyond chromosome-oriented cytogenetics. Studies at the Kihara Institute, for example, involved testing various conditions affecting plant growth, including low and high temperature, day length, plant hormones, chemicals in the soil, fertilizers, and insecticides, as well as the genetic background of plants.

As applied goals encouraged multidisciplinary approaches, Japanese geneticists began voicing the importance of nonchromosomal factors (i.e., the cytoplasm and internal and external environmental factors) for genetic studies, and of aiming their research direction specifically toward "physiological genetics" as was proposed by the German geneticist Richard B. Goldschmidt (1878–1958). Goldschmidt had visited Japan twice (in 1914 and again in 1924 for 2 years) and had close ties with Japanese biologists. In 1938, he again attracted much attention in Japan because that year he published a new book, *Physiological Genetics*.[46] Goldschmidt's goal was to connect genes and an organism's development by elucidating the gene's biochemical and physiological actions.

Kihara began advocating physiological genetics and also emphasizing his study of the cytoplasm (instead of chromosomes). Others also referred to Goldschmidt and suggested that genetics in the future should be developed into physiological genetics.[47] Scott Gilbert has described Goldschmidt as a "leader and prophet" of future genetics.[48] The Japanese took Goldschmidt's prophecy seriously.

Some genetic studies were indeed heading toward the direction of physiological genetics in and out of Japan at the time and began revealing functions of genes.[49] In 1941, the Japanese geneticist Kikkawa Hideo (1908–1990), who was an employee at a national sericulture experiment station, and the American team of George Beadle and Edward Tatum published independently on the chemical process of eye pigment formation in silkworm and fruit fly, respectively.[50] They proposed that in a series of biochemical steps (in the production of pigment precursors), each step

[45] Vavilov (1949/1950, p. 8).

[46] Goldschmidt (1938). See Richmond (2007).

[47] For example, Tanaka (1942).

[48] Gilbert (1988, p. 340).

[49] Earlier studies include Caspari (1933), Kühn et al. (1935), and Beadle and Ephrussi (1936). For historical studies see Rheinberger (2010, Chap. 6) and Sapp (1987, Chap. 5).

[50] Beadle and Tatum (1941) and Kikkawa (1941).

was controlled by a different gene, mediated by an enzyme. For example, "*v*+hormone" (kynurenine) in flies was a product of an oxidation of tryptophan, a process catalyzed by an enzyme controlled by "*v*+gene." This type of study was what Goldschmidt considered to be "physiological genetics" because it would be an important step in connecting genes and development.

By 1941 when their paper appeared in print, Beadle and Tatum had already switched their research material from flies to a completely different kind or organism, bread mold (*Neurospora*). According to Jonathan Harwood, the American team was as a result no longer studying "even *part* of a developmental process in a higher organism," and their research was only about the chemistry of gene action.[51] This switch would lead them to the famed one-gene–one-enzyme hypothesis.

In contrast, Kikkawa's work appeared to be much more deeply committed to the organism's biology and to a multidisciplinary approach. Kikkawa wrote that the problem of tryptophan metabolism went beyond genetics and required knowledge in other disciplines such as biochemistry, photochemistry, "protoplasma study," physiology, and developmental biology.[52] For example, photochemistry was essential for him because he was interested in the physiological functions of pigments. Genes and enzymes were necessary to produce pigments but why did insects need pigments, he wondered. He was conducting experiments to determine the spectrum of light absorbed by a pigment, hoping to understand the relation between insect biology (especially phototaxis) and pigments. Kikkawa hoped to cover everything about pigments, from the genes to the biological functions in the insect, because he was interested in the organism more than the genes.

This type of biological research was remarkably similar to the approach to "breeding as a science" that we observed in Kihara's laboratory. With the general emphasis on applied aspects in Japanese science, many geneticists engaged in multidisciplinary work. As the overall approach became multidisciplinary and sought full understanding of the organism's biological processes, physiological genetics was preferred as the approach to genetic research, and many of the leading geneticists in Japan believed that Goldschmidt had correctly identified the way in which future genetics had to develop. In addition, many including Kihara began shifting the emphasis of their interest from chromosomes to the cytoplasm. Kikkawa would also develop a new model of gene expression immediately after the war and incorporated the cytoplasm as an important factor in the model.[53]

"Breeding as a Science" in the Postwar Years

The fine balance between basic and applied science characterized the development of genetics through wartime and became the basis for further development of genetics in the postwar years. After Japan's defeat in 1945, Japanese geneticists

[51] Harwood (1993, pp. 92–93).

[52] Kikkawa (1943, p. 324).

[53] Ishidate (1980) and Kikkawa (1947).

immediately started rebuilding the research environment. One of their biggest accomplishments was the establishment of the NIG in Mishima in 1949.[54] Most research projects at the newly established Institute had relevance to agriculture. Among the initial eight laboratories at the NIG, two were run by the former members of the Kihara Institute, who worked on agricultural crops. Another principal investigator, Sakai Kan-ichi (1910–1999), conducted theoretical/quantitative study of genetics for the purpose of plant breeding of crops including rice, red pepper, barley, plants in the genus *Abelmoschus* (such as okra), and eggplant.[55] In the early 1950s, NIG researchers often used agricultural organisms for their genetic study: silkworm, virus infecting silkworms, various crops, and *Aspergillus* (a type of fungus used in the fermentation process of soy sauce, rice wine, and soybean paste).

In addition, the NIG began collaborative applied projects with agricultural industry. Shortly after the establishment of the NIG, poultry breeders asked for the improvement of strains for higher egg production, and the Japan Monopoly Corporation of Tobacco and Salt requested the improvement of the varieties of tobacco plants.[56] Starting in 1951, Kihara supervised a large project, "basic research for improvement of tobacco strains." Lilienfeld also returned to Japan from the USA in order to work with Kihara in 1950, and joined the tobacco team.[57]

The characteristic feature of the projects conducted at the NIG was not only their use of agriculturally relevant organisms but also their approach to "breeding as a science." For example, the members of the tobacco project examined physiological and chemical characters of the secretion of the plant (thought to be contributing to the aroma of tobacco), the plants' resistance to diseases, various environmental effects on the plants, cytogenetic studies, and creation of mutants by the use of polyploidy and X-ray.[58] This multidisciplinary approach was not just a manifestation of Kihara's own style. Two of the NIG laboratories used silkworm and both covered various aspects of silkworm biology, such as effects of day length and other environmental factors on silkworm physiology and development, transmission of virus infecting silkworms, viral development, and differentiation of silkworm embryos.[59] Such diverse knowledge was necessary to gain control of the organism.

Such "breeding as a science" was different from the mainstream approach to genetic studies outside Japan. During the planning of the International Genetics Symposia of 1956, for which Japan was selected as a host country, there were disagreements over the proposed program between the Japanese organizers and members of the International Union of Biological Sciences (IUBS), which was to support the symposium. In 1954, after reviewing the Japanese proposal for the symposium program, the IUBS responded that they could not approve it. The Japanese had proposed the following two sections: (1) "physical and chemical approach to the

[54] About the establishment of the Institute, see Iida (2010).

[55] Kokuritsu idengaku kenkyūjo (1952–1953, vol. 2 (1952), pp. 44–52; vol. 3 (1953), pp. 50–60).

[56] Kokuritsu idengaku kenkyūjo (1952–1953, vol. 2 (1952), p. 4).

[57] Majewski (1989).

[58] Kokuritsu idengaku kenkyūjo (1952–1953, vol. 2 (1952), pp. 81–89).

[59] Kokuritsu idengaku kenkyūjo (1952–1953, vol. 2 (1952), pp. 8–13, 52–59).

problem of chromosomes" and (2) "genetics of cultivated plants and domesticated animals (polyploidy breeding, resistance, microorganisms and viruses, breeding systems)."[60] These were, in short, nonagricultural and agricultural sections, respectively.

To the top members of the IUBS, this proposed program looked like a random package of "quite disparate and disconnected fields."[61] Before the official IUBS letter was sent to the Japanese, Claudio Barigozzi and I. Michael Lerner (the president and secretary of the Genetics Section of IUBS, respectively) internally exchanged their opinions earlier about the Japanese proposal. Barigozzi wrote: "I remark simply, that the topics chosen for the Symposium are of so little interest."[62] He wished to propose the inclusion at least of quantitative inheritance and of immunogenetics. Lerner also could not help but feeling that "what they want to have is not a Symposium but an unsystematic collection of topics." He wrote: "I see no relation… within the second division, I am wondering since when a virus has become either a cultivated plant or a domestic animal."[63]

What Barigozzi and Lerner did not grasp was that the Japanese style involved examining multiple biological problems related to an organism of interest in the agricultural context. The point was not that a virus was considered to be a domesticated animal or cultivated plant, but that viruses infected domesticated animals and cultivated plants that were being studied by Japanese geneticists. What appeared as a random or disconnected set of problems to them actually was thus logically connected in the Japanese context. Quantitative genetics and immunogenetics, which Barigozzi hoped to be included in the program, must have been already included in the "agricultural" section of the program to some extent. As seen in Sakai's theoretical work on various crops, the Japanese had used a quantitative approach for breeding studies.[64] Many veteran geneticists in Japan were revolving around a particular organism, rather than a particular scientific problem. Thus it made more sense for them to put what Barigozzi and Lerner thought to be "disparate" subfields of genetics together, in order to have comprehensive understanding of an organism.

However, it is notable that the Japanese Society of Breeding(*nihon ikushu gakkai*) branched off from the Genetics Society of Japan in 1951. At the founding meeting of the new society held at the Faculty of Agriculture of Tokyo University, 260 people gathered. Many older generations of geneticists, including silkworm

---

[60] IUBS rejection letter, Montalenti to Shinoto, 9 November 1954; the Japanese proposal in Barigozzi to Lerner, 8 February 1954, both in folder "Permanent International Committee on Genetics Congresses, Correspondence (1953–1954)," I. Michael Lerner Papers, American Philosophical Society (APS), Philadelphia, USA.

[61] Montalenti to Shinoto, 9 November 1954, folder "Permanent International Committee," Lerner Papers, APS.

[62] Barigozzi to Lerner, 8 February 1954, folder "Permanent International Committee," Lerner Papers, APS.

[63] Lerner to Barigozzi, 12 February 1954, folder "Permanent International Committee," Lerner Papers, APS.

[64] In fact, "breeding systems," included under the second section, was later replaced by "polygenic inheritance" (which was part of quantitative genetics) in the actual program of the symposia.

geneticists, former members of Fujii's laboratory, and Kihara and his former students, joined the new society as central members (board member, secretary, editor, or honorary member, etc.).[65] This shows that even Japanese researchers felt that a gap between genetic studies within "breeding as a science" and newer (nonagricultural) types of genetics was growing. In turn, this indicates that Japanese "genetics" had maintained a very close relation with breeding studies until around this time. It should be noted however that the emergence of the new breeding society by no means implied the split of genetic studies with agricultural relevance from "basic" genetic research. Many had membership in both societies and Kihara in particular was at this time the president of the genetics society. Moreover, as we saw, the NIG members continued projects that had agricultural relevance.

Even if some breeders began leaving the circle of genetics, what stayed was the approach of "breeding as a science." For example, the NIG acquired in 1952 a temperature-controlled room (suitable for "experiments with all kinds of temperature treatments"; likely for animals such as silkworms) and a "controllable greenhouse" (*chōshetsu onshitsu,* or "phytotron" in English), in which temperature, humidity, and light could be controlled "for physiological genetic research on germination and growth."[66] Because of their interests and needs in examining organisms' responses to various environmental effects, these two rooms must have been essential and were thus established at the same time as the other essential facilities such as a microbe laboratory, an optical and chemical laboratory, and facilities for electron microscopy and irradiation.

Furthermore, because Kihara served as the director of the NIG from 1955 to 1969, his idea of genetic research must have been influential. Kihara obtained a large sum of funding from the Rockefeller Foundation for a project, "research on the origin of cultivated rice," for a total of 8 years starting in 1957. (When the International Rice Research Institute in the Philippines was established in 1960 with support from the Rockefeller and Ford Foundation, Kihara became a member of the board of trustees.) Through their rice project, the NIG also acquired new facilities, including Japan's first phytotron designed specifically for rice (i.e., experimental rice fields were contained within a phytotron). Moreover, in the same year (1957–1958), the NIG established another phytotron, in which temperature, humidity, day length, and wavelength of light were adjustable for "research of physiological genetics."[67] The world's first phytotron, a new type of laboratory for the experimental study of whole organisms, had been built in 1949 under the direction of American plant physiologist Frits W. Went at the California Institute of Technology. Went, originally motivated by the perception that organism–environment relations were being neglected at Caltech, later came to see the multidisciplinary research done in the phytotron as helping to counter the divisive trends of molecular biology.[68] It would be highly interesting to know what type of social and cultural

[65] See *Ikushugaku zasshi* (or *Japanese Journal of Breeding*) 1, no.1 (1951).

[66] Kokuritsu idengaku kenkyūjo (1952–1953, vol. 3 (1953), p. 78).

[67] Kokuritsu idengaku kenkyūjo (1958, vol. 8, pp. 131–132).

[68] Kingsland (2009).

roles the multiple phytotrons at the NIG played in the 1960s. While the NIG rapidly incorporated various newer branches of genetics by sending younger scholars to the USA for training (such as Kimura Motoo in population genetics), both the agricultural connection and a holistic, physiological approach hardly ended in the new era.

Throughout this chapter, I have argued that Japanese genetics expanded by maintaining a close relation to practical goals in agriculture and horticulture. Applied goals created projects that were both basic and applied and nurtured a holistic understanding of organisms, cells, and genes. In the second half of the twentieth century, the scheme to attract funding to genetics would change, particularly with the growing connection between genetics and medicine. However, the role of genetics in agriculture would remain extremely important. How the relation between genetics and agriculture developed further with the increase of genetic knowledge and techniques and with various other issues such as food security, population growth, and war, and how that relation affected interdisciplinary interactions, approaches to genetic studies, agriculture, and our view of life would be themes for future research.

Acknowledgments I would like to thank Savithri Preetha Nair, Takehiko Hashimoto, and Hans-Jörg Rheinberger for making helpful comments on the earlier draft, and Sharon Kingsland for suggestions and encouragement. This work was partly supported by JSPS KAKENHI (grant numbers 22800020, 24700922).

References

Abe, Ayao. 1939. Meiji no makkigoro nihon iden kenkyū no omoide (Recollections about genetic research in Japan around the end of the Meiji period). *N.I.G. danwashitsu* 5:25–27.

Beadle, G. W., and Boris Ephrussi. 1936. The differentiation of eye pigments in Drosophila as studied by transplantation. *Genetics* 21:225–247.

Beadle, G. W., and E. L. Tatum. 1941. Experimental control of development and differentiation: Genetic control of developmental reactions. *American Naturalist* 75 (757): 107–116.

Blakeslee, A. F., and A. G. Avery. 1937. Methods of inducing doubling of chromosomes in plants. *Journal of Heredity* 28:393–411.

Caspari, Ernst. 1933. Uber die Wirkung eines pleiotropen Gens bei der Mehlmotte Ephestia kühniella Zeller. *Wilhelm Roux' Archiv für Entwicklungsmechanik der Organismen* 130:353–381.

Crow, James F. 1994. Hitoshi Kihara, Japan's pioneer geneticist. *Genetics* 137:891–894.

Curry, Helen Anne. 2010. Making marigolds: Colchicine, mutation breeding and ornamental horticulture, 1937–1950. In *Making mutations: Objects, practices, contexts,* ed. Luis Campos and Alexander von Schwerin, Preprint 393, pp. 259–284. Berlin: Max Planck Institute for the History of Science.

Dobzhansky, Theodosius. 1937. *Genetics and the origin of species*. New York: Columbia University Press.

Eigsti, O. J. 1957. Induced polyploidy. *American Journal of Botany* 44:272–279.

Fujihara, Tatsushi. 2012. *Ine no daitōa kyōeiken: Teikoku nihon no "Midori no kakumei"* (The Greater East Asia Co-Prosperity Sphere of rice: Imperial Japan's "Green Revolution"). Tokyo: Yoshikawa kōbunkan.

Fujii, Kenjiro. 1918. Tōdai rika daigaku shokubutsugakka ni okeru idengaku kōza no shinsetsu: Nomura-ke no bikyo (Establishment of the new genetics laboratory at the Department of

Botany, Science College of Tokyo Imperial University: Commendable contribution of the Nomura family). *Botanical Magazine* 32:277–278.
Fujii, Kenjiro. 1920. On the conceptions of 'id' and the question of its transmutability. *Botanical Magazine* 34:99–125. [in Japanese].
Gilbert, Scott F. 1988. Cellular politics: Ernest Everett Just, Richard B. Goldschmidt, and the attempt to reconcile embryology and genetics. In *The American development of biology*, ed. R. Rainger, K. R. Benson, and J. Maienschein, pp. 311–346. New Brunswick: Rutgers University Press.
Goldschmidt, Richard B. 1938. *Physiological genetics*. New York: McGraw-Hill.
Harwood, Jonathan. 1993. *Styles of scientific thought: The German genetics community, 1900–1933*. Chicago: University of Chicago Press.
Hoshino, Yuzo. 1902. Kanseishi ni okeru ryōsei no bunkai ni tsuite. *Sapporo nōgakkaihō* 3:106–113.
Iida, Kaori. 2010. Practice and politics in Japanese science: Hitoshi Kihara and the formation of a genetics discipline. *Journal of the History of Biology* 43:529–570.
Inoue, Katsuo. 2006. Sapporo nōgakko to shokumingaku no tanjō. (The birth of Sapporo Agricultural College and colonial study). In *'Teikoku' nihon no gakuchi.* (Knowledge of 'Imperial' Japan), ed. Tetsuya Sakai, et al., vol. 1, pp. 11–41. Tokyo: Iwanami shoten.
Ishidate, Mieko. 1980. Hideo Kikkawa's works on eye pigment in *Bombyx* (1937–1950). *Kagakushi kenkyū* (Japanese journal of history of science) (series II) 19:129–139. [in Japanese].
Kihara, Hitoshi. 1930. Genomanalyse bei *Triticum* und *Aegilops*. *Cytologia* 1:263–284.
Kihara, Hitoshi. 1942a. Rakuseishiki (Inauguration ceremony). *Seiken zihō* 1:103–104.
Kihara, Hitoshi. 1942b. Aisatsu (Speech). *Seiken zihō* 1:107–109.
Kihara, Hitoshi. 1944. Futsū komugi no ichisosen taru DD-bunsekishu no hakken, yohō. (The discovery of a variety with DD, an ancestor of common wheat, preview). *Nōgyō oyobi engei* (Agriculture and horticulture) 19 (10): 13–14.
Kihara, Hitoshi. 1951. *Komugi: Ichi-seibutsugakusha no kiroku* (Wheat: A record of one biologist). Tokyo: Chūō kōron.
Kikkawa, Hideo. 1941. Mechanism of pigment formation in Bombyx and Drosophila. *Genetics* 26:587–607.
Kikkawa, Hideo. 1943. Konchū no toriputofan taisha o meguru shomondai (2). (Various problems on tryptophan metabolism in insects (2)). *Kagaku* 13:319–325.
Kikkawa, Hideo. 1947. Idenshi no honshitsu to sayō. (The substance and function of genes). *Kagaku* 17:12–16.
Kimmelman, Barbara A. 1987. A Progressive Era discipline: Genetics at American agricultural colleges and experiment stations, 1900–1920. Ph.D. Diss., University of Pennsylvania.
Kimmelman, Barbara A. 2006. Mr. Blakeslee builds his dream house: Agricultural institutions, genetics, and careers 1900–1915. *Journal of the History of Biology* 39:240–280.
Kimura, Motoo. 1986. Kihara Hitoshi hakushi o shinobu. (Remembering Dr. Kihara Hitoshi). *Kagaku* 56:725–728.
Kingsland, Sharon E. 2009. Frits Went's atomic age greenhouse: The changing labscape on the lab-field border. *Journal of the History of Biology* 42:289–324.
Kokuritsu idengaku kenkyūjo, ed. 1952–1953. Kokuritsu idengaku kenkyūjo nenpō (National Institute of Genetics Annual Report). vol. 2–3, pp. 44–52; 50–60.
Kokuritsu idengaku kenkyūjo, ed. 1958. Kokuritsu idengaku kenkyūjo nenpō (National Institute of Genetics Annual Report). vol. 8, pp. 131–132.
Kühn, Alfred, E. Caspari and E. Plagge. 1935. Uber hormonale Genwirkungen bei Ephestia kühniella Z. *Nachrichten von der Gesellschaft der Wissenschaften zu Göttingen, Mathematisch-Physikalische Klasse* 2:1–29.
Kuwada, Yoshinari. 1910. A cytological study of *Oryza sativa* L. *Botanical Magazine* 24:267–281.
Kyoto Imperial University, ed. 1943. *Kyoto teikoku daigakushi.* (The History of Kyoto Imperial University). Kyoto: Kyoto Imperial University.

Kyoto daigaku nōgakubu 70-nenshi henshū iinkai, ed. 1993. *Kyoto daigaku nōgakubu 70-nenshi.* (The 70-year history of the Faculty of Agriculture of Kyoto University). Kyoto: Kyoto daigaku nōgakubu.
Majewski, Tomasz. 1989. Flora Lilienfeldówna (1886–1977). *Wiadomości Botaniczne* 33:101–108.
Matsubara, Yoko. 2000. Menderu idengaku no juyō: Nihon no baai. (Reception of Mendelian genetics: The case of Japan). *Seibutsugakushi kenkyū* 66:89–94.
Matsubara, Yoko. 2004. The reception of Mendelism in Japan, 1900–1920. *Historia Scientiarum* 13:232–240.
McFadden, E. S., and E. R. Sears. 1944. The artificial synthesis of Triticum spelta. *Records of the Genetics Society of America* 13:26–27.
Moriwaki, Yasuko. 2010. Toyama Kametarō to meijiki no sanshigyō ni okeru kaiko no 'shurui kairyō'. (K. Toyama and silkworm breeding in Japan, from 1891 to 1913). *Kagakushi kenkyū* (series II) 49:163–173.
Nihon kagakushi gakkai, ed. 1965. *Nihon kagaku gijutsushi taikei.* (Survey of the history of science and technology in Japan). vol. 15 (biological sciences). Tokyo: Dai-ichi hōki.
Noguchi, Yakichi. 1978. Nōgyō ni okeru idengaku. (Genetics in agriculture). In *Nihon nōgyō hattatsushi.* (The history of the development of agriculture in Japan), ed. Nōgyō hattatsushi chōsa kai, vol. 9, 2nd ed., pp. 241–318. Tokyo: Chūō kōronsha.
Onaga, Lisa. 2010. Toyama Kametarō and Vernon Kellogg: Silkworm inheritance experiments in Japan, Siam, and the United States, 1900–1912. *Journal of the History of Biology* 43:215–264.
Paul, Diane B., and Barbara A. Kimmelman. 1988. Mendel in America: Theory and practice, 1900–1919. In *The American development of biology,* ed. Ronald Rainger, Keith R. Benson, and Jane Maienschein, pp. 281–310. Philadelphia: University of Pennsylvania Press.
Rheinberger, Hans-Jörg. 2010. *An epistemology of the concrete: Twentieth-century histories of life.* Durham: Duke University Press.
Richmond, Marsha. 2007. The cell as the basis for heredity, development, and evolution: Richard Goldschmidt's program of physiological genetics. In *From embryology to evo-devo: A history of evolutionary development,* ed. Manfred D. Laublichler and Jane Maienschein, pp. 169–211. Cambridge: MIT Press.
Sakamura, Tetsu. 1918. Kurze Mitteilung ueber die Chromosomenzahlen und die Verwandtschaftsverhaeltnisse der *Triticum*-Arten. *Botanical Magazine* 32:150–153.
Sapp, Jan. 1987. *Beyond the gene: Cytoplasmic inheritance and the struggle for authority in genetics.* Oxford: Oxford University Press.
Shinotō, Yoshito, ed. 1967. *Idengaku no ayumi: Menderu iden hōsoku 100-nen kinen.* (The progress of genetics: Centennial anniversary of Mendelian laws of heredity). Tokyo: Shōkabō.
Stebbins, G. Ledyard. 1980. Botany and the synthetic theory of evolution. In *The evolutionary synthesis: Perspectives on the unification of biology,* ed. Ernst Mayr and William B. Provine, pp. 139–152. Cambridge: Harvard University Press.
Tahara, Masato. 1914–1915. Cytological studies on chrysanthemum I-IV. *Botanical Magazine* 28:489–494; 29:5–17, 45–51, 92–103. [in Japanese].
Tahara, Masato. 1941. Sonokoro no omoide (Some recollection of that time). *N.I.G. danwashitsu* 7:2–3.
Tanaka, Kōji, and Ryōichi Imai. 2006. Shokuminchi keiei to nōgyō gijutsu: Taiwan, Nanpō, Manshū (Colonial management and agriculture: Taiwan, the South, and Manchuria). In *'Teikoku' nihon no gakuchi* (Knowledge of 'Imperial' Japan), ed. Tetsuya Sakai, et al., vol. 7, pp. 99–137. Tokyo: Iwanami shoten.
Tanaka, Yoshimaro. 1916. Genetic studies on the silkworm. *Journal of the College of Agriculture, Tohoku Imperial University* 7:129–255.
Tanaka, Yoshimaro. 1942. Idengaku no shōrai (The future of genetics). *Shokubutsu oyobi dōbutsu* (Botany and zoology) 10:59–60.
Tanaka, Yoshimaro. 1961a. Nihon idengaku no yoake (1). (The dawn of Japanese genetics (1)). *Iden* 15 (1): 39–42, 35–37.

Tanaka, Yoshimaro. 1961b. Nihon idengaku no yoake (2). (The dawn of Japanese genetics (2)). *Iden* 15(2): 39–42, 35–37.

Toyama, Kametarō. 1906. Studies on the hybridology of insects: I. On some silkworm crosses, with special reference to Mendel's law of heredity. *Bulletin of the College of Agriculture, Imperial University of Tokyo* 7:259–393.

Vavilov, Nikolai I. 1949/1950. The origin, variation, immunity and breeding of cultivated plants: Selected writings of N. I. Vavilov. Trans. K. S. Chester. *Chronica Botanica* 13 (1): 6.

Yamamoto, Mihoko. 2011. Taiwan ni watatta hokudai nōgakubu sotsugyōsei tachi (Graduates of the Faculty of Agriculture of Hokkaido University who went to Taiwan). *Annual Report of Hokkaido University Archives* 6:15–41.

Yamashita, Kōsuke. 1942. Menka-han no kaisetsu (Establishment of the cotton team). *Seiken zihō* 1:97–98.

Kaori Iida is a lecturer at the Graduate University for Advanced Studies (Sokendai), Hayama, Japan. Her current research examines the history of Japanese genetics with particular focus on the work of Kihara Hitoshi, who among early Japanese geneticists was one of the best known internationally. She has a Ph.D. in genetics from Pennsylvania State University, and a Ph.D. in history of science from Johns Hopkins University.

Chapter 22
Speeding Up Evolution: X-Rays and Plant Breeding in the United States, 1925–1935

Helen Anne Curry

Introduction

On a hot mid-summer day in 1925, a young researcher tended to an experiment he had started on the grounds of the Missouri State Agricultural Experiment Station. Lewis J. Stadler, a professor in the Department of Agronomy, maneuvered a General Electric (GE) portable X-ray outfit out into the open cornfield. Moving along the rows of experimental maize, he stopped in front of one of the plants. He oriented the X-ray tube so that the radiation would fall on the tassel at the very top, where pollen grains were developing, before turning on the device. After a couple of minutes, he cut the power, then wheeled the machine towards his next experimental subject and repeated the procedure.[1] Stadler hoped that the ionizing radiation produced by the X-ray would affect the genes or chromosomes of the maturing pollen—the same pollen that would soon be shed to fertilize the female flowers growing below and, in time, produce the seeds of a next generation in the form of a cob of corn kernels. This was no small ambition. As Stadler explained, "The object of the experiments is to devise methods for affecting heredity by external treatments. This is a problem of fundamental importance in biological science and has very important applications in animal breeding and plant breeding."[2] He thought that radiation might be used to alter genes, and that this in turn might both answer some pressing questions in genetics and be a useful tool for agriculture, especially in breeding new varieties.

In conducting this research, Stadler was following in a longer tradition in genetics, attempting the experimental manipulation of heredity both to answer questions

[1] A description of the experimental procedure is found in Stadler (1928).

[2] Stadler to Henke, 21 March 1925, records of the Department of Agronomy (University Archives, University of Missouri-Columbia), 3C/33/4, Box 6, Folder "June 1925, H-O."

H. A. Curry (✉)
Department of History and Philosophy of Science, University of Cambridge, Free School Lane, Cambridge CB2 3HR, UK
e-mail: hac44@cam.ac.uk

D. Phillips, S. Kingsland (eds.), *New Perspectives on the History of Life Sciences and Agriculture,* Archimedes 40, DOI 10.1007/978-3-319-12185-7_22

about basic biological processes and to find possible applications. He was also about to become part of one of that tradition's landmark achievements: the demonstration of artificially induced mutation. In December of 1927 Stadler presented the findings of his X-ray experiments, some of which he felt demonstrated the effects of the X-ray in causing genetic mutations and other chromosomal changes.[3] His announcement followed close on the heels of the announcement of a similar—and better remembered—set of experiments conducted by the geneticist Hermann Muller. In July of that year, Muller had published results indicating that exposing *Drosophila* flies to large doses of X-ray radiation caused genetic mutations to appear in abundance in subsequent generations.[4] It was this work that geneticists and historians would take as the defining moment within the history of research on induced mutations, which in turn has had consequences for our understanding of why research in induced mutation was so celebrated in the 1920s and afterwards.[5]

Muller's experiment was significant in part because it encouraged geneticists and other biologists to induce mutations that they then used in their experimental procedures, whether this was producing genetic maps, understanding biological pathways, investigating evolution, or other research.[6] These were not, however, the only endeavors in which induced mutations were felt to be of value in the 1920s. Stadler's 1925 explanation of his X-ray experiments offers a case in point—there would be potential practical uses of this research as well as important implications for basic research. And Muller agreed. As he noted in his first-ever paper on the subject of X-ray-induced mutation, not only experimental biologists but also practical breeders, long "compelled to remain content with the mere making of recombinations of the material already at hand," would benefit from a tool that would produce variation on demand.[7]

For a time, these geneticists, and others, thought that X-ray radiation might be used to create through genetic mutation new and useful variations in agricultural crops and animals. This idea—and the research and discussions of it that followed—is the subject of this chapter. I focus on efforts and aspirations linked to the improvement of plants in particular, as biologists and breeders made greater efforts with X-rays in plant improvement than they did in animal improvement.[8] In the 1920s and 1930s especially, breeders attempted to induce mutations in a

---

[3] Stadler (1928); see also Stadler (1930).

[4] Muller (1927).

[5] Carlson (1981), Schwartz (2008). For an account of Muller's experiments that includes a reflection on how and why this came to dominate historical memory of induced mutation, see Campos (2006, Chap. 5).

[6] Carlson (1981) discusses some of the early follow-up research, esp. Chap. 11. For a more extensive history of mutations as objects of biological research, see contributions to Campos and Schwerin (2010).

[7] Muller (1927), p. 84.

[8] Research on animals was hampered by the greater sensitivity of many species to gross genetic disruption of the kind produced by X-rays (not to mention the other harmful effects of intense irradiation treatment). Plants by comparison are more resilient, tolerating changes in chromosome number, broken or altered chromosomes, and in many cases showing a range of unusual forms in

range of economic crops, and Americans were encouraged by many reports to believe that this research augured a new era in agricultural production. Some of the researchers who applied X-rays in breeding and many others who read about the procedure saw it as enabling an unprecedented ability to alter organisms through the direct manipulation of genetic material. As such, it gave rise to sweeping aspirations such as perfecting the ability to produce new varieties of plants "to order" through technological innovation, and better aligning American agricultural production with industrial ideals.

My account complements the now significant body of literature that addresses the intersections of agricultural interests and genetics (and vice versa) in the early decades of the twentieth century. Most historical narratives about genetics and agriculture in the United States have emphasized how Mendelian genetics confirmed and then refined practices already in use at the turn of the century.[9] Here I stress the hopes which followed on this initial Mendelian moment and persisted in subsequent decades: that continued research in genetics would provide entirely new capabilities to breeders, that technological innovation would result in dramatic changes in breeding practices and therefore in agriculture more generally. I draw particular attention to the ways in which these hopes were picked up and amplified by the popular press, giving rise to a shared enthusiasm for new technologies of manipulating living organisms.

To explore this aspect of the history of plant breeding and American agriculture, it is helpful to consider research in genetics, evolution, and agriculture in the period leading up to 1927. To that end, I begin with a consideration of the period immediately before X-ray-induced mutation was discovered, to suggest that both breeders and biologists had long been interested in a tool that would induce mutation. I then turn to events of 1927, and especially to early claims that X-rays would prove useful in agricultural breeding, and to the many attempts made to demonstrate this usefulness in subsequent years. Finally, I examine the coverage of these activities in the popular press to explore why the notion of induced mutation proved exhilarating to so many Americans "scientists and laypersons" alike.

Mutation, Evolution, and Breeding

The early years of the twentieth century witnessed the emergence of new understandings of biological variation and inheritance, most notably in the development of the discipline of genetics. As a number of historians have described, the field of study that became known as genetics was deeply intertwined with agricultural interests from the start. It both confirmed the long experience of breeders in crossing and selection while also promising to legitimize and further improve these methods

---

response to mutagenic treatment; in fact, many plant breeders and biologists sought actively these chromosomal changes as useful to their work and research.

[9] For example, Paul and Kimmelman (1988), Fitzgerald (1990), esp. Chap. 1, and Cooke (1997).

in a "scientific" manner.[10] Much has been written about Mendelian genetics at the turn of the twentieth century and its relation to agriculture; for the purposes of this chapter, which pursues these links between genetics and agriculture well into the 1930s, it is important to understand in particular shared interests among geneticists and agriculturists related to mutation, variation and heredity, and evolution.

Inheritable variations are the key raw material with which any plant breeder works. Just as evolution requires there to be genetic variability on which natural selection can operate, the improvement of a plant presupposes that there are a range of inheritable traits in any given population. The job of the breeder is to gather into a variety or breed the right combination of traits. For much of the eighteenth through twentieth centuries, this meant searching—sometimes around the globe—for plants bearing desirable characteristics. When such a plant was found, it could then be hybridized with established varieties, in order to transfer the trait into this existing line, or else it alone could become the basis of a new variety via selection.[11] One imagined alternative to this process was to be able to produce the traits desired in a particular agricultural plant, or to influence the rate at which variations in traits appeared. To many researchers, especially in the early twentieth century, the ability to control the appearance of variation promised to free the breeder from what had always been a significant constraint to the improvement of any agricultural or horticultural crop—the extent of existing variation. If one could influence the appearance of variation, then it would no longer be necessary to wait for nature to turn out such variation or to search the globe in hopes of finding it.

One of the most famous proponents of the idea of induced variation was the Dutch botanist Hugo de Vries, best known for his mutation theory of evolution published in 1901–1903 (in German; the English edition appeared in 1909–1910). As de Vries described in 1904, "We want to share in the work of evolution, since we partake of the fruits. We want even to shape the work, in order to get still better fruits." Accordingly, he encouraged biologists to study more closely the nature of obvious and sudden variations that sometimes occurred in nature—he called them "mutations"—and to seek their causes. If scientists could learn to control the process by which inherited variations occurred, he argued, "New and unexpected species will then arise, and methods will be discovered which might be applied to garden plants and vegetables, and perhaps even to agricultural crops, in order to induce them to yield still more useful novelties."[12]

When de Vries introduced the term "mutation" into debates about evolution and heredity around 1900, he used it to denote a change in an inherited character significant enough to be considered the cause of immediate speciation. He had first applied the term in referring to clearly visible changes in his experimental plant *Oenothera,* the evening primrose: A mutation was what set a parent plant apart

[10] On the entanglement of genetics research with agricultural interests in the American context, in addition to the references in no. 9, see Kimmelman (1983, 1987, 1992), Rosenberg (1997), Chap. 13, Allen (2000).

[11] On the long history of techniques of plant breeding, see Murphy (2007), Kingsbury (2009).

[12] Vries (1905), quotation p. 48.

from an offspring so different from the parent that the two did not interbreed. De Vries's observations of *Oenothera* became the basis of his "mutation theory," a revision of evolutionary theory that emphasized the process of rapid speciation via sudden leaps or saltations, instead of the accumulation of small variations over time as proposed by Darwin. The theory quickly captured international attention.[13] "Mutation," too, caught on as a biological concept indicating a change in an inherited character, except that the meaning of the word quickly, well, mutated, to encompass less dramatic changes that did not necessarily entail speciation. By the 1920s, "mutation" more commonly referred to inherited changes in specific traits, sometimes referred to as "unit factors" or "Mendelian unit factors" by geneticists. These mutations were most often assumed to be linked to otherwise invisible alterations of genes and chromosomes, but were not limited to events that distinguished one species from another.[14] This concept of mutation, though distinct from de Vries's formulation, would remain a critical component of ideas about evolutionary change, for spontaneous mutations, as inheritable changes in the fundamental units of heredity, were thought by some biologists to be the source of variations on which natural selection operated.[15]

Breeders' interest in finding a way to control the appearance of new or different traits in order to facilitate more efficient plant breeding mapped onto these changing ideas about mutation and evolutionary change. For example, part of the significant popular interest in de Vries's mutation theory had stemmed from the hope that understanding mutation might in turn lead to control of evolution and by extension the improvement of cultivated types.[16] In 1912, a news report detailing a lecture by de Vries appeared in various newspapers under headlines such as "How to Increase World's Foods" and "Grow Larger Grain." It described de Vries's hope in the existence of mutations that would generate higher yielding rice and wheat varieties. The director of the New York Botanical Garden, W. A. Murrill, summed up de Vries's research as "paving the way… Someday…some one will apply, in a practical way, the principles which Professor de Vries is now laying down." He envisioned that breeders, rather than wait to discover mutations as they spontaneously appeared, would be able to generate them through new experimental methods.[17]

By the 1920s, it was fairly standard in agricultural literature to aspire to the ability to control the production of mutations (which by this time typically referred to changes in Mendelian factors) in order to facilitate plant improvement—and there were many experiments conducted along these lines. Still, scientists had not yet achieved this much-hoped-for goal of understanding and replicating mutation. In

[13] On de Vries's mutation theory and its reception, with particular attention to the hopes it engendered for experimental evolution, see Kingsland (1991). Other accounts include Allen (1969), Bowler (1978).

[14] For a discussion of "mutation" contemporary to this period, see Morgan (1919), Chap. xx.

[15] See Bowler (1989), Chap. 11.

[16] Kingsland (1991). An account of mutation theory that gives particular attention to its popular reception is Endersby (2013).

[17] Some instances of this report include: Anonymous (1912a), p. 4; Anonymous (1912b), p. 4; Anonymous (1912c), p. 7.

a 1925 textbook, the geneticists Edmund Sinnott and L. C. Dunn discussed experiments using X-rays, radium, and biological injections in an attempt to produce inheritable changes, but concluded that this research had contributed little to the knowledge of why mutations occur. In their estimation, the origin of mutation "constitute[d] one of the most interesting unsolved problems of genetics".[18] A subsequent chapter on practical breeding underscored that this was also a fundamental problem facing breeding in achieving all that genetics promised. Control of the inheritance of specific traits, Sinnott and Dunn argued, "presupposes the power not only to manipulate inheritance by breeding operations, but also the power consciously to induce new heritable variations"[19]. Although the first was increasingly possible, the latter remained a dream for the future. Like many of their colleagues (and their predecessors), Sinnott and Dunn saw that if some way could be found to control the appearance of genetic mutations, and therefore the appearance of new characters, a significant bottleneck in breeding could be eliminated.

It was in this context that Muller's announcement in 1927 of X-ray-induced mutation, and the presentation of Stadler's findings and similar research on plants that quickly followed, came to assume immediate significance in the agricultural community. Induced mutation was already understood to be a feat with direct and important applications.

X-Ray-Induced Mutation, in the Laboratory and Beyond

Hermann Muller counted among those scientists who had long perceived control of mutation to be of critical importance, both to knowledge of genetics and evolution and to the application of this knowledge. In the 1920s, he was one of the leading participants in *Drosophila* research, a field that he had in fact helped to pioneer while working under Thomas Hunt Morgan in the Columbia University fly group.[20] As early as 1916, Muller had expressed a belief that understanding mutation was critical to the future of biology, in particular because of its potential to allow greater control over the evolution of new forms and potentially even human evolution—genetic mutations, after all, were (in his assessment) the key to evolutionary change. Motivated by this interest, he took up the study of mutation, its causes and mechanisms, more directly. It was this work that eventually led him to apply X-ray radiation to his stocks of *Drosophila* and then to the announcement of having successfully induced genetic mutation by this process in 1927.[21]

Muller had not been the first to attempt such an experiment. Others had in the preceding years similarly used radiation, whether X-ray or radium radiation, as well

[18] Sinnott and Dunn (1925), pp. 307–308.

[19] Sinnott and Dunn (1925), p. 369.

[20] On the early history of the *Drosophila* research community, including Muller, see Kohler (1994). Biographical accounts of Muller include Carlson (1981), Schwartz (2008).

[21] A concise overview of Muller's interest in mutation and the control of evolution can be found in Pauly (1987), pp. 177–183.

as chemicals, temperature changes and other environmental shocks in attempts to provoke genetic mutation. A few appeared to have in fact produced inheritable alterations, but demonstrating that these were unequivocally linked to the particular treatment and to changes in genes proved a challenge.[22] Convincing his fellow biologists of both of these points was critical to Muller's success, and he was able to do so in part because the experimental setup he devised enabled him to produce and detect mutations in enormous abundance.[23] The rates of mutation that Muller claimed to have obtained by administering high doses of radiation from an X-ray tube to cultures of fruit flies in his laboratory were spectacular: a rise in the mutation rate of some 15,000 % in germ cells that had received the heaviest radiation treatment.[24]

The work had clear implications for biological research, especially in genetics. For example, as Muller suggested in 1927, biologists could now take advantage of X-rays to create "in their chosen organisms a series of artificial races" to use in mapping the genes that governed any particular trait to a location on a specific chromosome.[25] And this is just what happened. In Muller's fly lab and elsewhere, X-rays became a standard tool for producing changes in the genes and chromosomes of experimental organisms.[26]

At the time that Muller published his first studies on this topic, the same genetic effect of X-ray treatment was under investigation by at least two plant biologists—a circumstance that among other things underscores how important this effect was perceived to be. Lewis J. Stadler of the University of Missouri had been studying whether X-rays could be used to alter observed linkage rates (a measure of the tendency of certain traits to be inherited together) among certain genes in corn. In the process of this research, he discovered that the X-ray treatment in some cases appeared to produce genetic mutations. As described above, he was at work confirming this finding when Muller made his announcement in 1927.[27] Another biologist, Thomas Harper Goodspeed of the University of California-Berkeley, had for

[22] The former was hard to demonstrate especially in small-scale experiments, given that most organisms were known to have mutations spontaneously appear without apparent provocation. It was also difficult to demonstrate that the inherited changes resulted in fact from alterations of Mendelian unit factors—in other words, that they resulted from changes in the still-invisible genes and not gross alterations of the chromosome or other damage to the cell. On the history of earlier efforts to induce mutation, see Campos (2006), Chaps. 3 and 4.

[23] Descriptions of the experiment and why it produced convincing evidence where previous efforts had failed to do so include: Carlson (1981), pp. 146–150, Campos (2006), pp. 363–364.

[24] Muller (1927), p. 85.

[25] Muller (1927), p. 87.

[26] Well-known examples of the use of X-ray-induced mutations in biological research include the mutations in the bread mold *Neurospora* that led the geneticists George Beadle and Edward Tatum toward their one-gene, one-enzyme hypothesis and the X-ray mutations in maize that proved essential to the advances cytologist Barbara McClintock made in genetic studies of that organism. See Berg and Singer (2003), Chap. 9; Kass and Chomet (2009).

[27] In subsequent accounts, Stadler would be noted as having just missed priority for the discovery. See, e.g., Carlson (1981), p. 151. Biographical accounts of Stadler include Rhoades (1957), Rédei (1971).

many years studied the evolutionary history of the genus *Nicotiana* (tobacco plants) using cytological analyses especially. His foray into radiation studies came at the encouragement of a Berkeley colleague, the chemist Axel Ragnar Olson, who had seen the X-ray produce unusual changes in bacteria. With Olson's help, Goodspeed irradiated some of his tobacco varieties and subsequently discovered an incredible display of plasticity of forms, more than he had ever encountered in his years of investigating *Nicotiana*.[28]

Geneticists considered the work of Goodspeed and especially Stadler to be an important confirmation of the results produced by Muller. Together their experiments contributed to a new enthusiasm for X-ray and other radiation experiments in biology. Within a year, a Committee on Effects of Radiation on Living Organisms had been assembled under the auspices of the National Research Council, with the mandate to solicit and distribute funds for research in this area.Goodspeed, Muller, and Stadler numbered among its grantees, as did more than 40 other researchers by 1934. Their investigations included a whole range of studies in the physiological, developmental, and genetic effects of X-rays (and other radiation such as radium rays and ultraviolet) on various plants, animals, and microorganisms. The research aims of the committee were many, but included especially support of "pure science" and "fundamental problems."[29] As Winterton Curtis, who had been a key figure in the organization of the committee, described, although the "trend of modern biology" was to explain biological processes in physical and chemical terms—in hopes of achieving control over them—physical and chemical experiments tended to damage or destroy living organisms. In the case of irradiation, however, "the organism may be profoundly changed and yet continue its existence in a normal state. We can, as it were, shoot some of its molecules to pieces, destroying cells or parts of cells by radiations, and thus modify the organism by a procedure infinitely more delicate than anything previously available." This would be "the real beginning of a physicochemical analysis of life processes."[30] It was clear to Curtis and his fellow biologists that the induced-mutation research, as well as other studies in radiation effects, had above all revealed an important experimental tool.

But what were nonscientists to make of all this X-ray research? The science journalist Frank Thone laid bare the implications of Muller's investigation for readers of *Science News-letter* in 1927: "[T]his is what Prof. Muller's experiments signify: evolutionary changes, or mutations, can be produced 150 times as fast by the use of X-rays as they can by the ordinary processes of nature. This means that man can force the production of new and desirable plant varieties far more rapidly than he

[28] Accounts of this research include Goodspeed (1927), pp. 226–227, 256–257, Goodspeed and Olson (1928), Goodspeed (1929). A biographical account of Goodspeed is Baker et al. (1967).

[29] See Annual Reports of the Committee on the Effects of Radiation on Living Organisms, Division of Biology & Agriculture, National Research Council (Archives of the National Academy of Sciences, Washington, DC). Quotation in "Cumulative report, 1928–1934."

[30] Curtis to Quinn, 11 September 1928, Records of the Committee on the Effects of Radiation on Living Organisms, Division of Biology & Agriculture, National Research Council (Archives of the National Academy of Sciences, Washington, DC), Folder "Requests for Support Foundations."

has hitherto been able to get them"[31]. In his appeal to a more general audience for the significance of this research, Thone put Muller's finding not in the context of basic genetic science, but in that of potential agricultural application. And if there were excitement over the implications of what Muller had done with flies, it was redoubled or more in the cases of the economically important species that Stadler and Goodspeed worked with, especially when it came to the popular press and accounts of the research found there. According to a reporter for the *New York Times,* Stadler had demonstrated that X-rays could be used to create "entirely new species of grain." "Here we may have a new method for Burbanking flowers and plants for man's benefit," the reporter declared, referring to the world-famous horticulturist Luther Burbank who had dazzled the public in preceding decades with his production of new fruits and flowers.[32] According to another report, Goodspeed and his colleague Olson had produced over 200 new varieties of tobacco simply by applying X-rays to the plant. This was offered as evidence to support above all the notion that X-rays "may provide a method of producing new and possibly improved types" of economically valuable plants and animals for agricultural use.[33]

These assessments may have sensationalized the as-yet-unconfirmed potential of X-ray-induced mutation in breeding, but they were accurate in reflecting the ideas of many of the experimental biologists engaged in this research. The Committee on Effects of Radiation on Living Organisms, committed though it was to "pure science" and not its applications, described in its 1930 "Popular Report of Research" that research such as Muller's indicated that radiation could be used "to increase the percentage of mutations…which the breeder can use as starting points in the development of new types, thus 'speeding up' the evolutionary process that proceeds so slowly with domesticated forms."[34] Muller likewise encouraged this perspective in his public declarations. In a statement published in *Scientific American* he noted that the "extension of such work to other organisms, especially to domestic animals, mammals that can be bred in the laboratory, and crop plants in general" was urgent, because their ability to tolerate radiation had not been established; if it were, then "the method [of x-ray breeding] should become a practicable one."[35] Believing in 1928 that the mutations generated by X-rays were in many cases no different from spontaneous or "natural" mutations, Muller encouraged researchers working with important crops to take up mutation research immediately.[36] He apparently even collaborated on a pecan-breeding project with H. P. Traub of the United States Department of Agriculture (USDA).[37]

Stadler and Goodspeed, both more familiar than Muller with practical breeding, also made efforts to extend their research into the realm of plant improvement.

---

[31] Thone (1927), p. 243.

[32] Anonymous (1927a), p. 6.

[33] Anonymous (1927b), p. 19.

[34] "Popular report of research supported during the year 1929–1930," Committee on the Effects of Radiation Upon Living Organisms, National Research Council Reports (Copies), MS-2469 (University of Tennessee Special Collections, Knoxville, Tenn.).

[35] From opening note, "A statement by the discoverer," in Thone (1928), p. 235.

[36] E.g., McKay and Goodspeed (1930).

[37] Traub and Muller (1934).

Stadler was skeptical that X-rays would ever be as useful in breeding as was being claimed, for even early on he suspected that they mostly produced not "true" mutations of specific genes (that is, the conversion of a gene from one allelic form to another) but rather damage to the chromosomes and subsequent loss or destruction of genes.[38] He was nonetheless willing, in the early years of his induced mutation research, to consider the use of X-rays in very specific breeding projects. These included introducing variation into highly inbred lines, such as the pure lines used in hybrid corn, and in the production of new somatic mutations in fruit breeding—both areas where traditional techniques of introducing variation through crossbreeding did not apply.[39] He subsequently collaborated with two prominent USDA corn breeders, X-raying prized strains of inbred corn in hopes of producing variations in disease resistance and other traits that could be further improved via selection.[40] He also collaborated with the horticulturist A. E. Murneek in the late 1920s on the possibility of inducing bud variations in apples through the use of X-rays[41]. Goodspeed at Berkeley, who was like Muller more confident that radiation treatment resulted in gene mutations, continued his research on induced mutation in tobacco for a number of years, and also aided in a research project that attempted to demonstrate the possibility of creating useful mutations in cotton. Both seemed to him to have produced clearly useful types by the 1930s; for example, he noted that one of the X-rayed tobacco strains had resulted in a true-breeding type with desired traits such as "a more compact habit of growth, almost no suckering, increase in leaf number and leaf size," and with a colleague he had observed a cotton plant in which the boll was not attached to the seed (and therefore did not require ginning).[42]

And it was not just these geneticists who considered whether their findings might be extended. In the wake of the demonstration of X-ray-induced mutation and especially the claims made for its importance to practical concerns, experimenters in a range of venues took up the study of the potential application of X-rays in plant breeding. Those interested in exploring the use of X-rays in breeding included, predictably, agricultural experiment station workers and USDA researchers. A pair of cotton breeders from the Texas Agricultural Experiment Station set out to assess the value of X-rays in producing so-called progressive mutations, i.e., changes that could be considered beneficial from an evolutionary perspective and hopefully also from an agricultural perspective.[43] A handful of US agricultural station research-

[38] Stadler to Sax, 17 December 1931, Lewis J. Stadler Papers (Western Historical Manuscript Collection, Columbia, MO), Folder 5. This is also discussed in Stadler (1930).

[39] Stadler (1930), pp. 18–19.

[40] Stadler to Slosson, 7 June 1928, Lewis J. Stadler Papers (Western Historical Manuscript Collection, Columbia, MO), Folder 1; Stadler to Richey, 2 September 1929, Records of the Bureau of Plant Industry, Soils, and Agricultural Engineering (US National Archives and Records Administration, College Park, MD), RG 54, 66/31/136, Folder "L. J. Stadler 1918–1930 Sorted."

[41] Bishop (1954); Stadler (1930), p. 19.

[42] For a description of the tobacco work, see "Cumulative Report, 1928–1934," Committee on the Effects of Radiation Upon Living Organisms, National Research Council Reports (Copies), MS-2469 (University of Tennessee Special Collections, Knoxville, Tenn.). On cotton see McKay and Goodspeed (1930).

[43] Horlacher and Killough (1933a, 1933b).

ers also used X-rays to induce mutations through the 1930s. In 1936, the USDA *Yearbook of Agriculture* listed three other agricultural stations pursuing X-ray-induced mutation in cotton, though it mostly catalogued their failures.[44] Reports on other crops noted similar efforts and gave general attention to X-rays as a tool of breeding.[45] One report noted that sorghum breeders had used the X-ray to produce unusual changes in milo, a small drought-resistant sorghum.[46] Breeding methods described by tobacco experts included both attempts to double the chromosome complement with X-rays, and the use of X-rays to induce mutations of single characters; Goodspeed at California received attention as having produced "seven derivative pure-breeding types" from a single irradiation.[47]

These investigations into the potential of X-ray breeding in genetics departments and at experiment stations were complemented by those carried out at commercial operations such as the Maui Agricultural Company, producer of sugarcane, and W. Atlee Burpee & Co. seed company, a commercial seedhouse specializing in flower and vegetable seeds for home growers. Both explored the possibilities that X-ray-induced mutation might enhance their plant breeding efforts in the late 1920s (with Burpee continuing its experiments with radiation much longer).[48] And even as agriculturists and horticulturists attempted to exploit the potential of this new technique by bringing X-ray apparatus into their fields, farms, and laboratories, the American technology conglomerate GE brought the field into their laboratory in hopes of the same. GE established a small greenhouse and farm site at their research laboratory in the early 1930s, which a pair of GE researchers used to cultivate a wide variety of plants from irradiated seeds and bulbs. These included ornamental flowers, food crops including especially fruits, and tree species prized for lumber. The goal of the research was both to produce novel types and to make the X-ray—at the time, an important area of GE innovation and production—a more precise tool of genetic manipulation and therefore more valuable to research scientists and breeders alike.[49]

X-Ray Visions

Many of the individuals and institutions catalogued above conducted experiments in X-ray-induced mutation because they wanted to create improved varieties of plants, or to explore whether the X-ray machine might help them do so. When word

[44] Ware (1936), pp. 742–743.

[45] For example, X-rays featured among options offering "unusual possibilities for breeding" in the future. See Anonymous (1936), p. 183.

[46] Martin (1936), p. 538.

[47] Garner (1936), p. 804, 828.

[48] For a report from the Maui Agricultural Company, see Foster (1929). Discussions of Burpee's interest in X-rays to produce floral novelties include Burpee and Taylor (1940), Anonymous (1942a), p. 29.

[49] Descriptions of this research program include Hawkins (1932), Haskins (1932a, 1932b), Anonymous (1932), Haskins and Moore (1934).

of their activities hit the popular press, these sometimes took on a larger significance—scientists had discovered how to accelerate evolutionary change and, more important, how to originate new varieties and species on demand. In many cases, these interpretations were encouraged by the researchers themselves. Take Caryl Haskins, one of the engineers-turned-biologists at the GE research laboratory, for example. As he described to one reporter, "The possibilities, as applied to man's welfare, of being able to modify, in controlled fashion, the heredity of farm stock and crops are practically endless." Among other things, he speculated that a mutation for cold tolerance in oranges or in sequoia would enable these to be grown as crops even in his hometown of Schenectady, New York.[50] Although this was likely inspired by Haskins's knowledge of mutations that enhanced cold tolerance in other crops, it was of course hardly a modest aspiration. His description suggested that even small genetic changes could mean large changes in agricultural production, which in turns helps to explain why some reporters thought that X-ray-induced mutation would, through enhanced control of heredity, "revolutionize agriculture."[51]

In some ways, this was nothing new. The science of genetics had since its earliest days inspired claims that humans had gained control over the direction of evolutionary change, hence the enthusiasm of farmers and breeders, not to mention the support of eugenicists, for this developing discipline at the turn of the century. But the demonstration of X-ray-induced mutation generated a novel set of ideas and hopes, which extended beyond those initially envisioned by the champions of Mendelian genetics. These focused not only on controlling the direction of evolution but also its pace. Many Americans now envisioned that breeders would be "speeding up" biological evolution by producing on demand the very mutations that were thought to be the basic material of evolutionary change.

Prior to 1927, breeders were sometimes declared to have accelerated the evolution of varieties through selection and hybridization, and this was especially true of the ever-popular Luther Burbank. The famous Burbank, mentioned above, had garnered significant media attention just before the turn of the century for his introduction of many varieties of fruits, vegetables, and flowers. These included some spectacular alterations of form, such as white blackberries, pitless prunes, and spineless cactus plants, alongside more typical improvements in colors, flavors, hardiness, and other traits. To many observers his achievements seemed to indicate greater human control over evolutionary change.[52] "Darwinism taught us that species arose only through slow ages of change by the gradual process of natural selection accumulating its effects for thousands and even millions of years; but Luther Burbank shows that *man* can progress species and do it in a dozen summers!" exclaimed one popular magazine of Burbank's methods.[53]

---

[50] Quotation in Gray (1932), p. 3, 13.

[51] Early (1932), p. 19. For Haskins on mutations for cold tolerance, see Haskins (1932b), p. 471.

[52] For biographical accounts of Luther Burbank, see Dreyer (1985), Smith (2009). On his enduring popularity, see Pandora (2001).

[53] Serviss (1905), p. 64.

After the demonstration of X-ray-induced mutation, this rhetoric shifted dramatically. Where Burbank had once been described as taking plant evolution into his own hands, deftly reorienting it to the whims of his own imagination, he now became the quintessential picture of slow and old-fashioned methods. According to popular lore, Burbank had planted 1000 seeds of each plant he hoped to improve, even though only one in that number would have a new trait helpful to his breeding projects. "You can imagine how much work that was for Mr. Burbank and how many miles of land he needed to grow his experimental plants," one reporter declared, and she was not alone in her characterization.[54] In the era of the X-ray breeder, all this would change. Where Burbank had had to wait for mutations to appear among his many varieties, taking the slow pace of natural evolution as a necessary constraint on his ambitions, breeders now had the option of causing mutations to appear "at will." No longer did they need thousands of plants and miles of productive land. "If Burbank had known what we know now, he could have done his work on a city plot," claimed a breeder who had taken up his own set of plant experiments in imitation of Muller, Goodspeed, and others.[55] It seemed logical to surmise that one mutation in a thousand individuals would now be one in ten, and a 100 years of slow change would be effected in just a couple of seasons.

The benefit to speeding up the pace of evolution was that it would offer humans greater opportunities—in the form of greater variety among individual animals and plants—for crafting organisms to meet the ideals they imagined. Journalists compared X-ray breeding to the streamlined processes of the factory floor, implying that a previously slow and haphazard process had been rendered acceptable to the machine age.[56] With the aid of X-rays, the breeding process would be as easy as building a new house, for according to one account, "the X-rays act like a carpenter with a saw and a hammer entering the strings of the chromosomes.... The rays cut some of these chromosomes out of their natural positions and fix them elsewhere."[57] *Popular Mechanics* likened the process to one of machine assembly: "As shifting of wheels, nuts, screws and bolts changes a machine, so the X-ray changes the life form and there is a new creation."[58] Another headline declared in short, "New Life Made to Order."[59]

Most of the X-ray researchers acknowledged that their method was not yet as precise as these journalists suggested in their discussions of making plants "to order." Instead, they described the process of X-ray-induced mutation as generating random change and often damage to or destruction of a plant.[60] An early report quot-

[54] Early (1932), p. 19. Similar assessments appeared elsewhere, e.g., Thone (1927), Curtis (1928), p. 147.

[55] Quotation in Early (1932).

[56] An example that references the future application of X-rays in animal breeding is Anonymous (1927c), p. 21A.

[57] Anonymous (1935), p. N1.

[58] Anonymous (1930), p. 567.

[59] Gray (1932).

[60] Stadler was among the clearest on this, but almost all researchers, even those more enthusiastic about mutation breeding, acknowledged the extent of damage done to both tissues and cell structures in the process of X-raying, and the large number of deleterious outcomes.

ed Muller's description of the "shot-gun method" in which researchers generated many random mutations and selected interesting ones from among these; according to the report researchers would rely on this technique "until some as yet unthought-of-improvement in selective technique can be made."[61] In 1929, Muller continued to be enthusiastic that agriculturists would through induced mutation "greatly improve and alter the forms and functionings of...domestic animals and plants" but cautioned that scientists were "almost as far as ever from producing to order the exact mutations which we want."[62] Other scientists working in the field felt that achieving controlled or selective mutation—though certainly not yet possible—was within closer reach. The researchers working on X-ray manipulation of plants at GE counted among the most optimistic on this front. As the GE employee Caryl Haskins readily admitted, although X-rays could "shatter" the pattern of genes, he and his coworker had no idea what would emerge from this process. He remained confident, however, that once physicists determined how radiation affected the cell, and biologists had mapped the position of genes, their combined knowledge would lead to use of X-rays to generate a precise and predictable outcome. In his words, "Someday we'll be able to eliminate the element of chance, will know just where to aim and what intensity and degree of bombardment to use to attain the exact result."[63]

Visions of life "made to order" obviously flourished in the press in spite of researchers' hesitations and caveats. This is perhaps not surprising given the aims of the popular press—to sell publications—and the celebratory tone of much science reporting of the 1920s and 1930s.[64] That these visions flourished in the late 1920s and persisted through the 1930s also reflected in part American attitudes towards industrialization and factory production. The achievements of industrial scientists and engineers in the early twentieth century in developing and managing large-scale technological enterprises inspired faith, in the words of the historian Thomas Hughes, "not only that they could create a new world, but also that they could control it."[65] This enthusiasm for mechanization and rationalization extended even to agricultural production, where the imposition of an industrial logic—in the form of new business and management models as well as new machines—was reshaping agriculture to more closely conform to the industrial ideal.[66] In this context, it might have been natural to assume that the creation of new plants and animals, or the

[61] Thone (1930).

[62] Muller (1929), p. 505.

[63] Gray (1932), p. 3; see also Anonymous (1933b), p. 19. Haskins's subsequent pursuit of research into genetic and other biological effects of radiation and his championing of research into physical and chemical methods of altering genes and chromosomes suggest that these statements were not simply promotional hot air.

[64] The early decades of the twentieth century saw a significant expansion in reporting on scientific and technological achievements. On the history of science journalism in the United States, with a particular focus on the Science Service news agency (which played a significant role in publicizing the X-ray research discussed here), see LaFollette (1990). See also Burnham (1987), Chap. 5. For a more focused reflection on the content of science journalism in the 1930s (related to chemistry), see LaFollette (2007).

[65] Hughes (1989), esp. Chaps. 5 and 7. Quotation on p. 8.

[66] Fitzgerald (2003).

improvement of existing varieties, could be as regularized and rationalized as any other American industry, if only the appropriate techniques and technologies could be developed. With the help of X-ray breeding especially, "agriculture is beginning upon a boom that will give it rank with the other great industries of the country," declared the director of the news agency Science Service, Edwin Slosson, in June of 1928.[67] "Speeding up evolution" did not mean that it would accelerate beyond human control. On the contrary, it would bring the creeping natural pace of this process into sync with the more demanding tempo of modern American society.

Conclusion

The demonstration of induced mutation by Muller in 1927 was not of interest to experimental biologists alone but rather caught the attention of a great many observers. The X-ray tube was understood as the first proven technology for generating permanent inheritable changes in living things. Although the exact mutations that would be produced by radiation treatment were unpredictable, it appeared that any organism and its future offspring could be made to change simply through exposure to ionizing radiation. At a time when the chemical substance and physical structure of the gene had yet to be determined, this effect could understandably be characterized in the popular press as "magic," as it frequently was, for even among scientists there was no truly satisfactory biological explanation.[68] But it was also a real effect, one that produced visible alterations that persisted through many generations. As such, it had no comparable technology in 1930 save for exposure to radium, a far more expensive and less accessible tool.[69]

Given the longer history of interest in controlling mutation among agriculturists, it was a small leap for many to assume that this was a discovery that might upend standard practices in plant breeding. Enthusiasm for experimenting in X-ray breeding was encouraged by contemporary understandings of genetic mutation and its role in evolution and heredity. Mutations occurred all the time in nature: in woods, fields, and gardens. Using X-rays would simply be a technological intervention to accelerate the natural process of change on which breeders already relied. Interest in X-ray mutation was also at times bolstered by the idea that it might be merely the first step towards the achievement of a far more precise improvement process. This freed some researchers to see themselves as working on the improvement not only of a specific type of plant but also of X-ray technology as a generalized tool of genetic modification. This was a particularly influential idea for those at the GE research laboratory, perhaps because they worked in a context of industrial innovation where technological devices remained of surpassing interest. But it had not escaped

[67] Anonymous (1928a), p. 10.

[68] Although radiation-induced changes provided a further means to study the physical nature of the gene, neither Muller nor his contemporaries knew in the late 1920s the mechanism of change. See Pauly (1987), pp. 177–183. For descriptions of the mutation effect as "magic," see for example, Anonymous (1927b), Thone (1930), Anonymous (1933a).

[69] For a comparison between X-rays and radium at this time, see Campos (2006), pp. 319–320.

other researchers. For example, even Stadler made note of efforts to make the various processes of gene and chromosome mutation "more efficient."[70]

As the history offered here suggests, research in genetics in the 1920s and 1930s included among its objects the aim of creating a method for inducing genetic change more efficiently, and perhaps even predictably. This biotechnological aspiration was shaped early on by agricultural needs and ambitions, especially by the idea that the extent of variation in a species formed a significant constraint to crop plant improvement and that artificially induced mutation would offer a way around this bottleneck. Efforts to realize this goal were reported, discussed, and celebrated in the popular press, engaging an audience that extended well beyond the realm of experimental and agricultural research. In other words, even in its earliest decades, research in genetics was thought by many Americans to hold the key to distinct new technologies for agricultural improvement. Our knowledge of this expectation serves as a reminder of the long history of hopes that genetic science would produce technological interventions for the manipulation of life.

Of course, the X-ray tube did not turn out to be the revolutionary tool for plant improvement initially imagined. However much hoped for, useful genetic alterations proved elusive to those working with crops, such as tobacco, maize, and cotton, whether they worked at academic departments, experiment stations, or commercial and industrial operations. Changes produced in the course of irradiation treatment were, as Stadler had predicted, mostly deleterious. Even when something promising did emerge, it still required years of selection to stabilize the trait or to breed it into an existing line. It became clear by the mid 1930s that the rate of evolution of crops could not actually be speeded up that much through X-ray radiation. In the United States, crop varieties bred from irradiated stock and incorporated into large-scale cultivation counted only one by 1956, a variety of field bean primarily grown in Michigan and known as the Sanilac bean.[71] Horticulturalists had better luck making use of radiation-induced changes, a result of the comparatively fewer constraints on garden or specialty-flower improvement (as compared to mass cultivated crops such as maize or tobacco) as well as the strong appeal of novelty in many areas of horticultural production. For example, in the early 1940s, Burpee Seed marketed a pair of calendula varieties, the "X Ray Twins," plants the company claimed had been produced through X-ray-induced mutation, and the flower breeder Frank Reinelt was reported to use X-rays in producing an improved line of delphiniums.[72] But here, too, commercial successes remained few and far between.

One might think, given the evident failure of X-ray breeding by 1940, that the experience would have dashed hopes for a quick technological fix to the problem of producing genetic variation through induced mutation. On the contrary, Americans remained ever optimistic on this subject, turning their attention to other tools such as chemicals and radioisotopes in the decades that followed.[73] This historical trajec-

[70] Stadler to Slosson, 7 June 1928, Lewis J. Stadler Papers (Western Historical Manuscript Collection, Columbia, MO), Folder 1.

[71] On the development of this bean, see Andersen (1972).

[72] Anonymous (1942b); Taylor (1947), p. 55.

[73] The longer history of mutation breeding in American agriculture is discussed in Curry (2012).

tory further speaks both to the felt limitations of Mendelian genetics and traditional plant breeding methods in the middle decades of the twentieth century, as well as to a continuing faith in technological solutions to the challenge of controlling heredity.

References

Allen, Garland E. 1969. Hugo de Vries and the reception of the 'Mutation Theory'. *Journal of the History of Biology* 2:55–87.
Allen, Garland E. 2000. The reception of Mendelism in the United States, 1900–1930. *Comptes Rendus de l'Académie des Sciences – Series III – Sciences de la Vie* 323:1081–1088.
Anonymous. 1912a. Grow larger grain. *Mills County Tribune*, October 31.
Anonymous. 1912b. How to increase world's grains. *Evening Tribune*, September 25.
Anonymous. 1912c. Larger grain. *Wellsboro Gazette*, November 14.
Anonymous. 1927a. Feathers on birds determined by skin. *New York Times*, December 31.
Anonymous. 1927b. Scientists grow peculiar tobacco plants with X-ray. *Charleston Gazette*, October 30.
Anonymous. 1927c. X-ray to breed super animals. *Frederick Post*, November 19.
Anonymous. 1928. Farm boom beginning, 4-H Club campers told. *Washington Post*, June 26.
Anonymous. 1930. Secret of life sought in mystery ray. *Popular Mechanics*, October:564–567.
Anonymous. 1932. A symposium of some activities of the research laboratory of the General Electric Company. *Transactions of the American Institute of Chemical Engineers* 28:31–54.
Anonymous. 1933a. Magic-ray farmers. *Popular Mechanics*, July:58–61, 120A, 124A.
Anonymous. 1933b. X-rays are found to alter heredity. *New York Times*, September 27.
Anonymous. 1935. Better regal lily produced by X-ray. *New York Times*, September 1.
Anonymous. 1936. Unusual possibilities in breeding. In *Yearbook of agriculture* 1936, 183–206. Washington, DC: USDA.
Anonymous. 1942a. Burpee introduces his new calendulas. *New York Times*, January 27.
Anonymous. 1942b. Science: New flowers by X-rays. *Time*, February 2.
Andersen, A. L. 1972. The Sanilac story. Paper presented at the Mutation Breeding Workshop, University of Tennessee Agricultural Campus, Knoxville, TN. http://www.osti.gov/bridge/servlets/purl/4297499-OQeKLV/4297499.pdf. Accessed 1 July 2014.
Baker, Herbert G., Adriance S. Foster, and Johannes M. Proskauer. 1967. Thomas Harper Goodspeed, botany: Berkeley. In *University of California: In memoriam 1967,* ed. University of California Academic Senate, 43–46. Berkeley: University of California.
Berg, Paul, and Maxine Singer. 2003. *George Beadle, an uncommon farmer: The emergence of genetics in the 20th century.* Cold Spring Harbor: Cold Spring Harbor Laboratory Press.
Bishop, C. J. 1954. Mutations in apples induced by x-radiation. *Journal of Heredity* 45:99–104.
Bowler, Peter J. 1978. Hugo de Vries and Thomas Hunt Morgan: The Mutation Theory and the spirit of Darwinism. *Annals of Science* 35:55–73.
Bowler, Peter J. 1989. *Evolution, the history of an idea*, rev. ed. Berkeley: University of California Press.
Burnham, John C. 1987. *How superstition won and science lost: Popularizing science and health in the United States*. New Brunswick: Rutgers University Press.
Burpee, David, and Frank J. Taylor. 1940. So we shocked mother nature. *Rotarian*, July 15–17.
Campos, Luis Andres. 2006. Radium and the secret of life. PhD Diss., Harvard University.
Campos, Luis, and Alexander von Schwerin, ed. 2010. *Making mutations: Objects, practices, contexts*, Preprint 393. Berlin: Max Planck Institute for the History of Science.
Carlson, Elof Axel. 1981. *Genes, radiation, and society: The life and work of H.J. Muller*. Ithaca: Cornell University Press.

Cooke, Kathy J. 1997. From science to practice, or practice to science? Chickens and eggs in Raymond Pearl's agricultural breeding research, 1907–1916. *Isis* 88:62–86.
Curry, Helen Anne. 2012. Accelerating evolution, engineering life: American agriculture and technologies of genetic modification, 1925–1960. PhD Diss., Yale University.
Curtis, Winterton C. 1928. Old problems and a new technique. *Science* 67 (1728):141–149.
Dreyer, Peter. 1985. *A gardener touched with genius: The life of Luther Burbank*. Berkeley: University of California Press.
Early, Eleanor. 1932. X-rays to produce giant flowers, fruits, and animals, *Galveston Daily News*, January 24.
Endersby, Jim. 2013. Mutant utopias: Evening primroses and imagined futures in early twentieth-century America. *Isis* 104:471–503.
Fitzgerald, Deborah Kay. 1990. *The business of breeding: Hybrid corn in Illinois*, 1890–1940. Ithaca: Cornell University Press.
Fitzgerald, Deborah. 2003. *Every farm a factory: The industrial ideal in American agriculture*. New Haven: Yale University Press.
Foster, J. P. 1929. Artificially induced mutations. *Planter and Sugar Manufacturer* 82 (2):21–23.
Garner, W. W. 1936. Superior germ plasm in tobacco. In *Yearbook of agriculture 1936*, 785–830. Washington, DC: USDA.
Goodspeed, T. H. 1927. The x-ray in evolution. *The California Monthly*, December, 226–227, 256–257.
Goodspeed, T. H. 1929. The effects of x-rays and radium: On species of the genus nicotiana. *Journal of Heredity* 20:243–259.
Goodspeed, T. H., and A. R. Olson. 1928. The production of variation in nicotiana species by x-ray treatment of sex cells. *Proc Natl Acad Sci of the United States of America* 14:66–69.
Gray, George W. 1932. New life made to order. *Charleston Daily Mail*, August 21.
Haskins, C. P. 1932a. X-ray and cathode-ray tubes in biological service. Part I. *General Electric Review* 35:367–70.
Haskins, C. P. 1932b. X-ray and cathode-ray tubes in biological service. Part II. *General Electric Review* 35:465–471.
Haskins, C. P., and C. N. Moore. 1934. X-ray and cathode ray tubes in the service of biology. *Radiology* 22:330–333.
Hawkins, L. A. 1932. Electricity and animation. *General Electric Review* 35:247.
Horlacher, W. R., and D. T. Killough. 1933a. Progressive mutations induced in Gossypium hirsutum by radiations. *American Naturalist* 67 (713):532–538.
Horlacher, W. R., and D. T. Killough. 1933b. Radiation-induced variation in cotton: Somatic changes induced in Gossypium hirsutum by x-raying seeds. *Journal of Heredity* 22:253–262.
Hughes, Thomas Parke. 1989. *American genesis: A century of invention and technological enthusiasm 1870–1970*. New York: Viking.
Kass, Lee B., and Paul Chomet. 2009. Barbara McClintock. In *Handbook of maize: Genetics and genomics*, ed. Jeff Bennetzen and Sarah Hake, 17–52. New York: Springer.
Kimmelman, Barbara A. 1983. The American Breeders' Association: Genetics and eugenics in an agricultural context. *Social Studies of Science* 13:163–204.
Kimmelman, Barbara A. 1987. A Progressive Era discipline: Genetics at American agricultural colleges and experiment stations, 1900–1920. PhD Diss., University of Pennsylvania.
Kimmelman, Barbara A. 1992. Organisms and interests in scientific research: R. A. Emerson's claims for the unique contributions of agricultural genetics. In *The right tools for the job: At work in twentieth-century life sciences*, ed. Adele E. Clarke and Joan H. Fujimura, 198–232. Princeton: Princeton University Press.
Kingsbury, Noël. 2009. *Hybrid: The history and science of plant breeding*. Chicago: University of Chicago Press.
Kingsland, Sharon E. 1991. The battling botanist: Daniel Trembly MacDougal, mutation theory, and the rise of experimental evolutionary biology in America, 1900–1912. *Isis* 82:479–509.
Kohler, Robert E. 1994. *Lords of the fly: Drosophila genetics and the experimental life*. Chicago: University of Chicago Press.

LaFollette, Marcel C. 1990. *Making science our own: Public images of science, 1910–1955*. Chicago: University of Chicago Press.
LaFollette, Marcel C. 2007. Taking science to the marketplace: Examples of science service's presentation of chemistry during the 1930s. In *The public image of chemistry,* ed. Joachim Schummer, Bernadette Bensaude-Vincent, and Brigitte Van Tiggelen, 259–296. Singapore: World Scientific.
Martin, John H. 1936. Sorghum improvement. In *Yearbook of Agriculture 1936,* 523–60. Washington, DC: USDA.
McKay, J. W., and T. H. Goodspeed. 1930. The effects of x-radiation on cotton. *Science* 71 (1851):644.
Morgan, Thomas Hunt. 1919. *The physical basis of heredity*. Philadelphia: J.B. Lippincott.
Muller, H. J. 1927. Artificial transmutation of the gene. *Science* 66 (1699):84–87.
Muller, H. J. 1929. The method of evolution. *Scientific Monthly* 29:481–505.
Murphy, Denis J. 2007. *Plant breeding and biotechnology: Societal context and the future of agriculture*. Cambridge: Cambridge University Press.
Pandora, Katherine. 2001. Knowledge held in common: Tales of Luther Burbank and science in the American vernacular. *Isis* 92:484–516.
Paul, Diane B., and Barbara A. Kimmelman. 1988. Mendel in America: Theory and practice, 1900–1919. In *The American development of biology,* ed. Ronald Rainger, Keith R. Benson, and Jane Maienschein, 281–310. Philadelphia: University of Pennsylvania Press.
Pauly, Philip J. 1987. *Controlling life: Jacques Loeb and the engineering ideal in biology*. New York: Oxford University Press.
Rédei, G. P. 1971. A portrait of Lewis John Stadler, 1896–1954. In *Lewis. J. Stadler Genetics Symposium 1,* 5–20. Columbia: University of Missouri-Columbia.
Rhoades, M. M. 1957. Lewis John Stadler, 1896–1954. *Biographical Memoirs of the National Academy of Sciences* 30:328–347.
Rosenberg, Charles E. 1997. *No other gods: On science and American social thought*, rev. ed. Baltimore: Johns Hopkins University Press.
Schwartz, James. 2008. *In pursuit of the gene: From Darwin to DNA*. Cambridge: Harvard University Press.
Serviss, Garrett P. 1905. Transforming the world of plants. *The Cosmopolitan,* November, 63–70.
Sinnott, Edmund W., and L. C. Dunn. 1925. *Principles of genetics: An elementary text, with problems*. New York: McGraw-Hill.
Smith, Jane S. 2009. *The garden of invention: Luther Burbank and the business of breeding plants*. New York: Penguin.
Stadler, L. J. 1928. Genetic effects of x-rays in maize. *Proceedings of the National Academy of Sciences of the United States of America* 14:69–75.
Stadler, L. J. 1930. Some genetic effects of x-rays in plants. *Journal of Heredity* 21:3–20.
Taylor, Frank J. 1947. How your flowers are remodeled. *Saturday Evening Post,* March 1, 54–56.
Thone, Frank. 1927. X-rays speed up evolution over 1,000 per cent. *Science News-Letter* 12 (340): 243–246.
Thone, Frank. 1928. Science prize winner makes discovery believed by many to have immense practical and scientific significance. *Scientific American* 138:235.
Thone, Frank. 1930. New plants and animals developed by magic x-ray. *Corpus Christi Times*, June 20.
Traub, H. P., and H. J. Muller. 1934. X-ray dosage in relation to germination of pecan nuts. *Botanical Gazette* 95:702–706.
Vries, Hugo de. 1905. The aims of experimental evolution. In *Carnegie Institution of Washington Year Book No. 3 (1904),* 39–49. Washington: Carnegie Institution of Washington.
Ware, J. O. 1936. Plant breeding and the cotton industry. In *Yearbook of agriculture 1936,* 657–744. Washington, DC: USDA.

Helen Anne Curry is a lecturer in the Department of History and Philosophy of Science at the University of Cambridge and a fellow of Churchill College. Her research considers the histories of biology and biotechnology, agriculture, and environmental change in the twentieth century. She is currently writing a book on the history of early genetic technologies in American agriculture and imagination.

Chapter 23
Watching Grass Grow: The Emergence of *Brachypodium distachyon* as a Model for the *Poaceae*

Christopher W. P. Lyons and Karen-Beth G. Scholthof

Introduction

In January 2006, President George W. Bush in his State of the Union Address, announced to the country the "advanced energy initiative." A main goal of the initiative was to facilitate "a 22% increase in clean-energy research at the Department of Energy to push for breakthroughs in energy efficiency in homes and offices and advancing technology for more efficient fuel for automobiles." Specifically, Bush's call to action was an appeal to support the funding of "additional research in cutting-edge methods of producing ethanol, not just from corn but from wood chips and stalks or switch grass [sic]," with a goal "to make this new kind of ethanol practical and competitive within 6 years."

Only weeks before, a "Biomass-to-Biofuels Workshop" had been held by the Department of Energy (DOE) to "define barriers and challenges" that limited the production of cellulosic ethanol from grasses and crops and "to determine ways to speed solutions through concerted application of modern biology tools."[1] During this December 2005 meeting, an objective of developing new crops—"green crops" or "bioenergy crops"—"plants specifically designed for industrial processing to biofuel"—was outlined for the coming decade. Yet, there was a significant barrier in bringing not only the candidate of choice, switchgrass, but also other crop plants into this equation. Switchgrass (*Panicum virgatum*), a plant native to North America, had not "been bred extensively for the characteristics most desirable in energy crops."

[1] US Department of Energy (2006).

K.-B. G. Scholthof (✉) · C. W. P. Lyons
Department of Plant Pathology and Microbiology, Texas A&M University,
College Station, TX 77843, USA
e-mail: kbgs@tamu.edu

C. W. P. Lyons
e-mail: christopherwplyons@tamu.edu

D. Phillips, S. Kingsland (eds.), *New Perspectives on the History of Life Sciences and Agriculture,* Archimedes 40, DOI 10.1007/978-3-319-12185-7_23

Fig. 23.1 It fits in the palm of your hand. *Arabidopsis thaliana* (erecta mutant; *left*) and *Brachypodium distachyon* (Bd21 line; *right*) shown. For any organism to be considered a model, it must exemplify certain characteristics: it must have a small genome, grow in the laboratory, and be petite in stature. (The *Arabidopsis* photograph (Meyerowitz and Chang 1988) is used with permission of Springer; the *Brachypodium* photograph is used with permission of BrachyTAG.org)

Instead, the concentration had been on developing the plant for forage.[2] Were these planned advances for bioenergy practical, when it typically required 10 years or more to develop improved crop varieties? The strategy, as set forward by the DOE in June 2006, was to use model organisms as the key to "bringing the techniques of 21st century systems biology to bear on the complex problems." For this purpose, two species—*Brachypodium distachyon* (purple false brome) and *Populus trichocarpa* (poplar)—would prove to be the models for "rapid testing of strategies to improve the usefulness of grasses and trees as energy crops. Such tools would allow scientists to use modern molecular genetic methods to design superior energy crops."[3]

For scientists, having an organism raised to the consensus status of "model" often provides increased group funding, resources, and prestige. The 20th century saw the rise of model organisms that have included animals (e.g., mice, nematode worms, and zebra fish), plants (e.g., *Arabidopsis,* rice, and maize), and microbes (e.g., yeast, bacteria, viruses, and fungi). These experimental models met specific research needs and were used to bridge the gaps between classical Mendelian genetics, the well-established field of biochemistry, and the emerging discipline of molecular biology. The intent of this chapter is to follow the development of *B. distachyon,* a wild grass species (Fig. 23.1), as a model organism that has come

---

[2] Vogel et al. (2007).

[3] Stokstad (2006) described *P. trichocarpa,* or black cottonwoods, as "lab rats of the tree world." *Populus* was selected because it is a fast-growing tree valued by the timber industry for commercial products (plywood, pulp, and cardboard) and targeted by the DOE as a high-biomass energy crop for the 21st century. The poplar genome sequence was published in September 2006. It was the first of a tree and the third of a plant species, after that of *Arabidopsis* and rice.

from seemingly "different roots"—first, as a taxonomically interesting plant, then as a plant used to study host–pathogen interactions, and most recently as a plant intended to speed progress in developing bioenergy crops.

Philosopher of science Marcel Weber, in an analysis of model organisms, commented that "historians have found that it is far from clear that biologists deliberately choose certain organisms for solving specific problems. It may well be the other way around: The choice of certain organisms defines what comes to be viewed as a relevant research problem. This choice is mainly determined by locally contingent factors."[4] While we agree that the choice of *Brachypodium* did indeed reflect locally contingent factors, we also maintain that scientists deliberately chose it in order to solve specific research problems.

The historiography of model organisms has been well represented for several animal species and microorganisms;[5] we hope to expand the discussion of model organisms to include recent plant models. Although there is a small literature collection on *Arabidopsis thaliana* as a model plant in the modern genomic era, other plant models have received even less attention.[6]

In 2005, *Brachypodium* was identified by the Department of Energy's Joint Genome Initiative (DOE-JGI) as a model laboratory grass to develop genetic tools to improve switchgrass for the US bioenergy program. We became intrigued by the aforementioned push by the DOE to develop new models to advance plants as bioenergy crops, and this interest led to further questions about how a species in the "post-genomics era" achieves model status. How and why *Brachypodium* became a model organism for the improvement of switchgrass as a bioenergy crop, is best reconstructed by beginning with its most recent role as a tool for bioenergy and working back to its beginnings as a taxonomic curiosity.

In the late 1990s, *Brachypodium* was identified as a model organism that would be used to bring agriculture from the field to the lab bench to solve "real" problems.[7] This intent to develop a model plant for applied biology suggests that there were locally contingent and practical reasons to focus on a plant with specific goals of addressing the technological improvement of grasses, an improvement that would have significant economic impact. We have found that there was a further, perhaps stronger, interest in using *Brachypodium* to study plant pathogens, with the implication that it would be of relevance to improving crop resistance against disease. Specifically, *Brachypodium* was developed as a model for improving wheat resistance

---

[4] Weber (2005, p. 155).

[5] For detailed histories of model organisms, see Ankeny (2001), Creager (2002), Keller (1984), Kohler (1994), Rader (2004), Endersby (2007), Summers (1993), and Summers (1999).

[6] That is not to say that there is no record. The historiography of *Arabidopsis* is being developed by Sabina Leonelli and James Evans, both of whom have focused on the roles of Chris Somerville and Shauna Somerville (while at the Carnegie Institution, Stanford University, from 1994 to 2007) in developing *Arabidopsis* as a model organism. Leonelli (2007a, b) and Evans (2007, 2010).

[7] Khan (1984), Khan and Stace (1999), and Vogel and Bragg (2009). *Brachypodium distachyon* (purple false brome) is one of about 18 species in the genus Brachypodium found in the temperate latitudes of the Mediterranean, Middle East, and India, and is also occasionally found as an introduced weed in South Africa, Australia, America, and many parts of Asia. A sister species, *B. sylvaticum* (slender false brome), is a perennial plant with a larger genome than *B. distachyon.*

to fungal pathogens. This breakthrough work was done by scientists at the United States Department of Agriculture (USDA), an institution that played an important role in establishing the USA as a center for *Brachypodium* genome-based research.

Since the 1980s, plant biology has been dominated by the use of the weed, *Arabidopsis,* as a model organism. We will trace how researchers have brought a second weed, *Brachypodium,* in from the wild to the laboratory, and how this plant was constructed as the "right tool for the job" through several transformations of the practice of plant biology.[8] We will argue that *Brachypodium* was made relevant by deliberate choice on the part of scientists, and that the knowledge gained from *Arabidopsis* was key in the development of the *Brachypodium* community, a small community that has been advocating *Brachypodium's* suitability as a model organism.

Background: Model Organisms for Plant Biology and Agriculture

For decades, *Arabidopsis* has been the model organism for plant biologists. It has been accepted, even embraced, by the life sciences research community—although until the 1980s, it was mostly regarded as a mere weed.[9] Initially, *Arabidopsis* was collected by evolutionary biologists to study linkages between phenotype and genotype, work that was established in the 1940s by Friedrich Laibach in Germany, and continued in the 1960s by George Rédei in the USA. As narrated by Rédei, Elliot Meyerowitz, Chris and Shauna Somerville, and Maarten Koornneef, each now renowned researchers in the *Arabidopsis* community, the plant met the needs of plant biology—it was easy to cultivate, being short in stature, with a rapid seed-to-seed life cycle (8–12 weeks), and was a self-fertile diploid with a small chromosome number and genome size.[10] These last traits made the plant desirable to geneticists, as it would prove to be amenable to laboratory research. Laibach pronounced that the diminutive, widely distributed crucifer was a "botanical *Drosophila.*" Yet, and we echo the work of Sabina Leonelli here, *Arabidopsis* had not gained model status even at the beginning of the 1980s. As an object for research, the plant had only begun its journey in the burgeoning era of molecular biology. With the advent of biotechnology, to produce genetically modified plants, petunia and tobacco were initially the plants of choice to advance molecular breeding. The history painted by

[8] Clarke and Fujimura (1992).

[9] Baker (1965, 1974). A plant is a weed, as defined by the ecologist Herbert Baker, "if, in any specified geographical area, its populations grow entirely or predominantly in situations markedly disturbed by man (without, of course, being deliberately cultivated plants)." That is, a weed is a plant growing where it is not wanted.

[10] Somerville (1989, 2000, 2001), Somerville and Koornneef (2002), Somerville and Somerville (1996, 1999), Meyerowitz and Chang (1988), Meyerowitz (1987, 1989, 2001), Meyerowitz and Pruitt (1985), Pang and Meyerowitz (1987), Koornneef and Meinke (2010), Meinke et al. (1998), and Rédei (1975). These scientists have published (sometimes together) several perspectives on *Arabidopsis*.

Leonelli, and echoed by the Somervilles, Koornneef, and Meyerowitz, guides us through the genesis of *Arabidopsis* research and how, in the late 1980s, this plant became identified as a "model and specifically as a model of any plant."[11]

Delving into the development of *Arabidopsis,* Leonelli focuses on two scientists, Chris and Shauna Somerville. From her interviews, Leonelli found that, in their youthful exuberance, the Somervilles wanted to establish a system that was not burdened by "competition," and "the cumbersome legacy" left by the *Drosophila* workers.[12] Two other scientists, Maarten Koornneef (Wageningen University) and Elliot Meyerowitz (California Institute of Technology), played crucial roles in bringing *Arabidopsis* to model status. Meyerowitz, in particular, was a key figure because his body of work bridged the era of classical plant physiology and the newer early molecular era in which he and other young researchers would shine as visionaries. Meyerowitz had taken a postdoctoral fellowship with David Hogness at Stanford. Hogness made many seminal contributions to the study of the molecular genetics of *Escherichia coli, Drosophila,* and lambda bacteriophage and was instrumental in defining each organism as a model in its own right. And Hogness had used these models as reagents to build tools that were widely used for the development of recombinant DNA methods.[13] From his training with Hogness, Meyerowitz had become an expert in the techniques of genetic mapping and DNA cloning, and realized the possibilities of using phages as tools. In the mid-1980s, Meyerowitz laid the foundation of *Arabidopsis* molecular genetics, using these recombinant DNA technologies and, in 1985 with Caren Chang, submitted the first report of cloning a gene from *Arabidopsis.*[14] The Somervilles, likewise, came to molecular biology from backgrounds in plant physiology and plant pathology. As their own reflections as well as Leonelli's analysis suggest, an important goal was to build a community of *Arabidopsis* workers where sharing and collaborative interactions would be rewarded.[15] Koornneef, as a PhD student (which gave him a coappointment as an assistant professor), produced a suite of papers detailing a unified genetic linkage map of the five chromosomes of *Arabidopsis*—a key offering helping to bring this plant to model system status.[16] Each of these pioneering researchers were exposed to the success of other, older model systems and technologies—experiences which would influence the development of their own ideals of how to form a cohesive collaborative community of plant biologists that would focus on the molecular biology of a single tiny weed. They did not, however, have particular shared goals (e.g., goals associated with direct translational application).

[11] Leonelli (2007a, p. 198). Sabina Leonelli, and James A. Evans have detailed portions of the history, philosophy, and sociology of *Arabidopsis* and its workers.

[12] Leonelli (2007a, p. 201).

[13] David Hogness received the 2003 Thomas Hunt Morgan Medal.

[14] Chang and Meyerowitz (1986). Chang and Meyerowitz showed for *Arabidopsis* "the feasibility of isolating a dicot plant gene by homology with a monocot gene," in this case, the maize alcohol dehydrogenase gene 1 (*Adh1*). Their study also demonstrates that *Arabidopsis* was in its infancy as an experimental plant, having to rely upon maize for genetic guidance.

[15] Somerville and Somerville (1996).

[16] Koornneef et al. (1983). *Arabidopsis* was introduced to Wageningen by Will van der Veen in 1968 as a model for mutagenesis, flowering time, and seed germination.

This initial small group of *Arabidopsis* workers in the USA and Europe pushed for funding, standardization, and shared resource material (e.g., seeds, transgenic plants, and DNA sequences) and, in 2000, completed the first genomic sequence for a flowering plant. As we discuss below, by 2005, the *Arabidopsis* workers' choices and their outcomes, made with strong guidance from Chris Somerville and building on the foundations laid by Rédei, Meyerowitz, and Koornneef, would prove instructive for the *Brachypodium* workers.

Brachypodium: Rooting a Grass

In the twentieth century, *Brachypodium* was of interest to botanists because it was difficult to place taxonomically in the grass family (*Poaceae,* formerly *Gramineae*). Taxonomic and cytogenetic studies of the 1960s suggested that *Brachypodium* is an ancient (or relic) genus.[17] Grasses are approximately 200 million years (evolutionarily) distant from the divergence event that led to their dicotyledonous kin, *Arabidopsis*.[18]

Clive Stace, a botanist at the University of Leicester (England), was instrumental in the contemporary classification of *B. distachyon.* We suggest that he facilitated bridging the study of *Brachypodium* as a plant for classical botany/taxonomy and as a model of interest in the newer field of molecular biology in a manner quite similar to the events that occurred during the development of *Arabidopsis*.[19] In 1978, Ian Robertson, a student in Stace's botany group, showed the number of chromosomes present in seven different *Brachypodium* species. *B. distachyon* had the smallest number of chromosomes ($2n=10$), as well as, the smallest genome, thus confirming an earlier 1963 report from Bulgaria.[20] Stace and his graduate student, Mir Ajab Khan, continued this work in great detail to include cytology, enzyme assays, and interspecies hybridization of other known *Brachypodium* species. They concluded that *B. distachyon* was indeed an ancient grass, had a small genome, and was a taxonomic and phylogenetic anchor for species in the family *Poaceae*.[21] The taxo-

[17] Tateoka (1968). This became an important point when *Brachypodium* was promoted as a model system for the *Poaceae,* as it demonstrated that the cereals (wheat, barley, rye, rice, and maize) and grasses (switchgrass and sugarcane) were evolutionarily related to and more evolutionarily recent than *Brachypodium*.

[18] Wolfe et al. (1989).

[19] Catalán et al. (1995). In 2012, the taxonomy was again revised and a new species, *Brachypodium stacei,* was "dedicated to Prof. Clive A. Stace, who initiated the systematic and evolutionary studies of Brachypodium."

[20] Robertson (1981).

[21] Khan (1984) and Khan and Stace (1999). Khan produced a dissertation in 1984, although the results were not published in a peer-reviewed journal until 1998. This suggests that *Brachypodium* was not on the "radar" of plant biologists—it was a curiosity for taxonomists.

nomic study of the genus *Brachypodium* was continued by Ying Shi, also a graduate student with Stace. Following an introduction of the botany and taxonomy of the genus, Shi wrote of "the likely importance of *Brachypodium* in our understanding of the evolution of the grasses" and the importance of molecular taxonomy for placement of grass species. To reach such conclusions, she used DNA-based techniques to determine the degree of relatedness between *Brachypodium* species and confirmed its small genome.[22] These data were published with the assistance of Pilar Catalán, a systematic botanist, who was a postdoctoral fellow with Stace for about 18 months (1990–1991) preceding her faculty appointment at the University of Zaragoza (Spain).[23]

The stated goal of determining the DNA-based taxonomy of this still unimportant plant was to resolve "the controversial position of the genus Brachypodium… within the family *Poaceae*."[24] In 2000, Catalán and Richard Olmstead provided the first DNA sequence data used for taxonomic placement of *Brachypodium*.[25] By 2000, with some confirmatory DNA-based analyses, the taxonomy was settled enough for the botanists—*Brachypodium* was quite different from other genera in the *Poaceae,* and the species *B. distachyon,* an annual plant, deserved placement away from the perennial members of the genus, such as *Brachypodium sylvaticum*.[26] At this time, there were no indications that *Brachypodium* was going to be presented as a model organism for plant biology.[27]

---

[22] Shi (1991). The methods used to assemble phylogenetic trees included restriction fragment length polymorphism (RFLP) and random amplification of polymorphic DNA (RAPD) assays. Draper was cosupervisor for Shi's dissertation which, according to Mur, "got John [Draper] interested in Brachypodium."

[23] Catalán et al. (1995, 1997). Catalán also had research visits in the laboratories of Elizabeth Kellogg (Harvard University Herbaria, and later at the University of Missouri–St. Louis) and Richard Olmstead (University of Colorado Boulder, then at the University of Washington, Seattle). Kellogg is an expert on grasses—including wheat, maize, and rice—and had coauthored several papers with Jeffrey Bennetzen, a key figure in the colinearity hypothesis, including Bennetzen and Kellogg, "Do plants have a one-way ticket to genomic obesity?" Catalán, Kellogg, and Olmstead coauthored a paper on *Poaceae* phylogenetics, including *Brachypodium* species, in 1997.

[24] Catalán et al. (1995).

[25] Catalán et al. (1997, 2012) and López-Alvarez et al. (2012). The sequence of the chloroplast *ndhF* gene and the nuclear-transcribed spacer region (ITS) of ribosomal RNA was used to determine taxonomic placement of *Brachypodium*. The interest in the evolution and taxonomy (systematics) and *Brachypodium* has not abated. As new molecular tools and data analyses become available, such as "DNA barcoding" identification of new species, the taxonomy of *Brachypodium* has been reinvigorated.

[26] That *Brachypodium* was merely of taxonomic interest to only a few botanists is supported by the dearth of publications for a 10-year span from 1985 until 1995.

[27] Gressel et al. (1983) and Aronson et al. (1992, 1993). Field experiments on *Brachypodium* included studies of herbicide-resistant plants growing along the roadsides in Israel with concerns it could become a noxious weed. In particular, the question arose if pollen from herbicide-resistant genetically engineered plants might spread to weeds, such as *Brachypodium,* with unknown ecological and agricultural outcomes. Ecology studies of *Brachypodium* were made in Israel related to an interest in drought tolerance of grasses.

Proposing *Brachypodium* as a Model

The first hint that *Brachypodium* had features expected of a model plant emerged in 1991. An abstract was presented at the Third International Congress of Plant Molecular Biology in Tucson, Arizona, that profiled the plant's amenable characteristics and small, unique genome. The paper even went on to state that "…*B. distachyon* is very reminiscent of *Arabidopsis*. It is suggested that *B. distachyon* may be worth developing as a model system for molecular genetic studies in the temperate cereals belonging to the *Triticeae*."[28] However, despite this call, *Brachypodium* seems to have been ignored for nearly 10 years.

In the December 2001 issue of *Plant Physiology*, the editors introduced a special issue on of the importance of "Celebrating Plant Diversity: Biodiversity in the Age of Genomics."[29] They pointed out the need to find model organisms in addition to *Arabidopsis:*

> Arabidopsis is, and will continue to be, an immensely important model system in plant biology…. But *Plant Physiology* is not and never will be the *Journal of Arabidopsis Research.* Clearly, Arabidopsis is an inferior, or even an impossible, system for studying many important plant processes. In such instances, plant scientists should not hesitate to seek alternative model organisms even though the molecular biology of these alternative species is less completely known or absent.

In the same issue, a report on a workshop on the maize genome project was presented, driven by a "mandate from the maize (*Zea mays*) genetics community." The promise of completing the sequencing of the maize genome was to:

> facilitate improvements in maize and other crop species. These agronomic improvements will have enormous impacts on mankind through improving human health, increasing energy production, and protecting our environment. The production of novel compounds in plants, including industrial feed stocks, biofuels, and medicinal compounds will increase the demand for corn and thereby directly benefit the agricultural community.[30]

This statement anticipated the announcement, in April 2002, that sequencing of the rice genome had been completed—a plant once itself trumpeted as "the model grass."[31]

This push for the sequencing of the maize genome, combined with the fact that corn and other grasses were actively being marketed via research funding, lobbying, and legislation as sources for biofuel suggested a new role and apparent need for "applied" genomics in the plant biology community.[32] In this same issue of *Plant*

---

[28] Shi et al. (1991).

[29] Raikhel and Minorsky (2001).

[30] Bennetzen et al. (2001)

[31] Goff et al. (2002) and Yu et al. (2002).

[32] The US interest in "biomass energy" extended back to at least 1980, when the first Bioenergy World Congress and Exposition was held in Atlanta. In 1980, the primary source of biomass in the US was "waste" wood, but Brazil was already using ethanol as a source of fuel for more than 300,000 cars "providing jobs and reducing dependence on foreign oil." In the USA, kudzu and kelp were potential sources of "liquid fuel." S. David Freeman, of the Tennessee Valley Authority, presciently remarked that "corn-based alcohol may be very good business for the farm lobby,

Physiology, lead authors John Draper and Luis Mur at the University of Wales, Aberystwyth, suggested to the plant biology community that *Brachypodium* would be an ideal model organism.[33] By publishing a paper detailing the utility of *Brachypodium* in this special issue, the editors likely intended it to gain the attention of the plant biology community.[34] And yet, *Brachypodium* remained essentially unknown to plant biologists—specifically the *Arabidopsis* workers and plant breeding communities (based on a search of peer-reviewed publications in the literature). What was so compelling about *Brachypodium* that led the authors to take the (bold) step of proposing that this new plant system was deserving of resources and scientific attention?[35]

Brachypodium was of no economic importance for agronomists. There were no hints in the literature that *Brachypodium* was on its way to being presented as a model organism for plant biology. During a decade that saw the rise of *Arabidopsis,* concurrent with the rapid development of the field of plant molecular biology,[36] plus impressive improvements in the technology for more rapid and cheaper DNA sequencing of a wide variety of organisms (largely a result of gearing-up for the Human Genome Project), there was no mention of *Brachypodium* in the "genomics" community at all.

There was, however, one trial balloon sent up for *B. distachyon* in 1995. Within the context of the excitement of "getting on the genome list," many scientists began to suggest that "their" species was a "model" or "model molecular genetic system." *Brachypodium* was not immune to this "modeling" or "systems building" phenomenon. An article published in 1995, by Pauline Bablak, John Draper, Mark Davey, and Paul Lynch, noted that *B. distachyon* had "several features of its genome and growth habit reminiscent of [Arabidopsis] that may allow it to be developed as a model molecular genetic system representative of the temperate grasses."[37] The

but…digging into our bread basket poses the grave risk of driving up the price of fuel in a hungry world." See Smith (1980).

[33] Draper et al. (2001). In 2001, the journals of record for the plant biology community were *Plant Physiology, Plant Cell, Plant Journal,* and *Proceedings of the National Academy of Sciences (USA).*

[34] Draper et al. (2001). According to Mur, "in early 2001 John [Draper] and I were visiting a US company that brought in Natasha [Raikhel, University of California-Riverside] in an advisory capacity. She liked our story so much that she asked us to draw it all together" for publication in *Plant Physiology*. This paper was received by *Plant Physiology* for peer review on 23 February 2001, a revision was received on 3 May 2001, and it was accepted for publication on 1 June 2001, and held for the special issue published in December 2001.

[35] As this is a preliminary exploration of the development of *Brachypodium,* sorting out the lineage is, thus far, based on peer-reviewed publications, publically available documents from the DOE, and personal correspondence and/or discussions with John Vogel, David Garvin, Pilar Catalán, and Luis Mur.

[36] Kim and Jansen (1995). In 1995 the *ndhF* gene was proposed as a tool for evolutionary placement of plants, in part, because it showed high rates of sequence change between species, first showed for the *Compositae* (the sunflower family), a family that is notoriously difficult to identify at the species level.

[37] Bablak et al. (1995).

authors reasoned that one way they could demonstrate the utility of *Brachypodium* would be to regenerate fertile plants from tissue culture—a fundamental, routine procedure for established plant models including *Arabidopsis,* maize, and rice. For this work, they used three *B. distachyon* ecotypes (one collected in Turkey and two from Spain). Although the efficiency of regeneration was low, it was "comparable with recent reports of describing rice tissue culture and thus its [Brachypodium's] potential as a model system has been confirmed." For the next 5 years, however, *Brachypodium* work seemed to stagnate in a manner similar to the history of *Arabidopsis* research before the 1970s.

The already-mentioned instrumental *Plant Physiology* follow-up paper by Draper and Mur, in 2001, would finally make a concerted effort to detail the merits of *B. distachyon* as a model species.[38] Around 1998, Draper and Mur moved to the University of Wales, Aberystwyth, charged with forming a new molecular plant pathology group that was to focus on disease resistance in cereals. Draper was familiar with *Brachypodium,* having published the before-mentioned series of papers as a coauthor with Stace and Catalán. As Catalán explained, Draper "had noticed that *B. distachyon* ($2n=10$) not only possessed the smallest known genome of the monocots but also showed other important attributes that make it an ideal model plant (e.g., short life cycle, small stature, inbred species, easy to cultivate, etc.)." Although there was not any initial funding behind it, "Draper pushed this proposal [for the development of Brachypodium] as he envisaged its importance and potential for plant breeding experiments and for the transference of the outputs to the temperate cereals and the biofuel grasses."[39] This 2001 *Plant Physiology* paper urged the plant biology community to consider *B. distachyon* as "a new model system for functional genomics in grasses":

> Hopes that Arabidopsis could serve as an "anchor" genome to help locate important chromosomal locations in cereal species have not been substantiated by recent studies. Thus, it is clear that grass (*Poaceae*) model systems are a key requirement for the future identification of genes of agronomic interest from cereals and forage grasses.

This argument built on earlier work of Jeffrey Bennetzen and coauthors on colinearity of the grass genome, first published in 1993.[40] Draper and Mur's statement, pushing for *Brachypodium,* neatly worked around the sensibilities of the maize and rice communities—and the wealth of knowledge developed by each group. Through a combination of field, greenhouse, and laboratory work, the *Brachypodium* community would develop chromosome maps with the location of hundreds of genetic loci using tools previously used for maize, rice, and *Arabidopsis* research.

The publicized importance of completing the sequencing of the (relatively small) rice genome was that the information could be applied to cereals (wheat, barley, and maize) and bioenergy crops. Based on a colinearity hypothesis, the genetic

---

[38] Draper et al. (2001).

[39] The quotations attributed to Pilar Catalán are from personal e-mail communication to the authors (22 June 2013).

[40] Bennetzen and Freeling (1993) and Bennetzen and Kellogg (1997).

diversity of cereals was a result of 50–60 million years of gene duplication and rearrangement of an ancestral grass genome.[41] Therefore, it would follow that sequencing a basal species would allow for more robust predictions about the evolution of grasses, with the end result being the creation of generalized strategies for cereal crop improvement. In 1993, *B. sylvaticum,* a perennial wild grass, was suggested as being useful for these comparative genetics. This and other *Brachypodium* "simpler genomes" could be used to characterize and isolate genes, which, in turn, would be used to "study their counterparts in other grass species."[42] This strategy would provide molecular plant breeders with "a powerful tool for fundamental plant science as well as for cereal breeding."

Draper and Mur provided several examples to justify working with *Brachypodium* (and not any *Brachypodium* species, but specifically, *B. distachyon*), primarily focusing on what rice was not. For instance, they commented that "the value of rice as a model for the temperate cereals and forage grasses...may be, on occasion, questionable." Thus, *Brachypodium* was a good model because it was just like the cereal crops (wheat and barley) and different enough from rice. Using the limited data on *Brachypodium,* which included chromosome counts, phylogenetic tree groupings, and botanically defined characteristics to show its evolutionary roots, rice was now on the wrong side of the cereal evolutionary tree, while *Brachypodium* fell in the right spot.[43] Furthermore, rice did not have all of the "good" features that plant biologists valued in the quintessential laboratory plant—in short, it was not *Arabidopsis*.

Constructing the Tools for a Model Grass

In 2001, *Brachypodium* shared only a few characteristics with *Arabidopsis*—it had a small genome, was an annual plant with a petite stature (including the de facto requirement that it fit in the palm of the researcher's hand), and was amenable to growth in the laboratory (Fig. 23.1).[44] But *Brachypodium* lacked most of what *Arabidopsis* (and maize) workers had established as critical tools. If development of *Brachypodium* was to follow the *Arabidopsis* research philosophy, the availability of many open-source tools, ranging from inbred and gene knockout lines, DNA sequence (completed in 2010), *Agrobacterium*-mediated transformation, and species hybridization would need to be developed and constructed. Even in comparison

[41] Bennetzen and Freeling (1993)

[42] Moore et al. (1993)

[43] The announcement was apparently premature. For the plant molecular geneticist "many steps remain before it [*B. distachyon*] can be fully utilized." These steps included developing "homogeneous inbred lines from diverse diploid ecotypes" and "an efficient" plant transformation system using *Agrobacterium,* both of which were achieved in 2006 by USDA scientists under the leadership of John Vogel and David Garvin. Quotes from Vogel et al. (2006a).

[44] We are grateful to Kranthi K. Mandadi for his comment that the de facto definition of a model plant is that "it must fit in the palm of your hand."

to the long list of what rice and *Arabidopsis* were not, *Brachypodium,* at the time, had a seemingly insurmountable list of tools that needed to be developed for it to become anywhere close to being a model system for plant biology—let alone for phytopathology or crop improvement.

Yet, *Brachypodium* had been noticed. In December 2002, John Vogel joined the USDA Western Regional Research Center (Albany, CA) to work on "a new project that was supposed to address developing grasses for use as biomass crops to make biofuel." Initially, Vogel intended to "use rice as a model to study the grass cell wall" (having worked previously on *Arabidopsis* cell-wall architecture). His and others' *Arabidopsis* research experiences would prove to be important for the subsequent development of *Brachypodium*. With "some serious roadblocks" including quarantine measures for imported rice seed (delaying experiments), poor climate for growing rice (too cool outside and lack of suitable controlled growth rooms and space), and the fact that the "rice community in those days was not about sharing resources as I had grown accustomed to with *Arabidopsis*," Vogel "gave up on rice before [he] even got started."[45] From a literature search, he found the Draper and Mur paper of 2001. From his personal recollection, some of his Agricultural Research Service (USDA-ARS) colleagues thought *Brachypodium* "looked promising," but "no one had grown it or knew anyone who had grown it."

From the *Plant Physiology* paper, the *Brachypodium* ecotypes were given alphanumeric designations, for example, ABR1 (referring to "Aberystwyth" (A) and "Brachypodium" (Br)). These materials originally were obtained from Stace (ABR1 from Turkey via Stace's collection) or the USDA (ABR15, USDA 254867 collected from Iraq, in 1958, by Paulden Knowles, but indicated in the list by Jenkins et al. as unknown location).[46] In the Draper and Mur paper, readers were referred to a website for detailed information about *Brachypodium* and that "seed samples of different *Brachypodium* ecotypes were obtained from Brachyomics [a company founded by the authors] under a "Research Only" Materials Transfer Agreement [MTA]".[47] Draper eventually replied to a request from Vogel with an "offer to send seeds,"[48] however, for the USDA, "the terms…were too restrictive" and the MTA was not signed.[49]

[45] Vogel was apparently not the only scientist having difficulty growing rice. In a talk presented on 19 October 2011 at the First European *Brachypodium* Workshop (Versailles, France) titled "*Brachypodium's* Rise as a Model" Vogel showed a slide with text from a French scientist who had written: "it is really hard to get rice plants to flower in northern France. We would therefore consider the use of Brachypodium as an alternative plant model. Would you mind sending us seeds… to see if this species could fit our needs."

[46] McGuire et al. (2012) and Jenkins et al. (2003).

[47] The website for Brachyomics is no longer accessible. Brachyomics Limited was founded on 8 October 2001, just prior to December publication of the Draper and Mur paper in *Plant Physiology*. Draper was listed as director and research professor from 24 June 2002 to 12 May 2009, when the company was dissolved, based on information from www.companydirectorcheck.com.

[48] The quotations attributed to John Vogel are from personal e-mail communication to the authors (24 April 2013). David Garvin also received an MTA from Draper (Brachyomics) and similarly recalled that "it was so restrictive that [the USDA] ARS would never agree."

[49] The requirement for a signed MTA prior to receipt of *Brachypodium* seed was confirmed on the Brachyomics web page from 19 May 2003. This page was retrieved from the Wayback Machine

As discussed by Leonelli, in reference to the *Arabidopsis* workers and the community developed around that plant, a lack of open sharing between researchers would have been antithetical to "an ethos of collaboration and coordination of research efforts."[50] As Leonelli observed, this code was so engrained within the *Arabidopsis* community that, by 2004, graduate students with Chris and Shauna Somerville at the Carnegie Institution's Department of Plant Biology "do not even understand the competitive attitudes that they witness outside Arabidopsis research."[51] *Brachypodium* resource sharing would need to be established under that same rubric for the community to survive and thrive.

If restrictions remained, there was the very real possibility that Brachypodium would never take root. However, around the same time, David Garvin, also of the USDA-ARS (Minneapolis, Minnesota), had initiated his own project with *Brachypodium* after reading the Draper and Mur paper in December 2001. Garvin had moved to Minnesota in August 2001 from his previous USDA-ARS position at Cornell with a "mandate" to work on wheat with a focus on diseases. He also had requested materials from Draper, but while waiting, Garvin "hiked over to the Ag library" and "it became crystal-clear" that many of the ABR lines were renamed USDA germplasm accessions.[52] Garvin requested from the USDA the available *B. distachyon* germplasm (ca. 30 lines) and started growing it in his lab. Garvin's research interest was in developing grasses that would be faster to grow and easier to handle in the greenhouse or growth chambers with a focus towards improving wheat resistance to the fungus-causing stem rust.[53]

By December 2002, Vogel had searched the USDA Project Database and found that Garvin was working with *B. distachyon*.[54] Importantly, Garvin "had decided to make his own inbred lines that he planned to make freely available" and sent Vogel "some seeds from his early generations." For Garvin, a plant geneticist, "one important step that needs to be taken is to develop homogeneous inbred lines from diverse diploid ecotypes." These would then be the inbred reference lines for future genetic and molecular studies and would be representative of the smallest genomes within the *B. distachyon* species. For developing inbred lines, Garvin found the Draper and Mur paper "very helpful" for selecting the putative diploid accessions. He raised plants from the USDA collection and collected "seed of several individual plants for (which) each accession was harvested and kept separate" for the development of 27 pure lines. Each line would be assigned the now-iconic alphanumeric designation of "Bd"—for *B. distachyon*—and a number (e.g., Bd21 from PI 452867 of the USDA

---

hosted by The Internet Archive (archive.org).

[50] Leonelli (2007a, p. 202).

[51] Leonelli (2007b, p. 58).

[52] The quotations attributed to David Garvin are from personal e-mail communication to the authors (30 April 2013 and 2 May 2013).

[53] Mackintosh et al. (2006). Paul D. Peterson has written extensively on the historiography of stem rust. Peterson (2001, 2003).

[54] Vogel and Garvin immediately began sharing information and resources. By Garvin's count, in 2003 "about 95%" of his e-mail correspondence on *Brachypodium* was with Vogel.

collection). He immediately made the DNA content, ploidy (e.g., 2*n*), and all seed freely available upon request. This intent to share materials, especially seed, was reminiscent of the open collaboration that helped to establish *Arabidopsis* as "the model plant" even in its infancy.

By 2004, Garvin had five diploid lines and initiated a project on cold tolerance with Eric Stockinger, a plant geneticist at Ohio State University. For that project, Stockinger had made a bacteriophage lambda DNA library of Bd3–1 and, using barley, "fished out" clones for sequencing. Additionally, he calculated genomic DNA values for a number of the inbred lines. A poster was presented detailing this and other work at the Plant & Animal Genome Conference (PAG) in January 2005. Following this meeting, Garvin felt "there was enough interest in Brachy[podium] (based on an increasingly large number of seed requests, scientists starting to work with it, etc.)" to request a *Brachypodium* Genomics Workshop for the following year, which was approved by the organizers.

In 2006, Vogel and Garvin coauthored two papers detailing "everything they had on Brachypodium."[55] In addition to showcasing the inbred lines, these papers met another primary goal—to develop an efficient method for *Agrobacterium*-mediated transformation. As had been done previously with *Arabidopsis,* it would be essential to be able to produce transgenic plants in order to generate gene knockouts and to build a library (a genetic database) to facilitate functional genomics using *Brachypodium*.[56] As remembered by Vogel,

> Fortunately, there was a good baseline for tissue culture conditions established going back to Bablak's paper in 1995. Unfortunately, all the previous transformation work was done using biolistics and polyploid accessions that are now considered a separate species, *B. hybridum*.... My initial transformation work was successful, but not highly efficient, and the first diploid transgenics created using Agro[bacterium] were made in September 2003.

After working with most of the inbred lines developed by Garvin, a single line, Bd21 was selected, as it could be transformed and produced fertile progeny.[57] From Bd21, further single-seed lines were tested, with Bd21–3 being the most amenable to transgenic work. Bd21 was selected for sequencing by the *Brachypodium* community (discussed below) and a genetic (T-DNA) insertion library was made using Bd21–3. The fledgling community had achieved a critical goal in developing the tools needed to make the claim that *Brachypodium* was a model system for grasses. Garvin, in an essay written in 2005, but published in 2007, argued that future work would still involve using *Arabidopsis* as a "template" to "determine where our energies need to be directed to establish *B. distachyon* firmly as a new model species."[58]

The background of Vogel, Garvin, Draper, and Mur suggests that the development of *Brachypodium* could highlight the field of basic plant pathology and its intimate links to agricultural research. From the Draper and Mur papers, Vogel's expertise with *Arabidopsis,* and Garvin's interests in wheat pathogens, we find that

---

55 Vogel et al. (2006a, b).

56 Brkljacic et al. (2011).

57 Vogel et al. (2006a).

58 Garvin (2007a).

each individual needed a suitable plant to pursue their work on cereals and grasses. Yet at the time, there was no suitable model—wheat was too complex, rice was not amenable to laboratory conditions, and *Arabidopsis* was not a host for cereal pathogens. By circumstance, luck, and resourcefulness, these groups brought *Brachypodium* to the laboratory. But the name of the game in the early 2000s was to get on the genome sequencing list.

Brachypodium and the Joint Genome Initiative

According to Vogel, we find that "at this point Chris Somerville and DOE step in." Vogel had had a postdoctoral appointment with Shauna Somerville, and, in 2002, he joined the USDA-ARS (Albany, California). Chris Somerville had a joint appointment at the Carnegie Institution and the DOE lab in Berkeley. He was "transitioning into biofuels at the time and was working closely with DOE management to set priorities" and "was interested in Brachy[podium]" to the extent that he had collected *B. distachyon* seed from the Jasper Ridge Biological Preserve (Stanford University) "at some point."[59] Vogel explained that "nothing came of that since no one in his lab chose to work with it."

For Somerville, the key to supporting and heralding *Brachypodium* was the ability to make transgenic plants—which would make possible gene-specific mutations (or knockouts) and the ability to move genes from various cereals (or *Arabidopsis*) to *Brachypodium* and vice versa. As narrated by Vogel, Somerville "thought it would be a good idea to use Brachy[podium] as a model for biomass crops." And this was "crucial" in Vogel's view because "when Chris says something is a good idea, people, specifically people at DOE, listen." Garvin also reminds us that we "have to appreciate that just a few of us were in the midst of building up even basic genetic and other resources [for Brachypodium]."

In June 2005, Garvin, "despite it seeming utterly audacious," contacted the Department of Energy Joint Genome Initiative (DOE-JGI) about having *Brachypodium* added to the sequencing list. Timing was everything, as there was rumored internal interest in sequencing a cool season grass with a small genome—a plant that could serve as a model for cereals (wheat and barley) and bioenergy crops. At the DOE Biomass Workshop in December 2005, Somerville chaired the Feedstocks for Biofuels section.[60] This is when decisions were made about the genomes with priority for sequencing, and when *Brachypodium* was truly proposed as a model plant to those individuals who seemingly controlled the purse strings.

Somerville had invited Vogel to attend the workshop and present his research priorities for *Brachypodium,* which were subsequently integrated into the 2006 "Biomass to Biofuels" report. Based on this meeting "DOE pretty much went

[59] *Brachypodium distachyon* is on the plant list maintained by the Jasper Ridge Biological Preserve. Later, Vogel showed this was *Brachypodium hybridum*.

[60] U.S. Department of Energy (2006).

down the list and funded grants to do all the things we put on the list for developing Brachy[podium] as a model system," and the JGI decided that Brachypodium would be sequenced, under the auspices of the bioenergy program, thus linking *Brachypodium* to switchgrass and bioenergy production.

Garvin, in contrast, had brought *Brachypodium* as a model plant for crop improvement, specifically temperate grasses. Garvin had already asked Michael Bevan (John Innes Institute, UK) to join his efforts to promote *Brachypodium* as model grass to the DOE-JGI, under the support of the Community Sequencing Program (CSP) and sought out Bevan's experience (and expertise) with assembly of the *Arabidopsis* genome in 2001. Following this meeting, Vogel contacted Garvin about teaming up for the sequencing project. However, from Vogel, we learned that "at this time the aim of the CSP shifted and all projects had to address biofuels in some way," so the focus needed to be towards biofuels, such as switchgrass, and not to cereals *per se* or grasses as plant disease models for crop improvement.[61] "Thus began the convergence of biofuels and cereal crop improvement as dual drivers of a Brachypodium genome sequencing project," as remembered by Garvin.

In 2007, the DOE-JGI added Brachypodium to the CSP plans with John Vogel as the "Proposer."[62] In the plan outlining why *Brachypodium* had been chosen, it was compared to *Arabidopsis* as having features that were favored for model plants. Specifically,

> *Brachypodium distachyon* is another potentially important model for such highly productive grasses as switchgrass and *Miscanthus.* Interest in *Brachypodium* arises because its very small genome has a DNA content about 2.5 times larger than that of *Arabidopsis.* Additionally, its simple growth requirements, small stature, self-fertility, and ready transformability make it well suited to become a model organism. Because *Arabidopsis* is distantly related to and differs from grasses in a number of important respects (e.g., cell-wall composition), *Brachypodium* could become a powerful new model for cell and molecular biological studies of grasses. A high priority in facilitating the development of *Brachypodium* is sequencing its genome.

The rationale for selecting *Brachypodium* reiterated its "desirable attributes," characteristics that essentially parallel those of *Arabidopsis,* and "the burgeoning research interest in this species." Additionally, the practical significance of this project was made clear, that the genome sequence of *Brachypodium* and the associated tools for gene discovery "will be a cornerstone resource for a vigorous research community seeking to promote the development of new energy crops and to contribute to global food security." As far as the DOE was concerned, *Brachypodium* was to be used as a "powerful new model" to meet a specific goal for addressing bioenergy needs.

---

[61] At around the same time, Todd Mockler (Oregon State University) had submitted a proposal to DOE-JGI (for EST sequencing). The Garvin/Bevan and Mockler proposals were approved and bundled into the larger project with Vogel as principle investigator.

[62] The principle investigators for the *Brachypodium* Bd21 DOE-JGI Community Sequencing Program were (with their affiliations in 2007): John Vogel and David Garvin (USDA-ARS); Michael Bevan (John Innes Centre, UK); Todd Mockler and Jeff Chang (Oregon State University); Samuel Hazen (Scripps); and, Todd Michael (Salk).

Yet, there was a larger role for *Brachypodium*. It specifically had potential to advance wheat-breeding programs through the study of molecular genetic perturbations caused by fungal pathogens. And as broader projects were percolating, there was interest in having a model for grass biology, shown by the increased number of abstracts at key scientific meetings. *Brachypodium* was first introduced at the 2003 PAG meeting, using a rice sequence to prepare libraries of DNA from *Brachypodium* to probe wheat chromosomal DNA. Without going into details, suffice it to say that *Brachypodium* was adequately used as a novel tool to explore the evolutionary gap between rice and wheat.[63]

In 2005, an abstract by Garvin and Stockinger (mentioned above) provided a procedural progress report on developing *Brachypodium* as a model, noting, "many steps remain to be taken before Brachypodium can be exploited fully as a model species."[64] In 2006, ten *Brachypodium*-themed abstracts were presented at the PAG XIV Conference, plus a workshop organized by Garvin entitled "*B. distachyon.*"[65] In 2007, 15 abstracts detailed the advances in *Brachypodium* genomics resources and its use to study plant diseases, such as rice blast. Stress physiology research was also finding a niche with *Brachypodium,* such as studies of the effect of boron on plant growth that were intended to inform the work on barley genetics and crop improvement. And last but not least, there was also the first report of using *Brachypodium* "to identify genes controlling traits relevant to energy crops."[66]

Even as early as 2007, this crossing over to use *Brachypodium,* as the right tool for a diversity of jobs, reflected a maturing model system. In each instance, the rationale for these *Brachypodium* workers was to use it to solve specific problems for bioenergy, crop, and forage grasses. By 2008, there were 20 abstracts, ranging from cytogenetics to natural population diversity.[67] In 2009, there were also 20 abstracts. In 2010, when the genome map was published, there were 41 abstracts. By 2011, as the genome sequence was completed, 42 abstracts included "Brachypodium" as a subject for comparisons or development—exemplifying its multifaceted uses for plant biology, plant pathology, genetics, and plant breeding.

Brachypodium publications show it to be emerging from a fundamentally different background from *Arabidopsis*. *Brachypodium* was specifically proposed as a model to study grass biology and plant pathogen–host interactions, all the while functioning as a key plant for rapidly developing the genetics of grasses suitable for bioenergy-based programs. *Arabidopsis* was, and would continue to be, an outstanding tool for plant biology,[68] but there was a need to augment plant biology

[63] Moore (2003). An update and extension of this was provided, also by Moore, in 2004.

[64] Garvin and Stockinger (2005).

[65] Other plant workshop topics at the 2006 PAG meeting focused on banana, barley, sugarcane, *Lolium,* grasses, wheat, maize, rice, sorghum and millets, brassicas, cucurbits, citrus, *Compositae,* forage and turfgrass, forest trees, fruit and nut crops, legumes, soybeans, cotton, grape, *Solanaceae,* and sugar beet. This is an indication of the explosion of resources allowing for cheaper, better, and faster genome sequencing and compilation of data.

[66] Vogel et al. (2007).

[67] Gill et al. (2008).

[68] Raikhel and Minorsky (2001).

research with other plants,[69] perhaps to include developing a "model molecular genetic system typical of graminaceous species" specifically built for translational research.[70]

However, there are distinct differences in the reasons and justifications for each model's rise to prominence. In particular, *Brachypodium* was selected to further our understanding of temperate grasses, which comprise the major sources of food, forage, and bioenergy crops. The specific intent was to model a laboratory-friendly grass species for direct translation towards agricultural improvement. The DOE-JGI focused on energy and switchgrass use, yet there was much occurring outside even this venue. While still framing their funding arguments around energy and switchgrass, the *Brachypodium* workers were establishing the plant as a model organism for evolution, genetics, and pathology. *Brachypodium* is an "*Arabidopsis* for grasses," much as *Arabidopsis* had been "a botanical *Drosophila*." Perhaps *Brachypodium* has even more to offer: With its close links to agriculture and plant pathology, it may become the "working grass hero"—a plant that came into the lab from the wild for the expressed desire to improve economically important grasses.[71]

Having identified a plant that was good in the lab, and could be used to address practical problems was not enough, *Brachypodium* needed to stand-in as a resource for bioenergy crops. By taking on this "job," the plant went from being essentially unknown in 2006 to moving to a priority list for the DOE-JGI program for sequencing and development of the resources that had heretofore been the purview of *Arabidopsis*. And *Arabidopsis* had taken on additional "jobs," serving as the template to bring a petite grass to the status of "model organism" and essentially functioning as a "model to model a model."

In 2010, with the completion of the *Brachypodium* genome sequence, in a recorded interview made by the DOE, John Vogel explained that the resources were now in place to do experimental biology, making it possible now to shift "from essentially developing tools to allow everybody to use *Brachypodium* as a model system, to actually using it to answer some of the questions we are interested in."[72] With the available molecular tools and genome sequence, *Brachypodium,* now, can be probed and tested to determine if it meets the needs of both experimental and applied biologists.

[69] Scholthof (2001).

[70] Draper et al. (2001).

[71] Garvin (2007b). "Working grass hero" was used by Garvin in a review describing *Brachypodium* to agricultural scientists.

[72] "John Vogel on the *Brachypodium distachyon* genome," (www.scivee.tv/node/16140, DOI: 10.4016/16140.03, 10 February 2010). Resources include transgenic knockout lines, expressed sequence tag (EST) libraries, microarrays, and devoted websites, such as brachypodium.org. Conferences are evidence of having a core group of scientists. The First European *Brachypodium* Workshop was held in France (October 2011). The First International Brachypodium Conference was held in Modena, Italy (June 2013).

Conclusion

For the *Brachypodium* community, the idea that this plant could be developed as a model was based on its genetic links to grasses that were critical for food security—including rice, maize, and wheat—and its potentiality for use in bioenergy research. Second, it was necessary to demonstrate that the plant was amenable to lab work (tissue culture, growth conditions, readily accessible seed collections) and that a strong community of scientists with experience in model systems would support such work in a "free-use" capacity similar to that endorsed by the old *Arabidopsis* community. Third, that same community had to develop the tools expected of a model. And finally, and perhaps essentially, the push by a president in his State of the Union Address provided the imprimatur to ensure resources and funding to develop these tools would be available—tools and resources which would drive rapid advancement in the development of a new translational model organism for plant biology.

Acknowledgments We appreciate the critical comments provided by Herman Scholthof, Kranthi Mandadi, Jesse Pyle, Will Cody, Sharon Kingsland, and Denise Phillips. We would like to especially thank David Garvin, John Vogel, Luis Mur, and Pilar Catalán for their helpful comments and discussions as we prepared this manuscript. This research was funded, in part, by a Texas A&M Energy Institute Fellowship (CWPL), the Program to Enhance Scholarly and Creative Activities at Texas A&M (K-BGS), and the Texas Higher Education Coordinating Board's Norman Hackerman Advanced Research Program Grant (No. 00517–002–2009, K-BGS).

References

Ankeny, Rachel. 2001. The natural history of *Caenorhabditis elegans* research. *Nature Reviews Genetics* 2:474–479.

Aronson, James, J. Kigel, A. Shmida, and J. Klein. 1992. Adaptive phenology of desert and Mediterranean populations of annual plants grown with and without water stress. *Oecologia* 89:17–26.

Aronson, James, J. Kigel, and A. Shmida. 1993. Reproductive allocation strategies in desert and Mediterranean populations of annual plants grown with and without water stress. *Oecologia* 93:336–342.

Bablak, Pauline, J. Draper, M. R. Davey, and P. T. Lynch. 1995. Plant regeneration and micropropagation of *Brachypodium distachyon*. *Plant Cell, Tissue and Organ Culture* 42:97–107.

Baker, Herbert G. 1965. Characteristics and modes of origin of weeds. In *The genetics of colonizing species*, eds. Herbert G. Baker and G. L. Stebbins, 147–172. New York: Academic Press.

Baker, Herbert G. 1974. The evolution of weeds. *Annual Review of Ecology and Systematics* 5:1–24.

Bennetzen, Jeffrey, and Michael Freeling. 1993. Grasses as a single genetic system: Genome composition, collinearity and compatibility. *Trends in Genetics* 9:259–261.

Bennetzen, Jeffrey, and Elizabeth Kellogg. 1997. Do plants have a one-way ticket to genomic obesity? *Plant Cell* 9:1509–1514.

Bennetzen, Jeffrey, V. Chandler, and P. Schnable. 2001. National Science Foundation-sponsored workshop report. Maize Genome Sequencing Project. *Plant Physiology* 127:1572–1578.

Brkljacic, Jelena, E. Grotewold, R. Scholl, T. Mockler, D. F. Garvin, P. Vain, and T. Brutnell. 2011. Brachypodium as a model for the grasses: Today and the future. *Plant Physiology* 157:3–13.

Catalán, Pilar, Y. Shi, L. Armstrong, J. Draper, and C. A. Stace. 1995. Molecular phylogeny of the grass genus *Brachypodium* P. Beauv. based on RFLP and RAPD analysis. *Botanical Journal of the Linnean Society* 117:263–280.

Catalán, Pilar, E. A. Kellogg, and R. G. Olmstead. 1997. Phylogeny of *Poaceae* subfamily Pooideae based on chloroplast ndhF gene sequencing. *Molecular Phylogenetics and Evolution* 8:150–166.

Catalán, Pilar, J. Müller, R. Hasterok, G. Jenkins, L. A. J. Mur, T. Langdon, A. Betekhtin, D. Siwinska, M. Pimentel, and D. López-Alvarez. 2012. Evolution and taxonomic split of the model grass *Brachypodium distachyon. Annals of Botany* 109:385–405.

Chang, Caren, and Elliot M. Meyerowitz. 1986. Molecular cloning and DNA sequence of the *Arabidopsis thaliana* alcohol dehydrogenase gene. *Proceedings of the National Academy of Sciences USA* 83:1408–1412.

Clarke, Adele E., and Joan H. Fujimura. 1992. What tools? Which jobs? Why right? In *The right tools for the job: At work in twentieth-century life sciences,* ed. Adele E. Clarke and Joan H. Fujimura, 3–44. Princeton: Princeton University Press.

Creager, Angela. 2002. *The life of a virus: Tobacco mosaic virus as an experimental model, 1930–1965*. Chicago: University of Chicago Press.

Draper, John, L. Mur, G. Jenkins, G. Ghosh-Biswas, P. Bablak, R. Hasterok, and A. Routledge. 2001. *Brachypodium distachyon.* A new model system for functional genomics in grasses. *Plant Physiology* 127:1539–1555.

Endersby, Jim. 2007. *A guinea pig's history of biology: The plants and animals who taught us the facts of life*. Cambridge: Harvard University Press.

Evans, James A. 2007. Start-ups in science: Entrepreneurs, diverse backing, and novelty outside business. In *The sociology of entrepreneurship,* ed. Martin Ruef and Michael Lounsbury, 261–305. Amsterdam: Elsevier.

Evans, James A. 2010. Industry induces academic science to know less about more. *American Journal of Sociology* 116:389–452.

Garvin, David F. 2007a. *Brachypodium distachyon*: A new model system for structural and functional analysis of grass genomes. In *Model plants and crop improvement,* ed. Rajeev K. Varshney and Robert M. D. Koebner, 109–23. Boca Raton: CRC Press.

Garvin, David F. 2007b. Brachypodium: A new monocot model plant system emerges. *Journal of the Science of Food and Agriculture* 87:1177–1179.

Garvin, David F., and Eric Stockinger. 2005. Development of a U.S. *Brachypodium distachyon* collection and the evaluation of microsynteny with cool season grass crops. Paper presented at the Plant & Animal Genome XIII Conference, San Diego, California, 15–19 January 2005 (unpublished).

Gill, Bikram, L. Qi, and B. Friebe. 2008. "Capturing 65 million year's history of chromosome evolution in cereal crops," paper presented at the Plant & Animal Genomes XVI Conference, San Diego, California, 12–16 January 2008 (unpublished).

Goff, Stephen, D. Ricke, T. -H. Lan, G. Presting, R. Wang, M. Dunn, and J. Glazebrook. 2002. A draft sequence of the rice genome (*Oryza sativa* L. ssp. *japonica*). *Science* 296:92–100.

Gressel, Jonathan, Y. Regev, S. Malkin, and Y. Kleifeld. 1983. Characterization of an *s*-triazine-resistant biotype of *Brachypodium distachyon. Weed Science* 31:450–456.

Jenkins, Glyn, R. Hasterok, and J. Draper. 2003. Building the molecular cytogenetic infrastructure of a new model grass. In *Application of novel cytogenetic and molecular techniques in genetics and breeding of the grasses,* eds. Z. Zwierzykowski, M. Surma, and P. Kachlicki, 77–84. Poznan: Polish Academy of Sciences.

Keller, Evelyn Fox. 1984. *A feeling for the organism: The life and work of Barbara McClintock.* New York: Henry Holt.

Khan, Mir Ajab. 1984. Biosystematic studies in Brachypodium (*Poaceae*). Ph.D. Diss., University of Leicester, UK.

Khan, Mir Ajab, and Clive Stace. 1999. Breeding relationships in the genus *Brachypodium* (*Poaceae: Poöideae*). *Nordic Journal of Botany* 19:257–269.

Kim, Ki-Joong and Robert K. Jansen. 1995. *ndhF* sequence evolution and the major clades in the sunflower family. *Proceedings of the National Academy of Sciences USA* 92:10379–10383.

Kohler, Robert E. 1994. *Lords of the fly: Drosophila genetics and the experimental life*. Chicago: University of Chicago Press.

Koornneef, Maarten, and David Meinke. 2010. The development of Arabidopsis as a model plant. *Plant Journal* 61:909–921.

Koornneef, Maarten, J. van Eden, C. J. Hanhart, P. Stam, F. J. Braaksma, and W. J. Feenstra. 1983. Linkage map of *Arabidopsis thaliana*. *Journal of Heredity* 74:265–272.

Leonelli, Sabina. 2007a. Growing weed, producing knowledge: An epistemic history of *Arabidopsis thaliana*. *History and Philosophy of the Life Sciences* 29:193–224.

Leonelli, Sabina. 2007b. Weed for thought: Using *Arabidopsis thaliana* to understand plant biology. Ph.D. Diss., Vrije Universiteit, Amsterdam.

López-Alvarez, Diana, M. L. López-Herranz, A. Betekhtin, and P. Catalán. 2012. A DNA barcoding method to discriminate between the model plant *Brachypodium distachyon* and its close relatives *B. stacei* and *B. hybridum* (Poaceae). PLoS One 7:e51058.

Mackintosh, Caroline, D. Garvin, L. Radmer, S. Heinen, and G. Muehlbauer. 2006. A model wheat cultivar for transformation to improve resistance to Fusarium Head Blight. *Plant Cell Reports* 25:313–319.

McGuire, Patrick E., A. Damania, and C. Qualset. eds. 2012. *Safflower in California: The Paulden F. Knowles personal history of plant exploration and research on evolution, genetics, and breeding*. Agronomy Progress Report No. 313, Department of Plant Sciences, University of California-Davis.

Meinke, David, J. M. Cherry, C. Dean, S. D. Rounsley, and M. Koornneef. 1998. *Arabidopsis thaliana*: A model plant for genome analysis. *Science* 282:662–682.

Meyerowitz, Elliot M. 1987. Arabidopsis thaliana. *Annual Review of Genetics* 21:93–111.

Meyerowitz, Elliot M. 1989. Arabidopsis, a useful weed. *Cell* 56:263–269.

Meyerowitz, Elliot M. 2001. Prehistory and history of Arabidopsis research. *Plant Physiology* 125:15–19.

Meyerowitz, Elliot M., and Caren Chang. 1988. Molecular biology of plant growth and development. In *The molecular biology of cell determination and cell differentiation,* ed. Leon W. Browder, 353–366. New York: Plenum Press.

Meyerowitz, Elliot M., and Robert Pruitt. 1985. *Arabidopsis thaliana* and plant molecular genetics. *Science* 229:1214–1218.

Moore, Graham. 2003. The application of comparative genomics to wheat. Paper presented at the Plant & Animal Genome XI Conference, San Diego, California, 11–15 January 2003 (unpublished).

Moore, Graham, M. Gale, N. Kurata, and R. Flavell. 1993. Molecular analysis of small grain cereal genomes: Current status and prospects. *Nature Biotechology* 11:584–589.

Pang, Patty P., and Elliot M. Meyerowitz. 1987. *Arabidopsis thaliana*: A model system for plant molecular biology. *Nature Biotechnology* 5:1177–1181.

Peterson, Paul D. 2001. *Stem rust of wheat*: *From ancient enemy to modern foe*. St. Paul: APS Press.

Peterson, Paul D. 2003. The common barberry: the past and present situation in Minnesota and the risk of wheat stem rust epidemics. Ph.D. Diss., North Carolina State University, Raleigh, NC.

Rader, Karen. 2004. *Making mice: Standardizing animals for American biomedical research, 1900–1955*. Princeton: Princeton University Press.

Raikhel, Natasha V., and Peter V. Minorsky. 2001. Celebrating plant diversity. *Plant Physiology* 127:1325–1327.

Rédei, George P. 1975. Arabidopsis as a genetic tool. *Annual Review of Genetics* 9:111–127.

Robertson, Ian H. 1981. Chromosome numbers in Brachypodium Beauv. (Gramineae). *Genetica* 56:55–60.

Scholthof, Herman. 2001. Molecular plant-microbe interactions that cut the mustard. *Plant Physiology* 127:1476–1483.

Shi, Ying. 1991. Molecular studies of the evolutionary relationships of *Brachypodium* (*Poaceae*). Ph.D. Diss., University of Leicester, Leicester, UK.

Shi, Ying, J. Draper, C. Stace, P. Catalán, J. Moffat, and L. Armstrong. 1991. "*Brachypodium distachyon*: a species that might be developed as a model molecular genetic system representative of temperate cereals," paper presented at the Third International Congress of Plant Molecular Biology, Tucson, Arizona, 6–11 October 1991 (unpublished).

Smith, R. Jeffrey. 1980. Experts endorse biomass energy. *Science* 208:1018.

Somerville, Chris. 1989. Arabidopsis blooms. *Plant Cell* 1:1131–1135.

Somerville, Chris. 2000. The twentieth century trajectory of plant biology. *Cell* 100:13–25.

Somerville, Chris. 2001. An early Arabidopsis demonstration. Resolving a few issues concerning photorespiration. *Plant Physiology* 125:20–24.

Somerville, Shauna, and Chris Somerville. 1996. Arabidopsis at 7: Still growing like a weed. *Plant Cell* 8:1917–1933.

Somerville, Chris, and Maarten Koornneef. 2002. A fortunate choice: The history of Arabidopsis as a model plant. *Nature Review Genetics* 3:883–889.

Somerville, Chris, and Shauna Somerville. 1999. Plant functional genomics. *Science* 285:380–383.

Stokstad, Erik. 2006. Poplar tree sequence yields genome double take. *Science* 313:1556.

Summers, William C. 1993. How bacteriophage came to be used by the Phage Group. *Journal of the History of Biology* 26:255–267.

Summers, William C. 1999. *Felix d'Herelle and the origins of molecular biology*. New Haven: Yale University Press.

Tateoka, Tuguo. 1968. Phytogeographical notes on the genus *Brachypodium* P. Beauv. (*Gramineae*). *Boletin de la Sociedad Argentina de Botanica* 12:44–56.

U.S. Department of Energy. 2006. Feedstocks for biofuels. In *Breaking the biological barriers to cellulosic ethanol: A joint research agenda*. Report from the December 2005 Workshop, DOE/SC-0095, 57–83. Rockville, MD: U.S. Department of Energy Office of Science.

Vogel, John. 2010. John Vogel on the *Brachypodium distachyon* genome. www.scivee.tv/node/16140. doi:10.4016/16140.03.

Vogel, John, and Metin Tuna. 2008. "Diversity of Turkish populations of Brachypodium distachyon," paper presented at the Plant & Animal Genomes XVI Conference, San Diego, California, 12–16 January 2008 (unpublished).

Vogel, John, and Jennifer Bragg. 2009. *Brachypodium distachyon*, a new model for the Triticeae. In *Genetics and genomics of the Triticeae*, eds. Catherine Feuillet and Gary J. Muehlbauer, 427–449. http://link.springer.com/book/10.1007%2F978-0-387-77489-3. Accessed 1 July 2014.

Vogel, John, D. Garvin, O. Leong, and D. Hayden. 2006a. Agrobacterium-mediated transformation and inbred line development in the model grass *Brachypodium distachyon. Plant Cell, Tissue, and Organ Culture* 84:199–211.

Vogel, John, Y. Gu, P. Twigg, G. Lazo, D. Laudencia-Chingcuanco, D. Hayden, T. Donze, L. Vivian, B. Stamova, and D. Coleman-Der. 2006b. EST sequencing and phylogenetic analysis of the model grass *Brachypodium distachyon. Theoretical and Applied Genetics* 113:186–195.

Vogel, John, C. M. Tobias, D. M. Haydn, Y. Q. Gu, G. R. Lazo, D. F. Garvin, M. W. Bevan, T. C. Mockler, E. K. Chow, and O. D. Anderson. 2007. Mapping and marker development for energy crops: Progress on switchgrass and the model grass *Brachypodium distachyon*. Paper presented at the Plant & Animal Genomes XV Conference, San Diego, California, 13–17 January 2007 (unpublished).

Weber, Marcel. 2005. *Philosophy of experimental biology*. Cambridge: Cambridge University Press.

Wolfe, Kenneth, M. Gouy, Y. W. Yang, P. M. Sharp, and W. H. Li. 1989. Date of the monocot-dicot divergence estimated from chloroplast DNA sequence data. *Proceedings of the National Academy of Sciences USA* 86:6201–6205.

Yu, Jun, S. Hu, J. Wang, G. K. -S. Wong, S. Li, B. Liu, and Y. Deng. 2002. A draft sequence of the rice genome (*Oryza sativa* L. ssp. *indica*). *Science* 296:79–92.

Christopher W. P. Lyons is a Ph.D. candidate in the Department of Plant Pathology and Microbiology whose dissertation research focuses on the proteomics of *Panicum mosaic virus* (PMV) and its satellite virus (SPMV) and the model grass species, *Brachypodium distachyon*, as part of the continuing development of *Brachypodium* as a tool for translational research in the fields of crop and dedicated-biofuel crop improvement. He is interested in the history and philosophy of science, focusing on the development of agricultural model species in the postgenomics era.

Karen-Beth G. Scholthof is a professor in the Department of Plant Pathology and Microbiology and director of the Bioenvironmental Sciences Undergraduate Honors Program at Texas A&M University. Her scientific research focuses on the molecular genetics of *Panicum mosaic virus* (PMV) and the development of disease, using *Brachypodium* as a model organism. As a practitioner historian, her research interests include the historiography of *Tobacco mosaic virus* (TMV) in the early 20th century.

Index

D. Phillips, S. Kingsland (eds.), *New Perspectives on the History of Life Sciences and Agriculture,* Archimedes 40, DOI 10.1007/978-3-319-12185-7

MIX
Papier aus verantwortungsvollen Quellen
Paper from responsible sources
FSC® C105338

If you have any concerns about our products,
you can contact us on
ProductSafety@springernature.com

In case Publisher is established outside the EU,
the EU authorized representative is:
Springer Nature Customer Service Center GmbH
Europaplatz 3, 69115 Heidelberg, Germany

Printed by Libri Plureos GmbH
in Hamburg, Germany